MATERIALS THERMOCHEMISTRY

Sixth Edition of *Kubaschewski and Alcock's Metallurgical Thermochemistry*

To Victor,

With warm memories of our long-standing and fruitfull collaboration

C.B. Alcock (Ben)

Pergamon Titles of Related Interest

Books

ASHBY & JONES
Engineering Materials 1, 2 & 3

BARRETT & MASSALSKI
Structure of Metals, 3rd edition

BEVER
Encyclopedia of Materials Science & Engineering

BROOK
Concise Encyclopedia of Advanced Ceramic Materials

CAHN
Encyclopedia of Materials Science & Engineering, Supplementary Volumes 1, 2 & 3

GILCHRIST
Extraction Metallurgy, 3rd edition

HULL & BACON
Introduction to Dislocations, 3rd edition

KELLY
Concise Encyclopedia of Composite Materials

SCULLY
Fundamentals of Corrosion, 3rd edition

TAYA & ARSENAULT
Metal Matrix Composites

WILLS
Mineral Processing Technology, 5th edition

Journals

Acta Metallurgica et Materialia

Calphad

Canadian Metallurgical Quarterly

Chemical Engineering Science

Journal of Physics & Chemistry of Solids

Journal of The Mechanics & Physics of Solids

Materials Research Bulletin

Minerals Engineering

Scripta Metallurgica et Materialia

Solid State Communications

Solid-State Electronics

MATERIALS THERMOCHEMISTRY

Sixth Edition

by

O. KUBASCHEWSKI
DR PHIL. HABIL. DRS. H.C.
Technische Hochschule, Aachen

C. B. ALCOCK
Ph.D. A.R.C.S. D.Sc. F.R.S.C.
University of Notre Dame, Indiana

P. J. SPENCER
Ph.D.
Technische Hochschule, Aachen

PERGAMON PRESS
OXFORD · NEW YORK · SEOUL · TOKYO

UK	Pergamon Press Ltd, Headington Hill Hall, Oxford OX3 0BW, England
USA	Pergamon Press Inc., 660 White Plains Road, Tarrytown, New York 10591-5153, USA
KOREA	Pergamon Press Korea, KPO Box 315, Seoul 110-603, Korea
JAPAN	Pergamon Press Japan, Tsunashima Building Annex, 3-20-12 Yushima, Bunkyo-ku, Tokyo 113, Japan

1st edition 1951
2nd edition 1955
3rd edition 1958
4th edition 1967
Reprinted 1974
5th edition 1979
6th edition 1993

British Library Cataloguing in Publication Data

A catalogue record for this book is available from the British Library.

Library of Congress Cataloging in Publication Data

Kubaschewski, O. (Oswald), 1912–
Materials thermochemistry / by O. Kubaschewski, C. B. Alcock, P. J. Spencer.—6th ed., rev. and enl.
p. cm.
Rev. ed. of: Metallurgical thermochemistry. 5th ed., rev. and enl. 1979.
1. Thermochemistry. 2. Metallurgy. 3. Materials—Thermal properties. I. Alcock, C. B. II. Spencer, P. J. III. Kubaschewski, O. (Oswald), 1912– Metallurgical thermochemistry. IV. Title.
QD511.KB 1993
669'.9—dc20 92-31280

ISBN 0 08 0418899 Hardcover
0 08 0418880 Flexicover

Printed in Great Britain by B.P.P.C. Wheatons Ltd, Exeter

Foreword

DURING the later stages of the preparation of this 6th edition the book's prime mover and inspiring genius Dr O. Kubaschewski died in Aachen. He fully approved the changes in text and data which appear in this edition although his acceptance of joules instead of calories, with which he always claimed to be more comfortable, was reluctant. With the appearance of a new author, Dr P. J. Spencer, who worked in close collaboration with "Kuba" over many years, the content shows an obvious new vitality. The extension of the examples into areas which are remote from metallurgy suggested the new title of *Materials Thermochemistry* because the basic content is also applicable in a wider field than originally conceived in 1951. It should be admitted here that Kuba was not enthusiastic about the change of title, and would have preferred this to be the 6th edition of *Metallurgical Thermochemistry*. However, his fellow authors hope that this broader title will enrol a new readership to extend the world-wide acceptance that this book has won since its inception. There can be no warmer tribute to Kuba's scientific contribution and leadership than this continued dissemination of his principles.

The many scientists throughout the world who knew Dr Kubaschewski as a friend will mourn the passing of a worthy example of the best of German culture. He was one who shared his attention and intellectual interest over the spectrum from Goethe to Nernst. Always a man of intensely practical background, he combined a skill in the development of new measurement techniques with an encyclopaedic knowledge of the extant data in his chosen field. A most valuable aspect of his compendium of knowledge was his ability to see the "wood from the trees" and to make the critical appraisal which alone can codify the real worth of scientific progress.

His ability with languages, and the warmth of his personality made him friends, especially among young scientists, in many parts of the scientific community. For one who spent his adult life between England and Germany we offer two thoughts of this practical scientist:

> "He was a man, take him all for all,
> (we) shall not look upon his like again." (Shakespeare)
>
> "Grau, teurer Freund, ist alle Theorie
> Und grun des Lebens goldner Baum." (Goethe)

C. B. ALCOCK

Contents

Notation xi

1. The Theoretical Basis

1. The mass action law 1
2. Thermodynamic functions 3
   A. The equilibrium constant and thermodynamic functions 3
   B. Heat content and enthalpy of formation 4
   C. Temperature dependence of the enthalpy of reaction 8
   D. Entropy 13
   E. Gibbs energy 17
3. Solutions 27
   A. Partial molar Gibbs energy 27
   B. Vapour pressures and partial Gibbs energies 30
   C. The calculation of integral values from partial molar thermochemical data 34
   D. Dilute solutions 44
      1. Raoult's law 44
      2. Henry's law 45
   E. Atomistics and solution thermodynamics 47
   F. Regular solutions 50
      1. Calculation of activities using equilibrium diagrams 50
      2. Spinodal decomposition 51
      3. Order–disorder transformation 53
   G. Non-regular solutions 53
   H. Ternary solutions 56

2. Experimental Methods

1. Calorimetric methods 65
   A. Measurement of temperature 66
      1. Mercury-in-glass thermometers 67
      2. Platinum resistance thermometers 67
      3. Thermocouples 68
      4. Thermistors 68
      5. Optical pyrometers 68
   B. Determination of water equivalent 68
      1. Heat contents and heat capacities 70
      2. Enthalpies of fusion and transformation 79
      3. Enthalpies of reaction and formation 83
2. Equilibria with a gaseous phase 95
   The equilibrium constant for vaporisation reactions 96
   A. Static methods for the measurement of vapour pressure 97

1. Manometric methods 97
2. Static methods using radiation 98
B. Gas-condensed phase equilibria in a closed system 100
1. The dew-point method 100
2. The isopiestic method 100
3. Tensi-eudiometer measurements 102
4. Sievert's method 104
C. Dynamic vapour pressure methods 106
1. The boiling point method 106
2. The transportation method 108
D. Other heterogeneous equilibria 112
1. Reaction between a gas phase and a condensed phase 113
E. Methods based on rates of evaporation 120
1. The Knudsen effusion method 121
2. The Knudsen cell–mass-spectrometer combination 125
3. The Langmuir vaporisation method 133
4. Torsion effusion 136
5. Vapour transpiration 138
3. Electromotive forces 140
A. Liquid electrolytes 143
1. Aqueous solutions 143
2. Molten salt electrolytes 143
B. Solid electrolytes 146
1. Glass electrolyte 146
2. Solid oxide electrolytes 148
3. Other cells 159

3. The Estimation of Thermochemical Data

1. Heat capacities 163
A. Gaseous atoms and molecules 163
B. Solids 164
C. Liquids 167
D. Some average values 168
2. Enthalpies and entropies of transformation, fusion and evaporation 169
A. Evaporation 169
B. Fusion 169
3. Entropy and entropy changes 171
A. Standard entropies 171
1. Solids 171
2. Gases 174
B. Entropies of mixing of non-metallic solution phases 177
4. Enthalpies of formation 179
A. General 180
B. Homologous series 180
C. Empirical relations 182
1. Polyvalent metal oxides 182
2. Metal oxyhalide compounds 182
3. Double salts with the formula MX_aY_b 183
4. Oxides, carbonates, sulphates, hydroxides and nitrates 184
5. Halides 185
D. Enthalpies of formation of double oxides 185
1. Plots involving the ratio of ionic charge to ionic radius 185
2. Statistical analysis methods 187
3. Le Van's method 187
4. Comparison of data for similar compounds 188
E. Volume change and enthalpy of formation 189

F. Enthalpy of solution 190
G. The packing effect 192
H. Temperature increase during compound formation 194
5. Thermodynamic properties of alloys 197
A. Calculations using the Wigner–Seitz model of metals 197
B. Properties of mixing from "Free volume" theory 199

4. Examples of Thermochemical Treatment of Materials Problems

1. Iron and steelmaking 203
A. Deoxidation of steel 203
1. Deoxidation with silicon 203
2. Deoxidation with aluminium 204
B. The decarburisation of iron–chromium–carbon and iron–silicon–carbon liquid alloys 205
C. Chill factors 208
D. "Window" for liquid calcium aluminates in continuous casting 209
E. Precipitation of carbide and nitride phases from dilute solution in alloy steels 210
2. Non-ferrous metallurgy 212
A. Aluminothermic type reactions 212
1. The production of uranium by reduction of its fluoride with calcium and magnesium 212
2. The production of manganese and chromium by the aluminothermic process 214
B. The chlorination of metal oxides 215
C. Refining of lead 216
1. The removal of zinc from lead 218
3. Stability and production of ceramics 218
A. The free evaporation of an oxide ceramic *in vacuo* 218
B. Metal–refractory interaction 220
C. Electrochemical cells and the stabilities of ceramics 225
D. Equilibrium phase relations relevant to the production of oxide superconductor materials 226
4. Chemical vapour deposition (CVD) and physical vapour deposition (PVD) processes 227
A. CVD of ultra-pure silicon 227
B. Vapour phase transport of silicon carbide 228
C. Prediction of metastable phase formation during PVD of mixed coating materials 231
5. Corrosion 231
A. The oxidation of iron–chromium alloys 231
6. Environmental and energy problems 235
A. Calculation of hazardous emissions during sintering of ores 235
B. Incineration of waste in a molten iron bath 237
C. Thermodynamic conditions for formation of dioxin in waste incineration 239
D. Energy conservation in waste incineration 241
7. Assessment of standard values 241
A. A pure stoichiometric substance—silicon monoxide 241
B. A system exhibiting wide solution ranges—chromium–nickel 247
8. Calculation of metallurgical equilibrium diagrams 253

5. Thermochemical Data 257

Index 361

Notation

The symbols used in this monograph are listed in the following table for rapid reference. The fourth column in the table cites the page in which the symbol is introduced or its use explained.

Confusion often arises in thermochemical calculation with the sign used for the various thermochemical quantities. The convention used in this monograph is the prevalent one nowadays of regarding the heat of any reaction, ΔH, as the increase in heat content of the system, i.e. the heat absorbed. Thus the heat of an *exothermic* reaction is a *negative* quantity. The signs of the Gibbs energy and entropy are most conveniently deduced and remembered from the Gibbs–Helmholtz equation in the form $\Delta G = \Delta H - T\Delta S$.

Symbol	Meaning	Common Units or Value
a	activity (Raoult)	—
c	concentration	—
c_p	specific heat at constant pressure	J/g deg
C_p	molar heat capacity	J/mole deg
E	electromotive force (e.m.f.)	V
f	activity coefficient (Henry)	—
F	Faraday's constant	96,485 J/V
G	Gibbs energy (Gibbs' function or thermodynamic potential)	J
ΔG	Gibbs energy of formation or reaction	J
$\Delta \bar{G}$	partial molar Gibbs energy of solution	J
γ	activity coefficient (Raoult)	—
H	heat content, enthalpy	J
ΔH	enthalpy of formation or reaction	J
ΔH_{298}	enthalpy of formation or reaction at 25°C	J
$\Delta \bar{H}$	partial molar enthalpy of solution	J/mole
$K_{p,c,a}$	equilibrium constant expressed in terms of pressures, concentrations, or activities (reaction products in numerator)	
κ	electrical conductivity	$\text{ohm}^{-1}\ \text{cm}^{-1}$
ln	natural logarithm, $\log_e$	—
log	$\log_{10}$	—
$L_{t,f,e,s}$	latent heat of transformation, fusion, evaporation, sublimation	J/mole
m	mass	g
M	molecular weight	g
μ	chemical potential	J
n	number of moles	—
N_1, N_2	mole fractions	—
N	Avogadro's number	6.02×10^{23}

Symbol	Meaning	Common Units or Value
p	pressure	mmHg, atm, bar
r	radius, also atomic or ionic radius	cm, or Å
$\boldsymbol{R}$	gas constant	8.314 J/deg
$\boldsymbol{R} \ln 10$	—	19.144 J/deg
ρ	density	g/cm^3
S	entropy	J/deg
S_{298}	standard entropy	J/deg
ΔS	entropy of reaction	J/deg
$\Delta \bar{S}_1$	partial molar entropy of solution	J/deg mole
t	time	sec (min, hr)
T	absolute temperature	deg K
θ	temperature	deg°C
V	volume	cm^3, or litres
W	water equivalent of a calorimeter	J/K
z	electrochemical valency	—

Brackets around Symbols Denoting Molar Quantities

( )	gaseous
{ }	liquid
⟨ ⟩	solid
[ ]	dissolved (suffix denoting medium)

CHAPTER 1

The Theoretical Basis

1. The Mass Action Law

The problem which most commonly confronts the production metallurgist is that of the preparation of a pure metal by the reduction of one of its compounds; for example

$$Fe_2O_3 + 3C = 2Fe + 3CO$$

or to put the equation in a more general form,

$$M_aX_b + cR = aM + R_cX_b \tag{1}$$

where M represents the metal in question, R the reducing agent and X the radical to be removed. The yield of such a reaction depends firstly on the position of equilibrium between the reacting substances and secondly on the rate at which equilibrium is attained. This latter may be so slow that the yield is still far below the equilibrium value after a lapse of time of hours, days, or even years. Simple thermodynamic considerations cannot be applied to such incomplete reactions, though they govern to a certain extent the force driving a given reaction toward equilibrium.

At the high temperatures often encountered in materials production, reaction rates are usually so high that it may be assumed that equilibrium is attained during the process.

A numerical value for the equilibrium state of a reaction is given by the mass action law, first derived by Guldberg and Waage (1867) from kinetic considerations. Its thermodynamic derivation, by van't Hoff's well-known hypothetical equilibrium-box operation (1883) entails that the only components of the reaction which need be taken into consideration are those in the gaseous state or in solution. A pure condensed phase has of course a certain vapour pressure, but this is constant at any given temperature and need not be considered in the mass action relationship.

It is convenient therefore to use symbols to denote the state of aggregation of the components of any reaction. In this monograph we shall make use of various types of brackets. Thus diamond-shaped (angular) brackets will denote the solid state, $\langle NaCl \rangle$; braces the liquid state, $\{H_2O\}$; light parentheses the gaseous state, (O_2); while the state of dilute solution will be indicated by square brackets, $[NaCl]_{aq}$, the suffix denoting the solvent.

In making mass action calculations a distinction must be drawn between heterogeneous and homogeneous reactions. A homogeneous reaction is one which takes place completely in one phase, such as

$$(H_2) + (I_2) = 2(HI)$$

or

$$[CuSO_4]=[Cu^{2+}]+[SO_4^{2-}].$$

For these examples the mass action law is written

$$K_p=\frac{p_{HI}^2}{p_{H_2}p_{I_2}} \tag{2a}$$

$$K_c=\frac{c_{Cu^{2+}}c_{SO_4^{2-}}}{c_{CuSO_4}} \tag{2b}$$

K_p and K_c are the mass action constants at a certain temperature, p denoting pressure in atmospheres, and c, the concentration.*

For thermochemical calculations the unit of quantity should generally be the gram-molecule or mole, the symbol (H_2) then representing one mole of gaseous hydrogen. The number of moles of a substance which are considered to take part in a reaction appears in the mass action equation as the power of the partial pressure or concentration.

In heterogeneous reactions one or more condensed phases appear in addition to the gaseous or solution phase. The mass action relationship is then expressed as in the following three examples:

$$[FeO]_{slag}+[C]_{Fe}=\{Fe\}+(CO) \tag{α}$$

$$K_{p,N}=p_{CO}/N_C N_{FeO}$$

$$\langle Cr_2O_3\rangle+3(H_2)=2\langle Cr\rangle+3(H_2O) \tag{β}$$

$$K_p=\frac{p_{H_2O}^3}{p_{H_2}^3}$$

$$\langle SiO_2\rangle=[Si]_{Fe}+2[O]_{Fe} \tag{γ}$$

$$K_N=N_{Si}N_O^2.$$

This should of course be familiar to all chemists, and it is not intended to give fuller details either of the exact derivation or of the experimental proof of the mass action law.

In this monograph we shall follow the usual custom of writing the mass action constant with the active masses of the substances to the right of the chemical equation (the products) in the numerator, and those of the substances to the left (the reactants) in the denominator. The equation of any reaction may of course be written in either direction, but in practice the metal or other substance to be produced will be written on the right. Thus the most favourable conditions for reaction are indicated by a high value of K_p or K_N, while a value far below unity shows that conditions are unfavourable for producing the desired substance. In this latter case the yield can still be made large by the familiar device of removing one of the products, particularly a gas, as it is formed, thereby displacing the system continuously from equilibrium. Thus in reaction (β) described above, a continuous flow of hydrogen would effect this by removing the water formed.

*Concentrations will be expressed in this monograph as mole fractions N_i = number of moles of solute i per total number of moles $n_i/(n_i+n_2+\cdots)$; atomic per cent = mole fraction × 100; and weight per cent = number of grams per 100 g total solution: $c_1=[m_1/(m_1+m_2+\cdots)]\times 100$.

It follows from **Le Chatelier's principle** (1885) that the magnitude of the total pressure of a system will not affect an equilibrium if the number of moles of gas on either side of the chemical equation are equal. If the numbers differ, an increase in total pressure will displace the equilibrium in favour of the side with the smaller number of moles, and vice versa.

It should be emphasised here that the mass action law in the form given above will apply only in ideal cases where no interaction occurs between the molecular particles of the gas or the solute. This assumption is usually justified for gaseous reactions if the total pressure is not considerably higher than 1 atm. Generally it does not apply to reactions in the solution phase except with very dilute solutions. In more concentrated solutions the interaction of the dissolved components and the solvent affects the equilibrium conditions and K_N is no longer constant. A constant can be obtained, however, by replacing the concentrations N by activities a. The concept of activity will be discussed later.

2. Thermodynamic Functions

A. The Equilibrium Constant and Thermodynamic Functions

Generally the aim of any thermochemical calculation in materials production is to obtain values for the equilibrium constant under consideration. The direct method is to measure the relative concentrations of the substances themselves at some selected temperatures, and to derive values of these concentrations or of the constant at other temperatures by interpolation. As a rule, however, this direct determination of the constant either cannot be carried out or is precisely what the chemist is trying to avoid. The number of reactions which can require investigation is astronomical, and a simpler approach is necessary. The advantage of thermochemical calculations is that the vast number of equilibria encountered in materials processes may be computed from a much smaller number of functions characteristic of the separate compounds or elements which take part. These functions can often be determined much more easily than the equilibrium constants.

To illustrate the simple connection between these functions and the equilibrium constant, we may anticipate later discussion by stating that the function usually tabulated instead of the equilibrium constant is the change in standard **Gibbs energy** for any given process, ΔG°. Its relationship to K_p is expressed by

$$\Delta G^\circ = -RT \ln K_p \tag{3}$$

or

$$\Delta G^\circ = -19.144T \log K_p. \tag{3a}$$

The change in Gibbs energy is under certain conditions an additive function, which the equilibrium constant is not, and is related to the enthalpy and entropy change of reaction by the simplified **Gibbs–Helmholtz equation**

$$\Delta G^\circ = \Delta H^\circ - T\,\Delta S^\circ. \tag{4}$$

The following sections deal with the determination of enthalpy changes (ΔH) and entropy changes (ΔS) of reaction at various temperatures, and with the derivation and determination of Gibbs energy values.

The heat change in a reaction will be defined as the sum of the heat contents of the products of the reaction minus the sum of the heat contents of the reactants, and similarly for the entropy change.

B. Heat Content and Enthalpy of Formation

It may be deduced from the first and second laws of thermodynamics that at a given temperature every substance in a given state has a fixed, characteristic value of H and S. If H had not this constancy, then a cyclic process might be envisaged in which heat is produced from nothing, and this has not yet been experienced. The heat content exists in the substance as kinetic and potential energy of its atoms and molecules (energy of translation, rotation, and oscillation). Whatever may be the relative amounts of these constituent energies, their sum remains constant for a given amount of the substance in a given state.

Heat content is so defined that the heat evolved or absorbed during a change taking place in a system* at constant pressure is equal to the change in its heat content. If in the general reaction given in eqn. (1) the heat contents of the participants are respectively $H_{\mathrm{M}_a\mathrm{X}_b}$, H_{R}, H_{M} and $H_{\mathrm{R}_c\mathrm{X}_b}$, the summation

$$(aH_{\mathrm{M}}+H_{\mathrm{R}_c\mathrm{X}_b})-(H_{\mathrm{M}_a\mathrm{X}_b}+cH_{\mathrm{R}})=\Delta H_T \tag{5}$$

is equal to the amount of heat produced if the reaction takes place completely to form the substances on the right of the chemical equation from the reactants on the left. This difference ΔH_T is termed the heat or enthalpy of reaction and can be measured calorimetrically. In the more simple case of the formation of a single compound from its elements

$$a\mathrm{M}+b\mathrm{X}=\mathrm{M}_a\mathrm{X}_b$$

(e.g. $\langle \mathrm{Pb}\rangle+(\mathrm{O}_2)=\langle \mathrm{PbO}_2\rangle$ the difference in heat contents

$$H_{\mathrm{M}_a\mathrm{X}_b}-(aH_{\mathrm{M}}+bH_{\mathrm{X}})=\Delta H_T \tag{5a}$$

is called the enthalpy of formation. The enthalpies of formation of chemical substances are usually recorded and tabulated at a standard temperature of 25°C (=298.15 K) and are written as ΔH_{298}.

The numerical values of the enthalpies of formation or reaction will generally be negative, corresponding to an evolution of heat during reaction. This is illustrated diagrammatically in Fig. 1, which is self-explanatory.

If a substance is cooled from a high temperature T to room temperature the heat content will also decrease, the heat being lost by the system to its surroundings. The change represented by $M_T \rightarrow \mathrm{M}_{298}$ (e.g. $\langle \mathrm{Fe}\rangle_{773} \rightarrow \langle \mathrm{Fe}\rangle_{298}$), is accompanied by heat *evolution*

$$H_T-H_{298}=\Delta' H_{298}^T \tag{6}$$

*A system may be defined as any number and quantity of chemical substances considered separately from their surroundings.

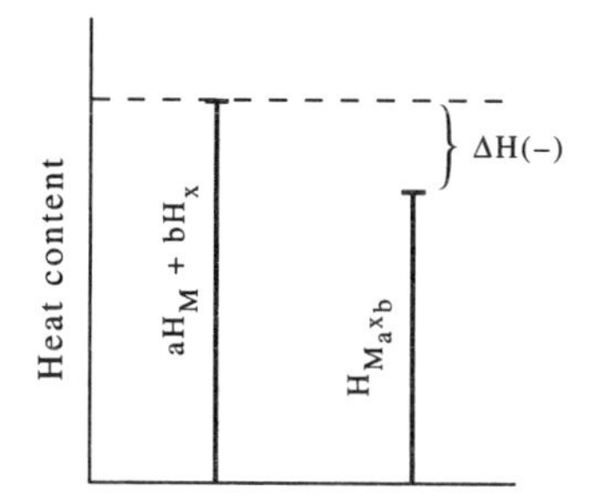

FIG. 1. The heat of compound formation.

Alternatively, it may be stated that the change (i.e. increase) in heat content of the systems is $-\Delta' H_{298}^T$. (The use of the notation Δ′ is explained in the footnote on p. 9.)

If the total heat contents of all substances were known they could be tabulated and used additively for the evaluation of the enthalpy of any reaction. These total heat contents would include the enthalpies of formation of atoms from the constituent elementary particles, protons, neutrons, electrons, etc. Since in any chemical reaction the nuclear energy and the energy of the inner electron structure of the atoms is in no way altered, the values of these quantities remain constant and can be omitted from thermochemical consideration. Furthermore, as we are principally concerned with heat changes be they as a result of chemical reaction or changes in temperature of a substance we can, without loss of rigour, select an arbitrary reference state where the substance has zero heat content, *by definition*, and make all of our calculations with respect to this reference or standard state.

Thus we know that the molar heat content of a substance at absolute zero is related to that at, say, 25°C by the relationship

$$H_{298} = H_0 + \int_0^{T=298} C_p \, \mathrm{d}T$$

where C_p is the molar specific heat of the substance at constant pressure, and we can define H_{298} as having the value zero for an elementary substance at 25°C. H_0 then becomes a negative quantity, since C_p must be positive, with reference to the element at 25°C as the standard state.

The heat content of compounds at 25°C by the convention which has been laid down is therefore equal to their enthalpy of formation from their elements at this temperature. Thus the enthalpy of formation of liquid water from gaseous oxygen and hydrogen at 25°C

$$(H_2) + \tfrac{1}{2}(O_2) = \{H_2O\}$$

is found to be $\Delta H_{298} = -285.8$ kJ which becomes the heat content of water, since those of hydrogen and oxygen are zero at this temperature. For this reason the enthalpies of formation of compounds are tabulated for this same temperature of 25°C.

One example may be given of the use of these values for calculating enthalpies of reaction:

$$\langle CoO \rangle + (CO) = \langle Co \rangle + (CO_2)$$

$$-237.7 \quad -110.5 \quad 0 \quad\quad -393.5 \text{ kJ.}$$

The respective enthalpies of formation (Table A) are written below each participant. The difference

$$\Delta H_{298} = -393.5 - (-237.7 - 110.5) = -45.3 \text{ kJ}$$

is the enthalpy of reaction. This additive property of enthalpies of reaction has long been known as **Hess' principle** (1840). Further use is made of it in evaluating enthalpies of formation of compounds by combining the enthalpies of various associated reactions. This is often applied in organic chemistry and to numerous inorganic substances. One example will suffice as illustration.

Roth measured the enthalpy of combustion of $\langle Al_4C_3 \rangle$ according to the equation

$$\langle Al_4C_3 \rangle + 6(O_2) = 2\langle Al_2O_3 \rangle + 3(CO_2)$$

and found it to be -4332.1 kJ/mole. (Probable error ± 31.4 kJ) The enthalpy of formation of $\langle Al_2O_3 \rangle$ is -1675.7 ± 1.3 kJ/mole for constant volume, and that of (CO_2) is -393.5 ± 0.13 kJ/mole. Then the enthalpy of forming $\langle Al_4C_3 \rangle$ may be calculated:

$$\Delta H^\circ_{298} = 3\Delta H^\circ_{CO_2} + 2\Delta H^\circ_{Al_2O_3} - 6\Delta H^\circ_{O_2} - \Delta H^\circ_{Al_4C_3}$$

$$-4332.1 = -1180.5 - 3351.4 - 0 - \Delta H^\circ_{Al_4C_3}$$

$$\Delta H^\circ_{Al_4C_3} = -4531.9 + 4332.1 = -199.8 \text{ kJ.}$$

One disadvantage of applying Hess's law is demonstrated by this evaluation. The errors of the separate ΔH determinations are additive, giving an overall error of $31.4 + 3 \times 0.13 + 2 \times 1.3 = 34.4$ kJ. The calculated enthalpy of formation $\Delta H_{Al_4C_3} = -199.8 \pm 34.4$ kJ is therefore uncertain to the extent of 17%. This shows how the evaluation of an enthalpy of formation as a small difference of large values makes it subject to a relatively great percentage error even if the original errors are quite small (a maximum of 0.7% in this example). For comparison, the accepted value of the enthalpy of formation of Al_4C_3 is -209.2 kJ/mole.

The determination of enthalpies formation will be described in the next chapter, but the principal methods are summarised below:

(1) Measurements of the change in temperature resulting from the reaction taking place in a system of known heat capacity (a calorimeter). Hess's law may be applied in connection with this method. For instance, the enthalpy of formation of an alloy may be obtained by measuring the enthalpy of solution of the alloy and of its constituents in a given solvent.

(2) Determination of the temperature coefficient of the equilibrium constant of reactions involving gases. The enthalpy of formation or reaction is calculated with **van't Hoff's isobar***

$$\Delta H^\circ_T = \boldsymbol{R}T^2 \frac{\mathrm{d} \ln K_p}{\mathrm{d}T} = -\boldsymbol{R} \frac{\mathrm{d} \ln K_p}{\mathrm{d}(1/T)}. \qquad (7)$$

In practice K_p is determined at two or more temperatures ($T_1, T_2, \ldots$) or two values K_{p1} and K_{p2} are taken from a curve of $\log K_p$ plotted against $1/T$. Then by integration, eqn. (7) becomes

*The term *isochore* is perhaps more familiar, but is a misnomer, so the relationship is referred to as an *isobar* throughout this monograph.

$$\Delta H_T^\circ = 19.144\ T_1 T_2 \frac{(\log K_{p1} - \log K_{p2})}{T_1 - T_2}\ \text{J} \tag{7a}$$

or

$$\Delta H_T^\circ = 19.144 \frac{(\log K_{p1} - \log K_{p2})}{1/T_2 - 1/T_1}\ \text{J}$$

which gives a mean value for the enthalpy of formation in the range T_1 to T_2.

If only one of the compounds taking part in the reaction is gaseous, then eqn. (7) can be written

$$\Delta H_T^\circ = \pm RT^2 \frac{\mathrm{d}\ln p}{\mathrm{d}T} \tag{8}$$

or in the integrated form

$$\Delta H_T^\circ = \pm 19.144\ T_1 T_2 \frac{\log p_1 - \log p_2}{T_1 - T_2}\ \text{J} \tag{8a}$$

(which has a positive numerical value if the gaseous component is on the right side of the chemical equation).

Equation (8) is a general relationship, which is perhaps better known in its more particular form as the **Clausius–Clapeyron** equation applied to the vapour pressure of a pure substance, when eqns. (8) and (8a) may be re-written

$$L_e = \boldsymbol{R}T^2 \frac{\mathrm{d}\ln p}{\mathrm{d}T} = -\boldsymbol{R} \frac{\mathrm{d}\ln p}{\mathrm{d}(1/T)} \tag{9}$$

$$L_e = 19.144\ T_1 T_2 \frac{\log p_1 - \log p_2}{T_1 - T_2}\ \text{J}. \tag{9a}$$

Here L_e is the average latent heat of evaporation over the temperature range T_1 to T_2.

(3) Equation (8) is also applicable to solutions provided that the dissociation pressures p are measured at various temperatures with the same solution composition. In this case the so-called partial enthalpy of solution of the volatile constituent is obtained (see pp. 27–33). It can be shown that the enthalpy of solution is related to the solubility of the substance (c_{sat}) in the solvent by

$$\Delta H_{soln} = \boldsymbol{R}T^2 \frac{\mathrm{d}\ln c_{sat}}{\mathrm{d}T} \tag{10}$$

or

$$\Delta H_{soln} = 19.144\ T_1 T_2 \frac{\log c_{sat(1)} - \log c_{sat(2)}}{T_1 - T_2} \tag{10a}$$

where ΔH_{soln} is the constant, partial enthalpy of saturated solution over the temperature range T_1 to T_2 and the concentration range $c_{sat(1)}$ to $c_{sat(2)}$. It is, of course, essential that the saturated solution should be in equilibrium with the same phase at all temperatures T_1 to T_2, i.e. either with the volatile solute of constant pressure (e.g. 1 atm) or with a compound formed by solute and solvent. Thus, in the special case of eqn. (10) under the restriction mentioned, enthalpies of solution can be calculated from equilibrium data obtained at varying concentration of solute.

Equation (10) is further restricted to the so-called regular solutions the properties of which will be discussed later (pp. 50–53). This restriction implies as a rule that the equation may only be used for rather dilute solutions in which the degree of interaction between the solute atoms or ions is not likely to be significant. It follows that eqn. (10) is useful mainly for low-solubility systems, to which it has frequently been applied, for instance to the metallurgically important solutions of hydrogen in various metals.

As an example, the low solubility of oxygen in liquid bismuth may be mentioned. From the solubility data (1953b) obtained at various temperatures, an enthalpy of +128,867 J/mole of oxygen is calculated by means of eqn. (10a). This solution is in equilibrium with solid bismuth oxide, and the corresponding chemical reaction is to be written as follows:

$$\tfrac{2}{3}\langle Bi_2O_3\rangle = 2[O]_{\{Bi\}} + \tfrac{4}{3}\{Bi\}; \quad \Delta H_{soln} = 128{,}867.$$

If (O_2) of 1 atm pressure were to be taken as the standard state, the standard enthalpy of formation of $\tfrac{2}{3}\langle Bi_2O_3\rangle$ from liquid bismuth and gaseous oxygen must be added to ΔH_{soln}.

(4) The measurement of the electromotive force of a cell in which the chemical reaction under consideration is the agent which produces the current. This provides information about certain thermochemical values, including the enthalpy of formation or of reaction, which is calculated from the Gibbs–Helmholtz equation (pp. 18–19)

$$\Delta H_T = zF\left(T\frac{dE}{dT} - E\right) \tag{11}$$

where F is Faraday's constant (96,485 J/V), E is the e.m.f. in volts, and z is the number of coulombs which pass if the electrolytic reaction of which ΔH_T is the enthalpy change proceeds to completion.* ΔH_T is expressed in Joules.

Over temperature ranges of less than 100°C, the temperature coefficient of the e.m.f. is usually practically constant. This allows eqn. (11) to be used in the form

$$\Delta H_T = z \cdot 96{,}485\left(T\frac{E_1 - E_2}{T_1 - T_2} - \frac{E_1 + E_2}{2}\right). \tag{11a}$$

C. Temperature Dependence of the Enthalpy of Reaction

For exact calculation consideration must be given to the variation of enthalpies of reaction with temperature. We shall derive the relationship by discussing as an example the reaction

$$\langle Fe_2O_3\rangle + 3\langle C\rangle = 2\langle Fe\rangle + 3(CO).$$

Given the enthalpy of reaction at room temperature, we shall derive its value at 727°C (ΔH_{1000}). This is given, as is seen from Fig. 2, by the equation

$$\Delta H_{1000} = \Delta H_{298} + \Delta' H_R - \Delta' H_L \tag{12}$$

*Thus for the reaction $ZnCl_2 + Cu = CuCl_2 + Zn, z = 2$.

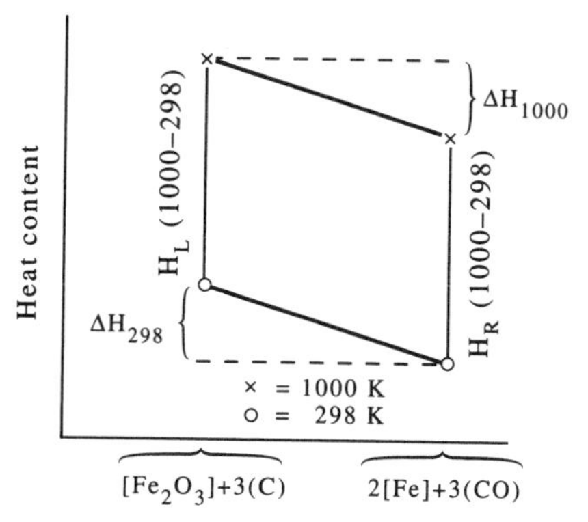

FIG. 2. Illustration of Kirchhoff's law.

where $\Delta' H_L$ and $\Delta' H_R$ are the changes in heat contents of the reactants and products respectively over the temperature range 25–727°C.* These quantities can be obtained by multiplying together the temperature change and the mean heat capacity values:

$$\Delta' H = (T_2 - T_1)\bar{C}_p. \tag{13}$$

In general, however, heat capacities are not tabulated as mean values for various temperature ranges, but as specific heat or atomic heat equations which include powers of T.

The atomic or molar heat of a substance is the amount of heat required to raise the temperature of one g-atom or one mole respecitvely by one degree. This equals the corresponding change in heat content, which we may express as $d(\Delta H)/dT$. In our example

$$\frac{d(\Delta H)}{dT} = 2C_{p(Fe)} + 3C_{p(CO)} - C_{p(Fe_2O_3)} - 3C_{p(C)} = \Delta C_p.$$

Integrating between temperatures T_1 and T_2,

$$\int_{\Delta H_1}^{\Delta H_2} d(\Delta H) = \Delta H_2 - \Delta H_1 = \int_{T_1}^{T_2} \Delta C_p \, dT. \tag{14}$$

This is simply one way of expressing **Kirchhoff's equation** (1858). For its evaluation the heat capacities of each participant must be known over the whole temperature range under consideration. Heat capacities are usually expressed in equations of the form

$$C_p = a + bT + cT^2 + dT^{-2} + eT^{-1/2}$$

(of which not all the terms need to be present in any single formula), so no difficulty in integration should be experienced:

*In thermochemical calculation, changes in heat content, entropy and Gibbs energy fall into two categories, and confusion can arise between the two. The first includes those taking place in an isothermal reaction, and these are denoted by the ordinary Δ symbol. The other describes the change in value of the thermodynamic functions of a substance due to change in temperature, and will be denoted by the symbol Δ'. Equation (12) illustrates the use of the two symbols and the necessity for keeping their identity separate though they represent similar physical quantities.

$$\int C_p \, \mathrm{d}T = aT + \tfrac{1}{2}bT^2 + \tfrac{1}{3}cT^3 - dT^{-1} + 2eT^{1/2} + k$$

where k is a constant.

To determine the change in heat content due to a change in temperature from T_1 to T_2, the integral becomes

$$\int_{T_1}^{T_2} C_p \, \mathrm{d}T = a(T_2 - T_1) + \tfrac{1}{2}b(T_2^2 - T_1^2) + \tfrac{1}{3}c(T_2^3 - T_1^3)$$

$$- \mathrm{d}(T_2^{-1} - T_1^{-1}) + 2e(T_2^{1/2} - T_1^{1/2})$$

in which the constant k disappears.

In making our calculation, we must first add up the total heat capacities of products and reactants.* Taking the above reaction, over the range 298–1000K:

$$\left.\begin{aligned} 2 \times C_p(\alpha - \mathrm{Fe}) &= 34.98 + 49.54 \times 10^{-3}T \\ 3 \times C_p(\mathrm{CO}) &= 85.23 + 12.30 \times 10^{-3}T - 1.38 \times 10^5 T^{-2} \end{aligned}\right\} +$$

$$120.21 + 61.84 \times 10^{-3}T - 1.38 \times 10^5 T^{-2}$$

$$\left.\begin{aligned} C_p(\mathrm{Fe_2O_3}) &= 98.28 + 77.82 \times 10^{-3}T - 14.85 \times 10^5 T^{-2} \\ 3 \times C_p(\mathrm{C}) &= 51.46 + 12.80 \times 10^{-3}T - 26.36 \times 10^5 T^{-2} \end{aligned}\right\} +$$

$$149.74 + 90.62 \times 10^{-3}T - 41.21 \times 10^5 T^{-2}.$$

The difference of the two sums gives

$$\Delta C_p = -29.53 - 28.78 \times 10^{-3}T + 39.83 \times 10^5 T^{-2}.$$

On integration between the temperatures 298 and 1000K we obtain

$$\int \Delta_{C_p} \, \mathrm{d}T = -29.53(1000 - 298) - 14.39 \times 10^{-3}(1000^2 - 298^2) - 39.83$$

$$\times 10^5 (1000^{-1} - 298^{-1})$$

$$= -24{,}468.$$

Now ΔH at 298K is given by*

$$\langle \mathrm{Fe_2O_3} \rangle + 3\langle \mathrm{C} \rangle = 2\langle \mathrm{Fe} \rangle + 3(\mathrm{CO})$$

$$-824{,}248 \qquad 0 \qquad 0 \qquad -331{,}373$$

$$\Delta H_{298} = -331{,}373 + 824{,}248 = +492{,}875$$

whence

$$\Delta H_{1000} = \Delta H_{298} + \int_{298}^{1000} \Delta C_p \, \mathrm{d}T = 492{,}875 - 24{,}468 = 468{,}407.$$

*Since it is the principle of this calculation that matters, the numerical values have not been altered compared with those in earlier editions although more recent thermochemical values have partly been adopted in the tables attached to this monograph.

If any of the components taking part in a chemical reaction undergoes change of state of order in the temperature range under consideration (it may be a transformation, fusion or evaporation) the heat effect must be taken into account. The enthalpy of transformation of a reactant must be subtracted from the total while that of a product must be added. In the above reaction, iron undergoes magnetic (order–disorder) transformation in the neighbourhood of 1000 K accompanied by an enthalpy of transformation of 2761 J/g-atom. The corrected enthalpy of formation, taking iron in its normal state at 1000 K, is then

$$\Delta H_{1000} = 468{,}407 + 5522 = 473{,}930$$

In calculating the change of enthalpy for higher temperatures, the data of the transformed material must be included. Kirchhoff's equation is modified to the form

$$\Delta H_2 - \Delta H_1 = \int_{T_1}^{T_t} \Delta C_p \, \mathrm{d}T \pm L_t + \int_{T_t}^{T_2} \Delta C_p \, \mathrm{d}T \tag{15}$$

L_t is the enthalpy of transformation which takes place at the temperature T_t. It is subtracted for the transformation of a reactant, and added for that of the product. Any further transformation, including melting and evaporation, must be allowed for in the same way.

Table I, pp. 259–325 gives the best values known to the authors of the heat capacities of a large number of substances. Even if this list were complete as far as the compounds are concerned, the research worker will often require to derive new formulae from more recent and accurate measurements. To illustrate the method used, the curve for the true heat capacity (atomic heat) of nickel over the range 0–900 K is given in Fig. 3. We see that the heat capacity increaes from zero at 0 K to about 29.5 J/K . mol at 450 K. Between 450 and 650 K there occurs a discontinuity in the C_p curve with a sharp maximum at 631 K corresponding to the magnetic transformation of nickel. These heat capacities are obtained by direct experimental methods.

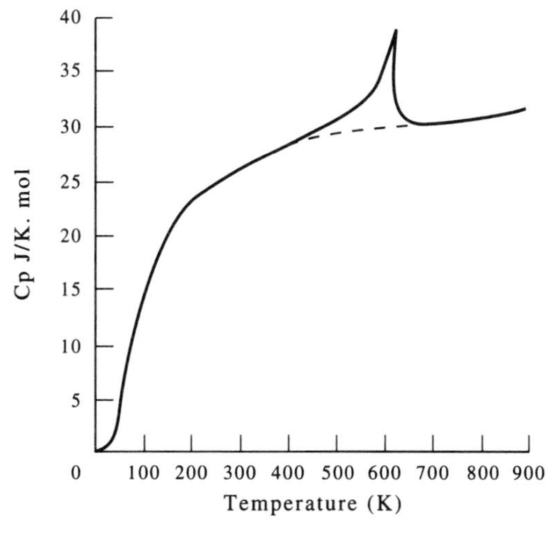

FIG. 3. Heat capacity of nickel from 0 to 900 K.

From such a curve the change in heat content between two temperatures is evaluated by graphical integration. The increase in heat content between 298 and 900 K for instance is given by the area under the curve between these two temperatures.

One may also derive formulae for the heat capacity from smoothed curves of experimental results. Straight lines are expressed by the general equation

$$C_p = a + bT \tag{16}$$

where a and b are constants.

To describe the heat capacity of a substance over a wider temperature range the most commonly used four-term expression is:

$$C_p = a + bT + cT^2 + dT^{-2}. \tag{16a}$$

Only in exceptional cases will a T^3 term be needed.

After choosing the type of C_p expression, numerical values are used to evaluate the constants of the equation. For nickel in the temperature range 298–631 K the equation $C_p = 11.17 + 37.78 \times 10^{-3}T + 3.18 \times 10^5 T^{-2}$ describes the experimental values satisfactorily. The discontinuity in the curve in Fig. 3 near 631 K is characteristic of magnetic or order–disorder transformation which occurs over a relatively broad range of temperature. In such cases the enthalpy of transformation can be obtained only by drawing the probable continuation of the C_p curve if there were no transformation, as is done in Fig. 3, and evaluating graphically the area between this curve and the one actually observed. The enthalpy of the magnetic transformation in α-nickel is thus found to be 586 ± 85 J/g-atom. Heat effects of order–disorder transformations are generally not very large.

Phase transformations on the other hand occur normally at definite temperatures. They entail a total rearrangement of the atoms or molecules in a new structure, or a breakdown of the lattice (melting or evaporation). Their enthalpy effects are not evaluated from C_p/T curves in the way described above but are determined by other methods to be described later (pp. 65–95).

True heat capacities are seldom obtained by direct measurements so much as from measurements of change in heat content between a series of high temperatures and room temperature. We shall consider one example of such an evaluation.

Values of several observers for the molar heat content of cadmium at various temperatures have been used to construct Fig. 4. A smooth curve has been derived from the most reliable experimental points. The height of the discontinuity at the melting point (321°C) is equivalent to the enthalpy of fusion.

Since the form of the C_p curve can be expected to conform to the empirical formula (16a), we must use the corresponding curve for the heat content

$$\Delta' H_{298}^{T} = A + BT + CT^2 + DT^3 + ET^{-1}. \tag{17}$$

For liquid cadmium, the values are not very accurate. Since they do not appear to fall on a definite curve, a straight line must be drawn through the points and a mean value obtained for C_p for that temperature range.

$$\frac{H_2 - H_1}{T_2 - T_1} = \bar{C}_p = C_p \tag{18}$$

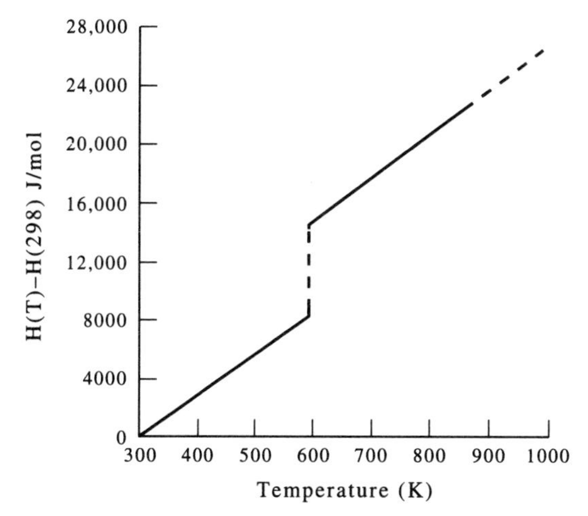

FIG. 4. Change of heat content of cadmium from 300 to 1000 K.

D. Entropy

The concept and nature of entropy follows from the **Second Law** of **Thermodynamics**. Its derivation, usually through the consideration of a reversible **Carnot cycle**, is a common feature of any textbook dealing with classical thermodynamics, so it will not be repeated here. It is not easily defined directly, but for the purpose of this monograph it may be interpreted in terms of *increase* in entropy, which is equivalent to the decrease in atomic order of a system undergoing any process; quantitatively, this is equal to the heat taken up isothermally and reversibly,* divided by the absolute temperature at which the process takes place (Boltzmann, 1866).

The true heat capacity (molar heat) of a substance is such a reversible heat effect, and the quantity C_p/T for any temperature is an entropy expression. Figure 5 shows the C_p/T curve derived from the C_p values for nickel in Fig. 3. Its form is characteristic of all such curves.

The equation for the entropy of a substance at any temperature T, corresponding to that for the heat content (p. 5), is

$$S_T = S_0 + \int_0^T \frac{C_p}{T}\,\mathrm{d}T. \tag{19}$$

It also follows from the Second Law of Thermodynamics that every substance in a given state at a given temperature has a certain unequivocal entropy as well as a heat content. Thus if a substance undergoes a cyclic process involving several intermediate changes, it will at the end have the same entropy as it had initially. We can therefore define an entropy of reaction ΔS_T, corresponding to the heat of reaction, replacing eqn (5) by the relationship

*A reversible process may be defined as one which is carried out under conditions such that the system at all times differs only infinitesimally from a state of equilibrium. A process produces maximum work, i.e. it works most efficiently, when done reversibly. Melting or evaporation at the melting or boiling point are examples of reversible processes. Since the true specific heat is measured in small steps of one degree it represents energy which is to all intents and purposes introduced reversibly.

$$\Delta S_T = (aS_{\mathrm{M}} + S_{\mathrm{R}_c\mathrm{X}_b}) - (S_{\mathrm{M}_a\mathrm{X}_b}) + cS_{\mathrm{R}}). \tag{20}$$

Combining this with eqn. (19) we obtain an expression for the entropy of reaction at any temperature T,

$$\Delta S_T = \Delta S_0 + \int_0^T \frac{\Delta C_p}{T}\,\mathrm{d}T. \tag{21}$$

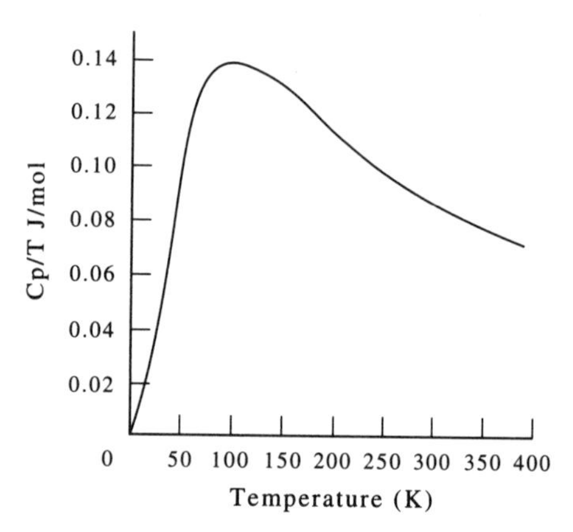

FIG. 5. C_p/T values for nickel between 0 and 400 K.

It was W. Nernst who first postulated (1906) that the entropy of reaction at absolute zero, ΔS_0, is nil, though at first this was not expressed in so simple a form. He and his co-workers established the validity of his assumption for reactions involving completely ordered crystallised substances only by measuring a number of heat capacities and reaction entropies down to very low temperatures.

Planck extended that deduction, and concluded (1912) that the entropy of any ordered crystalline substance at absolute zero S_0 should be zero. Applying this, eqn. (19) becomes

$$S_T = \int_0^T \frac{C_p}{T}\,\mathrm{d}T. \tag{22}$$

This conclusion has found general acceptance, and in this form it is the simplest formulation of the **Third Law of Thermodynamics**. Its significance to us is that, unlike the values of the heat contents (see p. 5), which are related to arbitary standards, those of entropies can be determined absolutely. They are obtained either from heat capacity measurements down to low temperatures and applying eqn. (22), or from reaction entropies as will be described later. For convenience in calculation, entropy values are tabulated for 25°C (=298.15 K). Termed S°_{298}, they are known as *standard entropies.*

The materials scientist will rarely be called upon to evaluate standard entropies from C_p measurements, since this is usually done by those who measure the heat capacities. The measurement of C_p in the neighbourhood of absolute zero is obviously rather difficult. If measurements are not extended much below 50 K, the **Debye equation** for the molar heat of a solid at constant volume, C_v, in terms of its absolute temperature in this low range:

$C_v = 465T^3/\Theta^3$, may be applied. In this equation Θ is a characteristic constant for each solid, and can be generally derived from the slope of the C_p curve above 50 K. This enables the C_p curve to be extrapolated to absolute zero, the relationships between C_p and C_v being known.

This equation is based upon the assumption that a crystalline solid has the properties of a perfectly elastic body. This applies to temperatures near zero and the agreement is good for temperatures up to a value of $T = \Theta/12$ (Θ for the elements varies between 50 [calcium] and 1840 [carbon]). Debye's equation cannot be applied at higher temperatures.

To clarify further the evaluation of ΔS_T from heat capacity curves, reference may be made to Fig. 6. The molar heat of a compound and the sum total molar heats of its constituent elements are plotted against temperature in Fig. 6a. The shaded area represents the term $\int_0^T \Delta C_p \, dT$. Plotting the same values as C_p/T against T, we obtain Fig. 6b, in which the shaded area represents the entropy of formation of the compound at the temperature T:

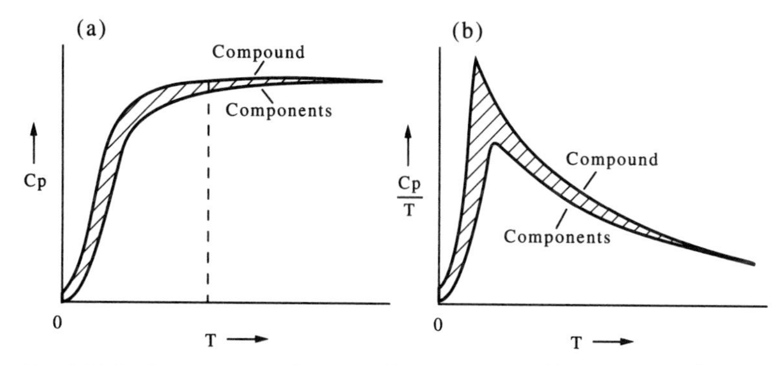

FIG. 6. Molar heat content and entropy of a compound and its component elements.

$$\Delta S_T = \int_0^T \frac{\Delta C_p}{T} \, dT. \tag{23}$$

If any phase change occurs in the temperature range considered, a transformation (t), fusion (f) or evaporation (e), its effect on the entropy must be taken into account. Such phase changes are thermodynamically reversible, so their associated entropy changes are obtained by dividing the respective latent heats by the temperatures at which they occur.

$$\Delta S_t = \frac{L_t}{T_t}; \quad \Delta S_f = \frac{L_f}{T_f}; \quad \Delta S_e = \frac{L_e}{T_e} \qquad \text{(24a–c)}$$

Equation (22) may then be expanded to the form;

$$S_T = \int_0^{T_t} \frac{C_p}{T} \, dT + \frac{L_t}{T_t} + \int_{T_t}^{T_f} \frac{C_p}{T} \, dT + \frac{L_f}{T_f} + \int_{T_f}^{T_e} \frac{C_p}{T} \, dT + \frac{L_e}{T_e} + \int_{T_e}^{T} \frac{C_p}{T} \, dT. \qquad \text{(22b)}$$

Nitrogen, for instance, has a transformation at $-237.6°C$ (35.5 K), its melting point at

$-209.9°C$ (63.2 K) and boiling point at $-195.8°C$ (77.3 K). Its standard entropy is therefore

$$S_{298}(N_2)=\int_0^{35.5}\frac{C_p}{T}\,dT+\frac{L_t}{35.5}+\int_{35.5}^{63.2}\frac{C_p}{T}\,dT+\frac{L_f}{63.2}$$

$$+\int_{63.2}^{77.3}\frac{C_p}{T}\,dT+\frac{L_e}{77.3}+\int_{77.3}^{298.1}\frac{C_p}{T}\,dT$$

and each heat effect must be determined in order to evaluate the standard entropy.

We have already mentioned that such evaluations are usually carried out by the observers of these heat effects, and do not require calculation by the materials scientist himself. Equation (22b) is, however, equally applicable to the determination of entropy values above room temperature when phase changes occur within the temperature range covered.

It was emphasized that eqn. (22) applies only to completely ordered, crystalline substances, and eqn. (23) to reactions involving such substances. Alloys are sometimes disordered at room temperature and may even remain so down to absolute zero. In such cases S_0 (eqn. (19)) has a non-zero value and must be taken into consideration in eqns (22) and (23). The zero-point entropy of completely disordered binary mixtures (e.g. alloys) is given by

$$S_0=-\boldsymbol{R}(N_1\ln N_1+N_2\ln N_2) \qquad (24)$$

N_1 and N_2 being the atomic fractions of the two components of a binary alloy. S_0 amounts to 5.77 J/deg for the atomic fraction $N_1=N_2=0.5$. Thus the formation of a disordered binary alloy phase is accompanied by an entropy change of

$$\Delta S_T=\int_0^T\frac{\Delta C_p}{T}\,dT-\boldsymbol{R}(N_1\ln N_1+N_2\ln N_2). \qquad (25)$$

The Third Law of Thermodynamics in addition does not apply to substances in a metastable state such as glass. It also leads logically to the postulate that every substance should be in an ordered crystalline state at absolute zero, if in equilibrium. In so many cases, however (e.g. the gold–silver alloys), this state is unattainable, since the disorder–order transformation temperature is too low to permit any observable atomic rearrangement within any practical length of time.

Entropies and entropy changes can be determined by several methods.

(1) The determination of entropies from low-temperature heat capacities, and of reaction entropies from equilibrium measurements has already been outlined.

(2) The reaction entropy of a chemical process which can be harnessed in a galvanic cell is obtained from the temperature coefficient of the resultant e.m.f. by the relationship

$$\Delta S=zF\frac{dE}{dT} \qquad (26)$$

or, for limited temperature ranges,

$$\Delta S=z\,96{,}485\,\frac{E_1-E_2}{T_1-T_2}\,\text{J}. \qquad (26a)$$

(3) From measurements of the equilibrium constant at different temperatures, reaction entropy is given by equation

$$\Delta S_T^\circ = \boldsymbol{R}\left(T\frac{\mathrm{d}\ln K_p}{\mathrm{d}T} - \ln K_p\right) \tag{27}$$

or, for a limited temperature range,

$$\Delta S_T^\circ = 19.144\left(T\frac{\log K_{p(1)} - \log K_{p(2)}}{T_1 - T_2} - \log K_p\right)\mathrm{J}. \tag{27a}$$

E. Gibbs Energy

While Berthelot and Thomsen were, over a century ago, carrying out their work on enthalpies of reaction in the belief that they were identical with the "driving force" of a reaction, it became clear from the work of Carnot, van't Hoff, Gibbs and Helmholtz that there was another thermodynamic function which conformed more exactly to the conception of a driving force. This quantity was the **Gibbs energy*** $\boldsymbol{G}$ which for any substance under given conditions is the maximum portion of its energy which can be converted into mechanical work. Any process which takes place is accompanied by a change in Gibbs energy of the system, and this change is equal to the net reversible work done by or absorbed during the process. It is a maximum when the process takes place reversibly. Thus the expression

$$\Delta G = G_{\text{products}} - G_{\text{reactants}} \tag{28}$$

represents the maximum mechanical work that can be carried out by a reaction at constant pressure.

The Gibbs energy change during a reaction is related to the various other functions associated with the reaction and these will be dealt with in some detail.

In van't Hoff's equilibrium-box experiments, Gibbs energy is derived as the maximum work ΔG done in a system when the reactants in a chemical process pass reversibly from their initial state to the one in which they are in equilibrium with and are transformed to the products, which are brought in turn to their final state. The Gibbs energy is then given by the relationship

$$\Delta G = -\boldsymbol{R}T\ln K_p + \boldsymbol{R}T\sum n\ln p \tag{29}$$

in which the term $\sum n \ln p$ defines the initial and final states of the system, n and p representing respectively the number of moles and the pressure of the participants in the reaction. (The n values for the reactants are negative, for the products positive.) It is sometimes written $\ln k_q'$, but K_q' is not an equilibrium constant. At any temperature the term $\boldsymbol{R}T\ln K_p$ is constant for any given reaction but $\boldsymbol{R}T\sum n\ln p$ is not, and the Gibbs energy change therefore varies according to the pressures of the gases in the initial and

*The term Gibbs energy (G) used in this monograph is also known as the thermodynamic potential and is related to the heat content H by the equation $G = H - TS$. There is another term (F) which is also described as Gibbs energy, but more precisely as Helmholtz function or the maximum work function which is related to the internal energy U by the equation $F = U - TS$. Thus $G = F + pV$. It should be noted that the symbol F is often used, particularly in American literature, to denote Gibbs energy.

final states of the system. If the pressures are all fixed at unity, the term $\boldsymbol{RT}\sum n \ln p$ vanishes. Thus defined, the term

$$\Delta G^\circ = -\boldsymbol{R}T \ln K_p \tag{29a}$$

becomes the standard Gibbs energy change of the reaction under consideration.

In a reaction involving only one gaseous component, the equilibrium contant consists only of its partial pressure p. This applies similarly to the Gibbs energy change: hence if the gas is a *product* of the reaction,

$$\Delta G^\circ = -\boldsymbol{R}T \ln p \quad \text{or} \quad \Delta G^\circ = -19.144\, T \log p. \tag{30}$$

The same equation applies to the Gibbs energy of evaporation of a single substance. At the boiling point or sublimation point p is unity and ΔG° is nil: there is no driving force to alter the equilibrium since it involves the participants in their standard states.

It should be noted that standard Gibbs energy changes are conventionally denoted as ΔG°.

There is another method, in addition to van't Hoff's equilibrium-box operation, by which a reaction can be made to take place reversibly and the associated mechanical work measured. This is to harness the reaction in an electrolytic cell and to measure the reversible electromotive force it produces. In practice this is measured by a null method; i.e. one in which there is no passage of current. Under these conditions, the reaction is in effect tending reversibly toward equilibrium, and its Gibbs energy is given by

$$\Delta G = -zFE \tag{31}$$

where F is the Faraday, zF the number of coulombs which pass if the reaction proceeds to completion, and E is the e.m.f. With ΔG expressed in Joules and E in volts, equ. (31) may be re-written

$$\Delta G = -z.\, 96{,}485\, E \text{ (J)}. \tag{31a}$$

This is the most direct method of measuring Gibbs energies of reaction, which can then be used of course for calculating equilibrium constants. Gases involved in e.m.f. measurements must have a pressure of 1 atm to fulfil the conditions discussed above. Electromotive force measurements do not only provide the values of the Gibbs energy of formation of equilibria, but also of complete condensed reactions.

For instance, the e.m.f. of the common Daniell cell gives the Gibbs energy of the complete reaction

$$\langle \mathrm{Zn} \rangle + [\mathrm{CuSO_4}]_{\mathrm{aq}} \rightarrow \langle \mathrm{Cu} \rangle + [\mathrm{ZnSO_4}]_{\mathrm{aq}}$$

or, more precisely,

$$\langle \mathrm{Zn} \rangle + [\mathrm{Cu^{++}}]_{\mathrm{aq}} \rightarrow \langle \mathrm{Cu} \rangle + [\mathrm{Zn^{++}}]_{\mathrm{aq}}.$$

The connection between the enthalpy and entropy of reaction and the maximum work was first derived by Gibbs (1875) and independently by Helmholtz (1882). The relationship is nowadays written in the form

$$\Delta G = \Delta H - T\Delta S \tag{32}$$

or, according to its initial derivation from Carnot's cycle,

$$\Delta G = \Delta H + T\frac{\mathrm{d}\Delta G}{\mathrm{d}T}. \tag{33}$$

Equation (32) is valid for values of ΔG, ΔH and ΔS at the one temperature T. Since values for ΔH and ΔS are usually known and tabulated for room temperature, the values of ΔG may be calculated directly for this temperature. For values at higher temperatures heat capacities must be considered, and equs (15 and (23) included, to obtain

$$\Delta G_T = \Delta H_{298} + \int_{298}^{T} \Delta C_p \,\mathrm{d}T - T\Delta S_{298} - T\int_{298}^{T} \frac{\Delta C_p}{T}\,\mathrm{d}T. \tag{34}$$

If any phase transition should occur in the range 298–T K its entropy and heat have also to be considered, as in eqns (15) and 22b). It may be permissible, as a first approximation, to neglect the heat capacity terms, thus:

$$\Delta G_T = \Delta H_{298} - T\Delta S_{298} \tag{32a}$$

but enthalpies and entropies of formation, fusion or evaporation are usually too large to be omitted.

If the values of both ΔG and the equilibrium constant at a given temperature are required, eqn. (34) must be applied. An example will be given of such an evaluation. For this purpose eqn. (34) is expanded into the form

$$\Delta G_T = \left(\Delta H_{298} - \int_0^{298} \Delta C_p \,\mathrm{d}T\right) + \int_0^{T} \Delta C_p \,\mathrm{d}T$$

$$- T\left(\Delta S_{298} - \int_0^{298} \frac{\Delta C_p}{T}\,\mathrm{d}T\right) - T\int_0^{T} \frac{\Delta C_p}{T}\,\mathrm{d}T. \tag{34a}$$

In practice, the numerical value of the term

$$\Delta H_{298} - \int_0^{298} \Delta C_p \,\mathrm{d}T = \Delta H_{(0)}$$

does not usually represent the enthalpy of reaction at absolute zero, since the C_p formulae used for these evaluations are generally *valid only for temperatures above* 298 K. Thus $\Delta H_{(0)}$ is a value used for calculation and has no fundamental or effective meaning. The same also applies to the term

$$\Delta S_{298} - \int_0^{298} \frac{\Delta C_p}{T}\,\mathrm{d}T = \Delta S_{(0)}.$$

The fictitious nature of these quantities is denoted by the parenthesised zero in the suffix.

An example will be given of such an evaluation,* that of the reaction

$$2\langle \mathrm{Al}\rangle + 1\tfrac{1}{2}(\mathrm{O_2}) = \langle \mathrm{Al_2O_3}\rangle. \tag{α}$$

*The C_p equations used in this example may not exactly agree with those in Table I. However, since the evaluation is simply an exercise, it has been deemed unnecessary to revise these data.

The heat of formation of Al_2O_3 is $\Delta H^\circ_{298} = -1{,}675{,}700$ J/mol and its entropy of formation is

$$\Delta S^\circ_{298} = S^\circ_{Al_2O_3} - 2S^\circ_{Al} - 1\tfrac{1}{2}S^\circ_{O_2} = 50.92 - 56.6 - 307.6 = -313.33 \text{ J/K.mol.}$$

The sum of the heat capacities of the reactants is given by

$$C^\circ_p(1\tfrac{1}{2}O_2) = 44.94 + 6.28 \times 10^{-3}T - 2.51 \times 10^5 T^{-2}$$

$$C^\circ_p(2Al) = 41.34 + 24.77 \times 10^{-3}T$$

$$\sum C^\circ_p = 86.28 + 31.05 \times 10^{-3}T - 2.51 \times 10^5 T^{-2}$$

which is subtracted from the molar heat of $\langle Al_2O_3 \rangle$

$$C^\circ_p(Al_2O_3) = 114.77 + 12.80 \times 10^{-3}T - 35.44 \times 10^5 T^{-2}$$

to give

$$\Delta C^\circ_p = 28.49 - 18.24 \times 10^{-3}T - 32.93 \times 10^5 T^{-2}.$$

Integration of this equation gives

$$\int \Delta C^\circ_p \, dT = 28.49T - 9.12 \times 10^{-3}T^2 + 32.93 \times 10^5 T^{-1} + k_1.$$

For $T = 298$,

$$\int \Delta C^\circ_p \, dT = 8491.4 - 809.6 + 11050 + k_1$$

$$= 18{,}732 + k_1.$$

Whence

$$\int_{298}^{T} \Delta C^\circ_p \, dT = 28.49T - 9.12 \times 10^{-3}T^2 + 32.93 \times 10^5 T^{-1} - 18{,}732.$$

The ΔC°_p equation divided by T and integrated gives in the same manner

$$\int \frac{\Delta C^\circ_p}{T} \, dT = 28.49 \ln T - 18.24 \times 10^{-3}T + 16.47 \times 10^5 T^{-2} + k_2$$

$$\int^{T=298} \frac{\Delta C^\circ_p}{T} \, dT = 162.34 - 5.44 + 18.58 + k_2 = 175.48 + k_2$$

whence

$$\int_{298}^{T} \frac{\Delta C^\circ_p}{T} \, dT = 28.49 \ln T - 18.24 \times 10^{-3}T + 16.47 \times 10^5 T^{-2} - 175.48.$$

Reverting to eqn. (34),

$$\Delta G^\circ_T = \Delta H^\circ_{298} + \int_{298}^{T} \Delta C^\circ_p \, dT - T\Delta S^\circ_{298} - T\int_{298}^{T} \frac{\Delta C^\circ_p}{T} \, dT$$

$$= (-1{,}675{,}700) + (28.49T - 9.12\times 10^{-3}T^2 + 32.93$$

$$\times 10^5 T^{-1} - 18{,}732) - (-313.33T) -$$

$$(28.49T \ln T - 18.24\times 10^{-3}T^2 + 16.47\times 10^5 T^{-1} - 175.48T)$$

$$= -1{,}694{,}432 - 65.61T \log T + 9.12\times 10^{-3}T^2 + 16.46$$

$$\times 10^5 T^{-1} + 517.3T. \qquad (34\alpha)$$

This is typical of the formulae usually obtained for change in Gibbs energy with temperature.

The temperature-independent term is equivalent to

$$\Delta H^\circ_{298} - \int_0^{298} \Delta C^\circ_p \, dT,$$

which is the fictitious standard change in heat content for the reaction at absolute zero, i.e. $\Delta H_{(0)}$. We may then write the general formula for the standard Gibbs energy change of a reaction as

$$\Delta G^\circ_T = \Delta H^\circ_{(0)} + aT \log T + bT^2 + cT^{-1} + IT. \qquad (35)$$

Continuing our evaluation of the Gibbs energy of formation of alumina, we note that the heat capacity values we have used are valid in the following ranges: $\langle Al_2O_3 \rangle$, 273–3000 K; (O_2), 298–1500 K; and $\langle Al \rangle$, 273–932 K (m.p.). Formula (34a) is therefore applicable at temperatures between 298 and 932 K. To evaluate the Gibbs energy of reaction above this temperature range, we must introduce the Gibbs energy of melting of aluminium

$$2\{Al\} = 2\langle Al \rangle. \qquad (\beta)$$

The reaction is written in this way so that its simple addition to eqn. 34α (α) will give

$$2\{Al\} + 1\tfrac{1}{2}(O_2) = \langle Al_2O_3 \rangle. \qquad (\gamma)$$

The Gibbs energy formulae will likewise be additive.

The enthalpy of reaction (β) is double the enthalpy of fusion of aluminium and opposite in sign

$$\Delta H^\circ_{932}(\beta) = -2\times 10{,}460 = -20{,}920.$$

Further,

$$\Delta S^\circ_{932}(\beta) = -20{,}920 \div 932 = -22.45.$$

The further calculation of $\Delta G^\circ_T(\beta)$ is similar to that of $\Delta G^\circ_T(\alpha)$

$$2C^\circ_p(\text{Al, solid}) = 41.34 + 24.77 \times 10^{-3}T$$

$$2C^\circ_p(\text{Al, liquid}) = 58.58 \qquad (932\text{–}1273\ K)$$

$$\Delta C^\circ_p = -17.24 + 24.77 \times 10^{-3}T$$

$$\int \Delta C^\circ_p\, dT = -17.24T + 12.39 \times 10^{-3}T^2 + k_3$$

$$\int \frac{\Delta C^\circ_p}{T}\, dT = -17.24 \ln T + 24.77 \times 10^{-3}T + k_4$$

$$\int_{932}^{T} \Delta C^\circ_p\, dT = -17.24T + 12.39 \times 10^{-3}T^2 + 5309$$

$$\int_{932}^{T} \frac{\Delta C^\circ_p}{T}\, dT = -39.71 \log T + 24.77 \times 10^{-3}T + 94.81$$

$$\begin{aligned}\Delta G^\circ_T(\beta) &= \Delta H^\circ_{932}(\beta) \\ &\quad + \int_{932}^{T} \Delta C^\circ_p\, dT - T\,\Delta S^\circ_{932}(\beta) - T\int_{932}^{T} \frac{\Delta C^\circ_p}{T}\, dT \\ &= (-20{,}920) + (-17.24T + 12.39 \times 10^{-3}T^2 + 5309) \\ &\quad -(-22.38T) - (-39.71T \log T + 24.77 \\ &\quad \times 10^{-3}T^2 + 94.81\ T) \\ &= -15{,}610 + 39.71T \log T - 12.39 \\ &\quad \times 10^{-3}T^2 - 89.66T. \end{aligned} \qquad (34\beta)$$

This is the equation for the change in Gibbs energy of melting of 2 moles of aluminium. As with all equilibrium phase changes, it is zero at the melting point itself, but it has a finite value which is additive in computing Gibbs energies above the melting point from values which are valid below that temperature. To obtain the Gibbs energy of reaction (γ), $\Delta G_T(\alpha)$ and $\Delta G_T(\beta)$ are added, giving

$$\begin{aligned}\Delta G^\circ_T(\gamma) &= -1{,}710{,}042 - 25.9T \log T - 3.27 \times 10^{-3}T^2 \\ &\quad + 16.46 \times 10^5 T^{-1} + 427.64T.\end{aligned} \qquad (34\gamma)$$

Since the C_p formula for liquid aluminium is valid up to 1273 K, eqn. (34γ) is reliable from the melting point of aluminium up to this temperature, i.e. from 932 to 1273 K.

The heat capacity equation which has been adopted for these calculations is of a standard form which successfully represents the behaviour of most substances from 298 K and higher temperatures. The term in T^{-2} dominates the curvature of the heat capacity in most circumstances only up to 600–700 K. Above this temperature range, the contribution to the heat capacity from this term is normally negligible, and the term which is linear in the temperature accounts for the variation of the heat capacity of a given phase in the given state. For thermochemical calculations we are more frequently interested in the Gibbs energy change of a reaction than in the heat capacities of the individual reactants and products. An approximate value for the standard Gibbs energy change of a reaction at temperatures a few hundred degrees above 298 K can be obtained using calorimetric data for ΔH°_{298} and low-temperature data for ΔS°_{298}. Thus, corresponding to equ. (32a):

$$\Delta G^\circ_T \text{ (approx.)} = \Delta H^\circ_{298} - T\Delta S^\circ_{298} \tag{36}$$

the difference between the real Gibbs energy change ΔG°_T and this approximate value is given by

$$\Delta G^\circ_T - \Delta G^\circ_T \text{ (approx.)} = (\Delta H^\circ_T - \Delta H^\circ_{298}) - T(\Delta S^\circ_T - \Delta S^\circ_{298}). \tag{37}$$

This deviation can be expressed in terms of the heat capacity change equation

$$\Delta C_p = \Delta\alpha + \Delta\beta T + \Delta\gamma T^{-2}$$

by means of the equation

$$(\Delta H^\circ_T - \Delta H^\circ_{298}) - T(\Delta S^\circ_T - \Delta S^\circ_{298}) = \int_{298}^{T} \Delta C_p \, \mathrm{d}T - T\int_{298}^{T} \frac{\Delta C_p}{T} \mathrm{d}T. \tag{38}$$

We define the function, the heat capacity deviation (h.c.d.) by the equation

$$\text{h.c.d.} = \frac{1}{T}\int_{298}^{T} C_p \, \mathrm{d}T - \int_{298}^{T} \frac{C_p}{T} \mathrm{d}T \tag{39}$$

for each reactant and product and the change in the h.c.d. function can be related to the heat capacity change thus:

$$\Delta\text{h.c.d.} = \frac{1}{T}\int_{298}^{T} \Delta C_p \, \mathrm{d}T - \int_{298}^{T} \frac{\Delta C_p}{T} \mathrm{d}T \tag{39a}$$

and substituting for ΔC_p and using ϕ for the fraction $298/T$, we find

$$\Delta\text{h.c.d.} = \Delta\alpha(-\phi + \ln\phi) + \Delta\beta T\left(\phi - \frac{\phi^2}{2} - \frac{1}{2}\right) + \Delta\gamma T^{-2}\left(\frac{1}{\phi} - \frac{1}{2\phi^2} - \frac{1}{2}\right). \tag{40}$$

Each of the bracketed functions will appear in the calculation of the Gibbs energy deviation for *any* reaction in which the recommended form of the heat capacity equation is employed, and the additional appropriate terms can readily be calculated for other heat capacity equations. If each term in Δh.c.d. is multiplied by the temperature under consideration for a given reaction, then the sum of these terms when added to the ΔG°_T (approx.) equation yields the correct standard Gibbs energy change at the temperature T.

Where a computer is not available, it is useful to tabulate the Gibbs energy deviation contributions since this table is of general applicability. Table I has been prepared for 200 K intervals from 400 to 2000 K, and if each term is multiplied by the appropriate value of $\Delta\alpha$, $\Delta\beta$ and $\Delta\gamma$ for a reaction of interest, then the deviation may be calculated. The reader can bring the values in the heat capacity table and those for the enthalpy of formation and entropy contents up to date as new results appear in the literature and, together with a suitable computer program for the Gibbs energy deviation or a tabulation

TABLE I. *Values of the Gibbs energy deviation contributions as a function of temperature (1966a, b)*

T (K)	$-T(1-\phi+\ln\phi)$ J	$-T^2\left(\phi-\frac{\phi^2}{2}-\frac{1}{2}\right)$ J	$-\frac{1}{T}\left(\frac{1}{\phi}-\frac{1}{2\phi^2}-\frac{1}{2}\right)$ J
298	0	0	0
400	67	21,765	0.63×10^{-3}
600	494	1.908×10^5	3.60×10^{-3}
800	1205	5.272×10^5	7.41×10^{-3}
1000	2130	10.309×10^5	11.59×10^{-3}
1200	3222	17.021×10^5	15.98×10^{-3}
1400	4452	25.401×10^5	20.42×10^{-3}
1600	5803	35.464×10^5	24.98×10^{-3}
1800	7259	47.196×10^5	29.54×10^{-3}
2000	8812	60.584×10^5	34.14×10^{-3}

as presented here, have a flexible calculation procedure to obtain standard Gibbs energy change of as high a precision as the current "data bank" will allow.

With reference to the values in Table I, it should be observed that the temperature-dependent parts of the heat capacities of substances have coefficients which are typically of the order $4–40\times10^{-3}$ for the coefficient β and $4–40\times10^5$ for γ, so these deviation terms only amount to a few kilo Joules under normal circumstances.

The errors in formulae such as (34γ) (p. 21) vary within wide limits, due mainly to the error in the experimentally determined value of the heats of formation at room temperature. A figure accurate to ±2000 J may be deemed a good one; errors of 4000–12,500 J are more the rule. The first consequence of this is that figures in equations such as (34α) and (34γ) may be rounded off without loss in accuracy. The next consequence may be demonstrated by drawing the $\Delta G/T$ curves of a number of substances, when it is seen that they do not differ much from straight lines. In most cases, therefore, no loss in accuracy is involved by drawing the ΔG vs T curve corresponding to an equation such as (34α), and using it to derive a simple equation of the type

$$\Delta G^\circ_T = \Delta H^\circ_x - T\Delta S^\circ_x \tag{36a}$$

ΔH°_x and ΔS°_x are two numerical values determining the $\Delta G^\circ/T$ relationship, and are not necessarily close in value to ΔH°_{298} and ΔS°_{298} for the reaction.

It will be demonstrated later (p. 170) that the change in molar heat contents in a reaction ΔC_p depends to the greatest extent on the change in number of moles of gaseous components when the reaction proceeds from one side to another. Thus in such a reaction as

$$\langle \mathrm{Ta}\rangle + 2\tfrac{1}{2}(\mathrm{Cl_2}) = \langle \mathrm{TaCl_5}\rangle$$

ΔC_p will have a relatively large value and there will be an appreciable curvature in the plot of ΔG_T against T over a range of a few hundred degrees. If, in addition, all the data used in evaluating ΔG_T are known with good accuracy, then one is justified in introducing a third term to express the $\Delta G_T/T$ relationship:

$$\Delta G^\circ_T = \Delta H^\circ_y + mT\log T + T\Delta S^\circ_y. \tag{35a}$$

This equation may be derived by drawing the curve and selecting three points on it, or by direct calculation from eqn (35a) using three temperatures. If the curves are to coincide at

300, 700 and 1200 K, the relationship between the constants of eqns (35) and (35a) are as folows:

$$\Delta H_y^\circ = \Delta H_{(0)}^\circ + 410 \times 10^3 b + 731 \times 10^{-5} c + 6.25d$$

$$m = a + 3.125 \times 10^3 b + 1.241 \times 10^{-5} c - 0.0224d$$

$$\Delta S_y^\circ = I - 8.78 \times 10^3 b - 4.37 \times 10^{-5} c + 0.0927d.$$

Of course, the availability of computer software now makes such ΔG calculations a rapid and reliable procedure.

Turning back to the Gibbs energy of dissociation of alumina, ΔG° values have been calculated from eqns (34α) and (34γ) and are given in Table II. Since the reactions involve only one gaseous component, oxygen, its equilibrium pressure can be calculated from eqn (30) which must be used in the form

$$\frac{2}{3}\Delta G_T^\circ = -19.144T \log \frac{1}{p_{O_2}} = 19.144T \log p_{O_2} \qquad (30\alpha,\gamma)$$

for formulae α and γ and for *one* mole oxygen. Pressures calculated from this equation are also tabulated in Table II.

TABLE II. *Oxygen pressure of alumina and its reduction with hydrogen*

°C	K	ΔG_γ	p_{O_2} (atm)	ΔG_δ	p_{H_2O}/p_{H_2}
727	1000	−1,358,920	4×10^{-48}	+781,280	2.4×10^{-14}
927	1200	−1,292,645	3×10^{-38}	+748,935	1.3×10^{-11}
1127	1400	−1,226,750	3×10^{-31}	+717,138	1.2×10^{-9}

To illustrate the use of these data we can consider whether the reduction of alumina with hydrogen is a practical proposition.

The relevant equation

$$\langle Al_2O_3 \rangle + 3(H_2) = 3(H_2O) + \{Al\} \qquad (\delta)$$

may be obtained by combining eqn (γ) with

$$3(H_2) + 1\tfrac{1}{2}(O_2) - 3(H_2O) \qquad (\varepsilon)$$

so that

$$\Delta G_T^\circ(\delta) = \Delta G_T^\circ(\varepsilon) - \Delta G_T^\circ(\gamma). \qquad (34\delta)$$

We may now turn to the simplified ΔG formulae obtained in the manner described above. These are

$$\Delta G_T^\circ(\varepsilon) = -718{,}602 + 56.23T \log T - 27.74T$$

$$\Delta G_T^\circ(\gamma) = -1{,}697{,}700 - 15.69T \log T + 385.85T$$

which lead by subtraction to

$$\Delta G_T^\circ(\delta) = +979{,}098 + 71.92T \log T - 413.59T.$$

With this the equilibrium constant of reaction (δ) can be calculated by means of eqn (29)

$$\frac{1}{3}\Delta G^{\circ}_{T}(\delta) = -19.144T\log\frac{p_{H_2O}}{p_{H_2}},$$

Table II also includes the values of this Gibbs energy change at 1000 to 1400 K. It is evident that this equilibrium is far towards the left of eqn (δ) and the reduction could be carried out only in a stream of extremely dry hydrogen, containing less than $3\times10^{-12}\%$ of water vapour, at 1000 K. Since this purity is hardly attainable in practice it must be concluded that alumina cannot be reduced by hydrogen at these temperatures (a fact which is already well known to the materials scientist). Even if this dryness of the gas were attainable, the reaction would be impractical owing to the enormous volume of circulation of hydrogen which would be required to remove the oxygen at its minute equilibrium pressure.

We have discussed means by which equilibrium constants may be derived from Gibbs energies and other thermal data. Conversely, equilibrium pressure measurements can be used to determine Gibbs energies of reaction or, by suitable combinations, Gibbs energies of formation of compounds. Such reactions as the following are typical:

$$\langle FeO\rangle + (H_2) = \langle Fe\rangle + (H_2O) \qquad (\eta)$$

$$\langle V_2O_5\rangle + (CO) = 2\langle VO_2\rangle + (CO_2) \qquad (\theta)$$

$$\langle CrCl_2\rangle + (H_2) = \langle Cr\rangle + 2(HCl) \qquad (\iota)$$

$$\langle WC\rangle + 2(H_2) = \langle W\rangle + (CH_4). \qquad (\kappa)$$

The Gibbs energy of formation of the metal compounds may be evaluated by measuring the equilibrium of the reaction and combining it with the Gibbs energy data of the complementary reaction,

$$(H_2) + \tfrac{1}{2}(O_2) = (H_2O) \qquad (\eta')$$

$$(CO) + \tfrac{1}{2}(O_2) = (CO_2) \qquad (\theta')$$

$$(H_2) + (Cl_2) = 2(HCl) \qquad (\iota)$$

$$\langle C\rangle + 2(H_2) = (CH_4). \qquad (\kappa')$$

For instance,

$$\Delta G^{\circ}_{T}(FeO) = \Delta G^{\circ}_{T}(H_2O) - \Delta G^{\circ}_{T}(\eta). \qquad (34\eta)$$

Methods of measuring Gibbs energies may be summarised as follows:

(1) Gibbs energies of substances or of reactions may be computed from heats of formation, entropies and specific heats, using the relationships expressed in eqns (34) and (32a).

(2) Gibbs energies of reaction or formation are obtained directly from measurements of the equilibrium constants of the reaction, applying the relationships in eqns (29) and (30).

(3) Gibbs energies of reaction are also directly obtainable from e.m.f. measurements if the reaction under consideration can be made to produce current in a galvanic cell, applying the relationship in eqn (31).

(4) Gibbs energies of reaction or formation may finally be obtained by suitable

combination of other Gibbs energy equations, as was shown in two examples: eqns (34δ) and (34η).

3. Solutions

A. Partial Molar Gibbs Energy

In the foregoing sections we have been dealing with elements or substances which are stable only at a fixed composition of simple atomic proportions, i.e. those to which **Dalton's rule** of simple and multiple proportions applies to a close approximation. This inflexibility of composition characterises organic and inorganic compounds in which the atoms are held together by the major valency bonds (homopolar and heteropolar). These bonds arise from the sharing or donation of certain of the electrons of the various atoms. Such compounds generally have an ordered structure in the stable solid state, and consequently they have no significant entropy at absolute zero.

The "metallic" binding forces which are found generally in alloy systems do not arise from such inflexible anisotropic exchanges of electrons. Rather, the constituent atoms are (to express it simply) kept in an ordered array by the so-called "electron gas". This binding does not require the constituent atomic species to be present in a simple stoichiometric proportion, and so intermetallic phases are often found to be stable within quite wide composition limits. These are defined by the relative ionic sizes and by the relative strengths of the metallic, homopolar and heteropolar components of the overall binding forces. These forces are rarely present singly in intermetallic phases.

Mixed binding forces also account for the relatively large ranges of homogeneity present in certain systems of oxygen, sulphur and nitrogen with transition metals, such as the titanium–oxygen, zirconium–oxygen, titanium–nitrogen, cobalt–sulphur and nickel–sulphur systems. It is therefore not surprising that "compounds" such as TiO, TiN, VN and MoS_2 exhibit metallic conductivity.

The thermochemical treatment of these phases of variable composition differs from that of pure, fixed compounds. Similar considerations are applicable to liquid mixtures, such as those of metals or of inorganic salts, which are often completely soluble. To illustrate the modified treatment we shall discuss an example of each of the extreme types of system: the system tellurium–tin (Fig. 7a) which contains a compound of definite composition and no significant range of solid solution; and the system silver–gold (Fig. 7b), which consists at all proportions of a homogeneous, disordered solid solution.

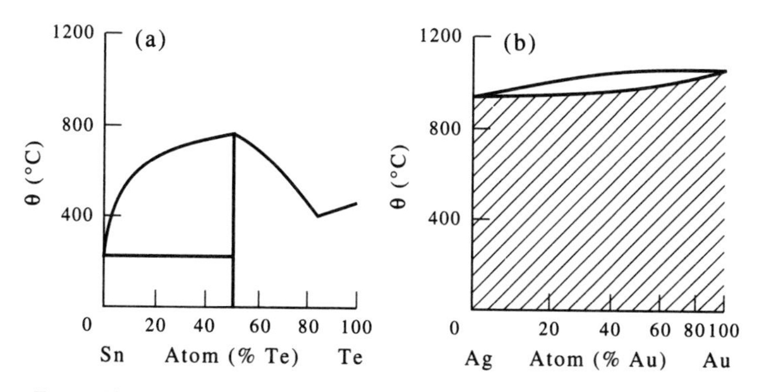

FIG. 7. Phase diagrams of the systems: (a) tin–tellurium and (b) silver–gold.

To compare these systems we must devise an experiment which shows the variation in Gibbs energy of formation over the complete range of alloys.

This is most conveniently done by constructing suitable galvanic cells, the e.m.f. of which is a measure of the Gibbs energies concerned. The negative electrode of such a cell is formed of the more electronegative of the two metals, and its positive electrode consists of the alloy itself. The cation of the electrolyte is the ionised form of the anodic metal. The cell can be represented schematically by

$$\ominus \mathrm{M^I}|\mathrm{M^I}-(\text{cation})|\mathrm{M^I_x M^{II}_y}\oplus.$$

When the cell is in operation, the metal M^I is transferred into the alloy $M^I_x M^{II}_y$. The potentiometric method used in measuring the e.m.f. involves no transfer of M^I, so in practice there should be no change in composition of the alloy. Actually, however, there is always a transportation of a small amount of material which must be considered in assessing the accuracy of the method (see p. 142).

The two alloy systems we are comparing have actually been investigated experimentally by this method.

The tin–tellurium cell (1936b) for all the alloy compositions between SnTe and Te may be represented by

$$\ominus\{\mathrm{Sn}\}|\{\mathrm{SnCl_2}\}|\langle\mathrm{Te}\rangle+\langle\mathrm{SnTe}\rangle\oplus.$$

The e.m.f. is constant over the whole of this range (Fig. 8a), being due to the formation of ⟨SnTe⟩ from liquid tin and solid tellurium. The Gibbs energy change

$$\Delta G^\circ = -2\times 96{,}485\ E\ \mathrm{J}$$

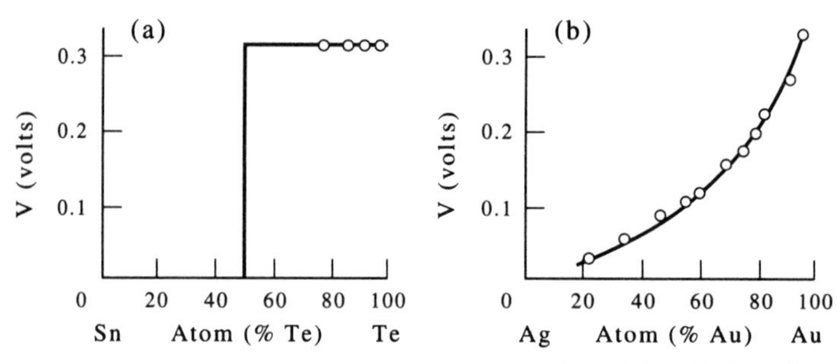

FIG. 8. E.m.f.s of cells with alloy concentration electrodes: (a) tin and tin–tellurium alloys, (b) silver and silver–gold alloys.

is that arising from the combination of one mole {Sn} with one mole ⟨Te⟩. Virtually no e.m.f. would be produced by an alloy of composition between SnTe and pure tin, since there would be not free tellurium present to combine with the tin.

The gold–silver cell which was studied (1948f) may be represented schematically by

$$\ominus\langle\mathrm{Ag}\rangle|[\mathrm{Ag^+}]\text{ in glass}|\langle\mathrm{Ag_x Au_y}\rangle\oplus.$$

Its e.m.f., arising from the transfer of ⟨Ag⟩ into the $\langle Ag_x Au_y \rangle$ solid solution, is found to vary over the whole alloy composition range (Fig. 8b). The Gibbs energy change calculated from each composition, from the formula

$$\Delta\bar{G}_{Ag} = -1 \times 96{,}485\, E\, J$$

is that of the assimilation of 1 mole $\langle Ag \rangle$ by an infinite amount of the alloy $\langle Ag_xAu_y \rangle$—(i.e. the composition of the solid solution is unchanged by the solution of the silver). It is to be expected that this Gibbs energy change will increase as the gold content increases, since the silver atoms liberated at the electropositive electrode will find a greater concentration of atoms of the other species for which they have an affinity in the gold-rich alloys. The Gibbs energy change is described more fully as the partial molar Gibbs energy of solution of $\langle Ag \rangle$ in $\langle Ag_xAu_y \rangle$. There are corresponding enthalpies and entropies and these partial molar thermodynamic functions are distinguished by a bar over the symbol, e.g. $\Delta\bar{G}$.

We may enlarge upon the comparison of integral and molar thermodynamic functions by considering a binary alloy to which both types of functions apply.

Integral Gibbs energies are those which are involved in the formation of an alloy from its constituent elements:

$$x\mathrm{A} + y\mathrm{B} = \mathrm{A}_x\mathrm{B}_y; \quad \Delta G(\Delta H, \Delta S). \qquad \text{[integral values]}$$

The quantities x and y can either be in atomic functions, N_A and N_B (e.g. $Mg_{0.6}/Bi_{0.4}$), when ΔG, ΔH and ΔS will be expressed in gram-atomic quantities, or denote mole quantities, n_A and n_B (e.g. Mg_3Bi_2), when ΔG, ΔH and ΔS will be expressed in terms of moles.

The partial molar Gibbs energies, etc, are related to the reaction

$$1\mathrm{A} + \mathrm{A}_m\mathrm{B}_n = \mathrm{A}_{m+1}\mathrm{B}_n; \quad \Delta\bar{G}(\Delta\bar{H}, \Delta\bar{S} \qquad \text{[partial values]}$$

where the ratio m/n is equal to the ratio x/y in the previous equation, but the magnitude of m and n is such that $(m+1)$ practically equals m. The partial molar thermodynamic functions are generally expressed in terms of 1 g-atom of the constituent A.

Taking the example of the silver–gold alloys, we can regard the partial Gibbs energy change as the difference between the Gibbs energy of the silver in the pure state and its Gibbs energy in solution in the alloy. These two quantities are often used and referred to (after Gibbs) as *chemical potentials*, denoted by μ. Using the term μ_{Ag} for the chemical potential of the pure metal and $\mu_{[Ag]}$ for that of the dissolved metal, we have

$$\Delta\bar{G}_{Ag} = \mu_{[Ag]} - \mu_{Ag}.$$

Chemical potentials are used often in theoretical discussions but rarely in the numerical calculations of the type with which we are concerned. They must be assigned an arbitrary value for substances in a standard state, so are no more absolute than Gibbs energies. We are only concerned with difference in chemical potential and we can relate this to the Gibbs energy G°_T of the substance concerned in its pure state. The authors favour the use of the symbol $\Delta\bar{G}$ in place of the appropriate term in μ so as to emphasise its connection with the other partial molar functions $\Delta\bar{H}$ and $\Delta\bar{S}$ (cf. ΔG, ΔH and ΔS).

In practice it will be found that e.m.f. measurements on concentration electrodes and vapour pressure measurements are primarily methods of deriving partial molar functions, while calorimetric methods usually provide the integral values.

In this section we have referred to measurements of e.m.f.s largely as an illustration of partial thermodynamic functions. We may finally summarise the relationship as follows:

$$\Delta\bar{G}_A = -z\,.\,96{,}485\ E_A \tag{41}$$

$$\Delta\bar{G}_A = \Delta\bar{H}_A - T\Delta\bar{S}_A \tag{42}$$

$$\Delta\bar{H}_A = z\,.\,96{,}485\left(T\frac{dE_A}{dT} - E_A\right) \tag{43}$$

$$\Delta\bar{H}_A = z\,.\,96{,}485\left(\frac{T_1+T_2}{2}\,\frac{E_2-E_1}{T_2-T_1} - \frac{E_1+E_2}{2}\right) \tag{43a}$$

$$\Delta\bar{S}_A = z\,.\,96{,}485\,\frac{dE_A}{dT} \tag{44}$$

$$\Delta\bar{S}_A = z\,.\,96{,}485\,\frac{E_2-E_1}{T_2-T_1}. \tag{44a}$$

These equations refer to the reaction $A + A_mB_n = A_{m+1}B_n$, as defined p. 29, and E is measured in volts.

The integral values of the thermodynamic functions of a solution are related to the partial values by the equations

$$\Delta G = N_A\Delta\bar{G}_A + N_B\Delta\bar{G}_B \tag{45}$$

$$\Delta H = N_A\Delta\bar{H}_A + N_B\Delta\bar{H}_B \tag{45a}$$

$$\Delta S = N_A\Delta\bar{S}_A + N_B\Delta\bar{S}_B \tag{45b}$$

where the integral values are those for a solution having mole fractions of A and B equal to N_A and N_B respectively.

B. Vapour Pressures and Partial Gibbs Energies

The Gibbs energy change when one mole of an ideal gas undergoes a change in pressure, at constant temperature, from p_1 to p_2 is given by

$$\Delta G = \boldsymbol{R}T\ln\frac{p_2}{p_1} \tag{46}$$

This expression may be applied to a parallel series of measurements which could be made with the tin–tellurium and silver–gold alloys which are studied by the electrochemical method above. Over each alloy composition there exists a gaseous phase which contains partial pressures of the vapours of each component of the alloy. The partial pressure of each component vapour changes with temperature and composition of the alloy, maintaining the vapour-condensed phase equilibrium. It therefore follows that the Gibbs energy change when one mole of a pure component is added isothermally to a large volume of the alloy so as not to change composition is equal to the Gibbs energy change when one mole of the vapour of the component undergoes the pressure change from p°, the vapour pressure of the pure component at the given temperature, to p the vapour pressure exerted by the component over the alloy. This process can be visualised as in Fig. 9.

Since these paths constitute a closed cycle the Gibbs energy changes must be the same by either route, hence

$$\Delta\bar{G}_A = \boldsymbol{R}T \ln \frac{p_A}{p_A^\circ}. \qquad (47)$$

It thus follows that the partial vapour pressure of a component over an alloy will change in exactly the same way as the partial Gibbs energy of solution of the component, its absolute value depending on the vapour pressure of the pure component at the given temperature. Moreover, as the *ratio* of vapour pressures between the pure and solution states is related to our useful thermodynamic function, we shall give this a special name, the *activity* and the symbol a. Thus

$$\Delta\bar{G}_A = \boldsymbol{R}T \ln a_A. \qquad (47a)$$

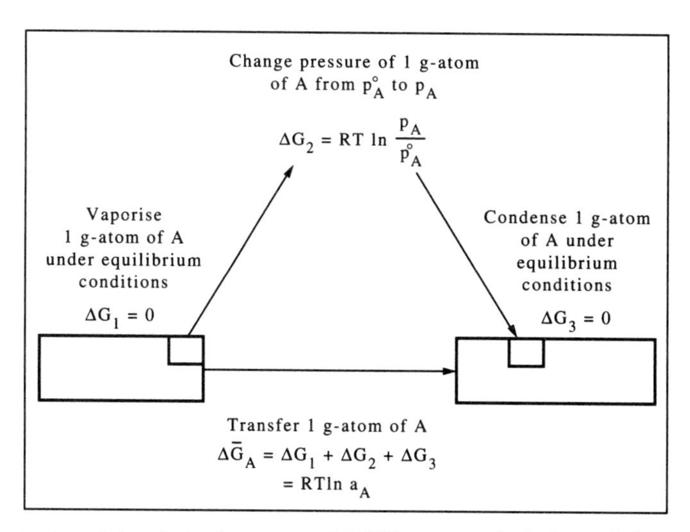

FIG. 9. Illustration of the relation between partial Gibbs energy of solution and change of vapour pressure of component A in an alloy.

This expression has no immediately clear relationship with the composition of the alloy but we may interrelate the two by an expression $a_i = \gamma_i N_i$ where γ_i is called the activity coefficient of the ith component of the alloy and the product $\gamma_i N_i$ must have values between the limits zero and unity. It therefore follows that

$$\Delta\bar{G}_i = \boldsymbol{R}T \ln \gamma_i + \boldsymbol{R}T \ln N_i. \qquad (48)$$

There are two particularly simple forms of this expression which are sometimes found for the behaviour of the partial Gibbs energy of a component as a function of its mole fraction. The first of these is when γ has the value unity and

$$a_i = N_i \tag{49}$$

This behaviour was first suggested as a general form by Raoult and the equality is a statement of **Raoult's law**. However, very few systems show this behaviour over wide ranges of composition.

The second simplified form is when γ_i is constant but not necessarily unity over a range of composition. This behaviour which is frequently shown by dilute solutions is called **Henry's law** and can be stated in the form

$$a_i = kN_i \tag{50}$$

where k is a constant for the system, and is represented by the tangents indicated in Fig. 10 as broken lines.

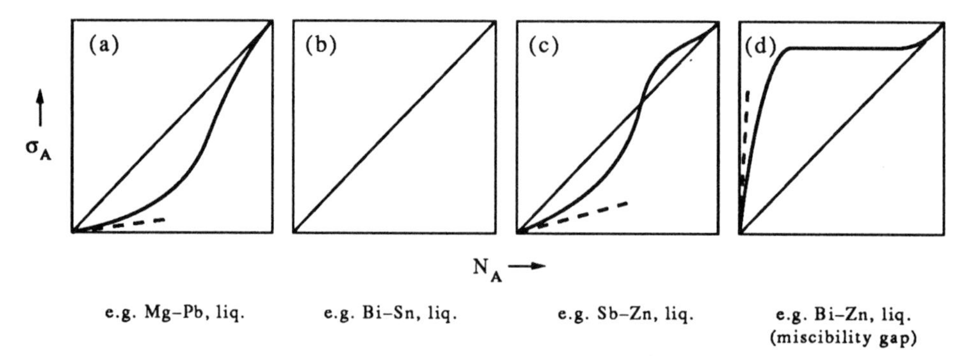

FIG. 10. Activities of single components of binary mixtures (schematic).

There is no simple law to express the vapour pressure of a component of a binary mixture at intermediate concentrations. Both Henry's law and Raoult's law are limiting generalisations, but they are often found to apply over quite large ranges of concentration. In fact they facilitate the construction of activity curves of substances with complete mutual solubility even when only a few measurements are carried out.

The activity of a substance can be derived by measuring the ratio of its vapour pressures over the solution and in the pure state. Its partial molar Gibbs energy of solution is simply related to its activity, as will be seen by considering as an example the zinc–aluminium alloys in the liquid state. We can make the following summation:

$$\begin{array}{lcr} + & \{\mathrm{Zn}\} = (\mathrm{Zn}); & p_{\mathrm{Zn}},\ \Delta G_0 \\ - & \{\mathrm{Al}_m\mathrm{Zn}_{(n+1)}\} = \{\mathrm{Al}_m\mathrm{Zn}_n\} + (\mathrm{Zn}); & p'_{\mathrm{Zn}},\ \Delta\bar{G}_1 \\ \hline & \{\mathrm{Al}_m\mathrm{Zn}_n\} + \{\mathrm{Zn}\} = \{\mathrm{Al}_m\mathrm{Zn}_{(n+1)}\}; & \Delta\bar{G}_{\mathrm{Zn}} \end{array}$$

Here, ΔG_0 is the Gibbs energy of evaporation of one mole (Zn), while $-\Delta\bar{G}_1$ and $\Delta\bar{G}_{\mathrm{Zn}}$ are the partial molar Gibbs energies of solution of gaseous and liquid zinc respectively in the alloy $\{\mathrm{Al}_m\mathrm{Zn}_n\}$.

Since

$$\Delta G_0 = -\boldsymbol{R}T \ln p_{\mathrm{Zn}}, \qquad \Delta\bar{G}_1 = -\boldsymbol{R}T \ln p'_{\mathrm{Zn}}$$

and

$$\Delta\bar{G}_{Zn} = \Delta G_0 - \Delta\bar{G}_1$$

then

$$\Delta\bar{G}_{Zn} = \boldsymbol{R}T \ln \frac{p'_{Zn}}{p_{Zn}} = \boldsymbol{R}T \ln a_{Zn} = 19.144\ T \log a_{Zn}. \qquad (51)$$

Activities can be derived from measurements of any equilibria involving a solution and a gaseous phase. Equilibria with hydrogen–steam mixtures, for example, are often used in experimental determinations. This may be illustrated by considering iron–nickel alloys (which form a continuous series of solid solutions at high temperatures). The activity of the iron is obtained by determining the H_2O/H_2 ratio of the vapour in equilibrium with FeO in contact with the alloys and pure iron:

$$- \qquad \langle FeO\rangle + (H_2) = \langle Fe\rangle + (H_2O); \qquad \frac{p_{H_2O}}{p_{H_2}}, \Delta G_0$$

$$+ \qquad \langle Fe_mNi_n\rangle + \langle FeO\rangle + (H_2) = \langle Fe_{m+1}Ni_n\rangle + (H_2O); \qquad \frac{p'_{H_2O}}{p'_{H_2}}, \Delta\bar{G}_1$$

$$\langle Fe_mNi_n\rangle + \langle Fe\rangle = \langle Fe_{m+1}Ni_n\rangle \ \Delta G_{Fe}$$

$$a_{Fe} = \frac{p_{H_2O}/p_{H_2}}{p'_{H_2O}/p'_{H_2}}$$

$$\Delta\bar{G}_{Fe} = \boldsymbol{R}T \ln a_{Fe} \qquad (51a)$$

Just as partial molar heats and entropies of solution were obtained from e.m.f. values (p. 33), so we may apply the **Gibbs–Helmholtz equation** to their derivation from activities.

$$\Delta\bar{G}_A = \boldsymbol{R}T \ln a_A = 19.144\ T \log a_A \qquad (51b)$$

$$\Delta\bar{H}_A = -19.144\ T^2 \frac{d \log a_A}{dT} \qquad (52)$$

$$\Delta\bar{H}_A = -19.144\ T_1 T_2 \frac{\log a_2 - \log a_1}{T_2 - T_1} \qquad (52b)$$

$$\Delta\bar{S}_A = -19.144\left(T \frac{d \log a_A}{dT} + \log a_A\right) \qquad (53)$$

$$\Delta\bar{S}_A = -19.144\left(\frac{T_1 + T_2}{2} \frac{\log a_2 - \log a_1}{T_2 - T_1} + \frac{\log a_2 + \log a_1}{2}\right) \qquad (53a)$$

Since the Gibbs energy of a substance in solution is measurable by its e.m.f. its activity can be determined in the same way.

$$\Delta\bar{G}_A = -zFE$$
$$\Delta\bar{G}_A = \boldsymbol{R}T \ln a_A$$

Hence

$$\ln a_A = -\frac{zF}{RT} E \qquad (54)$$

or, introducing the numerical values of the constants.

$$\log a_A = -\frac{5043\, zE}{T} \qquad (54a)$$

We have thus summarised the main sources of activity values. Two further important methods of deriving activities, from equilibrium diagrams and from equilibria in a melt, will be discussed at a later stage.

C. The Calculation of Integral Values from Partial Molar Thermochemical Data

Frequently the only experimental methods applicable to the study of a series of solutions are those which lead in the first instance to partial molar data. The integral values of the thermochemical functions are generally derived from these by graphical methods.

For simplicity we shall first consider a system the constituents of which, A and B, form a continuous series of solid solutions. Generally the partial molar values for *both* constituents of a solution will not be available from experimental measurements. Phase rule considerations show that the partial vapour pressures of the two components of a binary system are interdependent. From this we can deduce that their partial molar thermodynamic functions must also be related, and this relationship is given by the development by Margules of a generalisation of Gibbs and Duhem.

The **Gibbs–Duhem equation** can be used to calculate the change in partial molar Gibbs energy of B over the composition range N'_B to N''_B when the property is known for A over the same composition range. Thus

$$\ln \frac{a'_B}{a''_B} = -\int_{N_B}^{N_B} \frac{N_A}{N_B}\, \mathrm{d} \ln a_A \qquad (55)$$

Obviously the value of the activity of B can be calculated by this means provided that it is known at one of the compositional limits.

For the application of this equation it is important to realise that when no information is available concerning the properties of B, the partial molar functions of the other component A must be known across the whole range of concentration $N_A = 0$–1, and particularly accurately at low concentration of the solute—a condition not easily achieved experimentally. Since the partial molar entropies and Gibbs energies of disordered solutions tend towards infinity when zero concentration is approached, it is convenient to introduce the so-called excess functions $\Delta\bar{S}^E_A$ and $\Delta\bar{G}^E_A$ which are given by

$$\Delta\bar{S}^E_A = \Delta\bar{S}^{\text{xptl}}_A + \boldsymbol{R} \ln N_A \qquad (56)$$

and consequently

$$\Delta\bar{G}^E_A = \Delta\bar{G}^{\text{xptl}}_A - \boldsymbol{R}T \ln N_A = \boldsymbol{R}T \ln \gamma_A \qquad (57)$$

and correspondingly for a component B.

The Gibbs–Duhem equation can be used in the analogous form

$$\ln \frac{\gamma'_B}{\gamma''_B} = - \int_{N_B}^{N_B} \frac{N_A}{N_B} \, d \ln \gamma_A \tag{58}$$

since $dN_A = -dN_B$ in a binary system. Because the activity coefficient γ_A reaches a constant value for the dilute solution of A in B and has the value of unity close to pure A, the integral involving the excess Gibbs energy can be accurately evaluated whereas the integral which makes use of activity data directly must be estimated because a_A tends to zero as N_A tends to the same limit.

When it is desirable to compute the concentration dependence of the *integral* molar thermodynamic functions from measured values of the partial molar functions of one of the components, the **Duhem–Margules equation** may be used in the following form, where the symbol Z stands for H, S^E and G^E.

$$\Delta Z_f = N_B \int_0^{N_A} \frac{\Delta \bar{Z}_A}{N_B^2} \, dN_A \tag{59}$$

It is sometimes convenient to use the following formulation:

$$\Delta Z_f = N_B \int_0^x \Delta \bar{Z}_A \, dx \tag{60}$$

where $x = N_A/N_B$. The evaluation of the Duhem–Margules equation may be illustrated by one example. The e.m.f. values of Fig. 8b have been converted into excess Gibbs energies and are plotted in Fig. 11a against the atomic ratio N_{Ag}/N_{Au}, which represents the number of g-atoms of silver associated with 1 g-atom of gold at a given concentration. The integral excess Gibbs energy per g-atom of the solution, the composition of which is represented by x, is obtained by dividing the area under the curve between atomic ratio 0 and x by the quantity $(x+1)$. These integral values are plotted against the atomic fractions for the silver–gold system in Fig. 11b.

The integral heats and excess entropies of formation of the solution are obtained in the same manner from $\Delta \bar{H}_A$ and $\Delta \bar{S}_A^E$ respectively.

The integration of partial molar curves is more complicated if the system contains several homogeneous phases separated by heterogeneous fields. To clarify the problem we will first discuss the connection between partial and integral molar values of thermochemical functions in the two limiting types of system. The continuous lines in Fig. 12 show the integral excess Gibbs energies of (a) a binary system containing two compounds but with no significant range of solid solubility, and (b) a binary system forming a continuous series of solid solutions. The tangent to these lines at any point cuts the $N_A = 0$ and $N_A = 1$ ordinates at the values of the partial Gibbs energies of B and A respectively. While these vary continuously through the solid solution phase, they remain constant in heterogeneous phases of two constituents of varying proportion but of fixed composition.

A continuous series of solid solutions gives the continuous vapour pressure curve shown in Fig. 13. In the other type of system the vapour pressure changes in a stepwise manner with the formation of a new compound of higher content of volatile constituent.

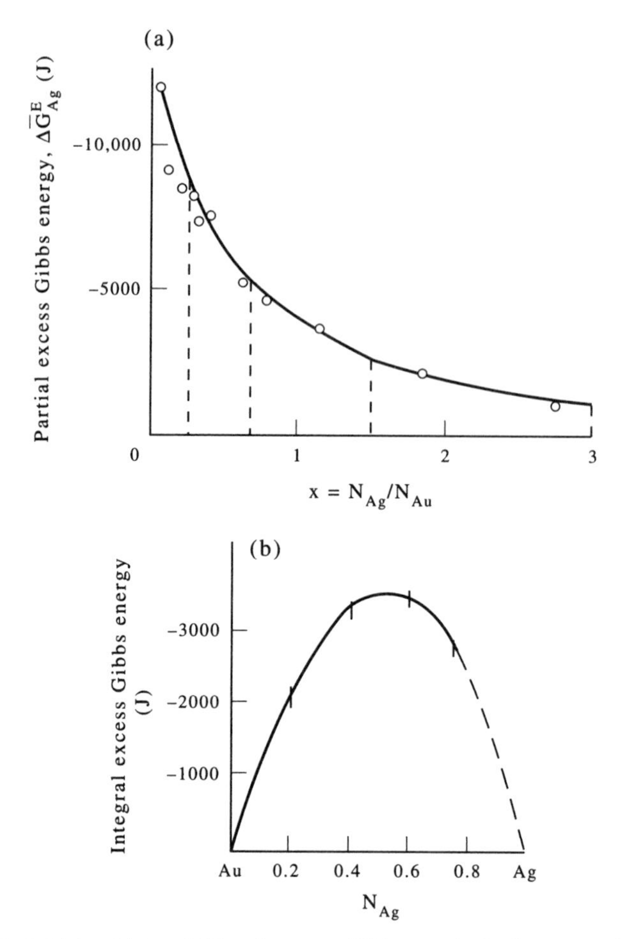

FIG. 11. Partial (a) and integral (b) molar excess Gibbs energies in the system silver–gold.

The magnitude of the pressure increase is a measure of the difference between the Gibbs energy of the phase richer in component A and the total Gibbs energy of component A plus the phase poorer in A (e.g. $G_{(A,B)} - G_{(AB+2A)}$).

It is often difficult to measure the reaction pressure of the lowest step on account of its very low value. The measurement must be carried out, however, since the integral thermochemical data of the whole system will be incomplete if the values associated with this low vapour pressure are not known. If a pressure measurement cannot be applied, the data must be obtained in some other way.

Consider a two-component system in which A is volatile and for simplicity monatomic, and B and the intermediate compounds AB_3, AB, A_3B are condensed phases.

Let the vapour pressures of AB_3, AB, A_3B and A be respectively p_1, p_2, p_3 and p_4.

The Gibbs energy change of each reaction step may be expressed as follows:

$$3B + A = AB_3; \qquad \Delta G_1 = 19.144\, T \log p_1/p_4$$
$$\tfrac{1}{2}AB_3 + A = 1\tfrac{1}{2}AB; \qquad \Delta\bar{G}_2 = 19.144\, T \log p_2/p_4$$
$$\tfrac{1}{2}AB + A = \tfrac{1}{2}A_3B; \qquad \Delta\bar{G}_3 = 19.144\, T \log p_3/p_4$$

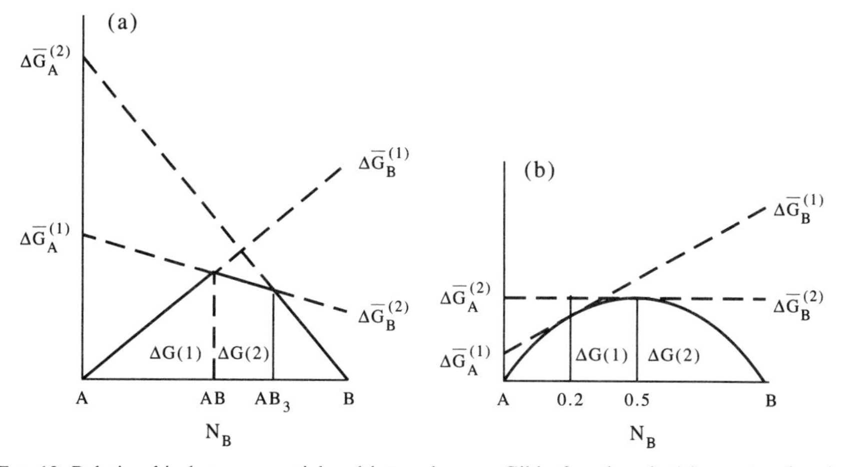

FIG. 12. Relationship between partial and integral excess Gibbs functions in (a) a system forming intermediate compounds, and (b) a system exhibiting complete mutual solubility.

Fig. 12a

$\Delta G(1)$: $0.5A + 0.5B = A_{0.5}B_{0.5}$
$\Delta G(2)$: $0.25A + 0.75B = A_{0.25}B_{0.75}$
$\Delta \bar{G}_B(1)$: $1B + \infty(xAB + yA)$
$= 1AB + \infty(xAB + yA)$
$\Delta \bar{G}_B(2)$: $1B + \infty(xAB_3 + yAB)$
$= 1AB_3 + \infty(xAB_3 + yAB)$

Fig. 12b

$\Delta G(1)$: $0.2B + 0.8A = A_{0.8}B_{0.2}$
$\Delta G(2)$: $0.5B + 0.5A = A^{0.5}B_{0.5}$
$\Delta G_B(1)$: $1B + \infty(A_{0.8}B_{0.2})$
$= 1A_4B + \infty(A_{0.8}B_{0.2})$
$\Delta \bar{G}_B(2)$: $B + \infty(A_{0.5}B_{0.5})$
$= 1AB + \infty(A_{0.5}B_{0.5})$

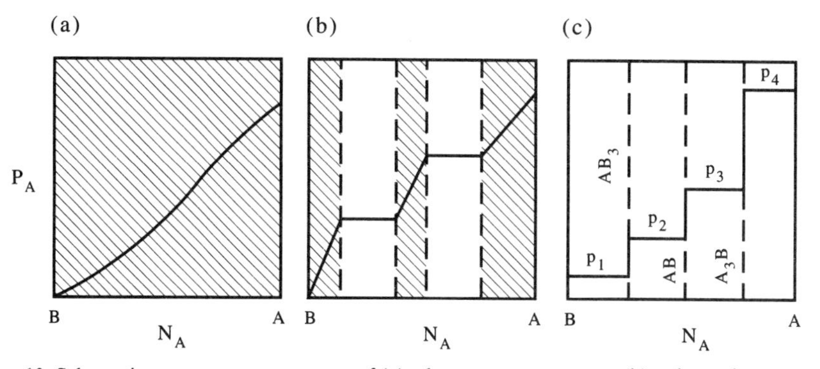

FIG. 13. Schematic vapour pressure curves of (a) a homogeneous system, (b) a three-phase system, and (c) a system containing three compounds.

Then the integral free energies are given by:

$$3B + A = AB_3; \qquad \Delta G_1$$
$$B + A = AB; \qquad \Delta G_2 = \tfrac{1}{3}\Delta G_1 + \tfrac{2}{3}\Delta \bar{G}_2$$
$$B + 3A = A_3B; \qquad \Delta G_3 = \Delta G_2 + 2\Delta \bar{G}_3.$$

Thus all the integral Gibbs energy values depend upon the determination of pressure p_1.

A similar limitation is found in the evaluation of Gibbs energy changes from curves of the type shown in Fig. 13. The accuracy of the integral thermochemical data derived from such curves depends upon the lowest concentration of A at which its vapour pressure can be measured. Below this concentration, an extrapolation must be made, and since its curvature can vary considerably, it is necessary that it should be as short as possible to minimise the error.

Methods for the determination of Gibbs energies (e.g. vapour pressure, e.m.f.) generally lead to *partial* thermodynamic functions. Calorimetric methods, on the other hand, normally produce *integral* enthalpies of formation. The best procedure for assessing the thermodynamic properties of binary systems with homogeneity ranges in such cases is to differentiate the enthalpy vs concentration curves, to calculate the partial entropies from the partial Gibbs energies and the differentiated enthalpies, and then to integrate the partial excess entropies. Unfortunately, the integral enthalpies have rarely been obtained with the high accuracy necessary for a reliable differentiation.

Differentiation can be carried out graphically as indicated in Fig. 12b. This method is not very precise, and the results must be checked by subsequent graphical integration. For the differentiation of relatively simple enthalpy (or excess Gibbs energy) vs concentration curves such as that in Fig. 12b, Guggenheim (1937c) has suggested the following form of the series expansion for the representation of the integral enthalpies, such as

$$\Delta H_f = N_A N_B [a_0 + a_1(N_A - N_B) + a_2(N_A - N_B)^2 + a_3(N_A - N_B)^3] \tag{61}$$

where a_0, a_1, a_2, and a_3 are coefficients to be derived from the experimental data. For the partial molar enthalpies one obtains:

$$\begin{aligned}\Delta\bar{H}_A = N_B^2[&a_0 + a_1(3N_A - N_B) + a_2(5N_A - N_b)(N_A - N_B)\\ &+ a_3(7N_A - N_B)(N_A - N_B)^2]\end{aligned} \tag{62a}$$

$$\begin{aligned}\Delta\bar{H}_B = N_A^2[&a_0 + a_1(N_A - 3N_B) + a_2(N_A - 5\mathrm{N}_B)(N_A - N_B)\\ &+ a_3(N_A - 7N_B)(N_A - N_B)^2].\end{aligned} \tag{62b}$$

Of course, for simple enthalpy curves one can reduce the number of terms. If $a_1 = a_2 = a_3 = 0$, eqns (61)–(62b) reduce to:

$$\Delta H_f = a_0 N_A N_B \tag{63}$$

$$\Delta\bar{H}_A = a_0 N_B^2 \quad \text{and} \quad \Delta\bar{H}_B = a_0 N_A^2. \tag{64}$$

Truly symmetric enthalpy vs concentration curves are, however, rarely encountered in practice.

From accurate experimental enthalpies of mixing in the liquid solutions of mercury and lead at 335°C, we have the following values:

ΔH_f	−117	−71	+63	+230	+393	+527	+602	+586	+410 J/g-atom
N_{Hg}	0.1	0.2	0.3	0.4	0.5	0.6	0.7	0.8	0.9.

Applying eqn (61) to four representative compositions (say, $N_{Hg} = 0.1, 0.3, 0.5$ and 0.7), we find the following coefficients;

$$a_0 = +1577, \quad a_1 = +3084, \quad a_2 = +67, \quad a_3 = +900.$$

There are many instances in the literature where values for the functions $\Delta H_f / N_A N_B$ and $\Delta G_f^E / N_A N_B$ of binary systems are described by the use of a simple power series. Thus

$$\Delta G_f^E/N_A N_B = \sum_{i=0} a_i N_B^i. \tag{65}$$

When the results are few for a given system, only a few terms are used in the power series development. However, when more data become available it is frequently found that a more precise description can be made using more terms in the series expansion.

For the simple power series each stage of adjustment usually means a drastic reworking of the expansion with new values for each coefficient a_i. This recalculation is not too time consuming if it is fitted to a flexible computer program, but the changes which occur in the values of a_i at each stage of the "refinement" should caution the theoretician not to make too much "theory" around the values at any given stage of development.

Bale and Pelton have shown that there are advantages in using the Legendre polynomials for the expansion of thermochemical data [1974a]. The power series discussed previously can now be replaced by the expansion

$$\Delta G_j^E/N_A N_B = a_0(1) + a_1(2N_B - 1) + a_2(6N_B^2 - 6N_B + 1) + a_3(20N_B^3 - 30N_B^2 - 12N_B - 1) + \cdots \tag{66}$$

The quantities in the brackets are the first four Legendre polynomials and the continuation of the series may be made by using the recursion formula for the nth polynomial

$$P_n = \frac{(2n-1)(2N_B - 1)}{n} P_{n-1} - \frac{(n-1)}{n} P_{n-2}. \tag{67}$$

It is found, in a number of typical cases, that refinement of the experimental data merely adds further polynomials to an already established expansion without changing the coefficients of the preceding terms very much. Thus the introduction of the fourth polynomial in an expansion which already described the data reasonably well with only three polynomials would not change the values of the coefficients a_1, a_2 and a_3 significantly.

In a complex system which included heterogeneous and homogeneous regions (e.g. brasses, bronzes and many sulphide systems) the reaction pressure vs concentration curve assumes the form shown in Fig. 13b. There is a continuous but not necessarily linear change in pressure (or activity) across the homogeneous phase fields. In the heterogeneous fields the reaction pressure of A is again determined by its pressure in the boundary phase richer in component A.

Examples of the two latter types of reaction pressure curves are given in Figs 14 and 15. The first shows the mercury pressure of the cerium amalgams (1928a). There is one compound, $CeHg_4$, in the range considered. The pressure does not increase in perfect rectangular steps, the "corners" on the curve being rounded. This may be due to a small variable concentration range of existence for the phase $CeHg_4$, but more probably results from the experimental method used in its determination. The volatilisation of mercury to fill the relatively large volume of the apparatus resulted in an impoverishment of the condensed phase which should have been of stoichiometric composition.

Figure 15 shows the partial molar Gibbs energy curve for oxygen in iron, which is described on p. 44.

The thermodynamic properties of the copper–zinc system are well established [M. Kowalski, private communication 1991a]. The partial Gibbs energy of solution of liquid

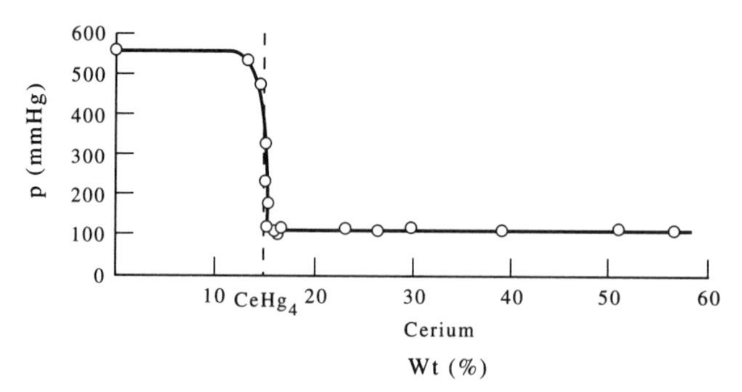

FIG. 14. Mercury pressures of cerium amalgams at 340°C.

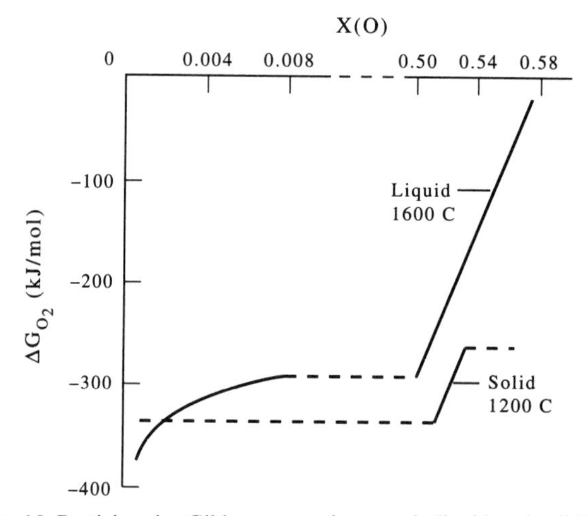

FIG. 15. Partial molar Gibbs energy of oxygen in liquid and solid iron.

zinc in the alloys at 500°C is shown in Fig. 16. As has been discussed above, the Gibbs energies of solution are represented by horizontal lines in the heterogeneous ranges, while $\Delta\bar{G}_{Zn}$ increases with the zinc content in the homogeneous ranges.

The partial enthalpies of solution of liquid zinc in the various 2-phase regions are also represented by horizontal lines, but these no longer form a continuous curve with the individual sections of curve of the homogeneous phases as in the ΔG_{Zn}–N_{Zn} diagram. In fact, the relative positions of the horizontal and curved ΔH_{Zn}–N_{Zn} lines determines the slopes of the phase boundaries with temperature.

The integral Gibbs energies are shown as a full line in Fig. 18. In the heterogeneous ranges the curve is a straight (dashed) line. Since there must be no reversal in the dG/dN curve, the integral Gibbs energy curve must not show any indentation; otherwise the phase would be unstable in that range.

This condition does not apply to the enthalpy vs concentration curve since an indentation can be compensated by a corresponding bulge in the entropy vs concentration curve. Such behaviour is illustrated in Fig. 19a.

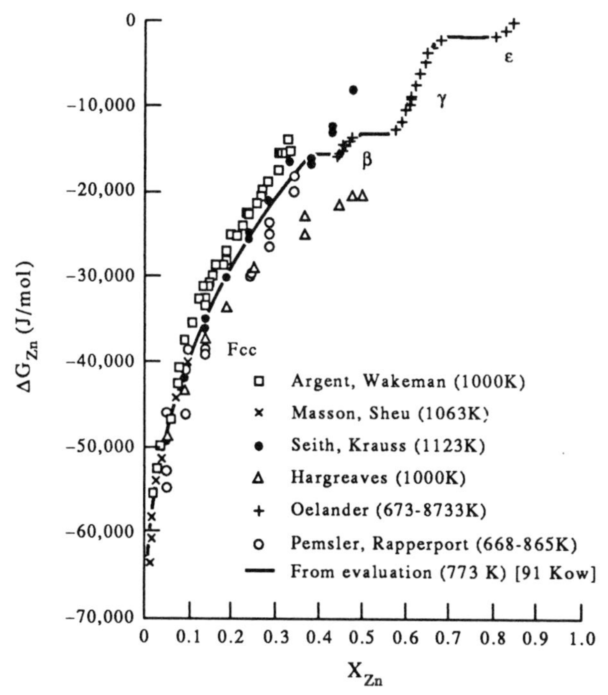

FIG. 16. Partial molar Gibbs energies of zinc in the solid copper–zinc system at 500°C. The liquid is the standard state for zinc.

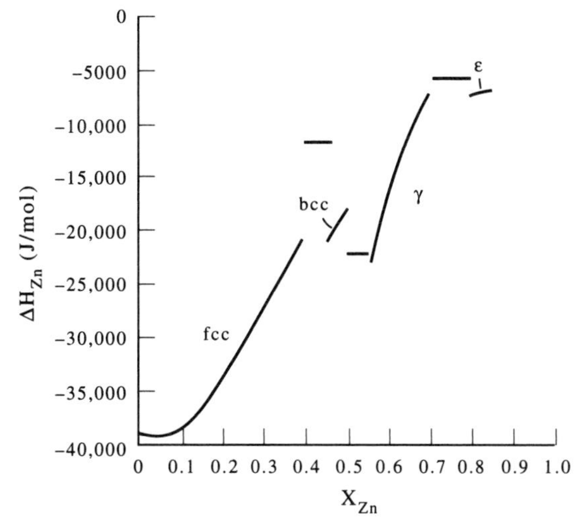

FIG. 17. Partial molar heats of solution of liquid zinc in solid copper–zinc alloys.

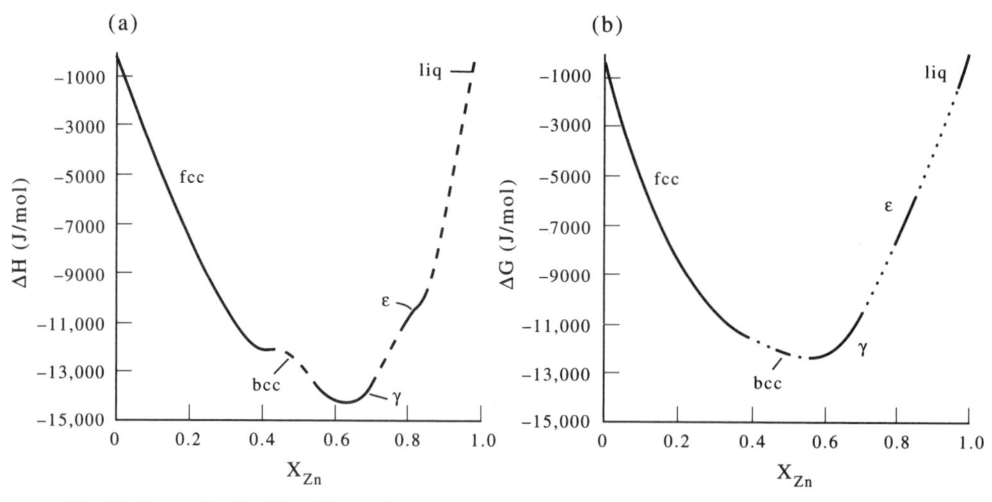

FIG. 18. Integral enthalpies of formation at 500°C and Gibbs energies of formation at 500°C of solid alloys in the copper–zinc system. The reference states are solid copper and liquid zinc.

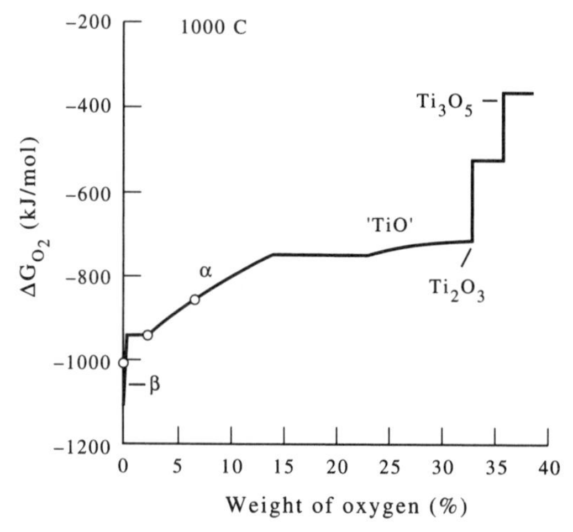

FIG. 19. Partial Gibbs energies of dissociation of 1 mole oxygen in the titanium–oxygen system.

Another system of practical significance is that of the titanium oxides. There are several phases in this system, some of them having quite a wide range of homogeneity. The partial molar Gibbs energy of assimilation of oxygen has been obtained for three compositions in the α and β phases by equilibration with Ca/CaO, Mg/MgO and Ba/BaO mixtures at 1000°C. These results were combined with the thermochemical data of the compounds to construct the $\Delta\bar{G}_{O_2}$ curve shown in Fig. 19. The branch of the curve in the concentration range Ti–TiO was obtained in the following manner. The experimental points (circles) within the solid solutions correspond to the Gibbs energies of dissociation of the three alkaline earths.

From enthalpies of formation, entropies and heat capacities, the value 794,960 J is obtained for the integral Gibbs energy of the reaction $2\langle TiO\rangle = 2\langle Ti\rangle + (O_2)$ at 1000°C. The connection between the partial and integral values is given by eqn (60), i.e.

$$\Delta G = \tfrac{1}{2} N_{Ti} \int_0^x \Delta \bar{G}_{O_2} \, dx$$

Thus, the area under the curve in Fig. 19 (with the abscissa expressed as atomic fraction $x = N_O/N_{Ti}$ instead of wt% oxygen, as shown) must be equal to the integral value given above. The experimental part of the curve is extrapolated to the phase boundary $\alpha/(\alpha + TiO)$, thus determining the horizontal part in the $(\alpha + TiO)$ range. To satisfy the condition that the area beneath the curve to be equal to ΔG for TiO, there can be no more than a slight drop in the value of $\Delta \bar{G}_{O_2}$ within the homogeneity range of TiO. It may be noted that the partial Gibbs energy of dissociation $\Delta \bar{G}_{O_2}$ of 2 "TiO" at 1000°C is 736,385 J, as compared with its integral Gibbs energy of dissociation of 794,960 J. Thus the dissociation pressure calculated from the last-named figure would not actually be observed at any but one composition in the range TiO–Ti. The actual dissociation pressures can only be obtained from the partial Gibbs energy curve.

The graph shows that a reducing agent that can overcome the affinity of oxygen to TiO_2, Ti_3O_5, Ti_2O_3 and even TiO may yet fail to produce pure titanium owing to the affinity of oxygen dissolved in the metallic titanium phase. For this reason, simplified thermochemical calculations which do not bring the solution phase into account may sometimes fail, reduction being less complete than would be expected from the calculations. Magnesium or lithium should be capable of reducing titania to titanium containing 2% oxygen by weight, but they will fail to overcome the affinity corresponding to the rapid increase of $\Delta \bar{G}_{O_2}$ in the solution phase towards pure titanium. Only calcium would be effective in this case, the corresponding equilibrium concentration of oxygen in titanium being 0.07% by weight at 1000°C.

Similar problems arise in other systems, especially those of oxygen or sulphur with metals such as titanium, zirconium, vanadium and niobium. Each of these systems contain several compounds, and in addition the oxygen and sulphur have a certain tendency for solution in the metallic lattice. The complete reduction of this metallic phase is essential to the production of the pure metal, and it is usually a very difficult step to overcome.

The corresponding data for the metallic solution phase in the liquid oxygen–iron system are more accurately known, since the partial molar Gibbs energy of solution of oxygen is much less in iron than in the metals just mentioned. It is more convenient to measure the oxygen pressure of the solution with the reaction

$$\langle FeO\rangle + (H_2) = \{Fe\} + (H_2O)$$

Dastur and Chipman (1949a, b) have determined the equilibrium constant of this reaction for various concentrations of oxygen in iron at 1600°C. By subtracting the Gibbs energy of water vapour formation from the partial molar Gibbs energies derived from their equilibrium values of p_{H_2O}/p_{H_2} we obtain the partial molar Gibbs energy of solution of oxygen in iron at various concentrations. These are plotted in Fig. 15. In the miscibility gap between $\{Fe\} + [FeO]$ and $\{FeO\} + [Fe]$, the value of $\Delta \bar{G}$, or the oxygen pressure, remains constant until the composition FeO is just exceeded.*

*The partial molar Gibbs energies within the FeO–Fe_3O_4 range are taken from the work of Spencer and Kubaschewski [1978b].

The activity of oxygen in a given solution in iron is given by the ratio of the oxygen pressure of the solution to the oxygen pressure of a saturated solution at the same temperature (i.e. the oxygen pressure of FeO itself).

The oxygen pressures calculated from Chipman's figures are given in Table III, the corresponding activities being listed in the fourth column. Then taking the saturation concentration $F_{FeO}=0.775\times10^{-3}$ to be unity and evaluating the other concentrations accordingly, we obtain the activity coefficients $f_{FeO}=h_{FeO}/c_{FeO}$ of solutions of oxygen in iron.

TABLE III. *Equilibrium of oxygen with ferrous oxide dissolved in liquid iron*

$N_{FeO}\simeq N_O$	$\frac{p_{H_2O}}{p_{H_2}}$	$\frac{1}{2}\log p_{O_2}$	f_{FeO}	h_{FeO}
1.05×10^{-3}	0.127	−4.896	0.135	0.990
2.10×10^{-3}	0.260	−4.585	0.260	1.015
2.20×10^{-3}	0.262	−4.582	0.262	0.975
2.35×10^{-3}	0.263	−4.580	0.263	0.918
5.66×10^{-3}	0.671	−4.174	0.710	0.968
6.44×10^{-3}	0.754	−4.123	0.800	0.962
6.65×10^{-3}	0.755	−4.122	0.800	0.932
6.86×10^{-3}	0.980	−4.010	1.035	1.168
6.86×10^{-3}	0.715	−4.146	0.756	0.854
7.00×10^{-3}	0.983	−4.008	1.041	1.128
7.31×10^{-3}	0.830	−4.082	0.878	0.928
7.38×10^{-3}	1.021	−3.991	1.081	1.130
7.75×10^{-3}	0.848	−4.072	0.898	0.897

D. Dilute Solutions

1. Raoult's Law

It has been found that Raoult's law, $N_A=a_A$ ($\gamma_A=1$), holds in many systems for high concentrations of the solvent (see Fig. 10). A simple illustration of why this should be is given by the following schematic consideration. If A represents the solvent and B the solute, then each ultimate particle of B in dilute solution is completely surrounded by A particles (Fig. 20). The attractive A↔B and B↔B forces, being thus hemmed in, as it were, do not affect the escape or addition of A particles from and to the solution. The behaviour of the A particles is therefore simply colligative.

The extent of the concentration range in which Raoult's law applies varies widely from one system to another, but rarely exceeds a lower limit of $N_A=0.85$, i.e. a maximum solute content of 15 molar %.

Raoult's law has at least one application that is useful to the metallurgist. On adding a substance B to a solvent A there is a lowering of the freezing point of A by an amount given by the Clausius–Clapeyron equation. The exact formulation of this equation, in terms of activities, is

$$L=RT^2(\mathrm{d}\ln a_A/\mathrm{d}T).$$

If we apply this equation to the liquidus line representing the equilibrium between a pure solid A and a liquid solution of B in A, we obtain, on integration,

$$\frac{L_f(T_0-T)}{RT_0T} = -\ln N_A \tag{68}$$

in which L_f is the enthalpy of transition of the solid ⟨A⟩ into the liquid, i.e. the enthalpy of fusion, T_0 is the melting point of the pure substance A, and T the liquidus temperature at composition N_A.

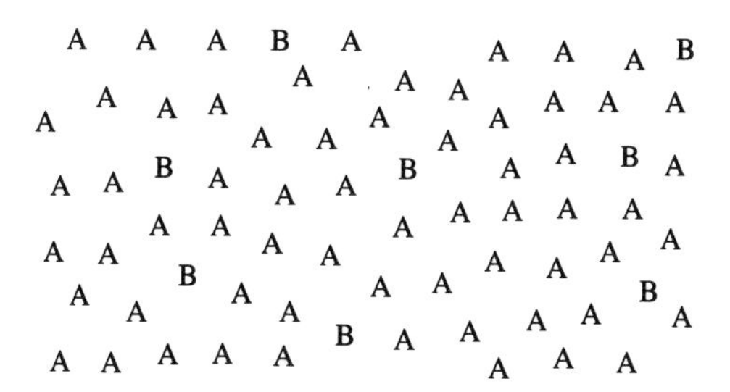

FIG. 20. Dilute solution of B in A.

The overruling proviso that there should be negligible solid solubility of B in A is often found to hold in simple eutectic systems. The relationship is used in all branches of chemistry to determine the molecular weight or degree of dissociation of a soluble substance, provided that the enthalpy of fusion of the solvent is known. Benzene and water are the commonest solvents for this purpose. The metallurgist makes rather more use of its converse application, that of determining an enthalpy of fusion from the freezing-point depression caused by a known molar concentration of solvent. The data required are of course directly available in equilibrium diagrams.

Kelley (1936c) has evaluated the enthalpies of fusion of a large number of elements by this method, and much still remains to be derived in this manner from the available data.

If there is appreciable solid solubility of the one component in the other, eqn (57) must be altered to

$$\frac{L_f(T_0-T)}{RT_0T} = -\ln \frac{N_A}{N'_A} \tag{69}$$

where N'_A is the solid solubility of A in B at temperature T and N_A is the liquidus composition at the same temperature.

Figure 21 shows a plot of N_{Pb} against the reciprocal liquidus temperature of alloys of lead with several other metals. The points fall into a straight line of slope -246 and the enthalpy of fusion thus obtained is

$$L_f, (\text{Pb}) = 4728 \pm 250 \text{ J}$$

which agrees very well with the value found calorimetrically: $L_f = 4770 \pm 125$ J.

2. *Henry's Law*

The law to which substances in dilute solution tend to conform is the more generalised

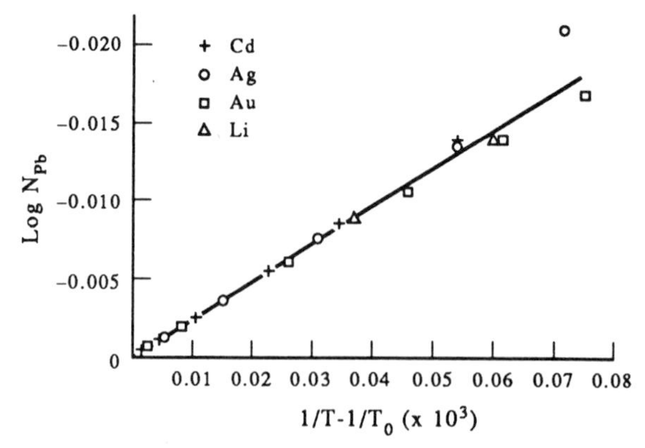

FIG. 21. Evaluation of the enthalpy of fusion of lead from freezing point data.

one derived from the postulate of Henry that the solubility of a gas in a liquid is proportional to its pressure (provided that its molecular state is the same in solution as in the gas). This will be found to lead to the relationship

$$a_A = \text{constant} \times N_A. \tag{50}$$

Like Raoult's law, that of Henry applies within a concentration range the extent of which varies from one system to another, and it is valid only at low concentration.

Henry's law is of greatest practical significance in its application to substances in dilute solution considered separately from the complete system, that is mainly in systems with a limited solubility range. On p. 31, it was expressed in the form

$$\Delta\bar{G}_i = \boldsymbol{R}T \ln \gamma_i + \boldsymbol{R}T \ln N_i. \tag{48}$$

Here, the first term on the right-hand side is in fact the excess Gibbs energy of solution

$$\Delta\bar{G}_i^E = \boldsymbol{R}T \ln \gamma_i. \tag{70}$$

For more accurate applications, the excess Gibbs energy should be split up into an enthalpy and an entropy term:

$$\Delta\bar{G}_i = \Delta\bar{H}_i + \boldsymbol{R}T \ln N_i - \Delta\bar{S}_i^E T. \tag{71}$$

If the solute element i is a diatomic gas i_2, one often finds that the solubility is not directly proportional to the pressure of the gas but to its square root owing to the dissociation of the gaseous species into two atoms upon solution. This observation is known as Sieverts' law, which is a special case of Henry's law, and may be written in the form

$$\Delta\bar{G}_{i_2} = \Delta\bar{H}_{i_2} + 2\boldsymbol{R}T \ln N_i - \Delta\bar{S}_{i_2}^E T. \tag{72}$$

In these forms, Henry's law (71) and Sieverts' law (72) respectively are to be used for practical calculations involving dilute solutions (see Chapter IV, e.g. examples (A–C). The constant enthalpy and entropy terms of eqn (72) are given in Table IV for the solution of diatomic gases (H_2, S_2, O_2, N_2) in a number of metals (e.g. ref. 1970m).

Some authors (see e.g. [1974b]), prefer to relate activities to reference states other than mole fraction. They express, for instance, concentrations as atomic percentages (or weight

TABLE IV. *Henrian constants for some solutions of diatomic gases in metals*

X = oxygen	$\Delta\bar{H}_{X_2}$ (J/mole)	$\Delta\bar{S}^E_{X_2}$ (J/K mole)	X = nitrogen	$\Delta\bar{H}_{X_2}$ (J/mole)	$\Delta\bar{S}^E_{X_2}$ J/K mole)
Solid metals			Solid metals		
Ti-β	−1,121,970	−186.1	V	+35,565	−16.5
Zr-β	−1,239,135	−168.9	Ta	−409,740	−148.6
V	−844,350	−201.4	Cr	+73,900	−20.0
Ta	−765,840	−191.8	Mo	+189,160	−41.5
Cr	−44,420	−132.5	Fe, α, δ	+65,145	−88.1
Co	−39,520	−216.0	Fe, γ	−5940	−119.0
Liquid metals			Liquid metals		
Cu	−173,300	−91.0	Fe	+17,480	−96.7
Ag	−29,850	−88.0	Co	+83,680	−97.9
Ga	−440,825	−131.4			
Ge	−326,350	−105.9			
Sn	−352,630	−112.3	X = hydrogen		
Pb	−246,220	−113.3	Solid metals		
Fe	−261,585	−64.3	V	−64,900	−121.8
Co	−135,225	−41.8	Ta	−72,760	−117.2
			Cr	+114,870	−57.0
			Fe, α, δ	+51,865	−102.0
			Fe, γ	+54,030	−94.5
X = sulphur			Co	+64,300	−91.2
Liquid metals			Cu	+97,990	−82.0
Cu	−76,100	+10.5	Al	+126,355	−96.1
Ag	−145,350	−93.7			
Sn	−208,780	−119.7	Liquid metals		
Pb	−193,720	−115.1	Fe	+69,705	−72.3
Fe	−251,040	−105.1	Co	+82,005	−67.8
			Cu	+87,030	−71.1
			Al	+118,305	−62.4

percentages) and take $a=1$ when $0=1$ at.% (or 1 wt%). Such a definition of activity, which also involves an activity coefficient, is confusing and is not recommended here.

E. Atomistics and Solution Thermodynamics

One of the objectives of statistical thermodynamics is to make a mathematical bridge between atomic models of systems and their macroscopic thermochemical properties. The highly simplified model of randomly mixed solution of two atomic species forms a convenient basis for discussing the measured values of the partial molar properties of, for example, binary alloys. Let us consider the mixing of n_A atoms of type A and n_B atoms of type B on a three-dimensional lattice which has the crystal structure common to both pure A and B and the whole range of composition of their continuous solid solutions. We will set

$$n_A + n_B = N = 6.06 \times 10^{23} \text{ atoms}$$

so that the discussion concerns 1 g-atom of material throughout. To begin with, let the lattice be completely void of atoms which are set apart separately, and randomly choose one atom to place on the first vacant lattice site. We may choose this atom in N ways from the total number. Now we take a second atom and place that on a neighbouring lattice site to the one which was first filled. This choice may be made in $N-1$ ways since one atom is already located on the lattice. The third atom may be chosen in $N-2$ ways and so on until the last atom can only be "chosen" in one way.

Clearly we have had a total number of ways of choosing the atoms from the original assembly to fill the lattice in

$$N(N-1)(N-2)\ldots 1$$

i.e. $N!$ ways. Each choice would have placed an atom on a given lattice site, so that n_A sites would have A atoms on them, and n_B sites would be occupied by B atoms. Such an array of sites filled with atoms we will call a configuration. But now if we select two sites occupied by A atoms we could not tell the difference between atom 1 of type A on site 1 and atom 2 of type A on site 2 and the reverse configuration with atom 2 on site 1 and atom 1 on site 2. These two configurations would be indistinguishable. We could only obtain a distinguishable configuration if the atoms were of different types, one of A and one of B. Thus the number of ways of producing distinguishable configurations will be smaller than $N!$ and will be given by

$$\text{Number of distinguishable configurations} = \frac{N!}{n_A!\,n_B!}$$

which is called also the **thermodynamic probability** (W). It can be deduced from an expression evolved by Boltzmann that the entropy change on mixing two species in this fashion is related to the thermodynamic probability by

$$\Delta S = k \ln W$$

where k is Boltzmann's constant, $k = R/N$. Hence

$$\Delta S = k \ln \frac{N!}{n_A!\,n_B!}.$$

According to Stirling's theorem, for a large number x,

$$\ln x! = x \ln x - x.$$

Applying this to the equation above, and making use of the fact that $n_A/N = N_A$ (the mole fraction of A) and $n_B/N = N_B$, we find that

$$\Delta S = -\boldsymbol{R}\{N_A \ln N_A + N_B \ln N_B\}.$$

Hence the configurational entropy of mixing the components of a binary alloy randomly has the value given above. From the general relationships given earlier between integral and partial entropies one sees that

$$T\Delta\bar{S}_i = \boldsymbol{R}T \ln N_i = \boldsymbol{R}T \ln a_i, \quad \text{if} \quad \gamma_i = 1.$$

In other words, in a binary alloy where the Gibbs energy of mixing arises simply from the random mixing of the component atoms on a fixed crystal lattice, that alloy would conform to Raoult's law.

Now let us turn to the energy of mixing of a binary alloy. To simplify this calculation, we will assume that each atom only interacts with its nearest neighbours, and that the energy of interaction between two atoms has a constant value depending only on the chemical identity of the atoms and is unchanged by the presence of other atoms. We may then define three interaction energies E_{AA} between the A atoms, E_{BB} between B atoms and E_{AB} between one atom A and one of type B. We will call the number of nearest neighbours which each atom has Z, the co-ordination number.

Now in the binary alloy which we constructed above the **statistical probability** that an A atom occupies any selected site is N_A. The probability that it will be occupied by a B atom is N_B. The probability that any selected neighbouring site has an A or a B atom on it is determined in the same way in a random mixture. Thus the probability that

the two sites are occupied by A atoms is N_A^2
the two sites are occupied by B atoms is N_B^2
the two sites are occupied by an A–B pair is $2N_AN_B$.

(The two here is because either site could be occupied by either species to produce an unlike pair.) There will be $\frac{1}{2}NZ$ of such pairs in the mixture as a whole and hence the total interaction energy will be

$$\tfrac{1}{2}NZ(N_A^2E_{AA}+NN_B^2E_{BB}+2N_AN_BE_{AB}).$$

Now to obtain the energy of formation of the alloy from N_A and N_B g-atoms of A and N_B g-atoms of B we must subtract

$$\tfrac{1}{2}NZ(N_AE_{AA}+N_BE_{BB})$$

from this quantity and thus

$$\Delta E_{\text{mixing}} \cong \Delta H_{\text{mixing}} = \tfrac{1}{2}ZN\{(N_A^2-N_A)E_{AA}+(N_B^2-N_B)E_{BB}+2N_AN_BE_{AB}\}$$

$$= NZN_AN_B\left(E_{AB}-\frac{E_{AA}+E_{BB}}{2}\right). \tag{73}$$

It should be noted that the interaction energies are negative quantities and thus when

$E_{AB} > \dfrac{E_{AA}+E_{BB}}{2}$ ΔH_{mixing} is negative

$E_{AB} = \dfrac{E_{AA}+E_{BB}}{2}$ ΔH_{mixing} is zero (Raoult's law)

$E_{AB} < \dfrac{E_{AA}+E_{BB}}{2}$ ΔH_{mixing} is positive.

Thus by taking into account the configurational arrangement of the atoms in a binary alloy we can arrive at three ideal laws of mixing.

(1). Raoult's law where $\Delta H=0$; $\Delta\bar{S}_i=-\boldsymbol{R}\ln N_i$.
(2) The strictly regular solution where $\Delta H=\alpha N_AN_B$; $\Delta\bar{S}_i=-\boldsymbol{R}\ln N_i$.
(3) Henry's law where $\Delta\bar{H}_i\rightarrow\alpha$ as $N_i\rightarrow 0$.

As well as these configurational terms there is also the possibility of thermal contributions arising from deviations from the Neumann–Kopp rule in the alloy. This states that

$$\Delta C_{p_{\text{alloy}}} = N_AC_{p_A}+N_BC_{p_B} \tag{74}$$

It is the presence of these terms as well as possible departures from random mixing, which will be discussed later, which causes real systems to deviate from these ideal laws.

F. Regular Solutions

Fifty years ago it was still assumed that all metallic solutions are approximately regular. In fact, one of the present authors in a monograph written in collaboration with the late Dr Weibke converted measured Gibbs energies of metallic solutions into enthalpies of solution by means of the equation

$$\Delta\bar{H}_A \equiv \Delta\bar{G}_A^E = 19.144T \log \frac{a_A}{N_A} = 19.144T \log \gamma_A \tag{75}$$

and presented all the thermochemical data merely in the form of enthalpy curves.

It will be seen in Section G that the underlying assumptions are not borne out by experimental evidence. However, there are a few aspects which have been mentioned in Section D and will be outlined in the present section where the regular solution concept retains some usefulness.

1. Calculation of Activities Using Equilibrium Diagrams

We may follow Rey (1949k) to demonstrate the calculation of chemical activities of the components of liquid solutions from the values represented by the liquidus and solidus lines in the equilibrium diagram. In the equations which follow, L_f denotes the enthalpy of fusion of the pure solvent substance A at its melting point T_0. The activities and corresponding mole fractions of substance A at the liquidus and solidus compositions at a lower temperature T_c are denoted respectively by a, a', N, N' (Fig. 22). T is an arbitrary temperature at which activities are to be calculated. The suffix A is omitted as all quantities refer to the solvent.

In systems where there is no solid solubility—that is, the solidus curve coincides with the pure solvent substance—then activities in the liquid mixture can be obtained by the

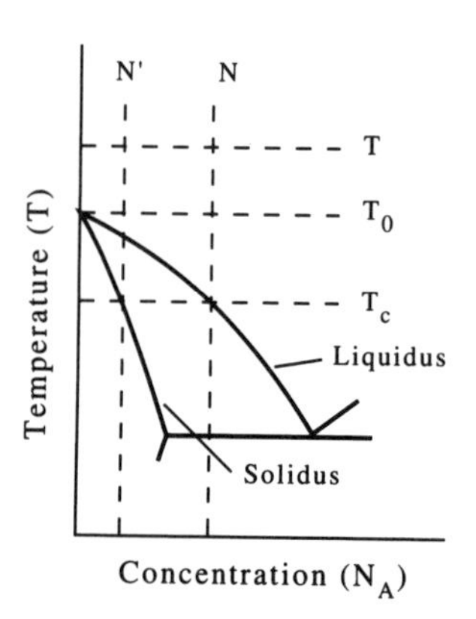

FIG. 22. Calculation of activities from equilibrium diagrams.

following considerations. Equation (74) may be taken to mean that $\boldsymbol{R}T \log (a/N)$ is independent of temperature. Hence

$$\log a_T = \frac{T_c}{T} \log a_{T_c} + \frac{T-T_c}{T} \log N.$$

Combining this with eqn (72),

$$\log a_T = -\frac{(T_0-T_c)L_f}{19.144T\,T_0} + \frac{T-T_c}{T} \log N$$

or, if there is a range of solid solubility over which Raoult's law is valid (i.e. the solution is ideal: $a' = N'$) the following equations are derived

$$\log a_T = -\frac{(T_0-T_c)L_f}{19.144T\,T_0} + \frac{T_c}{T} \log N' + \frac{T-T_c}{T} \log N \tag{76}$$

and

$$\log a'_T = \frac{T-T_c}{T}\left(\log N - \frac{L_f}{19.144T_0}\right) + \frac{T_c}{T} \log N'. \tag{77}$$

These or similar equations have often been used to derive activities, together with enthalpies of solution (see for example, Fig. 23). Changes in molar heats on solution are occasionally taken into account, and the results are sometimes presented in the form of integral enthalpies of solution.

There is, however, some objection to the low accuracy of the data so derived. The calculation involves an exponential relationship which requires that the enthalpy of fusion of the solvent should be known accurately, which is often not the case. The liquidus and solidus curves should be known with an accuracy well within $\pm 0.2°$ to guarantee a fair accuracy of the resulting thermochemical data; thermal measurements have, however, rarely been made with such precision. Deviations from regularity also affect the evaluations considerably; these will be discussed in some detail in Section G. The neglect of the change in heat capacities on melting introduces a further uncertainty, but this is probably of a minor order. All these uncertainties may amount together to considerable errors, and little evidence has so far been forthcoming to show that activities calculated from equilibrium diagrams without independent thermochemical evidence are reliable.

At present it may be regarded as a useful approximate means of filling gaps in our quantitative knowledge. Calculations of the activity curves of silicate melts by Rey (1949k) are an example. Owing to the objections raised above, these results are probably of low accuracy but they give quite a good indication of the general shape of the activity curves and illustrate the implications of the shape of the curves.

2. *Spinodal Decomposition*

In a strictly regular solution having a positive integral enthalpy of mixing there must exist a critical temperature above which the components are completely miscible and below which the system consists of a physical mixture of two solutions, one rich in component A, and the other rich in component B. The limiting compositions of these solutions may be calculated by means of the equations obtained above. Consider the

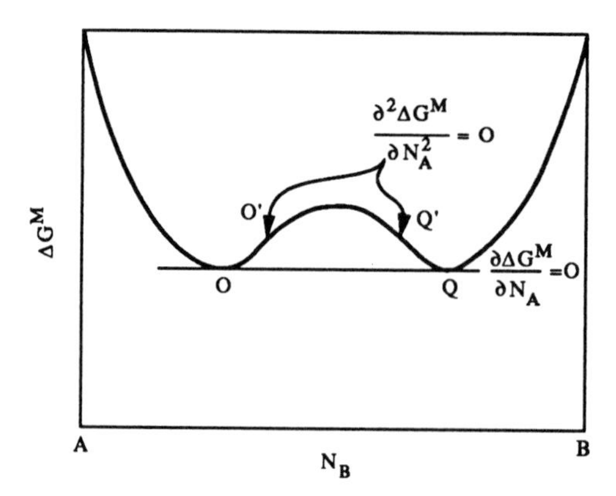

FIG. 23. First and second differentials with respect to mole fraction of A of the integral Gibbs energy of mixing of a binary A–B alloy. The curve is for a temperature below T_c.

Gibbs energy of mixing curve at a temperature below the critical value T_c as shown in Fig. 23.

The common tangent to the integral curves at the composition of the two limiting solutions passes clearly through OQ. These are the two points on the curve where $\partial G^M/\partial N_A = 0$. The two points O', Q', are those where $\partial^2 \Delta G^M/\partial N_A^2 = 0$ and these are called the spinodal points of the system, since it is at these compositions that the two new phases first appear when A–B alloys are cooled from a higher temperature, and not at the equilibrium compositions O and Q. The curve traced out as a function of temperature in the system by the points corresponding to O and Q are those showing the equilibrium compositions on the phase diagram. The spinodal points trace out the spinodal curves between these two limits.

From the expression for the Gibbs energy of mixing of a binary strictly regular solution, the phase boundaries are given by

$$\frac{\partial \Delta G}{\partial N_A} = \alpha(1-2N_A) + \boldsymbol{RT}\left[\ln \frac{N_A}{1-N_A}\right] = 0 \tag{78}$$

$$\frac{\boldsymbol{RT}}{\alpha} = (N_B - N_A) \ln \frac{N_A}{N_B}. \tag{79}$$

The critical temperature where immiscibility begins must occur at the equimolar composition at the temperature such that

$$\boldsymbol{R}T_c = 2\Delta H_{0.5}$$

Also,

$$\frac{\partial^2 \Delta G}{\partial N_A^2} = -2\alpha + \boldsymbol{RT}\left[\frac{1}{N_A N_B}\right] = 0$$

at the spinodal compositions.

Hence $N_A N_B = \boldsymbol{RT}/2\alpha$ along the spinodal curves.

3. Order–disorder Transformation

Some binary systems having a negative enthalpy of mixing show the order–disorder transformation. This is when, at a given composition, the system undergoes a change in the position of the atoms with increasing temperature. At the lower temperatures all the atoms of each component occupy well-defined sublattices, and above a critical temperature the atoms are randomly mixed on one common lattice. If the random mixture is also a strictly regular solution, the algebra of the transformation may be evolved in the following way.

In one g-atom of alloy of equiatomic composition there will be 0.5 g-atoms of A on one sublattice and 0.5 g-atoms of B on the other sublattice in the ordered low-temperature system. In the high temperature, disordered alloy, each sublattice contains 0.25 g-atom of each species. In the low-temperature system, each A atom is immediately surrounded by B atoms and vice versa. At high temperature the co-ordination of each atom is made of $Z/2$ A atoms and $Z/2$ B atoms. If S represents the fraction of the A sublattice which is occupied by the B atoms, then S takes the value $\frac{1}{2}$ in the random alloy. At some intermediate stage $(0<S<\frac{1}{2})$ the entropy of the alloy is $\Delta S=-\boldsymbol{R}[(1-S)\ln(1-S)+S\ln S]$ per g-atom and the energy of the lattice is $E=\frac{1}{2}NZ[S(1-S)E_{AA}+S(1-S)E_{BB}+(1-S)(1-S)E_{AB}]$.

The stable configuration at a given temperature can be obtained by minimizing the Gibbs energy with S as the variable

$$\frac{\partial G}{\partial S}=0=\tfrac{1}{4}NZ(1-2S)\left[\frac{E_{AA}+E_{BB}}{2}-E_{AB}\right]+\boldsymbol{R}T\left[\ln\frac{S}{(1-S)}\right]. \tag{80}$$

As $T\rightarrow T_c$ the critical temperature above which the alloy is random, $S\rightarrow\frac{1}{2}$, and the equation

$$(1-2S)\frac{T_c}{T}=\ln\left(\frac{1-2S}{S}\right) \tag{81}$$

is obtained for the value of S at any intermediate temperature.

G. Non-regular Solutions

Experimental evidence has shown that "regular solutions" are rare, but the concept retains its usefulness for didactic purposes. In order to describe deviations from regularity, the "excess Gibbs energy" and "excess entropy" functions have been introduced. By combination of the partial excess functions (p. 34) with those of Lewis (p. 30) one obtains the integral excess functions

$$\Delta S^E=\Delta S^{\mathrm{xptl}}+\boldsymbol{R}(N_A\ln N_A+N_B\ln N_B+\cdots) \tag{82}$$

$$\Delta G^E=\Delta G^{\mathrm{xptl}}-\boldsymbol{R}T(N_A\ln N_A+N_B\ln N_B+\cdots). \tag{83}$$

For a regular solution, ΔS^E is zero across the whole range of solution. In Fig. 24 are plotted the maximum or minimum excess entropies of mixing against the maximum or minimum enthalpies of mixing for a random selection of systems exhibiting complete mutual solubility, having different types of bonding. It may be seen that the condition for regular behavior, $\Delta S^E = 0$, is rarely fulfilled, strictly only for the "ideal solutions" for which $\Delta H = 0$. Actually, there seems to be a direct relationship between ΔH and ΔS^E as indicated by the curve which is quite empirical. In addition, it may be pointed out that the maxima in the enthalpy and entropy vs concentration curves is normally not observed at equiatomic composition as required by the regular solution model, but is shifted to higher or lower concentrations for both the enthalpy and entropy—apparently always in the same direction. As might be expected from these observations, a plot of the partial excess entropies against the partial enthalpies at high dilution also produces a direct relationship, as has been shown by Slough (1971q).

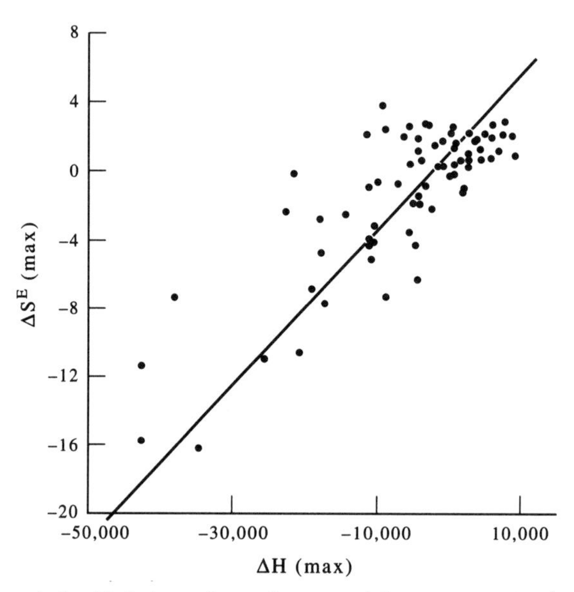

FIG. 24. The linear relationship between the maximum or minimum excess entropies and enthalpies of mixing for a number of binary alloy systems.

Kleppa (1949g) has drawn attention to the work by Scatchard (1937f) on the effect of a net volume change on the entropy of mixing. According to this, one could calculate the deviations from the ideal entropy of mixing if the volume changes are known, and Kleppa concluded under some simplifying assumptions that they should be approximately proportional, but could not confirm this relationship when comparing the experimental data. However, Table V shows that a rough connection does exist between the entropy and the volume changes on mixing liquid metals. According to Table V there also appears to be an approximate proportionality between the changes in volume and enthalpy of mixing.

Most liquid systems for which the changes on mixing of volume, entropy and enthalpy are known, involve one or two meta- or semi-metals, such as Bi, Sn, Pb, Tl, In, Cd and Zn, for which discussions of structure and other X-ray evidence have amply shown that covalent bonds occur (1949h, i; 1951d). If a more "metallic" metal is added to the more

TABLE V. *Maximum or minimum values of the integral changes of volume enthalpy and entropy of mixing of liquid binary alloys*

System 1–2	ΔV in %	N_2	ΔS^E in J/K . g-atom	N_2	ΔH in J/g-atom	N_2
Fe–Si	−27.0	0.52	−7.53	0.57	−37,865	0.48
K–Hg	−25.4	0.58	−6.93	0.47	−17,910	0.6
Mg–Pb	−7.0	0.33	−0.71	0.38	−10,125	0.38
Mg–Zn	−6.5	0.65	−2.47	0.55	−9205	0.55
Cu–Zn	−3.1	0.6	−3.33	0.5	−10,275	0.5
Mg–Al	−3.0	0.4	−0.88	0.4	−3370	0.5
Na–K	−1.0	0.33	−0.18	0.4	+735	0.6
Zn–Al	−0.5	0.45	+0.97	0.45	+2570	0.48
Pb–Bi	+0.3	0.33	+0.19	0.5	−1100	0.5
Zn–Cd	+0.6	(0.5)	+0.13	0.5	+2090	0.5
Cd–In	+0.8	0.43	+0.56	0.45	+1435	0.45
Cd–Pb	+0.95	0.38	+0.74	0.38	+2660	0.48
Cd–Tl	+1.0	0.44	+1.02	0.45	+2310	0.42
Sn–Bi	+1.1	0.33	+0.30	0.6	+105	0.5
Cu–Ag	+1.3	0.66	+0.64	0.7	+4245	0.55
Zn–Sn	+1.5	0.32	+2.18	0.4	+3190	0.4
Cd–Sn	+1.55	0.45	+1.18	0.45	+1810	0.46
Cd–Bi	+1.8	0.33	+1.58	0.35	+866	0.33

covalent one, this should result in a partial or complete break-up of the covalent bonds. This process would consume energy and increase disorder. In fact, Table V shows that the alloys concerned are generally formed endothermically and have positive excess entropies of mixing. Since the volume changes are also positive, it follows that the covalent "clusters" are not completely destroyed but are too rigid to allow a suitable close packing.

In Table V, the systems 1–2 are written in such a way that the more covalent metal stands in the second place. It is then seen that the maxima in the entropy and enthalpy curves are nearly always shifted to below equiatomic compositions. This is in line with the hypothesis just outlined: the covalent bonds are obviously more strongly affected when the respective metal is more dilute; in other words, when a covalent metal is dissolved in a less covalent one, more energy and order is consumed than in the reverse process.

We may take metallic bonds to be essentially additive and ascribe the enthalpies of formation of truly metallic phases to packing effects (1958f, g) (see p. 194). Then the mutual solutions of true metals should be nearly ideal as long as the component metals differ only little in atomic diameter. Although this is not always observed, regular behaviour is quite often approached in such solutions. As an example, the entropy curve of liquid silver–copper alloys, shown in Fig. 25, is near the ideal one. The entropy curve for zinc–tin, also depicted in this diagram, is higher owing to the destruction of covalent bonds and the resulting increase in vibrational entropy.

Alterations in the positional entropy can lead to a minimum ΔS of zero; alterations in the vibrational entropy may make ΔS negative. Suitable systems with which this question may be investigated are the liquid alloys of magnesium with bismuth and antimony. Both systems form a complete series of liquid solutions but solid Mg_3Sb_2 and Mg_3Bi_2 are stabilised partly by polar bonds which conceivably persist to some extent in the liquid state. The entropies of mixing of liquid magnesium and antimony are also shown in Fig. 25. It is seen that the curve is in fact negative over large ranges of composition with a minimum at approximately stoichiometric composition. This behaviour is undoubtedly

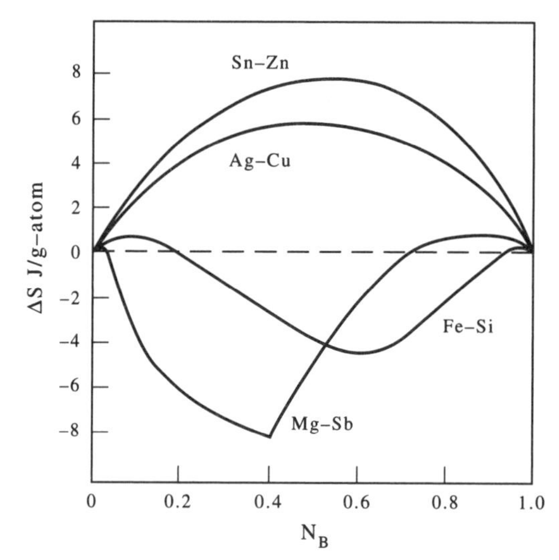

FIG. 25. Integral entropies of mixing in the liquid systems tin–zinc, silver–copper, iron–silicon, and magnesium–antimony.

due to strong polar forces in melts that consist partly of Mg^{2+} and Sb^{3-} or Bi^{3-} ions. These forces exert their influence mainly on the vibrational term of the entropy.

When strong covalent bonds are formed on alloying, the effect on the entropy is similar as with polar bonds. The example of the liquid iron–silicon alloys in Fig. 25 is probably due to this type of bond mechanism.

These are extreme cases, but they indicate the possible magnitude of the effect. Moreover, the alloys have strong negative enthalpies of mixing with minima at the stoichiometric compositions, namely -37.9 (Fe–Si), -33.9 (Mg–Sb) and -21.8 (Mg–Bi) kJ/g-atom. The corresponding curves of partial enthalpies of solutions are ~ shaped, as was predicted by Wagner on the basis of a similar consideration for the liquid system Ag–Te (1952j).

These examples illustrate the importance of the behaviour of the valence electrons in considerations of entropy- and enthalpy-changes on mixing. Even when two true metals are mixed together the conduction electrons will shift, provided that the metals are of different valence, and thus bring about bonding between the positive ions. This change in bond will influence the various thermodynamic functions. This is clearly indicated by the observation of Hume-Rothery that solutions for example of germanium, gallium and zinc in copper all become saturated at approximately the same concentration of valence electrons. Further examples of the significance of a change in the concentration of valence electrons are given by Wagner (1952j). These changes would affect the enthalpy as well as the entropy terms, but the quantitative elaboration is complicated.

H. Ternary Solutions

Three methods for solving the ternary Gibbs–Duhem equation have been given in the literature. The resulting equations make it possible in principle to calculate the activities of two components, say B and C, in an A–B–C ternary alloy, from a knowledge of the partial

Gibbs energy of component A over the whole ternary composition range for a given temperature. The consequence of this requirement of such detailed information is that attempts have been made to put workable approximations into the accurate equations in order to obtain the main features of ternary systems and to test the approximations with the limited amount of data which are at present established. Unfortunately, most of the experimental results relate to systems in which the interactions between the alloying components are relatively weak. Such systems do not provide a very rigorous test of the approximations and these are made to look more successful than they really are.

The methods of solving the ternary Gibbs–Duhem which are obtained by Darken and by Wagner have some degree of similarity; the principal difference resides in a sequence of integration and differentiation in Darken's procedure (1950b) which is reversed in Wagner's analysis (1952j). From many points of view, Wagner's approach is more readily applied to experimental data with reasonable accuracy than Darken's, but for the purposes of the present analysis we shall begin our discussion of the ternary systems with an equation which emerges during Darken's analysis. The method of solution of Schuhmann (1950j), which is also very powerful for experimental data of a limited scope, will not be referred to further, but the interested reader will find a useful discussion of this procedure in Richardson's monograph (1974k).

Returning to the Darken derivation, we will denote by α_B the function

$$\Delta\bar{G}^E_B/(1-N_B)^2$$

and then an important equation obtained by Darken may be written as

$$\Delta G^E_{B+A/C}=(1-N_B)\int_1^{N_B}[\alpha_B\,dN_B]_{N_A/N_C}-N_A\int_1^0[\alpha_B\,dN_B]_{N_C=0}-N_C\int_1^0[\alpha_B\,dN_B]_{N_A=0}. \tag{84}$$

The first integral on the right-hand side is obtained from values of α_B as a function of N_B and along an integration path with a constant atom fraction ratio of A and C, N_A/N_C (Fig. 26). The excess integral Gibbs energy on the left is for the ternary system containing component B at the atom fraction N_B, with A and C in the proportion N_A/N_C. The other two integrals oan the right-hand side are for α_B in the binary A–B and B–C systems, respectively.

Toop (1965f) has shown that this equation can be simply integrated for the ternary system in which all solutions are strictly regular, that is to say α_B remains constant along any given integration path with N_A/N_C fixed. The binary A–B and B–C systems must also be regular. Then,

$$\Delta G^E_{B+A/C}=\frac{N_A}{1-N_B}\,\Delta G^E_{A-B}+\frac{N_C}{1-N_B}\,\Delta G^E_{B-C}+(1-N_B)^2[G^E_{A-C}]_{N_AN_C}. \tag{85}$$

Kohler [1960f] obtained another solution which can also be obtained from this Darken equation by choosing different integration paths (Fig. 27) but again assuming strictly regular behaviour along these. The paths of integration are now along constant ratios N_A/N_B, N_B/N_C and N_CN_A to yield the equation

$$\begin{aligned}\Delta G^E_{A-B-C}=&(1-N_A)^2[\Delta G^E_{B-C}]_{N_B/N_C}+(1-N_B)^2[\Delta G^E_{A-C}]_{N_A/N_C}\\&+(1-N_C)^2[\Delta G^E_{A-B}]_{N_A/N_B}.\end{aligned} \tag{86}$$

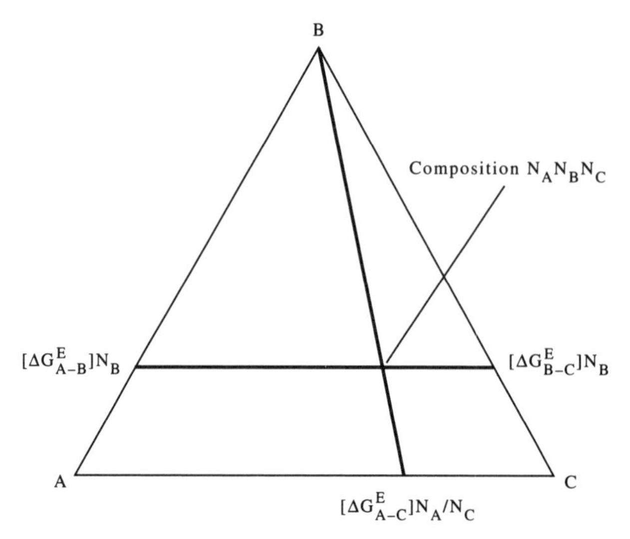

FIG. 26. Illustration of the binary terms used in the application of the Toop and Bonnier equations.

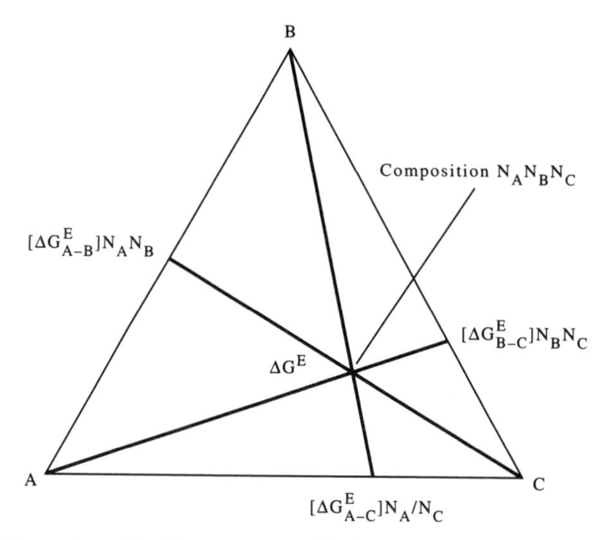

FIG. 27. Illustration of the binary terms used in the application of the Kohler equation.

A further equation has been obtained by Bonnier and Caboz [60]

$$\Delta G^E_{B+A/C} = \left[\frac{N_A}{1-N_B}\,\Delta G^E_{A-B} + \frac{N_C}{1-N_B}\,\Delta G^E_{B-C}\right]_{N_B} + (1-N_B)\,[\Delta G^E_{A-C}]_{N_A/N_C}. \qquad (87)$$

These equations have been surveyed by Ansara (1971a) and Spencer *et al.* (1972i) have compared calculated and experimental results from some ternary systems together with the equation of Margules (1946b) which contains an empirical ternary constant K_{ABC}, as well as the binary constants, K_{BA}, etc.

$$\Delta G^{E}_{A-B-C} = N_A N_B[N_A K_{BA} + N_B K_{AB}] + N_A N_C[N_A K_{CA} + N_C K_{AC}] + N_B N_C[N_B K_{CB} + N_C K_{BC}] + N_A N_B N_C K_{ABC}. \tag{88}$$

Although these equations have all been written for ΔG^E, they may also be applied in the calculation of ΔH and ΔS^E in the obvious manner.

For low-affinity solutions these methods mostly give similar values and agree quite well with the experimental ones, often within the experimental accuracy. For high-affinity systems, deviations are larger. Figure 28 shows as an example the experimental heats of formation of iron–cobalt–nickel alloys in comparison with the values calculated by means of the Kohler and Bonnier equations (1973m). However, it is difficult at the present stage to decide about the relative merits of the various modes of extrapolation, and further experimental values must be awaited for a systematic survey. Ansara *et al.* (1978a) have compared calculated and experimental phase equilibria in selected ternary alloy systems using the thermodynamic values derived from different models. The Margules equation has been employed by Kubaschewski and Counsell to describe the thermochemical properties of the system gold–palladium–platinum (1971j).

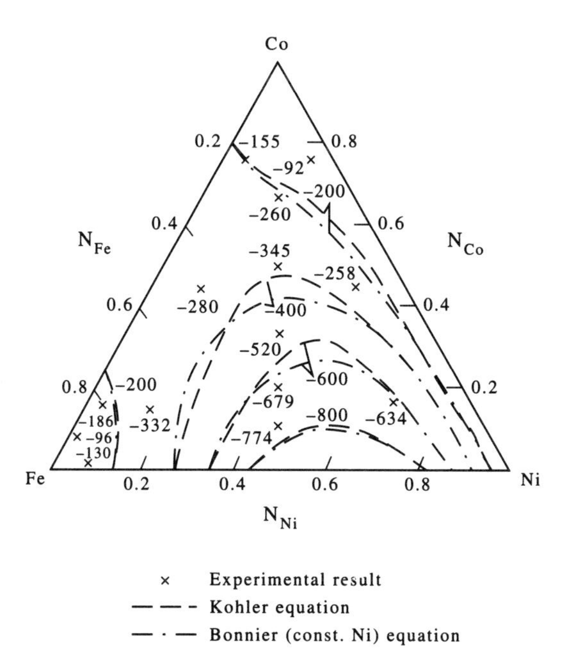

FIG. 28. Iso-enthalpies of formation (cals) of solid iron–cobalt–nickel alloys at 1473 K.

The application of these equations to the evaluation of phase boundaries in ternary systems will be described later (p. 265).

In the system where all solutions behave regularly, Darken's equation may be used to obtain the variation of the excess Gibbs energy of the dilute, Henrian, solution of B in the binary A–C solvent as a function of N_A/N_C. The solution is

$$\Delta\bar{G}^{E}_{B(A-C)} = N_A \Delta\bar{G}^{E}_{B(A)} + N_C \Delta\bar{G}^{E}_{B(C)} - \Delta G^{E}_{A-C}) \tag{89}$$

B(A–C) denotes B dissolved in the A–C alloy of mole fractions N_A, N_C and for this solution

$$N_A + N_C \cong 1.$$

Alcock and Richardson (1958a; 1960a) used an atomistic model in which the B atoms are surrounded randomly by A and C atoms and each A–B and B–C bond is of constant strength at all values of N_A/N_C. This constant pairwise bonding model which together with a fixed co-ordination number Z describes a strictly regular solution also yields eqn (89). The model system further suggests, however, that when, say, the A–B bond is much stronger than the B–C bond, then A atoms will tend to cluster around the B atoms to the exclusion of the less strongly bonded C atoms. The population of the Z atoms which surround any B atom is then calculated by the "quasi-chemical" approach to be in the proportion

$$\frac{n_A}{n_C} = \frac{N_A}{N_C} \exp(W/Z\boldsymbol{R}T)$$

where W is the exchange energy for the change of bonds $A-B+C \rightarrow B-C+A$ and therefore

$$W = \Delta\bar{H}_{B(C)} - \Delta\bar{H}_{B(A)} - \Delta\bar{H}_{C(A-C)} + \Delta\bar{H}_{A(A-C)}.$$

From these atomistic considerations, the equation

$$\left[\frac{1}{\gamma_{B(A-C)}}\right]^{1/Z} = N_A\left[\frac{\gamma_{A(A-C)}}{\gamma_{B(A)}}\right]^{1/Z} + N_B\left[\frac{\gamma_{C(A-C)}}{\gamma_{B(C)}}\right]^{1/Z} \tag{90}$$

was derived (1958a; 1960a). This equation contains the co-ordination number explicitly but approaches eqn (89) as Z becomes large. For a typical value in liquid metallic solutions $Z \simeq 10$ the effect of clustering is relatively unimportant in most alloy systems involving three metallic species. Results obtained by Jacob and Alcock for the dilute solution of indium in the solid copper–gold alloy where Z is known to be 12 are indicative of the size of the effect (1973e) (Fig. 29). The more significant effect which is found when the dilute solute is a non-metal such as oxygen has been accounted for by these authors in a further extension of the quasi-chemical model (1972f). If an oxygen atom attaches one or two electrons in a metallic alloy solvent, then this negatively charged species will tend to repel the conduction electrons in its immediate environment. In the language of bonds, the oxygen species will tend to cause the breaking of metal–metal bonds in its co-ordination shell. The assumption that oxygen and sulphur form strong bonds with four atoms in the co-ordination shell where half of the metal–metal bonds are broken, accounts for most of the published literature for the dilute solutions of oxygen and sulphur in liquid metallic alloys. The equation for oxygen in dilute solution in the A–C alloys is, then

$$\left[\frac{1}{\gamma_{O(A-C)}}\right]^{1/4} = N_A\left[\frac{\gamma^{1/2}_{A(A-C)}}{\gamma^{1/4}_{O(A)}}\right] + N_C\left[\frac{\gamma^{1/2}_{C(A-C)}}{\gamma^{1/4}_{O(C)}}\right]. \tag{91}$$

This equation resembles mathematically that given earlier for dilute metallic alloys, eqn (90) but differs essentially in the two values of Z which are used. Hydrogen in solution in liquid metallic alloys which might be expected to ionize in the normal "metallic" fashion

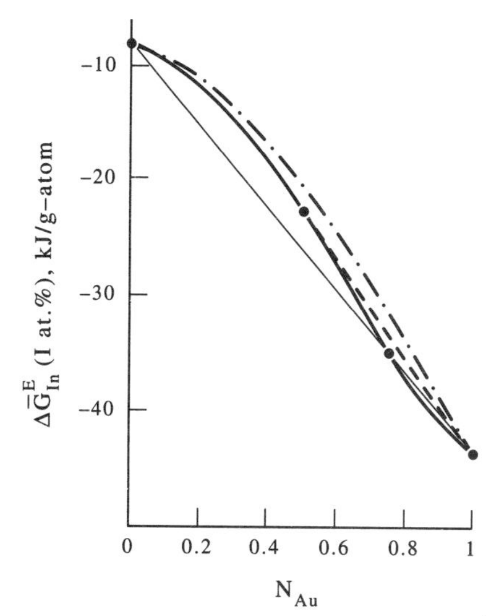

FIG. 29. Variation of the chemical potential of indium at 1 at.% concentration with N_{Cu}/N_{Au} ratio in Cu + Au + In alloys at 900 K: —·—·— regular solution model; ------ quasichemical model $Z = 12$; —●— experimental data.

to H^+, and an electron appears to follow eqn (90) thus differing from the electro-negative oxygen and sulphur.

The effect of an alloying element on the activity coefficient of a dilute solute in solution in a liquid metal is of considerable importance in the refining of metals. The atomistic models given above demonstrate that a number of factors determine this effect, not the least among these being the metal–metal interactions.

From the many experimental studies that have been made of these effects, collections of empirical data have been made covering many systems of industrial importance. These are usually reported as values of the so-called "interaction" coefficient, ε_B^C which is defined by the equation

$$\frac{\partial \ln \gamma_B}{\partial N_C} = \varepsilon_B^C$$

for the dilute solutions of B and C in the solvent metal A. This parameter only has a constant value over a limited range of composition, usually less than one atomic percent for each solute. As the concentration of C is increased while that of B is kept small, the change in ε_B^C will largely reflect the competing interactions of B with C and of the solvent A with C. Figures 30 and 31 show some collected data for the Fe–O–X and Cu–O–X systems, where X is the alloying element and the concentration of oxygen is low. For many uses it is convenient to express the concentration of X in weight percent and to use logarithms to the base ten. Then,

$$\frac{\partial \log \gamma_O}{\partial \text{wt\%X}} = e_O^X = \frac{M_{Fe}}{M_X}\frac{\varepsilon_O^X}{230.3} \tag{92}$$

when the solvent is iron and M_{Fe}, M_X are the atomic weights of iron and of X respectively.

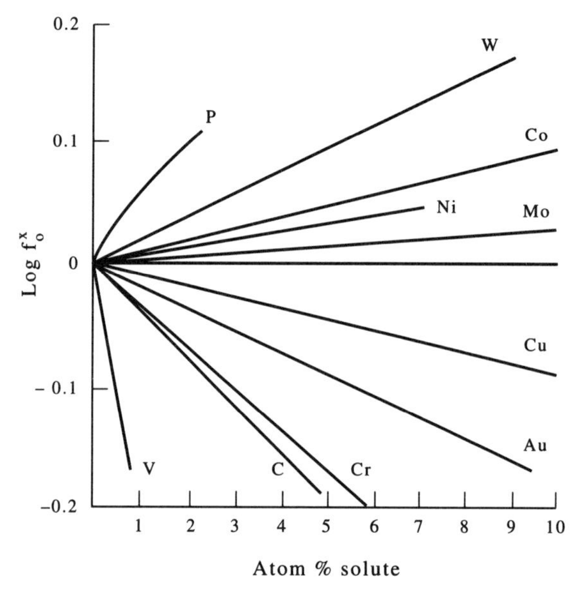

FIG. 30. The effects of some alloying elements on the activity coefficient of oxygen in liquid iron at 1600°C.

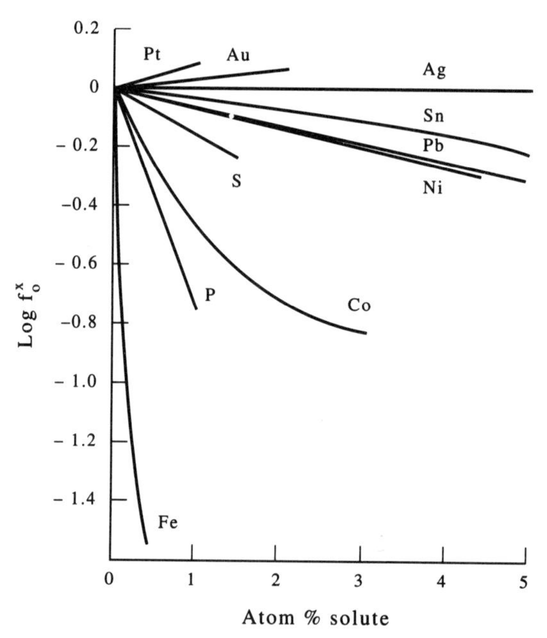

FIG. 31. The effects of some alloying elements on the activity coefficient of oxygen in liquid copper at 1200°C.

This simple relationship only applies for small concentrations of X which are usually of most importance in the industrial applications. The interaction coefficient ε_B^C is clearly related to the integral excess Gibbs energy of the ternary system and can be shown to be equal to the conjugate interaction parameter ε_C^B. This results from the fact that the sequence of differentiation of a homogeneous function, in this case ΔG_f^E, with respect to two variable N_B and N_C, does not affect the result. Thus

$$\varepsilon_B^C = \frac{\partial^2 \Delta G_f^E/RT}{\partial N_B\, \partial N_C} = \frac{\partial^2 \Delta G_f^E/RT}{\partial N_C\, \partial N_B} = \varepsilon_C^B. \tag{93}$$

TABLE VI. *Interaction coefficients for oxygen in liquid iron and copper*

Fe–X–O at 1600°C	ε_O^X	e_O^X	γ_X^0(Fe–X)
Fe–Ni–O	1.4	0.006	0.62
Fe–Mo–O	0.67	0.0035	Raoultian
Fe–Mn–O	−4.7	−0.021	1.33
Fe–Cr–O	−8.5	−0.04	0.87
Fe–Si–O	−15	−1.13	0.0013
Fe–C–O	−22	−0.45	0.57
Fe–V–O	−63	−0.3	0.18
Fe–Ti–O	−118	−0.6	0.01
Fe–Al–O	−433	−3.9	0.058
Cu–X–O at 1100°C			
Cu–Pt–O	+38	0.057	0.05
Cu–Au–O	+6.9	0.007	0.14
Cu–Ag–O	−0.2	0	3.23
Cu–Sn–O	−4.6	−0.0087	0.42
Cu–Ni–O	−7.0	−0.034	2.22
Cu–Pb–O	−7.4	−0.007	5.27
Cu–Co–O	−61	−0.31	1.54
Cu–Fe–O	−553	−2.73	19.5

The standard state for element X is the pure element in the equilibrium state at 1600°C for iron alloys and 1100°C for copper alloys. γ° is the activity coefficient at infinite dilution.

CHAPTER 2

Experimental Methods

The quantities conveniently recorded for the thermodynamic description of inorganic and metallic compounds are the enthalpy of formation at 298 K, the standard entropy, the heat capacity (for condensed substances up to the boiling point) and the enthalpies of transformation and fusion where applicable. All these data can in principle be obtained by means of suitable calorimeters. Of course, no single apparatus of this type can be employed to measure all these quantities. Application of each calorimeter is restricted to a certain range of temperature. There are calorimeters operating between 1–50 K and room temperature producing low-temperature heat capacities and standard entropies, provided the substances are in the well-ordered crystalline equilibrium state required by the third law. There are calorimeters specially designed to measure enthalpies of reaction and others for the determination solely of heat capacities and enthalpies of transformation and fusion. Some of the last-mentioned calorimeters can be made to produce not only heat capacities but also enthalpies of reaction or formation. Thus the calorimeter is the most important piece of equipment of the versatile thermochemist.

However, a calorimeter may not always give the desired results. A case in point is the evaluation of low-temperature heat capacities to provide standard entropies. Strongly ionic compounds, it is true, can normally be produced in the highly ordered form required for the application of eqn (22). Intermetallic compounds, the stability of which stems from a dense atomic packing, are also likely to have a near-zero entropy at 0 K, as has been demonstrated within the experimental error for $TaCr_2$ and other Laves phases by Martin *et al.* (1970n). Many phases of a predominant covalent–metallic nature, however, are not easily brought into a completely ordered state, and "frozen-in equilibria" prevail—in particular with the preparation of high-melting point compounds. In such cases it is necessary to find independent thermochemical methods to provide the desired information—and it is here where the usefulness of methods measuring Gibbs energies become apparent. Of such methods there is a great variety, entailing measurements of dissociation pressure, of condensed-state equilibria and of electromotive forces of suitable cells.

When such measurements are made over a range of temperature, enthalpies and entropies of reaction may be evaluated using the relevant equations recorded in Chapter 1. However, it cannot be emphasised strongly enough that such evaluations are unreliable when condensed phases take part in the reaction because the temperature coefficients are very sensitive to small experimental errors. The so-called second-law values are not *necessarily* inaccurate, but the trouble is that one never knows, how just unreliable!

This severe restriction has already been pointed out earlier for liquid and solid solutions of alloy systems. Whenever possible, Gibbs energy results should be combined with independently obtained enthalpies in assessing the standard values. It should not be overlooked that in cases where enthalpies of reaction are difficult to obtain by means of the available calorimeters for kinetic reasons (e.g. metal carbides), and Gibbs energy measurements seem to offer the only alternative, their temperature coefficients are even more affected by kinetic checks. In such cases and in alloy thermochemistry, it is sometimes preferable to record enthalpies and entropies not at 298 K but at some convenient higher temperature where the measurements have actually been made—a procedure followed, for instance, by Hultgren *et al.* (1973c, d) for the presentation of the thermochemical properties of alloys.

This is not necessarily a severe defect because the data are often required for industrial applications, calculations of equilibrium diagrams, etc, at temperatures which are not far from those of the original measurements. When standard enthalpies and entropies have been derived from the temperature coefficients of Gibbs energy measurements, they are given in the tables of the present monograph with relatively large limits of error. Nevertheless, when such values are incorporated in an actual calculation for practical ends, the result is likely to be more accurate at the temperature of application than the stated limits of error imply.

Naturally, as time goes on, more experimental information becomes available and the standard values can be improved. Unfortunately, there is at present a tendency among experimentalists to employ elaborate equipment that is not always justifiable because simpler thermochemical techniques may produce the same results with similar accuracy. A study of the following pages may therefore be recommended.

1. Calorimetric Methods

The first of the thermodynamic functions to be determined experimentally for a large range of substances was the enthalpy of formation. The foundations of this work were laid with the classic investigations of Berthelot in France and Thomsen in Denmark. At the time, both Berthelot and Thomsen regarded the enthalpy of formation of a compound as the measure of the affinity of its constituent elements for one another, and it was for that reason that their work was carried out on such an extensive scale.

Enthalpies of formation or of reaction are usually determined by calorimetric methods, which are quite simple in principle. The reaction to be investigated is carried out in an apparatus of known heat capacity, the calorimeter, so that the heat effect is measured by the change in temperature during reaction.

The heat capacity of the calorimeter is determined as its water equivalent. Theoretically this is the sum of the heat capacities of all the constituent parts of the calorimeter. In practice, however, various heat interchanges take place which are difficult to account for by calculation, and the water equivalent is always determined experimentally. The water equivalent W multiplied by the change in temperature then gives directly the heat effect of the reaction

$$q_T = W\Delta T. \tag{94}$$

The physics of heat conduction, capacity and interchange and their bearing upon the design of calorimeters will not be dealt with in great detail in this monograph. Readers are recommended to study sections on this subject in textbooks of practical physics, and particularly in the second volume of *Experimental Thermochemistry* (1962k).

The classification of calorimeters will be based here on the nature of the measurement to be made—heat content and heat capacity, enthalpy of fusion and transformation, enthalpy of formation and reaction. This classification is somewhat arbitrary, since some calorimeters may be used for more than one type of measurement. Nevertheless, the prime concern of the reader is likely to be a certain type of measurement rather than a certain type of calorimeter, and the following sections therefore present the necessary information in an appropriate manner for rapid reference.

Since some authors have classified calorimeters according to three main variables: the temperature of the calorimeter T_c, the temperature of its surroundings T_s, and the heat Q produced per unit time, a summary of the main types of calorimeter based on these variables is presented below.

Isothermal calorimeter: For this calorimeter $T_c = T_s$. Its best known representative is the ice calorimeter designed in its modern form by Bunsen.

Adiabatic calorimeter: For this calorimeter $T_c = T_s$ but is not constant. Such calorimeters have mainly been built for the determination of heat capacities, but also enthalpies of transformation and reaction. However, because the temperature of the enclosure must at all times be the same as that of the calorimeter, its application should be limited to relatively slow reactions, for instance the dissolution of metals, etc, in acids, or the formation of endothermic alloys.

Heat-flow calorimeter: For this calorimeter, which is somewhat simpler in design than the adiabatic calorimeters, $T_s - T_c = \text{constant}$. It is suitable for the direct determination of heat capacities and enthalpies of transformation, but not for the study of enthalpies of reaction because it cannot be "stopped" and held at constant temperature.

Isoperibol calorimeter: This most widely used type of calorimeter operates with the enclosure held at constant temperature ($T_s = \text{constant}$) and T_c being measured before, during and after a reaction. It is often wrongly categorized as an isothermal calorimeter.

The quantities to be determined in almost all calorimetric experiments are thus the change in temperature during the experiment, and the water equivalent of the calorimeter, which is found by calibration. In addition, it is essential to know exactly what amount of substance takes part in the reaction, and to ensure that no secondary reactions occur. The exact chemical analysis of the reaction products is, therefore, of prime importance in calorimetry. Without it, attempts to measure physical quantities, such as temperature change, with great accuracy become meaningless.

A. Measurement of Temperature

Temperature measurements in calorimetry usually involve the accurate determination of changes in temperature of a few degrees only. The importance of accuracy in these measurements is such that all those engaged on this work should study the rudiments of thermometry. Two short monographs by Hall (1953c) give a concise introduction to the subject with accounts of the main types of precision thermometers.

Mercury-in-glass thermometers, platinum resistance thermometers, and thermocouples have all been used extensively in calorimetry, and the account which follows is intended to indicate the salient practical features and advantages of each type.

1. Mercury-in-glass Thermometers

To measure temperatures with the necessary accuracy, these thermometers must have a fine-bore capillary and a large reservoir. The range of mean temperatures to be covered may be very wide, however, and since there are practical limitations to the length of a thermometer stem two methods have been developed to make calorimetric and similar work more flexible. The first is to use a set of thermometers, each calibrated over 6°C with overlapping ranges from 9° to 33° (1938a). These thermometers are very robust, and are ideal for the routine work carried out, for instance, in a fuel control laboratory.

The second method is to use an adjustable zero thermometer of the type introduced by Beckmann, which is also graduated over a 6°C range. Because this instrument is rarely used in thermochemistry nowadays and an experienced thermochemist made the point in a review of an earlier edition of the present book that he had never seen a Beckmann thermometer since his student days, the authors considered omission of its description. However, they found themselves very reluctant to do so because the instrument can and should continue to have many uses, being just as accurate as most electrical thermometers in the hands of a skilled observer and cheaper and simpler in operation. The "Beckmann" is, of course, restricted in temperature of application: the zero-point temperature may be varied from −10° to +100°C or more by confining superfluous mercury to a reservoir at the top end of the capillary. It is an instrument for individualists (like the late W. A. Roth) who through continual acquaintance have become thoroughly familiar with its idiosyncrasies and virtues.

Though graduated only to 0.01°C, with good Beckmann thermometers an accuracy in reading of ±0.0005°C is attainable with some practice. The thermometer must be immersed always to the same depth in the calorimeter liquid, and before each reading it should be tapped near the meniscus with the fingers or a rubber-covered rod. Whenever its temperature is changed by, say, 40°C or more, it should preferably be left for 24 hours before other measurements are made.

2. Platinum Resistance Thermometers

The accurate determination of temperature by the measurement of the resistance of a coil of platinum wire was developed initially by H. L. Callendar in the years 1887–92, and the principle of the method may be found in textbooks of heat. If the electrical measuring instruments are reliable and sensitive, these thermometers can be eight times more accurate than the mercury-in-glass type, so they are the obvious choice for work of high precision.

The conventional type of resistance element may prove too cumbersome for calorimetric work. Barber (1950a) has described a more compact thermometer the bulk of which is no larger than that of a Beckmann thermometer. It has a lag of 5 sec only, and its zero point is reproducible within 0.002° after temperature variations from −180° to 600°C. Much calorimetric work is carried out near room temperature, however, and Maier (1930d) has described a modified resistance thermometer for this range.

It may often be desirable to dispense with the normal self-contained resistance element and to incorporate the platinum resistance with the construction of the calorimeter itself. This has been done, for example, in an adiabatic calorimeter operated at 650°C (1939g), and in working with fluoride solutions (1925a) (which would have attacked glass). In constructing resistance thermometers for these special purposes, steps should be taken to observe as far as possible the principles followed in the design of the standard thermometers. The platinum element, once mounted, should be kept quite free from strain at all temperatures. It must be supported with a completely non-conducting material, and any possibility of condensation occurring on the wire must be rigidly excluded. In precision thermometry nowadays it is usual to measure the current flowing through the element and the potential difference across its ends. For much calorimetric work the earlier Wheatstone bridge method, involving the use of compensating leads, will be found sufficiently accurate and may be found simpler to set up with the equipment available in some laboratories.

3. Thermocouples

The sensitivity of measurements on a single thermocouple is too small for most calorimetric purposes, but may be increased by connecting a number of couples in series. In addition to multiplying the voltage, this can give an average temperature for a large body if the couple junctions are disposed judiciously. As many as 1000 couples have been combined in this way (1930c) to enable temperature differences as low as 10^{-7}°C to be measured. The number of thermocouples cannot be increased indefinitely, as they may introduce other sources of error, such as the thermal conductance of the connections. The usefulness of the method in giving a direct measure of temperature difference has, however, not yet been fully exploited. The use of ten to fifty thermocouples in series can be made normal practice. With ten copper-constantan couples 1 mV is equivalent to 2.34°C, and with the same number of chromel-constantan couples 1 mV$\equiv$1.58°C. This high precision may be offset in calorimetric work by the thermometric lag, which in most instances, due to the thickness of couple sheaths, will be relatively high.

4. Thermistors

In a thermistor the resistance element is a semiconductor, such as silicon carbide. Its high resistivity enables it to be made much more compactly than the usual metallic resistance winding. Also the temperature coefficient (which is negative) is much larger than that of metals, increasing the sensitivity. However, they are useful only over limited temperature ranges. Thermistors are theoretically capable of measuring temperatures with a precision of a few microdegrees without elaborate precautions. Their chief difficulty thus far has been an instability; after a temperature cycle they do not return to the same reading. This difficulty is probably being overcome.

5. Optical Pyrometers

Optical pyrometers may be mentioned because they are being employed to measure the temperature of the specimen before the drop in drop calorimetry combined with levitation melting, but a description is obviously not called for in the present context.

B. Determination of Water Equivalent

Ideally, determinations of water equivalent should always be made under conditions similar to those in which the reaction to be measured takes place. In small calorimeters, particularly, the apparent water equivalent varies with the magnitude of the temperature change. Consequently, the known heat effects used in calibration should be as near as possible in magnitude and in rate of evolution to those to be studied in subsequent experiments.

If the reaction to be investigated is endothermic, calibration is made by dropping into the calorimeter a quantity of metal of known heat capacity cooled initially to a known temperature below 0°C. The metal used may be annealed copper or silver, or mercury enclosed in a glass bulb.

For high-temperature heat content measurements a similar method is employed for the calibration as for the actual determination, using metals of known heat capacity. Aluminium oxide in the form of sapphire is recommended as the standard substance. Silver may also be used for heat content measurements, though it has the disadvantage of giving up its heat slowly and so prolonging the duration of measurement. This should be roughly equal for the calibration and the subsequent experiment. A coating of graphite on the silver does not materially affect the heat loss.

For exothermic reactions generally, the calibration is best carried out by electrical heating. A detailed description of the method is given by Roth (1947) for instance. An electric current is passed for a given time through a heating coil of known resistance inside the calorimeter, and the resulting temperature increase is determined. The input of energy is

$$q = RI^2t \text{ J} \tag{95}$$

where R is the resistance of the coil in ohms, I the current in amperes and t the time in seconds for which it is passed. Reliable electrical instruments must, of course, be used, their precise nature and mode of operation depending upon the accuracy it is intended to attain. The resistance of the heater may be measured at room temperature by a bridge circuit, and any correction due to change in resistance with rise in temperature is calculated from the temperature coefficient of the heater material. Current is best determined by measuring the potential drop across a standard resistance included in the heater circuit using a potentiometer. For accurate work the power being dissipated in the heater during an experiment may be measured by measuring the voltage across its ends as well as that across the standard resistance. The heat input is then given by the equation

$$q = \frac{E_S E_H}{R_S} t \text{ J} \tag{96}$$

where E_H and E_S are the potential drops across the heater and the standard resistance (R_S) respectively. In practice it will rarely be possible to carry out continuous observations with potentiometers during experiments since there are so many other observations to be made. The precise application of the electrical calibration method therefore demands a supply of perfectly constant voltage for the duration of the experiment (which should, however, not exceed the very few minutes taken by the thermal reaction subsequently to be studied).

For some calorimetric work sufficient accuracy in electrical energy measurement may be attained by the use of an ordinary voltmeter and ammeter. The water equivalent of the calorimeter and its contents is obtained by dividing the heat input by the corrected temperature increase,

$$W = q/\Delta\theta. \tag{97}$$

The design of an electrical heater will depend on the available space and on the voltage, which should in general be kept as low as practicable for reasons of safety.

It is important that heaters should have a low heat capacitance and a low heat lag, and for this reason the heater is often immersed unprotected in the calorimeter fluid. This method is impracticable in solutions of electrolytes, and with this in mind Murphy (1949j) has described the fabrication of heater containers of aluminium (and possibly tantalum), electrically insulated by a sealed anodised coating.

1. Heat Contents and Heat Capacities

The Drop Calorimeter. Heat capacities are often determined by measuring the change in heat content between room temperature and a number of higher temperatures. Isoperibol or isothermal calorimeters are used for this purpose. The method entails heating the substance to the desired temperature and dropping it into the calorimeter in which the heat effect is to be measured—hence the name "drop method" is applied to this technique.

It may be necessary to heat the substance in a suitable container, either to prevent its oxidation or decomposition, or because it is in powdered or liquid form. The hot sample is dropped either directly into a calorimetric fluid (water, aniline, paraffin, oil, etc) or into a suitable vessel surrounded by water. In the case of smaller specimens, the receiving vessel is surrounded by a thermopile.

Although there have been no fundamental changes in technique, in more recent years the drop method has been extended to higher temperatures using modern methods of temperature measurement and heating.

Drop calorimetry is widely used, but it is recommended that the method be limited to metals and stoichiometric compounds which do not undergo phase changes in the temperature interval used. Such changes, in view of the rapid cooling involved, may produce undefined end-states in the specimen. The enthalpies obtained may then be useful for calculating changes in enthalpies of reaction at different temperatures, but would yield unreliable heat capacity data upon differentiation.

Representative drop calorimeters are described below.

The Diphenyl Ether Calorimeter: This calorimeter makes use of the principle of the Bunsen ice calorimeter in which the heat given up by the specimen is absorbed by a mantle of a substance in two physical states which are in equilibrium (liquid and solid). The resulting amount of solid which is caused to melt is measured by determining the change in volume accompanying the transformation.

Diphenyl ether is particularly suitable because its melting temperature (26.88°C) is close to ambient temperature and it displays a large volume change on melting. A 3-fold sensitivity increase over an ice-water mixture, and long-term stability under working conditions, even after years of operation, account for the popularity of this particular substance for calorimetric studies. A particularly careful description of the properties of the diphenyl ether calorimeter has been given by Davies and Pritchard (1972d).

The construction, calibration and application of such a calorimeter has been described, for example, by Hultgren *et al.* (1958e) and by Allegret (1976a). The apparatus of the latter author is illustrated schematically in Fig. 32.

The heart of the calorimeter is a thermal diffuser with a copper–nickel fin construction.

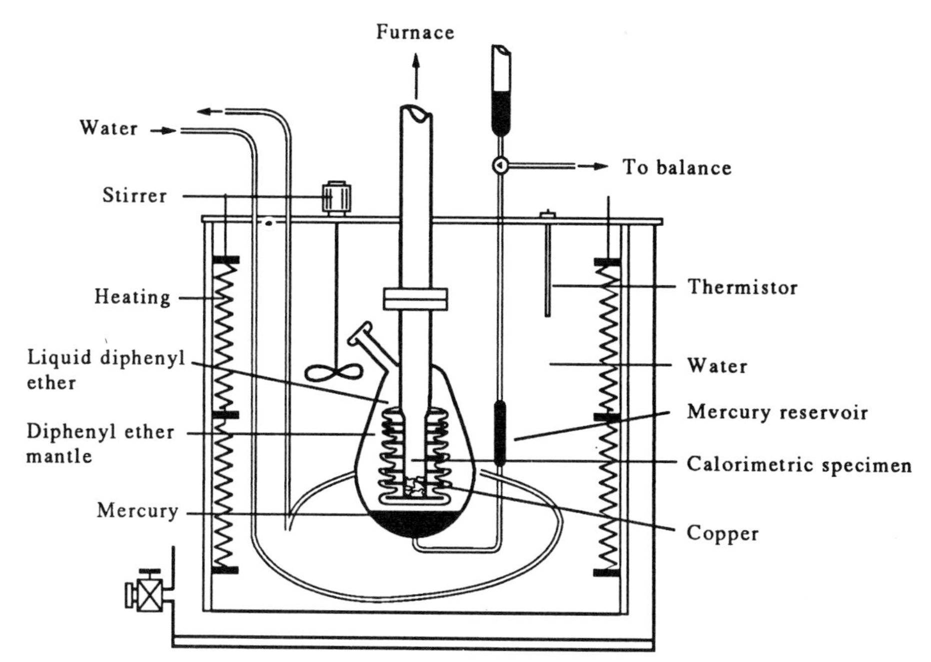

FIG. 32. The diphenyl-ether drop calorimeter due to Allegret [1976a].

This ensures that the heat effect due to the dropped specimen is transferred to the solid diphenyl ether which forms a mantle around the fins and is in thermal equilibrium with its liquid. These central components are contained in a glass vessel of approximately 2.5 litres capacity, which can be refilled via the side-arm illustrated.

The lower part of the vessel contains mercury in contact with the liquid diphenyl ether and connecting to a capillary via a glass tube. The capillary allows the mercury to flow in a regular manner into a beaker located on a recording balance.

The glass vessel is contained in a stainless-steel water thermostat, the temperature of which can be kept constant to $\pm 0.001°C$.

Specimens are dropped down the central tube of the apparatus from a furnace located above the calorimeter and separated from it by water-cooled shutters. Calibration is carried out either electrically or by use of platinum samples, and measurements of heat content of alloys and inorganic compounds have been made for temperatures up to about 1050 K.

High-temperature Drop Calorimetry: Klein and Müller (1987g) have described the adaptation of a commercial high-temperature calorimeter to studies of the enthalpy of mixing in the liquid $CaO–B_2O_3$ system using a drop method. The calorimeter itself has been described by Gaune-Escard and Bros (1974c) and its features were investigated and discussed by Pool *et al.* (1979f). The apparatus is illustrated in Fig. 33.

Two alumina crucibles are stacked one above the other as shown. The upper crucible, which has an inside diameter of 8 mm and a height of 95 mm is lined at its lower end by a platinum crucible (height 40 mm, wall thickness 0.2 mm) which receives the dropped samples during a run and acts as the working cell. The lower crucible (inside diameter 9.3 mm, height 28 mm) serves as the reference cell of the differential calorimeter. The temperature difference between the crucibles is measured using a thermopile composed of

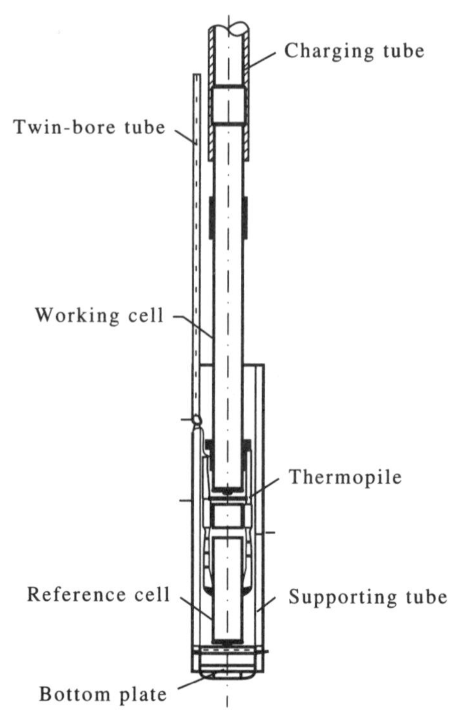

FIG. 33. Diagram of the high-temperature calorimeter.

(18 + 18)Pt–30%Rh/Pt–6%Rh couples interconnected in series. The crucible and the thermopile are enclosed by an alumina support tube, which in turn is suspended from three twin-bore alumina tubes, and adjusted in a constant temperature zone. Samples are introduced into the calorimeter through the charging tube, the lower end of which encloses tightly the top end of the working crucible.

A graphite heating element is used to heat the calorimeter and the temperature is measured by a Pt–30%Rh/Pt–6%Rh thermocouple positioned close to the working crucible. Signals from the thermocouple and the thermopile are fed into a data processing system. The e.m.f. signal of the thermopile is amplified, digitized and stored in a computer. After compensation using an electronic cold junction compensator, the e.m.f. of the thermocouple is measured by a digital voltmeter with a high input resistance.

Calibration is carried out by dropping platinum spheres held at room temperature into the calorimeter operating at a measurement temperature of 1725 K or 1775 K. The masses of the spheres are chosen so as to produce heat effects as similar as possible to the heat effects of the oxide samples.

Levitation Calorimetry: Most of the results obtained by drop calorimetry to date are for temperatures below 1800 K—the maximum operating temperature of a platinum resistance furnace. At higher temperatures two problems become paramount: the chemical reactivity of the specimens and the attainment and maintenance of high temperatures.

By use of electromagnetic levitation it is nevertheless possible to heat to very high temperatures and melt even the most refractory metals without contact with a crucible material. Enthalpy measurements for molybdenum at temperatures up to 3000 K, for example, have been carried out by Chekhovskoi *et al.* (1970d) and Berezin *et al.* (1971b). Their basic apparatus is illustrated in Fig. 34.

The apparatus consists of two parts—an isoperibol block calorimeter and a furnace chamber in which the sample is levitated and heated. The latter, which is of particular interest here, is constructed from a seamless drawn copper tube and has a volume of about 13 l. Use is made of various purpose-designed inductors differing both in size and shape. A removable upper brass flange incorporates two copper sleeves in which are mounted a quartz prism and a sight glass respectively: both are shielded from incident radiation and deposition of sublimation products by means of moveable metal screens. The prism allows the sample temperature to be measured using a disappearing filament pyrometer.

A moveable, water-cooled stainless steel disk holding a heavy copper pan is located underneath the levitation coil. This serves both to catch samples which fall out of the coil before heating is completed and also to shield the calorimeter from radiation from the sample. The lower flange of the furnace chamber is shielded from radiation by 0.1 mm thick stainless steel foil screens.

The samples used in the experiments were turned on a lathe to achieve a particular geometry, such that their position in the inductor is stabilized and the blackbody model used for temperature measurements faces vertically upwards. In assessing the temperature

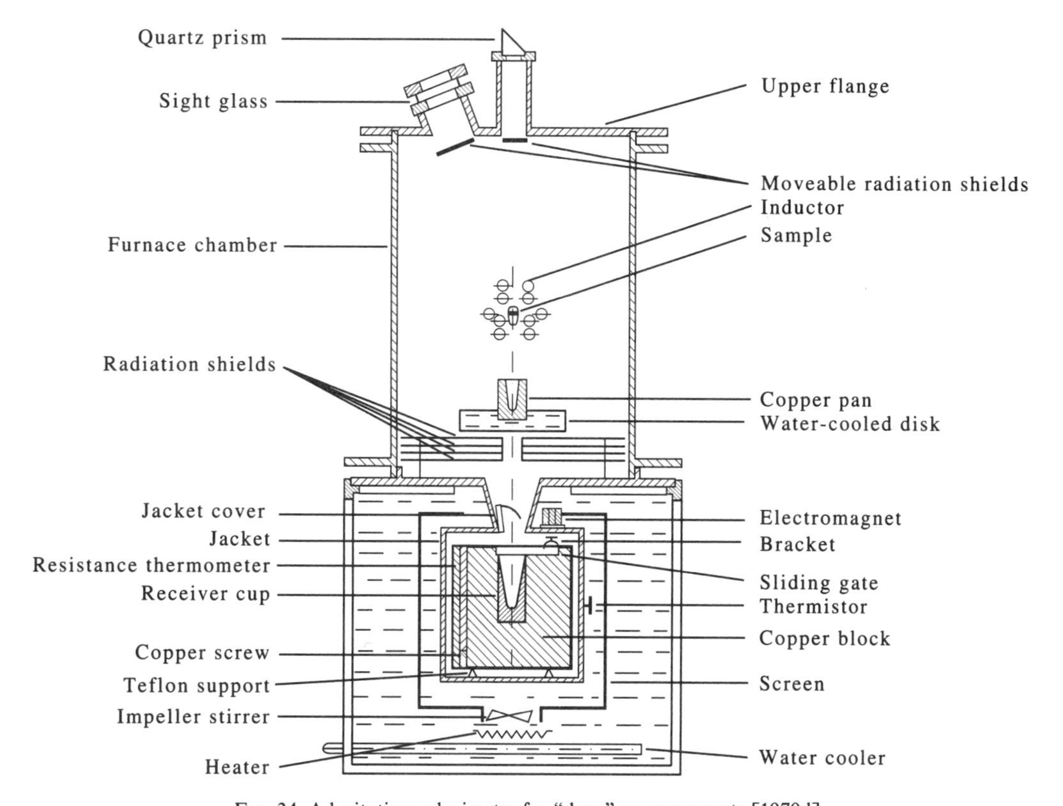

FIG. 34. A levitation calorimeter for "drop" measurements [1970d].

of the sample, absorption in the prism and imperfection of the blackbody model are taken into account.

At the beginning of an experiment, the sample is held in the inductor by means of a manipulator. After the required vacuum or argon pressure has been established in the apparatus, the temperature variation of the calorimeter is recorded for 15–20 min in electrical resistance units. The generator is then switched on and the sample becomes suspended in the inductor, at which point the manipulator is removed. The sample temperature is controlled by varying the power fed to the inductor. Thermal equilibrium is achieved in 1–2 min. After the sample temperature has been recorded, the gate and the cover of the calorimeter are opened, the disk is removed from beneath the inductor, and the sample drops into the calorimeter.

Experimental conditions are always selected so as to minimize the correction for heat transfer between the calorimeter block and the jacket. In order to prevent liquid specimens from spattering on hitting the bottom of the receiver cup, the latter was provided with a tightly fitting, helically wound, 3 mm copper strip. The distance between the cup bottom and the lower edge of the insert was 10 mm and the height of the insert about one-third of the depth of the cup.

Although heat capacities are often calculated from measurements of heat contents, a transformation occurring in the temperature range over which the heat content has been measured, especially one with a small heat effect, such as the magnetic change in nickel, necessitates the direct measurement of heat capacity. That is, the heat required to raise the temperature of the substance by a few degrees at a time must be measured.

Adiabatic Calorimetry. Adiabatic methods of measuring heat capacities at temperatures above room temperature have been developed, among others, by Moser and Sykes.

With the apparatus shown in Fig. 35 Moser (1936a) obtained an accuracy of about 0.5% in measurements in the range 50–670°C. A sample piece (v) is placed in well-fitting silver cylinder (k_1), 23 mm in diameter, which has a removable base. The cylinder is heated by a platinum heater set in a helical groove in a surrounding magnesia sleeve. The rate of heat input is determined from the heating current and voltage, using a calibrated ammeter and voltmeter. The resistance of the heater is 3 ohms at 25°C. The outer cylindrical silver container can be evacuated, and is filled with argon at a pressure of 30 mmHg for each experiment. To eliminate errors that may arise from induced eddy currents, a copper winding is provided outside cylinder (k_2), having the same number of turns as the platinum heater and carrying the same current in the opposite direction.

Constantan wires joining the cylinders to a common lead (d_3) and silver leads (d_1, d_2) constitute thermocouples. These are used to measure the temperature rise of the inner cylinder and to ensure that the outer cylinder is heated to the same temperature so that there is no heat exchange between them. The true specific heat c_p of a sample of mass m is calculated from the relationship

$$c_p = \frac{1}{m}\left[\frac{EI-a}{\Delta\theta} - W\right]\frac{1}{1+b}$$

where W is the water equivalent of the empty calorimeter and a and b are corrections for the loss of heat during measurement and for difference between the increase of temperature of the calorimeter and of the sample.

Further particulars of the method are available in the original paper. Moser's apparatus

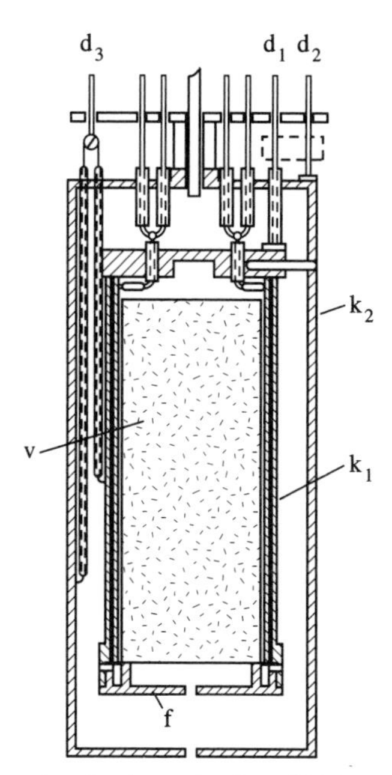

FIG. 35. Moser's calorimeter for the determination of high temperature heat capacities.

has become the prototype for a number of adiabatic calorimeters for the measurement of heat capacities, enthalpies of transition and enthalpies of reaction. Sykes and Jones (1936f) applied it to the investigation of various order–disorder transformations and to age-hardening phenomena.

The specimen in the form of a closed hollow cylinder is placed inside and thermally insulated from a closed copper cylinder. This is kept in an inert atmosphere in a silica tube which can be heated by a resistance furnace. The specimen is heated also by a small nichrome coil mounted inside it. Potentiometer leads are provided to measure the heat input by this auxiliary heater. The temperatures of specimen and block being T_S and T_B respectively, the measurements taken are those of T_B and $(T_S - T_B)$. In carrying out a determination, T_B is increased at a steady rate of the order of 1°C per min. Meanwhile the current through the auxiliary heater is adjusted at intervals so that the curve $(T_S - T_B)$ varies between plus and minus 0.1°C. Where this curve intersects the line made by plotting T_B against time, heat interchange block and specimen is zero. Consequently, at these points

$$Q = m_p\left(\frac{\mathrm{d}T_S}{\mathrm{d}t}\right)_{T_S = T_B}$$

T_S is given by the measurements $[(T_S - T_B + T_B]$.

Q is the power input of the auxiliary heater, m the mass of the specimen and c_p its specific heat. The heat capacity of the coil and thermocouples are usually negligible compared with that of the specimen.

Backhurst has extended the range of the adiabatic method to 1600°C for solid and liquid metals (1958b). The cylindrical specimen and its crucible are separated by three concentric radiation screens from an outer enclosure made of molybdenum. The outer enclosure is heated by tungsten spirals evenly distributed over the inner surface of a box formed by six ceramic blocks. The apparatus stands in an evacuated container. Heat is supplied to the specimen at a constant rate by means of an internal heater. Meanwhile the outer enclosure is kept at the same temperature as the crucible. The temperature of the specimen is recorded on a chart, and the specific heats are calculated using the gradients of this curve and the power supplied to the internal heater.

Sale (1970q) has constructed a spherical adiabatic calorimeter for operation in the temperature range 650 K to 1750 K (Fig. 36). The spherical geometry was chosen to avoid the problems of thermal assymetry often encountered in adiabatic calorimetry. The design of the calorimeter is based on the use of two spherical furnaces and associated heat shields as the adiabatic environment. A large, high thermal mass, low response furnace (180 mm diameter) supplies the majority of the thermal energy required to maintain the system at elevated temperatures, and a small, low thermal mass, high response furnace (110 mm diameter) is used to follow the changes in specimen temperature. Pt/Pt–13%Rh

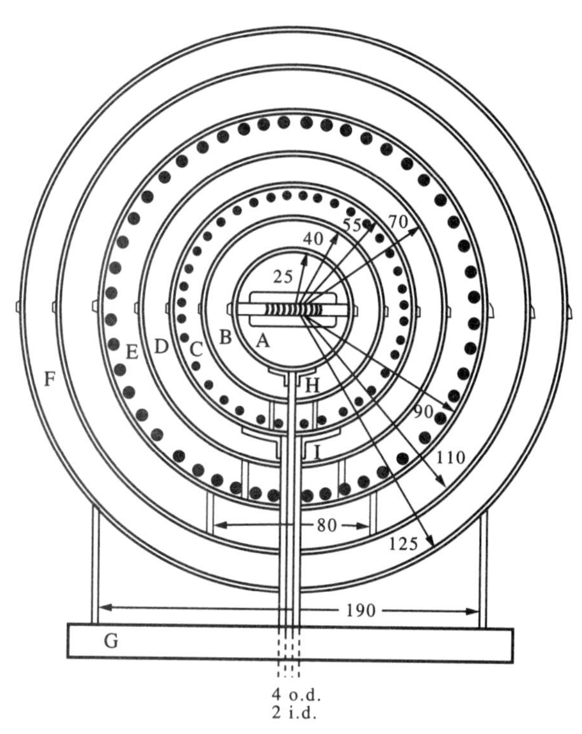

FIG. 36. Arrangement of the spherical heat shields and furnaces.

A—Tantalum specimen case
B—Molybdenum heat shield
C—Tantalum-wound inner furnace
D—Molybdenum heat shield
E—Tantalum-wound outer furnace
F—Radiation shield pack (tantalum spheres packed with molybdenum foil)
G—Water-cooled base plate
H,I—Molybdenum shoulders
All dimensions in mm.

differential thermocouples with junctions situated on the specimen and the 50 mm diameter sphere, and the 110 mm and 140 mm diameter spheres, are used to control the inner and outer furnaces respectively.

For a specimen heating rate of 4 K per minute and at temperatures around 1200 K, the departure from true adiabatic conditions was of the order of ± 0.005 K. The temperature distribution on the inner heat shield was measured to be less than 0.2 K at temperatures up to 1400 K.

Tantalum and molybdenum were used for the furnaces and heat shields respectively, and recrystallized alumina for the support plinths, the furnace-element supports and the electrical insulation on all measuring leads.

Figure 37 illustrates the specimen heater, which consists of a 30 mm long coil of 0.15 mm diameter Pt–13%Rh wire wound around a 1.2 mm diameter alumina tube. The whole assembly is contained within a 3 mm outer diameter alumina sheath. The shape of the specimen is selected to be a magnification of the cylindrical heater so that all points on the surface of the specimen respond to input energy at approximately the same time.

The electrical energy dissipated by the heater is recorded automatically by a system consisting of a scaler-timer, a digital voltmeter, a serialiser and a strip printer coupled

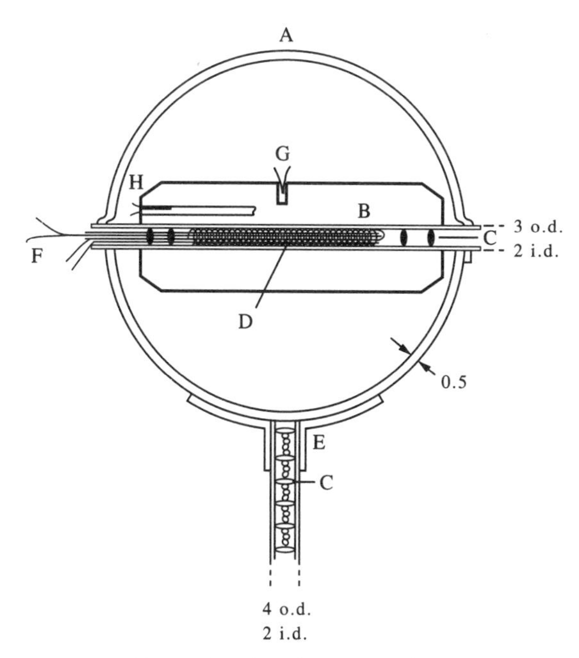

FIG. 37. Specimen heater assembly.

A—Tantalum specimen case
B—Specimen
C—Radiation shields
D—Pt-Rh heater
E—Molybdenum shoulder
F—Potential and current measuring leads
G—Differential thermocouple
H—Specimen temperature measurement thermocouple
All dimensions in mm.

together by suitable interface equipment. Approximately 88–90% of the energy supplied to the heater is used in heating the specimen. The remaining 10–12% is used in heating the heater assembly and thermocouples.

The apparatus has been used, for example, to determine the heat capacity of α, γ, and δ-Fe as well as the enthalpies of the $\alpha \rightarrow \gamma$ and $\gamma \rightarrow \delta$ transformations.

High-speed Methods. Because most experimental techniques for the determination of heat capacity require relatively long periods of time (minutes or even hours) for the measurements to be performed, problems quickly arise at higher temperatures due to increased heat transfer, reaction of the specimen with the container, evaporation, loss of mechanical strength, etc. The high-speed pulse technique for measurement of heat capacity has been developed in an attempt to overcome these problems.

Cezairliyan (1969a) has reviewed the experimental work carried out using the technique of pulse calorimetry, which has been applied particularly to the determination of heat capacities of electrical conductors (specifically metals) and semiconductors at high temperatures. The method generally involves rapid resistive heating of the specimen by the passage of large currents and the apparatus used by Cezairliyan is illustrated in Fig. 38.

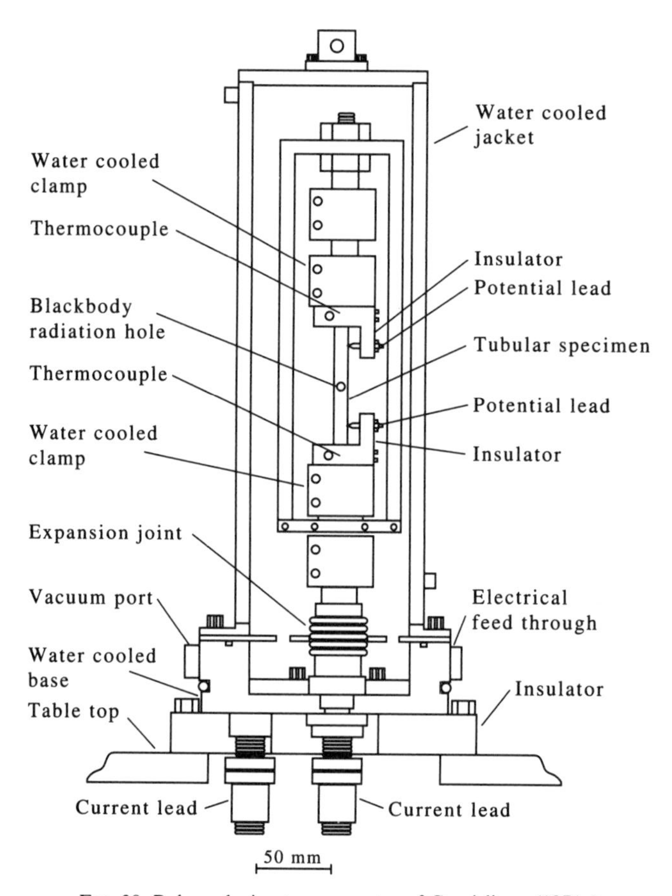

FIG. 38. Pulse calorimetry apparatus of Cezairliyan (1971c).

The power imparted to the specimen, and the specimen temperature, are both recorded as a function of time. The power is determined by measurement of the current flowing through the specimen, and of the potential difference across the effective specimen as a function of time, while temperature measurements are made using thermocouples or optical pyrometry. Power losses are usually estimated or obtained from data taken during the period immediately following the heating of the specimen. The duration of a high-speed measurement is generally of the order of 0.001–1 sec.

The problem of measuring the specimen temperature accurately during a measurement has been coped with by Cezairliyan *et al.* (1970c) using a high-speed pyrometer designed by Foley (1970f), which permits 1200 specimen temperature evaluations per second. This pyrometer has the advantage that it is capable of comparing the radiances from the specimen with that of a reference lamp during the pulse experiment. Problems arising from photomultiplier fatigue, drift, and changing environmental conditions are thereby eliminated.

Due to improvements in high-speed digital recording techniques, it is claimed that the accuracy of the best high-speed pulse calorimetric measurements is as good as can be obtained by conventional methods at temperatures up to 2000 K, and above 2500 K the accuracy of pulse calorimetry is superior. Typical error limits for the measurements are given as 2% at 2000 K and 3% at 3000 K with a precision of 0.5% between these temperatures.

The ability of the method to be used at temperatures beyond the scope of most other calorimetric techniques is demonstrated by the work of Affortit and Lallement (1968a) in their studies of the heat capacity of Nb and W at temperatures up to 3600 K, while Cezairliyan (1972c) has determined the melting point and electrical resistivity (above 3600 K) of tungsten using the pulse heating method.

Cooling Method. Another method of determining true specific heats is that first introduced by Dulong and Petit (1819) in the course of their well-known investigation of heat capacities. It involves the analysis of cooling curves, which depend on the specific heat of the substance concerned. Though at first sight this may appear a simple matter, in practice both the experimental method and the numerical evaluation require the utmost care to obtain dependable results. A comprehensive discussion of an application of this method is that of Knappwost (1943a). For various reasons gases alone should be used as the heat conducting medium, and Knappwost used a 2 mm layer of air or hydrogen between the cylindrical test sample and its enclosing crucible.

Oelsen *et al.* (1955c) simplified the evaluation by determining the rate-of-heat-loss vs time curve of the specimen in addition to the cooling curve. The specimen, in a thin-walled container provided with a thermocouple, was heated, then placed on pinpoints inside another container, also thin-walled, immersed in water in a room-temperature calorimeter. The true specific heat at any given temperature was calculated from the gradients of the two curves at the corresponding time. This simple method was used at temperatures between 100° and 1050°C, the errors being surprisingly small.

2. *Enthalpies of Fusion and Transformation*

Some methods of determining enthalpies of fusion have already been described in connection with specific heats. Other techniques available are described briefly below.

Differential Thermal Analysis (DTA): Differential thermal analysis (DTA) is a widely-used experimental method for investigating the enthalpy changes associated with different types of process and various commercially-available apparatus are available for such measurements. These have been discussed, for example, by Wilhoit (1967q) and by Wendlandt (1972l).

Examples of DTA apparatus used for measuring enthalpies of fusion are those of Speros and Woodhouse (1963g) for temperatures up to 960°C and of Lugscheider (1967r) for temperatures up to 1200°C. An apparatus for measuring enthalpies of fusion at higher temperatures is described below.

In DTA, the specimen and a reference body are heated in as closely similar a thermal environment as possible and the temperature difference between the two is measured and recorded. By suitable calibration, enthalpy effects occurring in the specimen can be quantitatively determined. The great advantage of this type of calorimetric study is that it can be performed simply and rapidly and is easy to automate. However, the accuracy of the measurements is generally not high compared with other methods.

Two representative DTA apparatus are described below.

The "tandem" calorimeter: This calorimeter was constructed by Hoster and Kubaschewski (1980c) for measurement of enthalpies of transformation and of alloying in the temperature range 800 to 1600 K. It is of very simple design, is mechanically stable and is significantly lower in cost than "conventional" calorimeters. Its principle of operation consists in measuring the temperature differences between an "active" and a reference metal specimen. The apparatus is illustrated in Fig. 39.

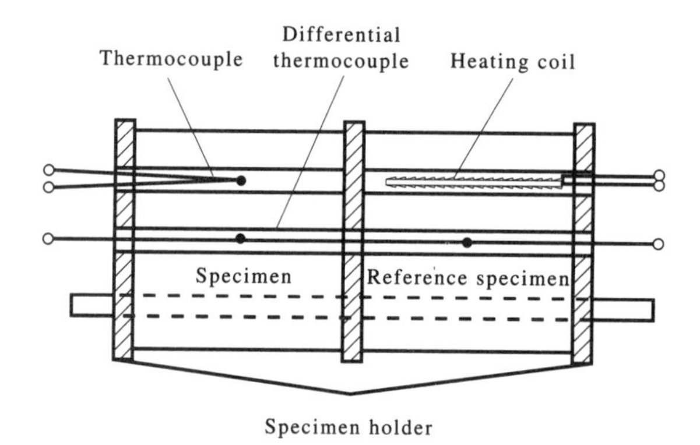

FIG. 39. The "tandem calorimeter" of Hoster and Kubaschewski [1980c].

The cylindrical specimens have a diameter of ca. 20 mm and a length of ca. 30 mm. These dimensions represent sufficient mass to produce significant thermal effects and a reasonably small heat-exchange with the environment. The weight of the two specimens is such that their heat contents are similar at the temperature of measurement. The junctions of a Ni/NiCr differential thermocouple are located in the centre of each specimen. A second thermocouple records the actual temperature in the test specimen. The assembly is situated in the uniform temperature zone of a furnace, through which a flow of gas consisting of 50% argon and 50% hydrogen is passed to prevent oxidation.

Exothermic or endothermic temperature effects in the specimen under investigation are registered as a function of time by a recorder (Fig. 40) and calibration is effected

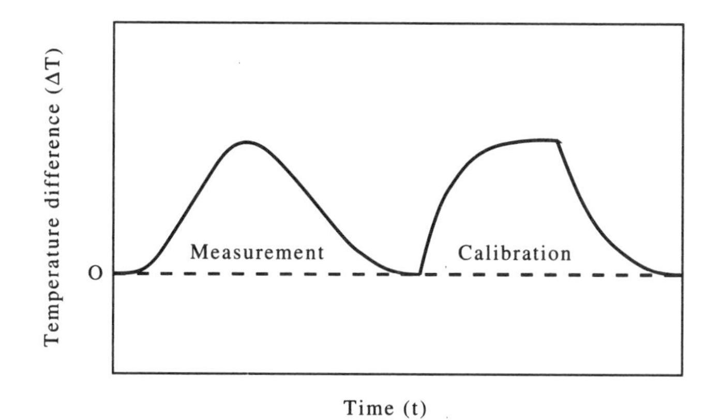

FIG. 40. Temperature difference—Time-Diagram.

electrically by means of a small heating coil situated in the reference specimen. In practice, the heating current and time are chosen such that the areas under the measurement and calibration curves are roughly equal. In each experiment, calibration is carried out in the same temperature interval as that in which the actual effect is recorded.

High-temperature DTA: Predel and Mohs (1970o) have described a calorimeter (Fig. 41) which is based on the principle of quantitative DTA and which operates at temperatures up to 1600°C in a vacuum of 1.3×10^{-3} Pa or in an inert gas. The apparatus has been used to measure, for example, the enthalpy of mixing of liquid Co–Ni alloys and the enthalpies of fusion of iron, cobalt and nickel.

The calorimeter is situated in a water-cooled jacket and employs two independently-controlled heating systems. The main heater (2) is a molybdenum cylinder, which enables temperatures up to 1400°C to be reached using a 5.5 kW power supply. The secondary heater (6) is situated in the molybdenum calorimeter block, and with a power supply of 1.5 kW, the temperature can be increased to 1600°C. The use of this dual heating system allows small temperature changes to be effected quite easily.

The central molybdenum block (3), which weighs 3.7 kg, contains the specimen crucible, and the reference thermocouple (7) is situated in a hole drilled in the base of the block. A double row of bore-holes near to the outside of the block contain the secondary heater, which is in the form of molybdenum wire insulated by ceramic tubing. The furnace and the block are surrounded by a nest of radiation shields (5).

In performing a measurement, the temperature difference between the specimen and the molybdenum block is continuously recorded, and calibration, which is carried out frequently and at different temperatures, is effected using a Pt–Rh heater.

Differential Scanning Calorimetry (DSC). A widely used method for studying transformations of different types in a wide variety of materials is differential scanning calorimetry. Various commercial instruments have been developed for this purpose. In this method, comparison is made of the energy required to heat the specimen under investigation and a reference sample of known thermal characteristics through the same temperature interval.

The advantages and disadvantages of the method have been discussed, for example, by O'Neill and Gray (1971n).

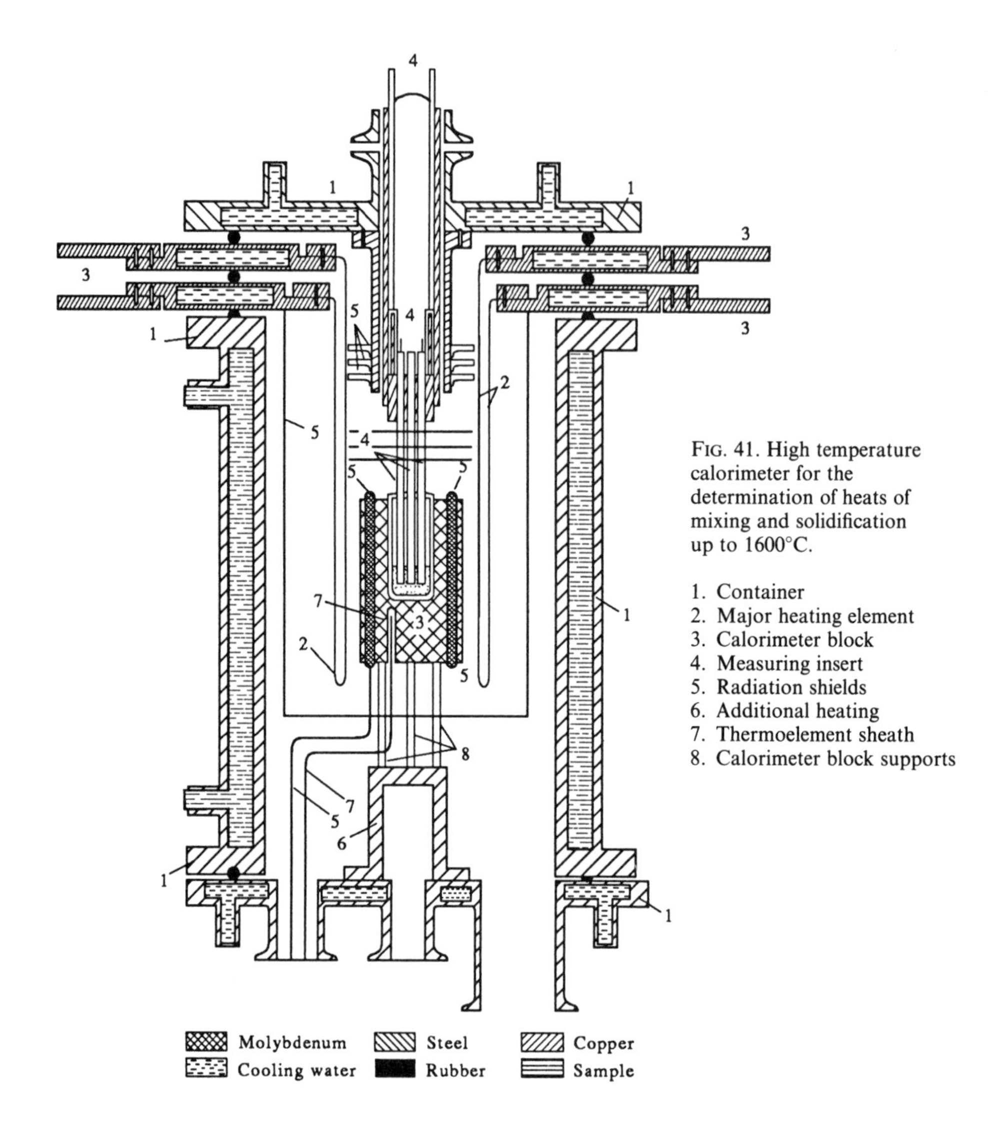

FIG. 41. High temperature calorimeter for the determination of heats of mixing and solidification up to 1600°C.

1. Container
2. Major heating element
3. Calorimeter block
4. Measuring insert
5. Radiation shields
6. Additional heating
7. Thermoelement sheath
8. Calorimeter block supports

The method employs a direct differential power measuring technique and is thus a more directly quantitative calorimetric method than differential thermal analysis, (DTA), which employs thermometric measurements. The apparatus has a low total thermal mass, and it is possible to carry out calorimetric experiments which require extremely rapid changes in temperature (either heating or cooling), or which involve precise isothermal temperature control.

The construction of the sample holders in a differential scanning calorimeter provides considerable design problems. Ideally, the holders should have a minimum thermal mass yet be of high thermal conductivity and be thermally stable. Each holder must contain a miniature heating element and a highly sensitive temperature sensing element. Unless the sample and reference holders are perfectly matched, variations in the program voltage and hence heater power can have a differential component which is, in effect, amplified by the high-sensitivity sensors and will appear as noise or fluctuations on the differential power record. Hence a very high standard of performance is imposed on the temperature programmer.

Another problem encountered in the direct power DSC design arises from the fact that the surroundings of the sample and reference holders are essentially at ambient temperature. Consequently the temperature gradient between the holders and the surroundings can be several hundred degrees when the calorimeter is operating at the highest temperatures. Nearly all the heater power provided to the holders is thus lost to the surrounding by conduction, convection or radiation. Although this results in the rapid cooling ability of the calorimeter, it also means that the matching of the sample and reference holders and the control of the environmental temperature should be excellent if the heat losses are not to produce differential power noise or baseline drifts. In practice, the difficulty of achieving perfect matching and environmental control is revealed by curvature of the baselines and problems in baseline repeatability.

A typical precision for DSC measurements is of the order of $\pm 0.2\%$. The accuracy varies according to the type of measurement and the temperature range concerned, but heat capacity values can generally be determined to $\pm 2\%$.

3. Enthalpies of Reaction and Formation

Combustion Bomb Calorimetry. The combustion bomb calorimeter, introduced by Berthelot, is probably the best-known piece of equipment used in thermochemistry and the application of combustion bomb calorimetry to the field of metallurgical research has been reviewed, for example, by Huber and Holley (1970j).

The technique involves carrying out a combustion inside a calorimeter with a gas (which may be introduced at a pressure of up to 25 atm) as one of the reactants. The amount of heat liberated from the combustion of a sample is compared with a known amount of electrical energy. This may involve the use of an intermediate substance, for example a well-characterized reference material, for calibration purposes.

The bomb must remain gas-tight during the reaction, which may occur with explosive violence. A stout shell of high tensile steel is therefore used and this is closed with an appropriately sealed heavy screw cap. The specimen is provided in the form of powder or thin foil and is placed in a crucible supported by thin platinum or iron wires about 0.25 mm in diameter, which also serve to carry the electric current used to start the reaction. The combustion gas is introduced at pressure and is retained by a needle valve or a spring-loaded non-return valve. The capacity of the bomb is usually between 350 and 500 cm. During an experiment, it is immersed in a water-filled calorimetric vessel which is itself contained in an outer jacket maintained at constant temperature. For all other work than the combustion of organic materials, the bomb should be lined with enamel, platinum, gold, or some similar material, to protect the steel from attack.

The combustion bomb is generally calibrated by the enthalpy of combustion of pure benzoic acid, although electrical calibration may be preferable.

Combustion gases used are oxygen, fluorine, chlorine and nitrogen (although hydrogen, nitrogen dioxide and bromine have also found application). Combustion with oxygen is suitable for metals, lower metallic oxides, carbides, nitrides, borides, hydrides, etc, and indeed bomb calorimetry is still the only method by which reasonable enthalpies of formation of carbides and borides have been obtained. Combustion in fluorine permits more refractory-type materials to be studied, for example hafnium and zirconium borides [Johnson *et al.* (1967e)] and non-stoichiometric nitrides of uranium [O'Hare *et al.* (1968g)].

During combustion of a metal in oxygen, the temperature may exceed 2000°C momentarily, resulting in melting both of the metal and the product oxide. To avoid interfering side-reactions, the combustion should, in such cases, be carried out on ceramic discs or in crucibles of the product oxide [Huber and Holley (1970j)]. Some metals form non-stoichiometric oxides of variable composition or form more than one oxide. It is therefore essential to determine what reaction has occurred on combustion, for example by obtaining an X-ray pattern of any solid product. This is illustrated by the fact that combustion of uranium in oxygen produces U_3O_8 deficient in oxygen, whereas combustion of a mixture of U with UO_2 is reported to produce stoichiometric U_3O_8 [Huber *et al.* (1952g)].

Non-metallic impurities present in metals, for example carbon, hydrogen and nitrogen, are a serious problem in combustion experiments because they contribute to the enthalpy of combustion, and in addition, their oxides may react with the product oxide or the crucible. Indeed, the enthalpy of solution of impurities in the parent metal or the enthalpy of solution of the metallic oxide in the product oxide can be significant in some cases.

In performing combustion experiments, it is preferable always to work in the same temperature region and over the same temperature range, so that errors in calibration will tend to cancel those in the experiments. A check on the reliability of the data obtained may be gained by varying the amount of sample burned, for the results should, of course, be the same when converted to the same weight basis. Efforts should also be made to equate approximately the burning times of the calibration sample and the sample under investigation, so that errors arising from the different shape of heating curves obtained with faster or slower burns are avoided. Of greatest importance perhaps, is the requirement that conditions should be selected such that the completeness of combustion is maximised, therefore reducing errors introduced in correcting for incomplete combustion.

Direct Chlorination Calorimetry. Gross (1955a) determined the enthalpy of formation of titanium tetrachloride by chlorinating titanium under pressure in a pyrex glass combustion vessel, shown in Fig. 42.

Inside the reaction vessel (A) (about 100 ml) was suspended a beryllia crucible (1) containing titanium pieces. Tube (4) permitted the vessel to be evacuated after introduction of the titanium. A small bulb (10), containing a known amount of liquid chlorine and kept at constant temperature (−75°C) was joined to the slide tube (5). When the vessel had reached equilibrium in a calorimeter the glass barb (6) was broken by allowing a glass tube (7) containing mercury to fall down the slide tube. Reaction began spontaneously after a few minutes and was generally complete within 10–15 min. The titanium tetrachloride, saturated with chlorine, collected on the bottom of the vessel forming a liquid seal at the capillary 5. The chlorine continued to bubble through this liquid thus making the solution homogeneous and bringing it into equilibrium with Cl_2 vapour. The enthalpy of solution of titanium tetrachloride in liquid chlorine was determined separately in the same apparatus.

Kubaschewski and von Goldbeck (1950f) have also used a direct chlorination method in determining the enthalpy of formation of vanadium tetrachloride.

Aqueous Solution Calorimetry. The measurement of enthalpies of solution of substances in dilute acid solvents at room temperature is relatively simple, and by applying Hess's law

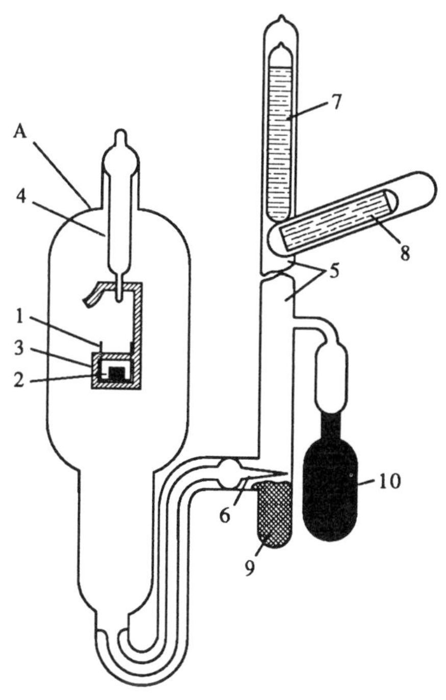

FIG. 42. Pyrex glass chlorination vessel (Gross, 1955a).

of constant heat summation, they may be used to evaluate enthalpies of formation or reaction that are difficult to determine directly.

To take a simple example, the enthalpy of formation of hydrated copper sulphate $CuSO_4 . 5H_2O$ from the anhydrous sulphate is obtained by measuring the enthalpies of solutions of the two salts in water. Many of the enthalpies of formation of the metallic compounds which we know today were obtained by this method, which figured prominently even in Thomsen's work carried out almost a century ago (1908d).

Strictly speaking, the term "solution" should be applied to the formation of a single homogeneous phase from two or more substances. In thermochemical work the term is used for experimental purposes in a wider sense to include certain processes in which not only solution but reaction and formation of a second phase, gaseous or possibly solid, occur. For example, the enthalpy of formation of an intermetallic compound of magnesium and zinc is determined by measuring the following enthalpies of solution in hydrochloric acid.

$$\langle Mg\rangle + 2[HCl] = [MgCl_2] + (H_2);\ \Delta H_a \tag{a}$$

$$2\langle Zn\rangle + 4[HCl] = 2[ZnCl_2] + 2(H_2);\ \Delta H_b \tag{b}$$

$$\langle MgZn_2\rangle + 6[HCl] = [MgCl_2] + 2[ZnCl_2] + 3(H_2);\ \Delta H_c. \tag{c}$$

From these equations we obtain

$$\langle Mg\rangle + 2\langle Zn\rangle = \langle MgZn_2\rangle;\ \Delta H_x$$

$$\Delta H_x = \Delta H_a + \Delta H_b - \Delta H_c. \qquad \text{(d)}$$

Here, again, the enthalpy of formation of a compound is the difference between its enthalpy of solution and that of its constituent elements in a given dissolving medium.

The validity of this experimental method depends upon solution taking place uniformly and upon the final state of the various solutions being the same.

No subsidiary reactions in addition to those considered must take place, and the solvent must be carefully chosen with this object in view. If hydrogen is evolved during solution, for instance, it must either be completely evolved in gaseous form or else completely oxidised by suitable reagents: otherwise the results will be inconclusive. The other requirement is that solution should take place neither too slowly nor too rapidly.

These experimental requirements can be met by careful attention to certain details. Table VII gives a number of solvents which have been found suitable for the compounds listed and their constituent elements. Hydrochloric acid of various concentrations is used extensively as hydrogen is evolved from it quantitatively. It is applicable to many metals and intermetallic compounds, to chloride, oxides (MgO), nitrides (CrN) and to sulphides (MgS). For the solution of sulphides the acid must first be saturated with hydrogen sulphide to eliminate a correction for additional thermal effects due to the formation and solution of this gas. Iodides may also be dissolved in hydrochloric acid (1941a), provided it is at a low concentration (N/20).

At such dilution it may be assumed that the nature of the anion is of no importance, and the enthalpy of formation of the iodides may be simply taken as the difference between their enthalpies of solution and that of the corresponding metal in this solvent.

Since the enthalpies of formation of the halides of hydrogen are frequently involved in the evaluation of results in solution calorimetry, their values are given in Table VIII which is taken from the NBS tables of chemical thermodynamic properties (1982b).

Metals too electropositive to be dissolved by hydrochloric acid should be investigated by oxidation with suitable agents in a calorimeter. Nitric acid should not be used, as it has been proved to be unsuitable as a calorimetric solvent (1901).

Sulphuric acid has not been used as frequently as hydrochloric acid or bromine- and bromide-containing solvents. It has proved useful for the solution of beryllium sulphide (at 80°C), and for manganese oxides. In the latter case it was mixed with potassium iodide and ferric ammonium sulphate (1943d, e), and the mechanism for obtaining the heat of formation of the oxide was quite complex, as may be seen from the following representation. Manganese and tri-manganese tetroxide were dissolved in N–H_2SO_4 containing potassium iodide:

$$3\langle Mn\rangle + 6[H^+] = 3[Mn^{++}] + 3(H_2); \qquad \Delta H = -681{,}046 \pm 594 \text{ J}$$
$$\langle I_2\rangle + [I^-] = [I_3^-]; \qquad \Delta H = +4753 \pm 67 \text{ J}$$
$$\langle Mn_3O_4\rangle + 8[H^+] + 3[I^-] = 3[Mn^{++}] + [I_3^-] + 4\{H_2O\}; \quad \Delta H = -327{,}369 \pm 259 \text{ J}$$
$$4(H_2) + 2(O_2) = 4\{H_2O\}; \qquad \Delta H = -1{,}143{,}370 \pm 167 \text{ J}$$
$$(H_2) + \langle I_2\rangle = 2[H^+] + 2[I^-]; \qquad \Delta H = -104{,}688 \pm 837 \text{ J}$$

$$3\langle Mn\rangle + 2(O_2) = \langle Mn_3O_4\rangle; \qquad \Delta H = -1{,}387{,}624 \pm 1925 \text{ J}$$

That solution takes place by this mechanism must, of course, be confirmed analytically. The heat effect of each step must be determined separately, and the errors are naturally additive.

TABLE VII. *Aqueous solvents for calorimetric investigations*

No.	Composition	Examples of application	Remarks	References
	Hydrochloric acid			
1.	$HCl:H_2O = 1:500$ to $1:1000$ (~N/20)	Mg, rare earth metals, their chlorides and iodides	Concentration of HCl to be increased progressively during solution	(1941a)
2.	$HCl:H_2O = 1:53.5$ (1N)	Mg, MgO, $MgCl_2$, $Mg(NO_3)_2$, $Ca(NO_3)_2$		(1943d, e)
3.	$HCl:H_2O = 1:18$ to $1:25$ (~2.5N)	ZnO, $ZnCO_3$, $CaSiO_3$, $CaCO_3$ Ca	50°C partially paraffin coated	(1928c)
		Ca–Zn, Ce–Mg, Ce_3Al		(1924a)
		Mg_3N_2, LaN, CeN, CrN		(1932e)
		MgS, CaS	solns. sat. with H_2S	(1943f)
4.	$HCl:H_2O = 1:8.8$ (~5N)	Mg, Ca, Na, Cd, Al, MgCd, CoAl,	accelerated with $PtCl_4$ or $PtCl_3$	(1924a)
		$CaAl_3$, $CeAl_4$, Na–Hg $FeAl_3$, La–Mg		(1923)
		Fe oxides	100°C; Fe mixed with Pt powder	(1930f)
5.	$HCl:H_2O = 1:3.1$ to $1:4$ (~10N)	Fe oxides, MnO, $MnCO_3$		(1941c)
6.	20% HCl, mixed with 20% HF	CaO, $CaSiO_3$, SiO_2, FeO, Fe_2SiO_4, slags	77°C	(1932f)
7.	40% HF solution	Al silicates		(1925a)
8.	30% $HCl:ICl_3:I_2$ = 10:10:1.5 by weight	AuSn, $AuSb_2$	90°C	(1934b)
9.	$Br_2:KBr:H_2O$ = 2:1:2 by weight	Cu–Cd, Cu_3Sn, Au–Zn, Au–Sn, $FeBr_2$, FeI_2, Ga, Ga halides		(1898, 1934b)
10.	$Br_2:HBr:H_2O$ = 2:1:2 by weight	AuSn, $AuSb_2$	90°C	(1934b)
11.	10 g $FeCl_3$ + 20 cm^3 HCl (conc.) + 30 cm^3 solution No. 12	Cu_3Sb, Cd–Sb		(1923)
12.	25 g KI + 12.5 g I_2 per 11 H_2O	cementite		(1917)
13.	35 g I_2 + 1 Citre 90% C_2H_5OH, with a few cm^3 HCl	cementite		(1917)
	Sulphuric acid			
14.	$H_2SO_5:H_2O = 1:50$ (2N) to $1:105$ (1N)	Te_2O, Mn, MnO		
15.	$H_2SO_4:H_2O = 1:4$ (12.5N)	BeS	80°C	(1943f)
16.	1N–H_2SO_4 + KI	Mn_3O_4		(1943d, e)
17.	2N–H_2SO_5 + $Fe(NH_4)(SO_4)_2$	Mn_3O_4, MnO_2		(1943d, e)
18.	3/4N–Na_2S	Sb_2S_3		(1942a)
19.	$KOH:H_2O = 1:272$ (N/5)	$Al_2(SO_4)_3$, $KAl(SO_4)_2$		(1945)

TABLE VIII. *Enthalpies of formation in kJ/mol of the hydrogen halides in aqueous solution at 25°C*

N	$[HF]_{aq}$	$[HCl]_{aq}$	$[HBr]_{aq}$	$[HI]_{aq}$
10	−318.967	−161.318	−116.955	−51.610
50	−319.306	−165.356	−120.160	−54.208
100	−319.407	−165.925	−120.562	−54.451
200	−319.482	−166.272	−120.809	−54.601
500	−319.754	−166.573	−121.026	−54.735
1000	−320.206	−166.732	−121.160	−54.836
5000	−322.75	−166.963	−121.361	−55.015
∞	−332.63	−167.159	−121.55	−55.19
Gas	−271.077	−92.307	−36.40	+26.48

N = moles H_2O per 1 mole of solute.

It is impossible to predict with certainty the suitability of any solvent for a given compound. The list in Table VII may be useful as a guide, but preliminary trials must be made to test the solvent chosen. The trial may suggest slight modifications, or it may show that the solvent is quite unsuitable.

In dissolving a compound and its component elements separately in a solvent, it is often found that one of the substances dissolves too slowly or rapidly relative to the others. This must be adjusted if at all possible.

The rate of solution may be increased by using the powdered material in a finer grain size; by carrying out the solution at higher temperatures; or by using more vigorous stirring. It must be remembered, however, that at extremely low particle sizes the surface energy can no longer be neglected.

The rate of solution is reduced by stirring more slowly or by covering the test pieces with an inert substance that is more or less impermeable to the solvent.

To decrease the violence of solution of magnesium in hydrochloric acid, Biltz and Hohorst (1922a) painted their metal sample with collodion. Calcium was coated with paraffin wax, leaving a small part of the surface unprotected. Sodium was dipped in liquid paraffin which forms a covering which is displaced sufficiently when the metal is immersed in acid to allow the reaction to proceed slowly.

The apparatus most commonly used for the determination of enthalpies of solution is either an ice calorimeter which is the more useful for slow solution, or one of the simple isoperibol type. This latter consists essentially of a Dewar flask containing the solvent, a stirrer, a thermometer and device used to introduce the substance to be dissolved. For some reasons, for instance to prevent atmospheric oxidation, it is sometimes necessary to enclose this test substance in a sealed bulb. The dead space in such a bulb must be as small as possible or else it must be evacuated; otherwise the air evolved on breaking the bulb may cause the loss of heat by evaporation of the solvent, which is increased if a gas is evolved during solution. The correction to be applied may be determined by a blank experiment in which dry air is bubbled through the solution and the decrease in temperature per unit volume of air is determined.

Stirring should be sufficiently rapid to prevent a "dip" or sharp concavity in the cooling curve. This results from a time lag in the dispersal throughout the calorimeter fluid of the heat generated by the reaction. Whether it is observed depends largely upon the position of the thermometer in the calorimeter, and it may perhaps be desirable in some calorimeters to place the thermometer fairly near the reaction vessel in a test experiment to check the efficiency of the stirring. When this "damming up" of heat occurs, there is a sharp drop in temperature from the maximum, before normal Newtonian cooling sets in, and correction with such a curve is rendered uncertain.

A solution calorimeter of good design which has been used for a number of salts (oxides, sulphates, nitrates and halides), has been described by Southard (1940e). A half-gallon Dewar flask (Fig. 43) is immersed in an oil bath maintained at 25° ±0.01°C. A stirrer is used which can provide vigorous agitation to accelerate the solution of substances such as gypsum. Temperature changes are measured by a specially constructed resistance thermometer. Metal parts are made acid-resistant with a coating of a Bakelite lacquer. The sample if introduced in a thin-walled glass bulb E, sealed to the end of a glass tube which passes through the hollow stirrer shaft to the outside of the calorimeter. At the appropriate time the operator breaks the bulb by jerking it upward against three prongs which project from the stirrer blade.

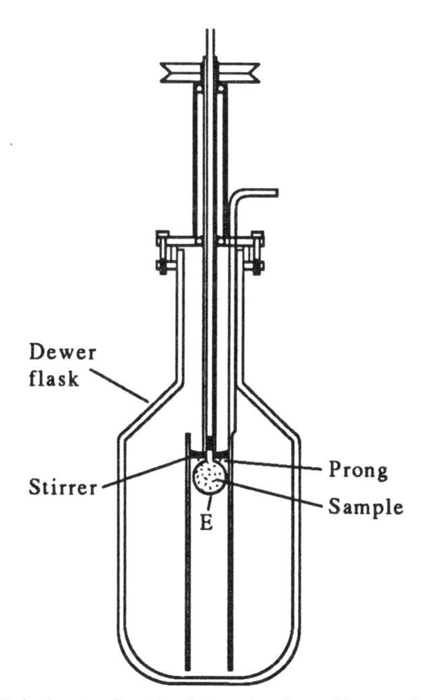

FIG. 43. Calorimeter for the determination of heats of solution.

The "Öfchen" Calorimeter. Kubaschewski and Dench (1955b) devised the apparatus shown in Fig. 44 for the investigation of the enthalpies of formation of transition-metal aluminides.

For the concentration of heat, the compact made of metal and aluminium powders (A) was suspended in a small furnace (B) made of molybdenum wire surrounded by ten radiation shields of 0.03 mm nickel foil (E). These were enclosed in an aluminium block (F) supplied with gas outlet holes (H) and suspended (J) within a brass vacuum container (K), which in turn was placed in a thermostat controlled at 25°C. Ten copper–Eureka thermocouples in series had their hot junctions (P) clamped to representative positions on

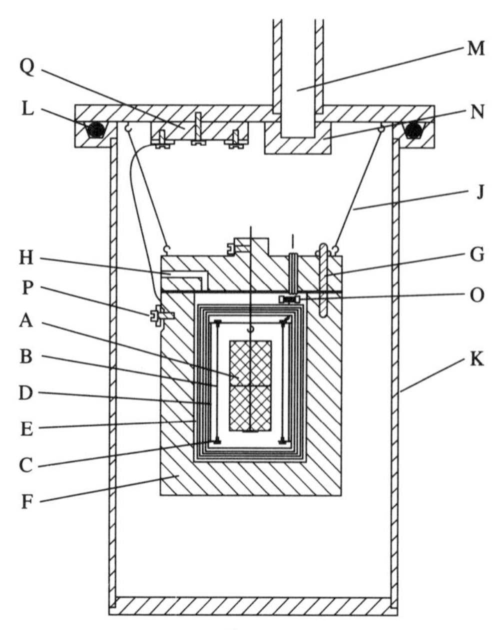

FIG. 44. The "Öfchen" Calorimeter.

the surface of the aluminium block and the cold junctions to the brass block (Q). The brass container could be evacuated through M, but instead of admitting an inert gas for the faster dissipation of heat, a somewhat novel evaluation was used. The furnace (B) was heated electrically until alloying took place rapidly, and the electrical energy supplied was measured by a watt-hour meter (accuracy: $\pm 0.2\%$). When the calorimeter was again at 25°C, an amount of electrical energy was put into the furnace so that the calorimeter block was raised to the same maximum temperature as in the reaction run: the difference between the input electrical energy in the reaction-run and calibration-run was then the energy evolved by the reaction. In this way the need to plot the cooling curve for a period of time was avoided, and the physical errors were restricted to the accuracy of the watt-hour meter. The water equivalent of the calorimeter need not be known. When the temperature of the compact rises above the melting point of the alloy formed (e.g. nickel–titanium alloys (1958f, g), the compacts must be encased, in molybdenum crucibles for instance. An attempt to react transition-metal mixtures with silicon failed in this calorimeter. It was found that at least one of the eutectic temperatures must be below 1100°C for reaction to take place.

High-temperature Adiabatic Calorimetry. Dench (1963b, c) has constructed an adiabatic calorimeter for use in the temperature range of approximately 773 K to 1673 K (Fig. 45). The calorimeter has been mainly used to determine enthalpies of formation of alloys (of the order of 4 kJ/g-at), although it is also suitable for measurement of heat capacities and enthalpies of transformation.

The specimen (A) in Fig. 45 is prepared by compacting a mixture of the powdered reactants into a cylindrical form, 19 mm diameter × 32 mm high. This is placed in the adiabatic enclosure (B) which consists of three concentric cylindrical radiation shields made from tantalum sheet. Around the adiabatic enclosure is the vacuum furnace (C),

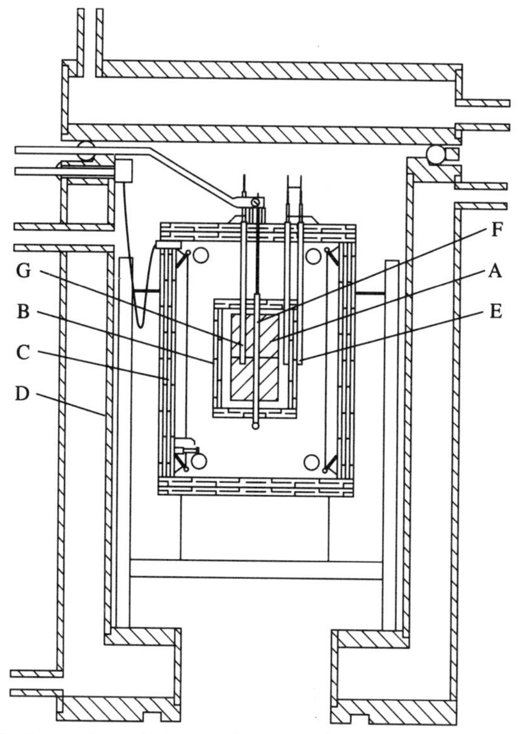

FIG. 45. Adiabatic reaction calorimeter for use in the temperature range 500–1400°C.

which has a tantalum winding and tantalum and molybdenum radiation shields. Part of the winding is arranged around the sides of the furnace, the remainder being formed into two coils—one near the top, the other near the bottom. The furnace is enclosed in a water-cooled brass vacuum chamber D. Three Pt–Pt/Rh differential thermocouples E in alumina sheaths are arranged round the middle of the adiabatic enclosure, each having one junction against the outside of the enclosure walls. Any temperature difference between the inside and outside of the enclosure is indicated by a deflection in a galvanometer to which the differential thermocouples are connected in series. The current in the surrounding furnace winding is continuously adjusted in order to keep the temperature difference at the small finite value corresponding to zero heat transfer across the wall of the enclosure. Heat may be supplied to the specimen via a tungsten resistance heating coil in an alumina sheath F in a hole down the axis of the specimen. The energy supplied to the heater is measured by means of a precision watt-hour meter. An alumina sheath G in a second hole in the specimen carries a Pt/Pt–Rh thermocouple by means of which the specimen temperature is measured.

In measuring enthalpies of formation, the following procedure is adopted. Preliminary experiments are carried out on small specimens to find the highest temperature at which the mixture of the powders to be reacted may be held without any appreciable proportion of the reaction occurring, the "safe temperature", and also to find the temperature range within which the mixture must be held to ensure complete alloying within an hour or so.

The specimen is then heated adiabatically from the safe temperature to the reaction temperature range using the internal heater which is eventually switched off. During the heating, the reaction will have begun, and, after switching off, the temperature of the specimen will continue to rise if the reaction is exothermic, or will begin to fall if it is endothermic. The rate of change of temperature will gradually decrease as the reaction proceeds, a constant temperature being eventually attained when it is complete.

The enthalpy of reaction is obtained by subtracting from the measured input energy the change in heat content of the reactants and empty calorimeter between the safe temperature and the final temperature, these heat contents being measured separately in the same apparatus. An estimated accuracy of ± 420 J/g-atom is attained.

Mean heat capacities may be obtained with an accuracy of $\pm 2\%$ by heating the specimen adiabatically over temperature intervals of approximately 25°C (1963b, c). Enthalpies of transformation may be measured by heating the specimen in a series of steps arranged so that one of the steps contains the complete transformation, the heat capacities of the phases just above and just below the transformation temperature being obtained by extrapolating the mean heat capacity curve. Modifications and improvements to this calorimeter made by Grundmann (1977b), by Hack (1979d) and by Nüssler (1979e) include the use of tungsten radiation shields, W–Re thermocouples, simplified specimen loading, a ceramic support structure for the heater, and in-built safety precautions.

The Tian–Calvet Calorimeter (contributed by F. Müller). The Tian–Calvet calorimeter may be described as an integrated heat-flow microcalorimeter in which enthalpy changes of a sample are determined by measuring the heat flux between the external surface of the calorimetric cell containing the sample, and the jacket surrounding the cell, which also acts as a thermostat heat reservoir, as a function of time. The principles of construction were developed by Tian and Calvet and have been explained in detail by Calvet and Pratt (1963a), and more recently by Hemminger and Höhne (1984b).

The Tian–Calvert calorimeter was originally devised for measurements near room temperature. However, during the past thirty years, considerable efforts have been made in France and the U.S.A. to develop this calorimeter for application at elevated temperatures. Progress has mainly been achieved by improvements to the design, by the application of more-refactory materials and by the use of electronic temperature control and data processing systems. As a result, the Tian–Calvet calorimeter has become very reliable for investigations in high-temperature chemistry, especially in metallurgy, ceramics and geochemistry. Modifications have been made to increase the stability of the baseline of the calorimeter, or to facilitate mechanical manipulations inside the calorimetric cells. Reviews have been written by Laffitte (1971k) and by Bros (1989a); brief descriptions of individual calorimeters have been published by Kleppa (1960e, 1971i, 1979b); Darby *et al.* (1966e); Laffitte (1971k, l); Gerdanian (1971g); Boureau and Gerdanian (1970b); and Vieth and Pool (1972k). Detailed information on the design and construction of a Tian–Calvet twin-calorimeter has been given by Mraw and Kleppa (1984e), and a computer-aided data processing system has been described by Dickens *et al.* (1980b).

Figure 46 shows a schematic diagram of the Tian–Calvet calorimeter due to Kleppa (1960e). This apparatus has been designed as a twin-calorimeter containing two nearly identical calorimetric cells. These are located at equal height in cylindrical wells in a massive thermostat jacket of aluminium, which is maintained at constant and uniform temperature by a furnace system composed of a cylindrical main heater and top and bottom heaters. One of the calorimetric cells is employed as the "measuring cell", while the other serves as "reference". The temperature difference between each cell and the surrounding jacket is monitored by a thermopile composed of 48 couples interconnected in series. The difference reading from the two independent thermopiles is meaured during an experiment. This difference voltage is amplified and the output of the amplifier is usually coupled to a recorder to obtain a graphical record of the experiment, and through an analog/digital transformer to an electronic data processing system which generates numerical values of the enthalpy effects to be determined (see below).

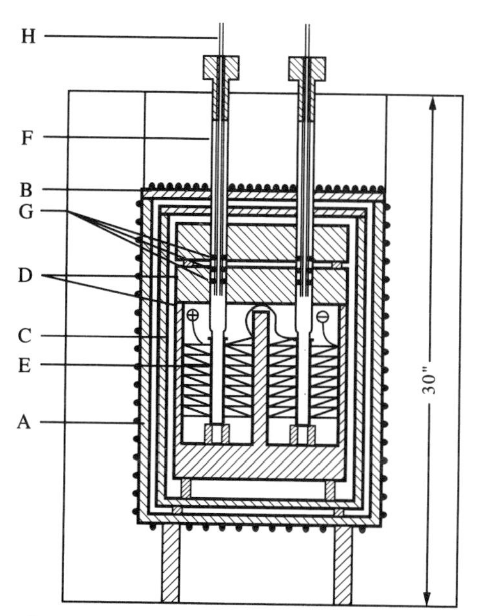

FIG. 46. General layout of Calvet type twin calorimeter. A, main heater; B, top heater; C, heavy shield; D, aluminium jacket; E, calorimeter; F, protection tube; G, radiation shields; H, manipulation tube; ⊕—⊖, thermopile.

When the calorimeter has attained a stationary state, the difference voltage, U, of the thermopiles displays a constant value, U°. When a heat change occurs in the calorimeter, the temperature of the measuring cell changes with respect to the temperature of the reference cell. As the heat exchange between the measuring cell and the jacket relaxes, the temperature returns to its initial value. In the graphical record of the experiment, U° determines the baseline, and the heat change generated in the calorimeter is represented by a peak-like area. The numerical value of the latter corresponds to the integral $I = \int \alpha(U - U^{\circ})\, dt$, the limits of which are fixed by the beginning and end of the measurement period; denotes the amplification factor. The total heat effect to be measured is proportional to the area under the calorimetric peak (1980b). Calibration is, in most cases, conducted by use of the drop method. A well-established heat effect is produced in the calorimeter by dropping weighed samples of an inert material (platinum or alumina) from a defined temperature, usually room temperature, into the measuring cell operating at the measurement temperature.

Referring to Fig. 46, some characteristic features and properties of the Tian–Calvet calorimeter may be considered:

(i) The thermopile which acts as a temperature sensing device also acts as a heat conducting bridge between the calorimetric cell and the jacket. Its influence on the sensitivity of the calorimeter is determined by the number of thermocouples as well as by its thermal resistance and configuration. Platinum–platinum rhodium couples have proved very suitable, and 50 to 100 couples have been found to be adequate (1979b) if provision is made for a proper distribution on, and a close thermal coupling of the thermo-junctions with the external surface of the calorimetric cell. Even so, in quite a number of cases, an attempt has been made to increase the sensitivity by using thermopiles with about 400 couples or even more.

(ii) The differential twin construction eliminates to a very large extent baseline shifts arising from small variations in the temperature of the jacket. Such variations alter the thermovoltage of each thermopile by nearly the same magnitude, but in the opposite sense, provided that the variations are identical at the internal surfaces of the jacket. The twin construction is especially useful, in combination with a good temperature control, for long-term measurements at high temperatures. In this respect the Tian–Calvet design is superior to the adiabatic calorimeter, which appears to be more suited to the study of fairly rapid processes. A single unit calorimeter of this type (employing only one calorimetric cell), when used in conjunction with a sensitive and precise temperature control system, can achieve nearly the same precision in short-term measurements as the twin type.

(iii) In an individual experiment, the precision of measurement may be improved by reducing the duration of the experiment. This can be achieved when the reaction rate is high and the time constant of the calorimeter small. This constant is largely proportional to the diameter of the calorimetric cells (through the heat capacity of the calorimeter). It is therefore preferable to keep the diameter of the cells as small as possible under the prevailing circumstances, but when mechanical manipulations must be performed inside the measuring cell, a minimum working diameter of 20 mm is required to accommodate the crucibles and manipulation devices (1979b).

The calorimeter shown in Fig. 46 may be employed at temperatures up to about 750 K, since the jacket is made from low-melting aluminium. Higher temperatures up to 1300 K can be achieved using more-refractory materials such as recrystallized alumina for the thermopile support, nickel (1971i), inconel (1972k), silver (1966e), recrystallized alumina (1979b) or boron nitride (1970b) for the calorimetric block, and nichrome (1960e), Kanthal (1971g), molybdenum (1971g) or platinum–40% rhodium (1979b) for the heating elements.

Some authors have succeeded in constructing calorimeters for operation at temperatures above 1400 K (1971k, l; 1979b; 1971g; 1970b; 1972k). Unfortunately, the number of scientific publications resulting from the use of these calorimeters has remained limited, but a detailed description of a typical calorimeter and of its operation has been given by Gaune-Escard and Bros (1974c), and its features and performance were investigated and discussed by Pool *et al.* (1979f) and by Arpshofen *et al.* (1979a).

Although this calorimeter uses the integrated heat-flow method as its principle of measurement, the construction of its heat detector is rather different from that of a Tian–Calvet calorimeter. Two recrystallized alumina crucibles are stacked one above the other, the upper crucible (inside diameter ≈ 9 mm, height ≈ 30 mm) containing the samples acts as the measuring cell, while the lower crucible serves as the reference cell of the differential calorimeter. Temperature differences between these cells are measured by a thermopile composed of $(18+18)$ Pt–30%Rh/Pt–6%Rh couples. One half of these temperature sensors encloses the sample crucible, the other half the reference crucible. Thus thermal effects are measured differentially along the longitudinal axis of the calorimeter, rather than radially as in the Tian–Calvet apparatus. The crucibles and the thermopile are enclosed in, and held by, an alumina tube located in a zone of constant temperature provided by an electric furnace employing a graphite tube as heating element.

This calorimeter offers several advantages (1986f; 1989c). It can operate continuously at temperatures up to about 1800 K with minimum measurable heat effects of about 2 J. Because its temperature can be changed rapidly, it can be employed not only isothermally, but also in a scanning mode. However, as a result of its temperature sensors encircling the sample crucible at equal height, the thermopile does not properly integrate the heat flux over the surface area of this crucible, and the calibration factor of the calorimeter depends on the sample mass contained in the crucible (1974c; 1979f, a). Various constructions have attempted to eliminate this deficiency, among which the versions due to Kleppa and Topor (1989b) and Wilsmann and Müller (1989c) are particularly noteworthy.

In these versions, the temperature sensors are distributed over a larger fraction of the surface area (1989b) or over the whole surface area (1989c) respectively, of the sample crucible. Proper integration of the heat flux is thereby ensured. The thermal coupling between the thermopile and the calorimetric cells is also better and the position of the thermojunctions with respect to the sample crucible is well-defined and remains virtually unaltered on changing the crucibles. This ensures a highly reproducible calibration factor in succeeding runs.

While the calorimetric detector constructed by Kleppa and Topor has been designed to be employed in its own furnace system using a Pt–40%Rh heating element, the version due to Wilsmann and Müller can be inserted into a commercial tubular graphite furnace. A new high-temperature calorimeter using the construction principles described by Wilsmann and Müller (1989c) has been found to be reliable and safe in test experiments (1992a).

The Tian–Calvet calorimeter and similar devices have mainly been employed in high-temperature reaction calorimetry, e.g. Bros (1989a), Kleppa (1971i; 1979b) and Navrotsky (1977e), as well as in the measurement of thermophysical properties, as demonstrated by Ziegler and Navrotsky (1986f). A further application of the calorimeter is for measurement of partial enthalpy values. Because of its high sensitivity, thermal effects which are virtually partial enthalpies of solution can be determined, e.g. oxygen in oxide phases (1971g; 1970b; 1974b) or hydrogen in metal–hydrogen systems (1969k), after admission of small amounts of the corresponding gas to the solid samples. The partial enthalpies of solution of the component oxides in liquid oxide mixtures have also been successfully determined by liquid–liquid or solid–liquid reaction calorimetry (1969k).

The experimental precision of high-temperature calorimeters based on the integrated heat-flow principle is remarkable. As a result, these calorimeters have become reliable research tools which enable heat effects as small as 2 J to be measured with a precision of 2 to 4% at temperatures of about 1300 K, while effects of 10 to 20 J can be determined with a standard deviation of about 1%.

2. Equilibria with a Gaseous Phase

As has been mentioned earlier, the standard Gibbs energy change of a chemical reaction can, in principle, be obtained by a combination of calorimetric data for the enthalpy and entropy change of the reaction. It may be seen from the preceding section that a number of calorimetric studies must be combined in order to yield the Gibbs energy change, and therefore in most circumstances it is preferable to obtain the standard Gibbs energy change at a given temperature from the value of the equilibrium constant at that temperature. The determination of the equilibrium constant for the reaction

$$A+B \rightarrow C+D$$

requires a measurement of the activities of the components in the equilibrium state since

$$K=\frac{a_C a_D}{a_A a_B}=\exp -\frac{\Delta G^\circ}{RT}. \tag{98}$$

The simplest forms of chemical reaction are those which involve a gaseous phase and pure condensed or phases only, or a gaseous phase and dilute condensed phase solutions when the equilibrium constants also take a simple form.

$$\langle A \rangle \rightarrow (A) \qquad K=p_A$$

$$\langle A \rangle + (B) \rightarrow \langle AB \rangle \qquad K=\frac{1}{p_B}$$

A and AB are both pure solids or liquids.

$$\langle A \rangle \rightarrow [A] \qquad K'=\frac{c_{[A]}}{p_A}$$

The value of the constant K' can only be obtained under conditions where the A atoms in solution do not interact with one another significantly, and this usually requires that the concentration of A in the solution is less than one atom per cent or so.

In high-temperature systems there are considerable experimental difficulties in the measurement of equilibria involving gaseous phases at pressures greater than one atmosphere, and therefore the following considerations will be limited to those situations where the total pressure of the gaseous phases, be they simple or mixtures, is less than or equal to one atmosphere pressure. Even with this limitation, there are a number of variants of each particular experimental technique which are designed to suit a specific reaction, and only typical examples will be referred to here. There now exist a number of more exhaustive reviews of high-temperature experimental techniques than will be attempted in this work, and the reader seeking further examples is encouraged to consult these (1967k; 1970p; 1973g).

1. The Equilibrium Constant for Vaporisation Reactions

When a pure solid or liquid vaporises, the equilibrium constant will only contain the partial pressure of the vapour species and, providing this is a gas of simple composition, then the equilibrium constant is equal to the saturation vapour pressure of the solid. Many metals vaporise to simple atomic species or to a gas phase which contains a negligible proportion of polyatomic molecules. For example, the vapour phases in equilibrium with copper, silver and gold consist mainly of the monatomic species and contain negligible amounts of Cu_2, Ag_2 and Au_2 gaseous molecules under most experimental circumstances. The measurement of the vapour pressure of these elements can therefore be made with simple apparatus under simple assumptions. Numerous examples are known, however, where the gaseous phase is a complex mixture of species such as MoO_3, which vaporises to give gaseous molecules of Mo_3O_9, Mo_4O_{12} and Mo_5O_{15} in significant proportions in the temperature range 800–1000 K. With a system showing this kind of behaviour, the *saturation vapour pressure* can be measured using techniques which yield the total pressure directly, but the equilibrium constants for the individual vaporisation processes

$$3MoO_3\ (s) \rightarrow Mo_3O_9\ (g)$$
$$4MoO_3\ (s) \rightarrow Mo_4O_{12}\ (g)$$
$$5MoO_3\ (s) \rightarrow Mo_5O_{15}\ (g)$$

cannot be obtained from such a simple measurement, and it is necessary to devise a technique which permits the measurements of each individual gaseous species. This is usually an extremely difficult and complicated task to carry out with a high accuracy which in the present context means $\pm 1\%$ of the correct value.

A. Static Methods for the Measurement of Vapour Pressure

1. Manometric Methods

The direct measurement of pressure by means of manometers has played a minor role in high-temperature thermodynamics, largely because of the problems associated with materials of construction.

Jenkins (1926c) used a static manometer to measure the vapour pressure of liquid zinc. The liquid metal was held in a quartz U-tube, one limb of which was closed and contained only the metal vapour in equilibrium with the liquid metal. The other, open, limb was joined to a "buffer" bulb containing nitrogen, the pressure of which could be measured relative to atmospheric pressure through a mercury manometer, and which could be varied to balance the vapour pressure of the zinc. The pressure was measured over the range 20–1500 mmHg and a temperature range of 900–1250 K.

Fischer *et al.* (1939c) constructed an apparatus for the direct measurement of vapour pressure in which a cylindrical molybdenum bell dipped into molten tin was contained in a closed silica vessel. The apparatus was used for the study of the vapour pressures of aluminium halides, a sample of which was placed inside the bell and then the apparatus was evacuated. On increasing the temperature, the force exerted by the contained vapour on the molten tin would raise the bell, the movement of which was taken up by a spring and indicated by an attached pointer. The apparatus was then filled with an inert gas until the pointer regained its original position, and the pressure of gas required to restore the pointer was read on a mercury manometer. Pressures from a few millimetres up to atmospheric could be measured by this technique with a reproducibility of 0.1 mmHg.

The vapour pressure of liquid tellurium was measured directly by the use of a quartz spiral manometer* by Brooks (1952b) and a Bourdon manometer by Machol and Westrum (1958h). In the spiral manometer, a flattened quartz tube is formed into a spiral, and one end of the tube is joined to a bulb containing the sample. The other end of the spiral is joined to a pointer which carries a mirror. This end is contained within another vessel in which the pressure can be regulated. As the apparatus is heated, the quartz spiral deforms and a pressure equal to that of the vapour pressure of the sample must be exerted

*The quartz spiral manometer was originally developed by Bodenstein (1908a). A detailed description of this instrument and its application to indium halides was given by Robert (1936e).

in the second vessel to restore the pointer to the null position. In the Bourdon manometer, a flattened thin-walled bulb replaces the spiral and the bulb containing the sample. Once again, the difference in pressure between the vapour pressure of the contained sample and that of the surrounding vessel must be reduced to zero to keep a pointer at the null position. The results for the vapour pressure of liquid tellurium which were obtained in these two studies agree well and cover a pressure range of 0.5–200 mmHg, and a temperature range of 750–1125 K. These techniques can be called static methods since there is no significant transfer of the material to be vaporised around or through the apparatus.

More recently, a membrane "null" manometer described by Oppermann *et al.* in connection with studies of the thermodynamic behaviour of Co_3O_4 (1980d) has been used to measure, for example, the total vapour pressures over liquid $MnCl_2$, $MnBr_2$ and MnI_2 [Oppermann and Krausze (1988e)] and to determine the barograms of the systems BiI_3–HgI_2 and BiI_3–I_2 [Oppermann and Witte (1991b)].

2. *Static Methods Using Radiation*

Measurements of vapour pressure in a sealed system may be made when a well-defined volume of the vapour can be characterised with respect to the number of atoms or molecules of the vapour species it contains. Any physical property of this mass of vapour which can be observed from outside can be used to determine the vapour pressure providing that some calibration point or an overlap with some other technique can be obtained. Optical absorption or emission have been used up to temperatures of about 1000°C when the condensed sample and its vapour can be contained in a quartz cell.

Hirst and Olson (1929b) determined the concentration of mercury atoms in atmospheres in equilibrium with thallium amalgams by measuring the absorption of light corresponding to certain of the mercury resonance wavelengths. In this way they were able to measure pressures of the order of 10^{-3} mmHg. An ionisation manometer was used by Poindexter (1926e) for the measurement of the vapour pressures of alkali metal amalgams in the range 10^{-2} to 10^{-8} mmHg.

Herbenar *et al.* (1950d) used an absorption spectrum method for measuring the zinc vapour pressure of six α-brasses over the temperature range 650–970°C. Millings of the various alloys were sealed in an absorption vessel of transparent silica (Fig. 47), which was located in a tubular furnace with transparent ends. Light from a spark source having two zinc electrodes passed through the optically flat part of the vessel which confined the zinc vapour within two planes with a separation of 0.8 mm. The transmitted light was analysed by a spectrograph and recorded photographically, a stepped intensity–density pattern being imposed by a tungsten filament source on each photographic plate for calibration purposes. Photometric observations were made of two resonance lines in the zinc spectrum; the 3035 Å line was not absorbed and was therefore used as a standard for measuring the absorption of the 3076 Å radiation. The relationship between absorption, measured by the logarithmic intensity ratio of the two spectral lines, and the vapour pressure p and absolute temperature T is given by the equation

$$-\log \frac{I_{3076}}{I_{3035}} = K \frac{p}{T} d$$

where K is a constant, I the absorption coefficient, and d the thickness of absorbing space.

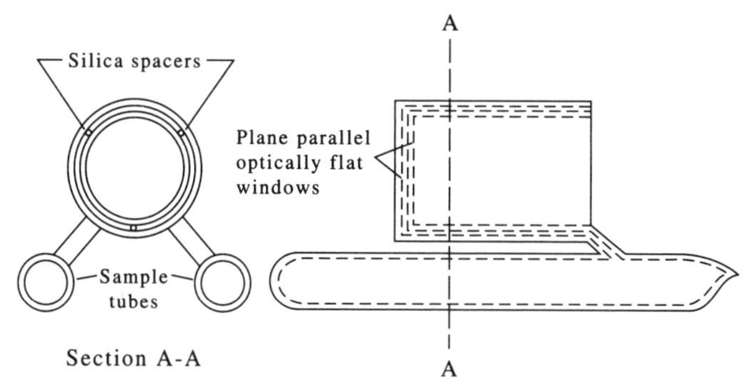

FIG. 47. Light-absorption cell for zinc vapour.

Preliminary measurements were made with liquid zinc over the temperature range 450–745°C to confirm the validity of this law and to determine the value of the absorption coefficient.

The determination of the partial pressure of an atomic species in the vapour above an alloy system by no means exhausts the potential applications of optical absorption studies. Mixtures of atomic and molecular species in the vapour phase above solids can, in principle, be measured separately since the absorption line corresponding to the excitation of the atomic species may be in another part of the spectrum from the band spectrum of the molecular species. Rice and Ragone (1965c) have made a theoretical and experimental study of the Bi–Bi_2 mixture over liquid bismuth. They included a sample of lead in the system to provide an independent check of the accuracy of the method.

The optical density of atomic bismuth was measured at 3067 Å from 770 to 1280 K, and this corresponded with a pressure range of 2×10^{-8}–10^{-3} atm. The absorption due to lead atoms was measured at 2833 Å in the same temperature range where the vapour pressure of lead changes from 3×10^{-8} to 1×10^{-3} atm. Finally, the Bi_2 band spectrum was observed starting at 2731 Å, and the optical density was measured between 1010 and 1280 K covering a pressure range of 10^{-5}–10^{-3} atm.

The fact that the vapour pressures have the Clapeyron dependence on temperature was invoked to obtain vapour pressures of these species directly from the optical density results for monatomic bismuth gas. This system obeys the simple absorption law only at low pressures, and deviates markedly at high pressures, but a mathematical treatment involving an extrapolation from the Beer's law region using the Clausius–Clapeyron equation enables the measured absorption curve to be interpreted. Alternately it was found that the presence of an inert gas, helium, at a constant pressure, simplified the absorption behaviour of the atomic species and the partial pressure p_{Bi} could be evaluated. The enthalpy of vaporisation of lead which was obtained in the study, $\lambda_e = 197{,}652 \pm 3556$ J was in good agreement with published data.

More elaborate studies have been made by Brebrick and Strauss (1964c) of the tellurides of lead and tin. These compounds evaporate congruently and thus the vapour pressures of PbTe and SnTe were established by conventional vapour pressure measurements (1959d; 1961b). The experimental technique involved the measurement of the optical densities of the vapour phase due to the presence of Te_2 and PbTe or SnTe molecules over a range of composition in solid alloys ($N_{Pb} = 0.05$–0.534; $N_{Sn} = 0.483$–0.504) at wavelengths of light which were selected by interference filters from the spectrum emitted by a mercury lamp. The pressure–optical density relationship for Te_2 was established in calibrating studies made with pure tellurium and that for the molecules PbTe and SnTe from the Russian studies of conventional vapour pressures. It appears to be the general case, with a few exceptions, that this relationship must be established between optical density and partial pressure at some point in the composition–temperature range in the study of a system by optical absorption, and that absolute measurements of the vapour pressure can only rarely be achieved by the use of this technique.

In the results which have been described above, the absorption cells were usually made of quartz, but this need not necessarily be the case. It is possible to operate a ceramic cell, with two orifices to allow the transmission of a light beam, in a vacuum furnace to function as the container of the equilibrium vapour. Some uncertainty then exists concerning the path length of light through the vapour phase because of the effusion of vapour out of the cell and into the surrounding vacuum. Such effects might yield a calibration procedure.

B. Gas-condensed Phase Equilibria in a Closed System

1. The Dew-Point Method

Hargreaves employed an elegant dew-point method (1939f) (originally proposed for liquids by Lescoeur, 1889) making use of an electric furnace with two independently heated portions. The brass test piece (A) (Fig. 48) is placed in an evacuated silica tube at the end situated in the hotter part of the furnace. Both ends of the silica tube are provided with small re-entrant tubes which carry thermocouples. The temperature of the whole furnace is raised to the required value, and then the end away from the test piece is slowly

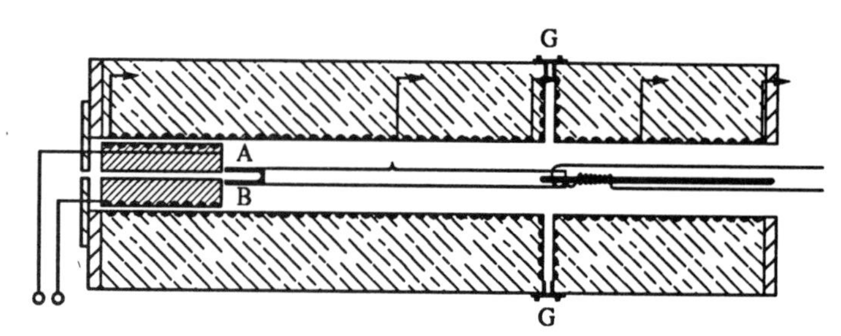

Fig. 48. Apparatus for the dew-point determinations on metals.

cooled, the coolest portion of the silica tube being observed through viewing windows at (G). At a certain temperature, measured by the thermocouple, pure zinc condenses in small droplets. By repeated heating and cooling of this part of the tube the "dew-point" of zinc can be determined within $\pm 1°C$. The vapour pressure of zinc at that temperature equals the zinc pressure of the brass at the temperature of A. This is, of course, a relative method. It can also be applied to other than metallic vapours, for example to the titanium tetrachloride pressures in a reaction such as $2\langle TiCl_3\rangle = \langle TiCl_2\rangle + (TiCl_4)$ for the investigation of which the vapour pressures of pure $TiCl_4$ must be known.

2. The Isopiestic Method

Whereas Hargreaves' method has nearly exhausted its usefulness, a similar method devised by Herasymenko (1956b) continues to find fairly wide application under the name *isopiestic method*. The principle again involves the establishment of an equilibrium vapour pressure between an alloy specimen held at the high temperature end of a sealed tube and the pure volatile component of the alloy held at the low-temperature end. In this technique, the temperatures of the hot and cold ends of the tube are fixed and the equilibrium composition of the alloy determined. The activity of the volatile component in the alloy at the temperature of the hot end can then be calculated provided the vapour pressure/temperature relation for the volatile component is known. As a time-saving device the hot end may be supplied with a row of several alloy specimens held in a temperature gradient rather than at constant temperature. Since the activity of the volatile component in the alloys varies with temperature, subsequent analysis of the alloy specimens produces a number of results per run.

Komarek and Silver (1962d) have made an elegant use of the method in their determination of partial molar Gibbs energies of solution of oxygen in titanium-group metals. For these studies specimens consisting of, for example, titanium and an alkaline earth, MgO, in close physical contact were placed within the temperature gradient of the sealed tube with an excess of the pure alkaline earth metal being held at the cooler end. Reaction was allowed to proceed until equilibrium was reached where the partial pressure of oxygen over the $[O]_{Ti}$ alloy was equal to the partial pressure of oxygen in the reaction $(Mg)+\frac{1}{2}(O_2)=\langle MgO\rangle$. By varying the vapour pressure of the alkaline earth metal from one run to another, the chemical potential of oxygen in the Mg + MgO system was changed and hence, in turn, the chemical potential of oxygen in the $[O]_{Ti}$ alloy. The oxygen content of the alloys was calculated from the mass gain of the specimens after equilibration.

Richardson and Webb (1955d) used an isopiestic technique to obtain the activities of PbO in liquid PbO–SiO_2 mixtures by equilibrating a slag sample placed in one crucible with a bead of liquid lead contained in a separate crucible. The two condensed phase systems were contained within a larger crucible, and equilibrium was achieved when the PbO pressures of the liquid silicate and PbO dissolved in the metal sample were equal. Since the latter solution is very dilute and hence obeys Henry's law, the activity of PbO in the silicate can be obtained from the ratio of the oxygen content of the metal divided by the saturation oxygen content of lead in equilibrium with pure PbO at the same temperature.

Krachler *et al.* (1982a) have investigated the thermodynamics of non-stoichiometric β'-AuMn using a specially adapted isopiestic method (Fig. 49).

Since manganese reacts both in the condensed and gaseous state with quartz, the equilibration tube and the crucibles containing the specimens were made of high-purity alumina. The specimens for one experiment were first weighed and placed in about 20 crucibles (10 mm o.d., 16 mm height), each provided with a slit, stacked one on top of the other inside a 16 mm o.d., 500 mm long alumina tube which was closed at the bottom and fitted with a precision-ground stopper at the top. The alumina tube with the crucibles, the bottom one of which contained pure manganese, was first placed horizontally in a quartz tube with a thermocouple protection tube and with the Al_2O_3 stopper fitted only very loosely. The outer quartz tube (see Fig. 49) was then evacuated to ca. 10^{-1} Pa and backfilled with titanium-gettered argon about three times before it was sealed under vacuum. The ground stopper was made to slide into its joint by raising the assembly into a vertical position. The seal was then vacuum-tight and prevented any manganese vapour from escaping out of the alumina tube.

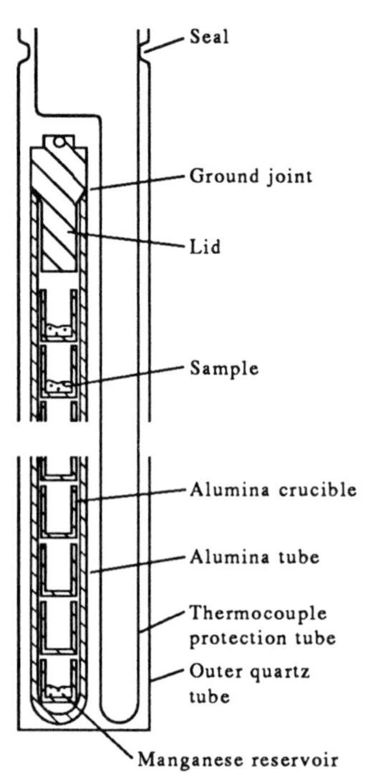

FIG. 49. Isopiestic equilibration tube; the hatched parts are made of alumina.

The fully assembled apparatus containing samples and reservoir was heated in a vertical two-zone resistance furnace, each zone of which could be independently controlled to within ± 1 K. The reservoir with the pure manganese was held at the lowest temperature in a constant temperature zone of about 3 cm, and the specimens were heated in a constant temperature gradient. The temperature was measured repeatedly by raising and lowering a Pt/Pt–10%Rh thermocouple in its protection tube. (It had previously been established that the lateral temperature gradient across the alumina tube was negligible.) After a specimen equilibration time of about 6 to 8 weeks, the quartz tube was removed from the furnace and cooled in air. The equilibrated specimens were reweighed and their composition determined from their gain or loss in weight, which was attributed entirely to gain or loss of manganese. The compositions of a series of specimens analysed by complexometric titration agreed within the limits of error with those calculated from the weight changes. The sample compositions were estimated to be accurate to ± 0.1 at.% and their temperatures to ± 1 to 2 K.

3. *Tensi-eudiometer Measurements*

The sulphur and phosphorus pressures of a large number of sulphides and phosphides were investigated by Biltz and his co-workers. Their long series of experiments began with the determination, by vapour pressure measurements, of the various steps in the progressive decomposition of inorganic ammonia complexes.

The apparatus used to measure the dissociation pressures of sulphides (1930a) (Fig. 50) has several refinements of experimental interest. The highest valency sulphide of the metal under consideration is enclosed in the bulb (A) (shown enlarged in the smaller diagram) which is heated to the required temperature. The experiment consists in measuring the vapour pressure of the sulphide after the removal of successive fractions of sulphur which are afterwards determined. The vapour pressure is measured by means of a quartz spiral manometer. In order to isolate the sulphur vapour at its equilibrium pressure, use is made of a so-called "sulphur valve", invented originally by M. Bodenstein. The side tube leading to the sulphur receiving tube is constricted at (C) to a diameter of 0.5 mm (Fig. 50a).

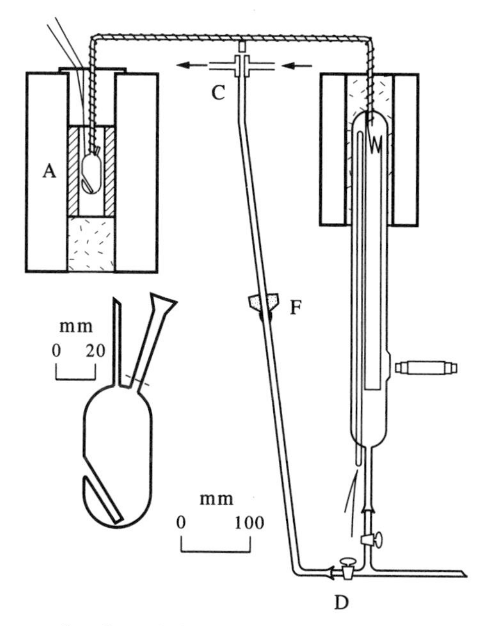

FIG. 50. Tensi-eudiometer for thermal decomposition and measurement of vapour pressure of sulphides.

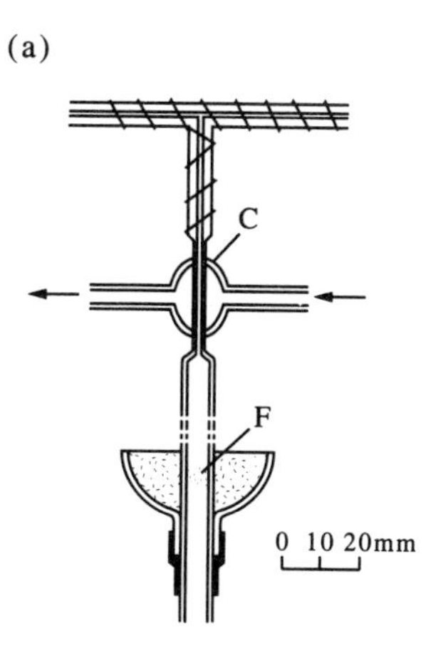

FIG. 50a. "Sulphur valve" in apparatus in Fig. 50.

Passage of cold water through a jacket surrounding (C) results in the condensation of a plug of sulphur, which is vacuum-tight when 1–3 mm long. When this plug is broken, sulphur distils into the receiving tube (CD). In order to measure the increments of sulphur withdrawn from (A), each one is collected at a different position along the receiver by means of the movable cooling device (F), which is filled with a solid carbon dioxide–alcohol cooling mixture. Condensation in other parts of the apparatus is prevented by heating them to a temperature higher than that of the heating vessel (A). At the end of each experiment the receiving tube is cut into portions, and the various deposits of sulphur are analysed. It may be noted that the apparatus is made of one piece, and this may necessitate the use of vitreous silica for a refractory sulphide such as that of platinum.

A similar arrangement has been used for phosphides. A full description of the method was given by Haraldsen (1932d) and a summary, together with the main experimental results, are available in a paper by Franke *et al.* (1934d).

Since phosphorus has not the same sealing properties as sulphur, the "phosphorus valve" must be narrower, 0.33 mm diameter, and the cooled portion at least 1.5 cm in length. For the collection of the phosphorus distillates a straight receiver tube is unsuitable, since one experiment may last several weeks. Instead, a comb form of reservoir is used (1939a) (Fig. 51). One "tooth" at a time is cooled, and is sealed off when the phosphorus valve is closed.

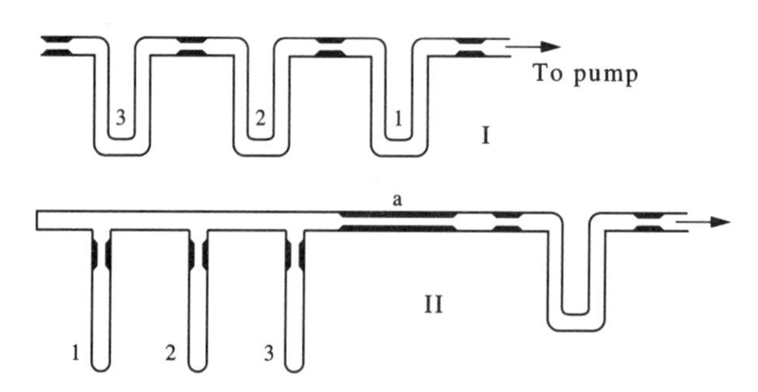

FIG. 51. Devices for condensing phosphorous fractions.

In all equilibrium measurements of the kind described in this section account must be taken of the molecular aggregation of the volatile component. If this exists in more than one form (e.g. S_2, S_4, S_6, S_8; P_2, P_4; Bi, Bi_2) its dissociation equilibrium at the temperature of measurement must be known and used in the evaluation of results. In these cases the mass spectrometer is an indispensable apparatus.

4. Sievert's Method

A considerable amount of work has been carried out by Sieverts *et al.*, Steacie and Johnson, Smithells, Ransley and others on the solubility of gases, more especially hydrogen, nitrogen and oxygen, in metals. In so far as these measurements refer to true equilibria they may be used to evaluate the thermodynamic data of the solution phase and even of the lowest valency compound of the metal with the gaseous element.

A method which is most generally applicable was developed by Sieverts (1907). The metal is heated in a silica or porcelain vessel connected to a burette (Fig. 52) which is also connected via a three-way stopcock to a pump and a gas supply. A known volume of gas is introduced to the reaction space, and the decrease in volume due to solution is observed by the level in the burette. The solubility of the gas at the operating pressure is thus given directly.

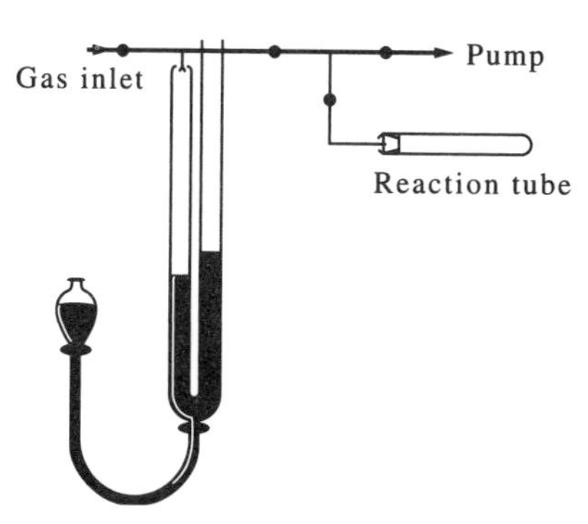

FIG. 52. Sievert's apparatus for determining the solubility of gases in metals.

Considerable advances have been made in the measurement of the solubility of gases in liquid metals and alloys with the advent of induction heating. Weinstein and Elliott (1963h) have described an apparatus for the measurement of the solubility of hydrogen in liquid iron and its alloys at temperatures around 1600°C (Fig. 53). Because the metal sample, which is both heated and stirred by the induction field, can be contained in a crucible mounted within a water-cooled quartz tube, the permeability of the containing vessel, which was an important source of error in Sievert's apparatus, can be reduced to negligible proportions. The hot volume of the assembly could be maintained at only 40 cm^3 by filling ouf the "dead" space, and equilibrium between the metal and gas was reached in 5 min.

The temperature of the melt was obtained by optical pyrometric measurement to within ± 2°C, and the solubility in pure iron at 1592°C which was found to be 24.47 ppm was obtained with a standard deviation of ± 0.24 ppm, from the results of fifteen different experiments.

In measuring the hot volume of the apparatus, an inert gas is substituted for the gas which is soluble in the liquid metal. In order that the temperature gradients in the apparatus are satisfactorily reproduced, the thermal properties of the calibration gas, the heat capacity and thermal conductivity, are matched as closely as possible to the gas under study. Thus helium should be used for calibration of an apparatus which is to be employed for the study of hydrogen solubilities in liquid metals, and argon is well suited as the calibration gas for studies of oxygen or nitrogen solubilities.

C. Dynamic Vapour-pressure Methods

1. The Boiling Point Method

The best-known thermochemical method that is truly dynamic is the determination of

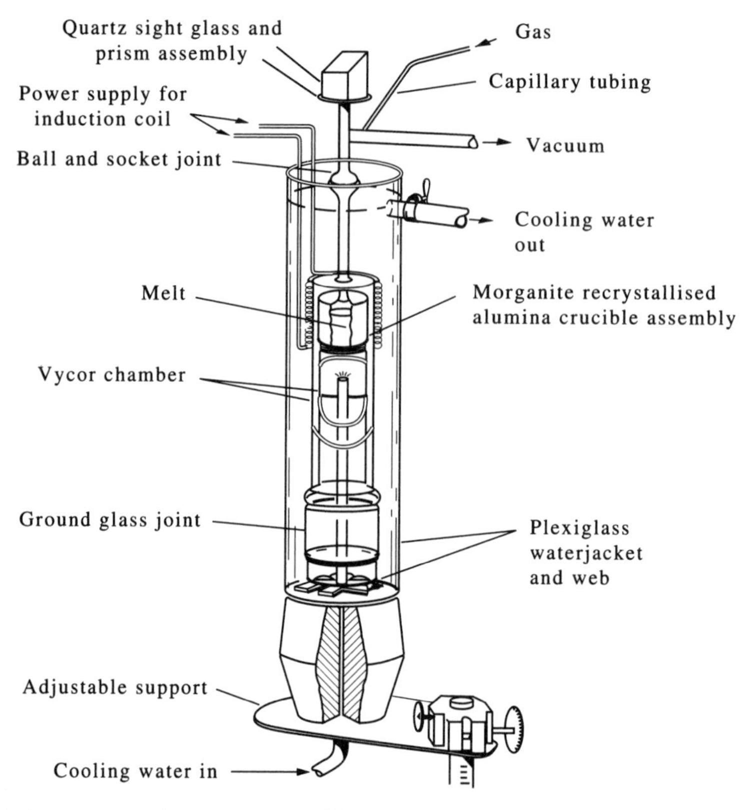

FIG. 53. Apparatus for the measurement of hydrogen solubility in liquid metals using induction heating [Weinstein and Elliott (1963h)].

the boiling point of a substance. This, of course, gives the temperature at which the vapour pressure of the substance equals the pressure of the atmosphere with which it is in contact.

Early applications of this method to metals were not very successful owing probably to the porosity of the containing vessel, which was commonly of graphite. As a result, the vapour pressure values obtained were too low. Fischer showed (1934c) that Ruff's method (1919) should be improved by using a completely impervious material for the crucible and by removing other lesser sources of error.

The principle of Ruff's method is to observe the temperature at which there is a sudden decrease in weight of a substance as it reaches its boiling point. The substance is enclosed in a crucible suspended from a spring balance inside a furnace. The boiling point is in effect determined from a discontinuity in the weight–temperature curve. The disadvantage of the method is that the change in slope of the curve at boiling is sometimes not sharply defined and, in addition, since the observations are taken at changing temperature, the observation of the actual temperature of the substance is subject to some error.

A modification of Ruff's method was therefore used by W. Fischer and Rahlfs (1932c), who worked with constant temperature and changing pressure. In this method the boiling point is much more clearly marked, as may be seen in Fig. 54. The apparatus which they used for aluminium halides is shown in Fig. 55. A bulb containing the substance was suspended from a steel spring, and an indication of the weight was given by the level of the mark M_2.

The methods used by Leitgebel (1931a) and Fischer (1934c) for observing boiling points

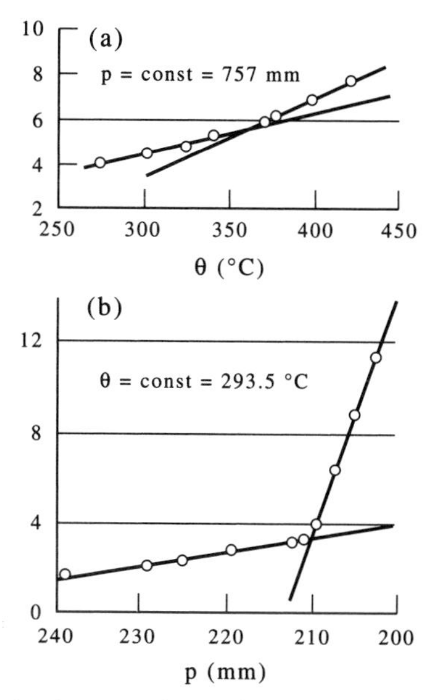

FIG. 54. Boiling point determinations: (a) with varying temperature and (b) with varying pressure.

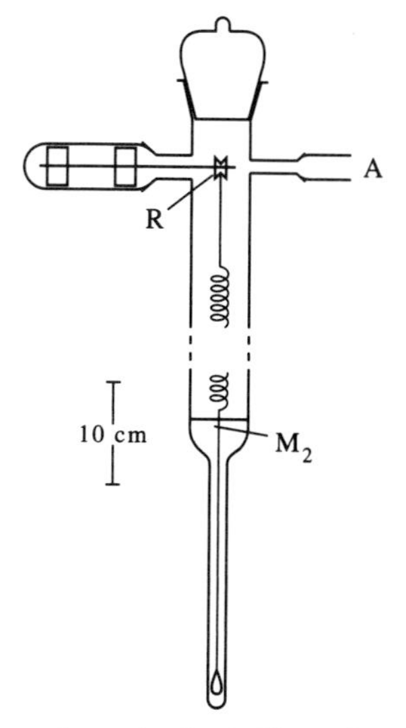

FIG. 55. Ruff type apparatus for evaluating boiling points (Fischer and Rahlfs 1932c).

at constant pressure are less responsive than Ruff's to porosity of the containing vessel. Both observers determined the boiling of metals and alloys in the range 700–1750°C. Leitgebel used a closed graphite crucible, and a thermocouple to indicate the temperature, while Fischer used an alumina crucible and an alumina tube terminating in a hollow graphite "black body" which was completely immersed in the melt for optical determination of temperature.

The boiling point is marked by a well-defined arrest in temperature increase. Its observed value can be high only as a result of superheating, and any such error, as far as metallic systems at least are concerned (1934c), is small and within the limits of accuracy of the temperature observation.

Even this small possibility of error can be eliminated by using the condensation method in which the vapour temperature is measured at various heights above the boiling liquid. Within a certain distance from the liquid, at which its vapour is condensing, the temperature is constant and is equal to the boiling point of the liquid at the pressure of the buffer gas. This method gives very reliable values but is applicable only at temperatures at which thermocouples can be used (i.e. generally below 1500°C). The apparatus can be quite simple; the one used by Fischer and Rahlfs (1932c) for aluminium halides is shown in Fig. 58. Hydrogen was used as the buffer gas. The temperature during boiling was constant within 3–6 cm above the liquid surface.

Gattow and Schneider (1959b) have studied and emphasized the usefulness of the simple boiling-point method devised by Bauer and Brunner (1934a) which consists in the determination of the boiling point of a substance at a given pressure. The buffer gas of the evaporation vessel is separated from a large volume of gas by a drop of mercury in a horizontal glass capillary. When the temperature is gradually increased the movement of the drop is accelerated the moment the boiling temperature is attained.

Barton and Bloom (1956a) took the added precaution of bubbling nitrogen through the melt in their studies of the vapour pressures of a number of metal halides. This procedure helps to remove the possibility of overheating which can occur in liquids of relatively low thermal conductivities. In this apparatus, the boiling point was measured by a thermocouple which was set just below the surface of the liquid attached to a double-bore sillimanite tube which carried nitrogen gas down to the bottom of the melt. The absolute vapour pressures of NaCl and KCl were obtained in this way over a range of temperatures and applied external pressures (15–500 mmHg). The results were combined with data from subsequent transportation studied to yield the monomer/dimer ratio of these vapours as a function of temperature, and hence the heats of dimerisation.

2. *The Transportation Method*

The more general dynamic method of determining vapour pressure is known sometimes as the transportation method. In the application to the measurement of the vapour pressure of a metal, for example, or of the volatile component of a binary metallic alloy, a steady, measured stream of inert gas is passed over the substance under investigation, which is maintained at a constant temperature. The gas removes the vapour or volatile component of the substance at a rate which is dependent upon the relative pressures and upon the rate of gas flow. The vapour is condensed or collected by absorption or chemical combination at a cooler portion of the apparatus. The rate of removal of vapour is measured at different rates of gas flow.

In a system which is completely isothermal, saturation is attained at a zero gas flow rate and the value of the vapour pressure can be determined by extrapolation to zero rate. The value so obtained should be equal to the value observed by the use of static methods. In the typical high-temperature system in which a temperature gradient exists between the sample, which is at the centre of the furnace, and the cold end of the furnace where the vapour is analysed, saturation can be achieved at low, but finite flow rates, and an error is introduced by extrapolation to zero flow rate of the transporting gas.

Alcock and Hooper (1960a) in a study of the vapour pressure of gold by this technique found that saturation could be achieved over a wider range of flow rates the more the evaporating surface could be made to fill the cross-section of the reaction tube. Their results shown in Fig. 56 indicate this effect.

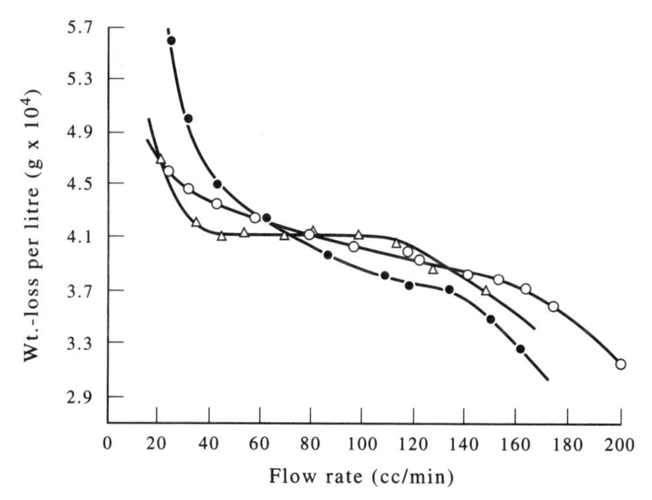

FIG. 56. The apparent vapour pressure of gold obtained by the transportation method, as a function of sample to reaction tube diameter ratio. △, large beads with plugs nearly filling the tube cross-section on either side of the sample; ○, small beads with plugs; ●, small beads without plugs.

At low flow rates incorrect results are obtained because the gold atoms diffuse down the temperature gradient at a different mean velocity than the argon atoms, resulting in apparently higher evaporation rates than the equilibrium rate.

At high flow rates the transporting gas does not have a long enough residence time in contact with the sample to reach saturation. Mertens (1959c) has analysed the transportation system for the case of laminar flow of the transporting gas over the condensed phase which covers the wall of the reaction tube along the length of the even temperature zone of the furnace. The conclusion from the experimental studies that a long "plateau" can be achieved, where the measured vapour pressure is independent of the carrier gas flow rate, when the condensed phase intercepts the flow pattern of the gas as much as possible, suggests that in a typical real system the gas is in turbulent flow around the sample, and such a theoretical analysis is of limited application.

The partial pressures being measured are calculated from the volumes of transported substance V_s and of transporting gas V_g at s.t.p., assuming the validity of Boyle's law, from

$$p_s = \frac{760 V_s}{V_g + V_s} \text{ (mmHg).}$$

This method is very commonly applied to organic and inorganic compounds at relatively low temperatures. It was first applied by v. Wartenberg (1913c) at the higher temperatures to metals, and his type of apparatus has been used with very little modification in subsequent work on inorganic salts and metallic systems by Jellinek and Rosner (1929c) and others. The diagram of the apparatus in Fig. 57 is taken from a paper by Schneider and Stoll [1941d].

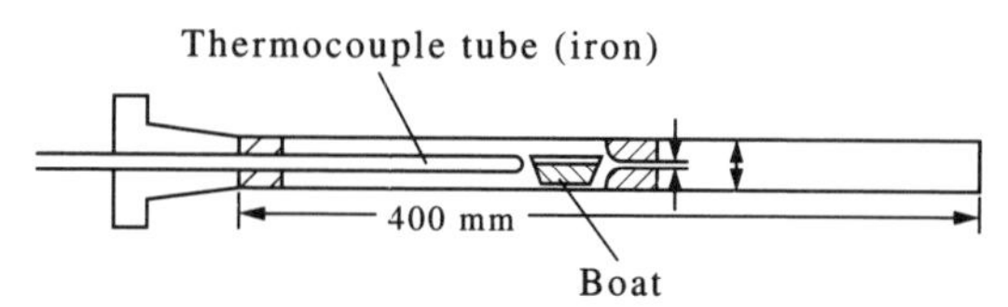

FIG. 57. Reaction tube for the measurement of vapour pressures by the transportation method.

Constrictions are made in the furnace tube on both sides of the sample to minimise counterdiffusion of the vapour. This may cause trouble especially when hydrogen is used, and whenever available the heavier gas argon is to be preferred. The transporting gas must be quite pure. The total gas flow and streaming rate are measured with a volume- rather than a flowmeter, since the fundamental quantity being measured is the total volume over a known time, not the rate at any instant.

The furnace is so mounted on wheels that it can be moved away from the test portion of the main tube. Before the experiment the furnace is moved well to the left (Fig. 57) and its temperature raised to about 50° higher than the desired final value. When it is pushed to its normal position the test sample is rapidly heated while the furnace temperature drops to within a few degrees of its desired value.

In the apparatus used by Fischer and Rahlfs (1932c) for the study of aluminium halides, the reaction and condensation chambers were connected by a heated tube. This allowed experiments to be done with a very slow gas flow which ensured complete saturation without any loss of transported vapour by condensation elsewhere than in the collecting vessel.

Another method of ensuring that all the vapour in the slowly moving saturated gas is quantitatively transferred into the condensation chamber is found in Fischer and Gewehr's apparatus (1932c). It was used to study the formation of aluminium chloride on passing chlorine or hydrogen chloride over heated alumina. The accuracy of the method was tested by measuring the transfer of pure mercury by an inert gas. The basis of the method is the use of a further "carrier" gas to take the saturated gas, diluted from the hot portion of the apparatus to the condensation chamber. The saturated gas leaves the reaction chamber (R) (Fig. 58) by a capillary tube (C). The carrier gas enters at (3), mixes with the saturated gas at the end of (C), and transports it to the chamber (A). The portion (B) is heated sufficiently to keep the temperature of (C) higher than that of (R).

When the condensed phase is a liquid, saturation of the carrier gas can be rapidly achieved if the gas is bubbled through the liquid phase. Morris and Zellars (1956d) first used this method in a measurement of the vapour pressure of liquid copper in the

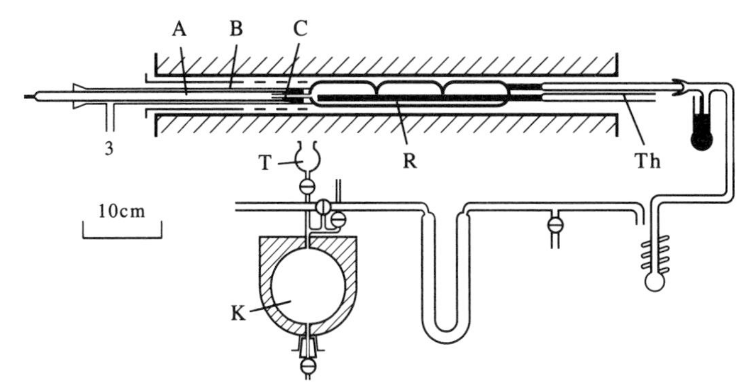

FIG. 58. Apparatus for the transportation method, utilising a carrier gas (vessel K is drawn on a reduced scale).

temperature range 1450–1600°C using both argon and helium as carrier gases. The mouth of the bubbling tube was placed just in contact with the surface of the liquid metal so as to create ripples on the surface. To sample the equilibrium gas, a sample mixture of known volume was drawn from immediately above the liquid metal through a narrow-bore tube. The quantity of copper in the measured volume of gas was determined from the mass of condensate within this sampling tube.

The transportation technique can be extended to use where the carrier gas undergoes a reaction with the condensed phase to form gaseous products. Again the most advantageous design of the reaction chamber brings the gas and the solid phases into intimate contact, thus establishing equilibrium at a finite flow rate, and the existence of a plateau in the apparent vapour pressure/flow-rate graph.

Typical examples of successful studies of this nature are those by Bookey (1952a) who studied the reaction

$$\langle Ca_4P_2O_9\rangle + 5(H_2) = 4\langle CaO\rangle + (P_2) + 5(H_2O)$$

$$K = p_{P_2}\left(\frac{p_{H_2O}}{p_{H_2}}\right)^5; \quad p_{P_2} = \tfrac{1}{5}p_{H_2O}$$

by passing hydrogen through a reaction chamber containing pellets of the solid reactants which occupied a large fraction of the cross-section of the reaction tube. Kitchener and Ignatowicz (1951c) passed hydrogen in nitrogen/hydrogen mixtures through a bed of zinc oxide granules and condensed a sample of the zinc transported as a result of the reaction

$$\langle ZnO\rangle + (H_2) \rightarrow (Zn) + (H_2O)$$

in a known volume of the carrier gas. Since $p_{Zn} = p_{H_2O}$ under these circumstances, the equilibrium constant, and hence the Gibbs energy of the reaction was obtained.

Oxygen was used by a number of workers to establish the thermodynamics of the gaseous oxides which are formed by the platinum group metals at high temperatures (1960a, k; 1943c). The oxide PtO_2, is now well established as a result of the studies in which oxygen was passed over heated platinum samples in a transportation apparatus, and, again, good plateaux could be obtained in the apparent vapour pressure/flow-rate curves (Fig. 59).

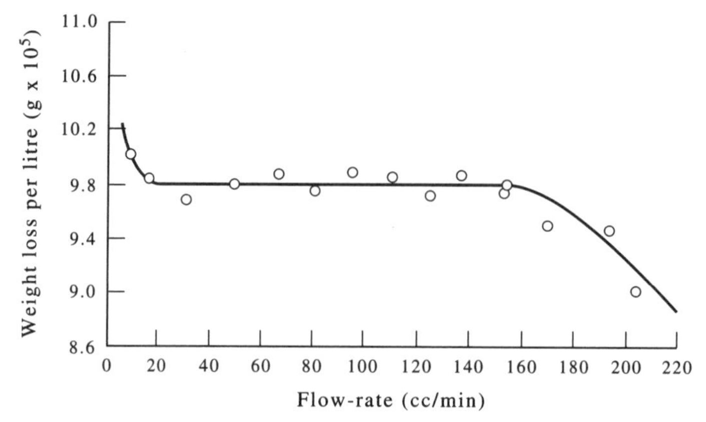

FIG. 59. Weight loss of a platinum sample as a result of (PtO_2) formation in oxygen at 1398°C as a function of the gas flow rate.

It is generally so, in the use of the gas transportation method as a technique for the measurement of vapour pressures, that the material which vaporises is collected in the condensed state. In order to apply the equation given above to the determination of the vapour pressure, it is necessary to know the molecular species in the vapour phase in order to calculate V_s from the mass which is condensed. In many cases for the metals, the monatomic species predominates, but in the case of substances which are evaporated as molecules, or vapours which are formed as the result of reaction between the transporting gas and the condensed phase, it is usually necessary to establish the composition of the vapour species in order to obtain the vapour pressure.

$$m\langle M\rangle + \frac{n}{2}(X_2) \rightarrow (M_mX_n)$$

the equilibrium constant will depend on the partial pressures of the gaseous species and the activity of the metal

$$K = \frac{p_{M_nX_n}}{a_M^m p_{X_2}^{m2}}.$$

The response of the vaporisation process to changes in the partial pressure of X_2, e.g. by dilution with an inert gas or by a reduction of the total gaseous pressure, will indicate the number of X atoms in the gaseous species. A study of the dependence of the partial pressure of M_mX_n upon the activity of M in an alloy with a less-reactive metal will then yield the number of M atoms in the molecule. This procedure is clearly only of any use when the vapour phase is relatively simple. When a number of polymers occur together in significant amounts, e.g. $(MoO_3)_n$ or a number of valency states for the metal e.g. MX, MX_2, MX_3, then another and much more difficult situation arises, and one which cannot usually be resolved by the use of this technique alone.

D. Other Heterogeneous Equilibria

The methods used in the study of systems involving two or more gaseous components are very similar to those described in the preceding section. The transportation method, for instance, may be used to determine the rate of evaporation of two volatile components. Their respective vapour pressures are evaluated and treated individually, and such a system lies completely within the scope of the foregoing section.

There are many factors peculiar to systems including more than one gas, however, and they may be considered separately. A large number of such systems have been investigated and some of the methods used are best described by a few examples. We shall limit these considerations to reactions between a gas and a condensed phase giving rise to at least one volatile product. The reactions will be classified according to the gas used as the reducing (or oxidising) medium in place of distinction between methods. Most of the equilibria measured hitherto have included hydrogen as the reducing agent.

1. Reaction Between a Gas Phase and a Condensed Phase

H_2–CH_4 Equilibria. Equilibria between hydrogen, metals and their carbides, and methane have been determined on a large scale by Schenck and his co-workers (1927c). Their method is simple and is in principle the one generally used for such equilibria with hydrogen. The partly carburised metal is heated in contact with a mixture of hydrogen and methane. The progress of reaction is observed by the change in pressure. When this has become steady, the gaseous and condensed phases are analysed. Sensitive methods of analysis must be used, and many have been developed by Schenck for this purpose. To investigate each system, experiments are carried out with a number of different mixtures of hydrogen and methane, and equilibrium is approached from the two directions by carburising or decarburising. The gas mixture is slowly circulated, and to ensure that the correct equilibrium is attained and measured, the cooler dead space in the furnace chamber is kept to a minimum with blocks of ceramic material. Even with the best materials, all heated furnace tubes are to some extent pervious to hydrogen at high temperatures, and a correction must be made for this. Since the method is static the results also require correction for thermal diffusion in the gaseous phase.

Systems involving binary iron alloys (with manganese or tungsten, for example) have been studied in addition to those which include only one metal. The kinetics of carbide formation, however, involve a number of incalculable difficulties, and thermochemical data obtained by this method are therefore not entirely satisfactory.

H_2–NH_3 Equilibria. Little work has been done on these equilibria, though of this work the investigation of the iron nitriding reaction

$$2\langle Fe_4N\rangle + 3(H_2) = 2(NH_3) + 8\langle Fe\rangle$$

by Emmett *et al.* (1930b) is of some importance. A weighed amount of ferric oxalate was placed in the reaction tube (A) (Fig. 60) and reduced in a stream of hydrogen. When the exit gas no longer contained water vapour, a mixture of hydrogen and ammonia of fixed composition was passed into the tube, which was maintained at a constant temperature. The ammonia content of the outgoing gas was periodically analysed, and when it became constant, the gas mixture was replaced by nitrogen. Finally, some of the nitrided iron powder was shaken into the thin-walled side tube in which it was sealed for subsequent X-ray analysis.

H_2–H_2O Equilibria. Since all metals are produced from their ores by reduction methods, the thermochemical data of the reduction of metallic oxides are of great fundamental importance, and have been investigated directly by numerous workers.

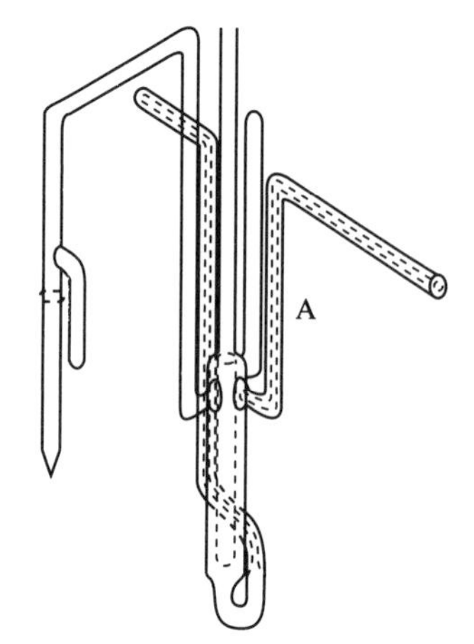

FIG. 60. Glass reaction vessel for the determination of NH_3/H_2 equilibria over iron.

A static method which is widely applicable was originally devised by Deville (1870) and used subsequently by a number of observers, notably by Eastman and Evans (1924b) to study the reduction of iron oxides by hydrogen. The apparatus (Fig. 61) contains water vapour at a pressure determined by the fixed temperature of the water in (A). This water vapour is allowed to react with iron powder in the boat (B) which is heated to the required temperature in the silica furnace tube. Measurement of the total pressure after attainment of equilibrium gives, on subtracting the water vapour pressure, the equilibrium pressure of hydrogen. Variation in the composition of the solid is obtained by successive high-temperature treatments in a stream of water vapour, beginning with the pure metal, or by controlled hydrogen reduction of the oxide. The composition is calculated from the change in weight during the treatment.

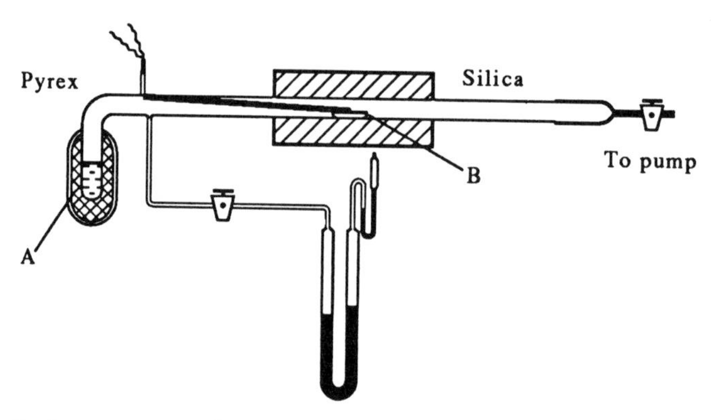

FIG. 61. Static measurement of H_2O/H_2 mixtures in equilibrium with iron and iron oxides.

A great source of trouble and error in this method is the evolution of occluded gases from the walls of the apparatus on heating. The apparatus must be thoroughly baked and evacuated and blank determinations carried out before each series of measurements.

The rapidity with which equilibrium is attained in Eastman and Evans' experiments decreased with increasing oxygen content of the iron. Over the range 0 to 10% O_2 a time of 30–34 min was taken at 990°C.

It was shown by Emmett and Shultz (1924c) for the equilibrium 'FeO' + H_2 = Fe + H_2O that the results obtained by various observers by static and dynamic methods are discordant and the difference was ascribed to the effect of thermal diffusion in static measurements. Shibata and Kitagawa have determined (1938b) this effect for H_2O/H_2 mixtures at various temperatures and various compositions. Their results can be expressed approximately by the empirical relationship

$$\log\left(\frac{p_{H_2O}}{p_{H_2}}\right)_{T_2} = \left(1 + 0.065 \log\frac{T_2}{T_1}\right)\log\left(\frac{p_{H_2O}}{p_{H_2}}\right)_{T_1} - 0.205 \times 10^{-3}(T_2 - T_1), \qquad (99)$$

which represents the experimental data (400–1021°C) within ± 0.005 on the logarithmic scale. T_2 denotes the higher, T_1 the lower temperature. Whalley (1951f) pointed out a mistake made by Shibata and Kitagawa in the discussion of their experimental results, and has recalculated certain values of theoretical importance. This does not affect the above equation, however, which for experimental purposes should be quite adequate for calculating H_2O/H_2 ratio in the hot zone of any apparatus from the ratios measured in a cooler part, thus making the static method as useful as the dynamic one.

These observations have been confirmed in an investigation (1949h, i) of the oxidation of iron–nickel alloys at high iron contents between 650° and 950°C. The method of Eastman and Evans was quite suitable for this system, since only the iron is oxidised in the temperature and concentration ranges covered. The simplicity of the method combined with its reliability makes it quite attractive. It must be determined in any series of experiments, however, whether the effect of thermal diffusion necessitates a correction.

Deville's method is not, however, well adapted to equilibria in which the hydrogen pressure is less than a few millimetres, or to those in which the equilibrium vapour pressure is very low. Dynamic methods are needed for such systems.

The reduction of cobalt and nickel oxides, for instance, proceeds with a very low hydrogen pressure. In a study of these equilibria (1929a; 1926d), steam and electrolytic hydrogen were mixed in fixed proportions and passed over the heated oxides. The volume of ingoing hydrogen was measured from the electrical power input, the exit steam was condensed and weighed, and the exit hydrogen determined with a volumeter.

At all proportions of water vapour to hydrogen, precautions must be taken to ensure that there is no thermal or back diffusion of the vapour before or after passage through the equilibrium chamber.

Equilibrium measurements at much higher temperatures, those of *molten* iron for example, require more elaborate equipment and refractory materials which must remain practically impervious to gases on heating. The most serious difficulty encountered in practice is that furnace tubes found to be gas-tight on first heating often become pervious after a few more thermal cycles.

The solution to this trouble may lie in the use of induction heating methods. These have been developed to a high degree of refinement by a succession of workers under the direction of Chipman at the Massachusetts Institute of Technology. The present basic form of apparatus was developed by Dastur and Chipman (1949a), and its elaborated version, which was used by Gokcen and Chipman (1952f) for the study of the silicon–oxygen equilibrium in liquid iron, is shown in Fig. 62. The iron alloy is melted in an alundum crucible supported on a tube which serves to carry the exit gas and to lower the crucible for rapid cooling when equilibrium has been established. The gas mixtures are passed into the furnace through a refractory tube wound internally with a platinum resistance heater which is maintained as nearly as possible to the temperature of the melt. The design of the furnace, and preliminary experimental work, were intended to eliminate thermal diffusion (see p. 109) and to ensure accuracy in the measurement of temperature. Apparent temperature readings taken by training an optical pyrometer on the surface of the melt were calibrated (1949b) by the use of crucibles fitted with re-entrant tubes in the base. True temperature readings were made both with the insertion of thermocouples in and with the sighting of a pyrometer into these re-entrant tubes.

As a result of these basic studies in which the equilibrium between oxygen dissolved in liquid iron and H_2–H_2O gas mixtures was established, it has been possible to extend the technique to understand the "equilibration" between these gas mixtures and levitated droplets of molten iron or its alloys. The advantages of this procedure over the conventional technique are, firstly, that there can be no metal-refractory interactions because no crucible is used, and, secondly, the whole droplet can be rapidly quenched for analysis by turning off the radio-frequency heating source and allowing the droplet to fall into a copper mould or "splat-cooling" device.

One disadvantage clearly results from the fact that the gas-levitated droplet is by no means a classical, isothermal, equilibrium system, and steep thermal gradients exist in the gas phase, thus emphasising thermal diffusion effects. Secondly, the temperature of the droplet cannot be measured by thermocouple as is possible when a crucible is present, and observations, usually by two-colour pyrometer, must be relied upon for temperature measurement.

As would be expected from the general pattern of thermal diffusion behaviour in gas mixtures, the errors due to this source are greater when H_2–H_2O mixtures are used to fix the oxygen than when CO–CO_2 mixtures are employed. In the latter case, reasonably good agreement was obtained with the results of conventional studies, and the effects of thermal diffusion in H_2–H_2O mixtures can be reduced by introducing argon into the gas mixture. The presence of this gas raises the average molecular weight of the gas, and an extrapolation of the values of log K of the equilibrium constant to the theoretical gas of infinite molecular weight yields a value of log K which is within 0.05 of the value obtained by the conventional technique (Fig. 63).

In an investigation of the reduction of chromic oxide Wartenberg and Aoyama (1927d) suspended the chromium test piece on thermocouple wires in a furnace supplied with hydrogen whose water content was fixed by cooling by suitable freezing agents. The test piece was heated slowly, and the temperature at which the surface oxidation disappeared was noted. Similarly, the temperature at which oxidation recommenced on cooling was recorded. By repeating these observations several times with slow temperature changes, the equilibrium temperature for the given water–hydrogen mixture was determined.

The visual observation of surface oxidation is not reliable, however, for surface colours can be misleading. The method was therefore improved by Grube and Flad (1939d), who dispensed with visual observations by substituting determinations of the change in weight of the test sample. The water content of the hydrogen after passing the sample was determined by absorption with phosphorus pentoxide. In these experiments the hydrogen must be very carefully purified.

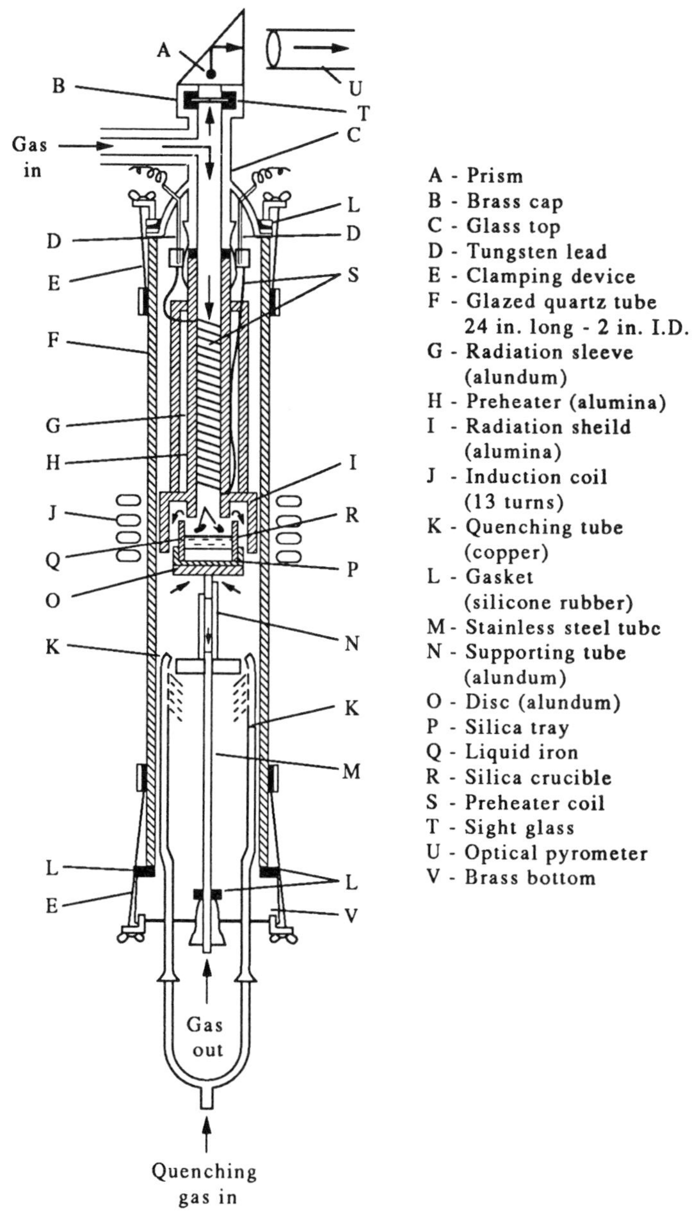

FIG. 62. Gokcen and Chipman's furnace tube for the measurement of H_2O/H_2 equilibria over molten iron.

Grube and Flad (1942b) extended their equilibrium pressure measurements to determine the thermodynamic data of alloy formation. By determining the water/hydrogen pressure ratios, (1) for pure chromic oxide and (2) for the oxide in intimate admixture with nickel powder, the free energy of chromium–nickel alloy formation is obtained from the relationship

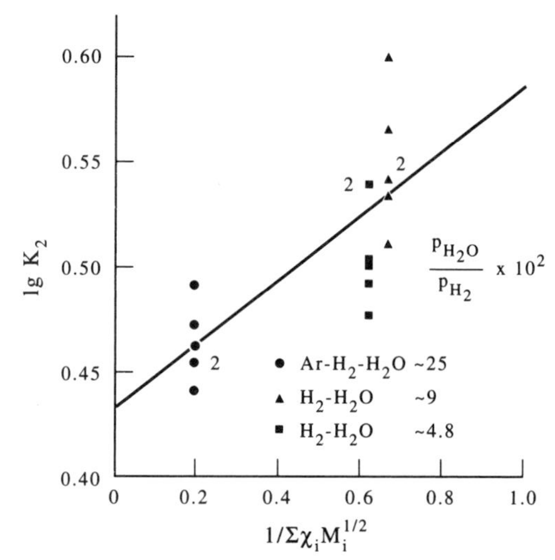

FIG. 63. The effect of thermal diffusion on the apparent equilibrium constant at 1700°C for the reaction $(H_2)+O \rightarrow (H_2O)$. χ_i is the mole fraction of i and M_i is the molecular weight of i. [After (1974h)].

$$\Delta \bar{G}_{Cr} = \tfrac{3}{4}\boldsymbol{R}T\left[\ln \frac{p_{H_2(1)}}{p_{H_2O(1)}} - \ln \frac{p_{H_2(2)}}{p_{H_2O(2)}}\right].$$

The reverse procedure, i.e. the determination of equilibria with alloys containing oxygen at known activities in order to obtain the dissociation pressure of a pure oxide, will rarely be used, presumably. Chiche (1952d), however, has done so. He first obtained the activities of copper in copper–gold alloys by an e.m.f. method and then made a direct determination of the dissociation pressures of Cu_2O–Cu–Au mixtures. He was thus able to calculate the dissociation pressure of cuprous oxide which is too low to be measured directly with convenience. Since the temperature coefficients of the e.m.f. of copper–gold alloys obtained by Chiche appear to be rather uncertain, it is better to take the assessed activities of copper in copper–gold (1973c, d) and to combine these with the dissociation pressures determined by Chiche. The resulting values for the dissociation pressure of pure Cu_2O then appear reasonable.

H_2–H_2S Equilibria. The field of heterogeneous equilibrium measurements discussed in this chapter was first explored by Pélabon who developed a static method, similar to that which Bodenstein used for homogeneous gas reactions, and applied it to sulphides and selenides. This work (1902a) formed the basis of later static measurements.

Pélabon enclosed the metal sulphide or a mixture of sulphide and metal with hydrogen or a hydrogen/hydrogen sulphide mixture in a tube heated to the required temperature. After allowing a certain time for equilibrium, the tube was quickly removed from the furnace, on the end of a wire, and rapidly aircooled by swinging it around. The gas, its composition retained at its high-temperature equilibrium value, was transferred to a eudiometer under mercury and analysed.

This method was used in an improved form by Keyes and Felsing (1920) and also by Watanabe (1933b). Schenck and Forst studied H_2/H_2S equilibria by a method similar to that used for H_2/CH_4 equilibria, described on p. 113. In the early investigations of

Schenck the effect of thermal diffusion was neglected, but precautions were taken against it in later work (1939i, j).

Equilibria between hydrogen/hydrogen sulphide mixtures and molten iron–sulphur alloys have been studied by Sherman *et al.* (1950k). The furnace equipment was similar in principle to that described on p. 117, but the melts were sampled periodically during each run to determine the approach to equilibrium by measurement of sulphur content.

Rosenqvist (1949m) has investigated, apparently very accurately, the equilibrium $Ag_2S+(H_2)=2Ag+(H_2S)$ for the solid and liquid states in the range 600–1280°C. The experimental system was closed and gas-tight, and the gas mixture was circulated by means of a glass propeller, driven by an external magnet, until equilibrium was established, which took about 5 hr. The circulation rate was high to minimise thermal diffusion, the gas being sufficiently preheated before it reached the metal. In its circulation the gas mixture passed through a chamber containing a buoyancy balance, where the density of the gas was determined by magnetic balancing. From the measured density, corrected for pressure and temperature, the H_2S content which varied from 1 to 25% was determined. This method has the advantage that it is more accurate than the usual chemical-analytical methods and that the attainment of equilibrium is being observed in a closed system. After each run, the sulphur content of the alloy was determined by analysis.

Another important feature of Rosenqvist's work is that a correction was applied for the S_2 content of the gas in the hot zone using the known dissociation constants of H_2S. This correction may be as high as 10% and has been overlooked by most previous investigators of equilibria with H_2S/H_2 mixtures.

Other Equilibria. While those equilibria in which hydrogen is the reducing agent have had the most attention, others, such as those with $CO–CO_2$ mixtures have also been investigated. By and large the static and dynamic methods which have been used are similar to those already described.

The attention of the reader may be drawn to a more detailed survey of the equilibria and transport phenomena involving gas mixtures and condensed phases by Schwerdtfeger and Turkdogan (1970r).

The heat change which accompanies a chemical reaction was employed in the study of the thermodynamics of sulphate formation from metal oxide, SO_2 and O_2 by Dewing and Richardson (1959a). In a particular example, a powdered specimen of MgO was packed into a spiral of Pt–13%Rh wire which had a thermocouple junction placed at its centre. A reference thermocouple junction was exposed to the gas phase, and the whole system was supported in an alumina boat. This was placed in the even zone of a furnace reaction tube down which a metered mixture of SO_2, O_2 and N_2 was passed. The furnace temperature was raised and $MgSO_4$ soon formed over the MgO sample. Heating was continued until a thermal signal from the thermocouple system showed that the temperature of the MgO sample lagged behind that of the gas phase. This indicates decomposition of the sulphate and at this temperature $MgO–MgSO_4$ is in equilibrium with the gas phase, which at the high temperature consists of an equilibrium mixture of SO_3, O_2, N_2 and SO_3. This composition can be calculated from the pertinent thermochemical data, and the constant for the reaction

$$MgO+SO_2+\tfrac{1}{2}O_2 \rightarrow MgSO_4$$

can be established at the temperature of the thermal arrest. On cooling the furnace, a heat evolution is observed accompanying the re-formation of the sulphate. By varying the input gas mixture, the value of the equilibrium constant can be obtained over a range of temperatures. Results have been obtained by the use of this technique for a number of metal sulphates, and the results for this technique, employing a flowing gas system, agree well with the static dissociation pressures for these sulphates reported by Kellogg (1964f) which have been obtained by Ingraham *et al.* The agreement indicates that thermal segregation errors in SO_2–O_2–SO_3 mixtures are not significant in this relatively low temperature range. This was further demonstrated by Ingraham and Kellogg (1963d) who showed that when the gases were circulated in an apparatus which could be used for static measurements, no significant change was observed in the total pressure of the system. The apparatus, used for the study of zinc sulphates, consisted of a furnace tube which was closed at one end by a pyrex bellows capable of transmitting the total pressure of the SO_2, O_2 and SO_3 mixtures generated by decomposition of the sulphate to a mercury manometer. When the gases within the furnace tube were recirculated by means of a pyrex centrifugal pump, the total pressure changed for a short time, and then returned to the original pressure exerted by the static gas. This indicates that the total pressure, which would change with a significant thermal diffusion contribution, was unaffected by recirculation and hence significantly free of this source of error.

E. Methods Based on Rates of Evaporation

Two principal methods have been developed of determining the vapour pressure of a substance by its rate of evaporation *in vacuo*: they are the methods of Knudsen (1909a) and Langmuir (1913a). The pressure p (atm) is calculated from the formula

$$p = \frac{m}{tA}\sqrt{\frac{2\pi RT}{M}} = 0.02256\,\frac{m}{tA}\sqrt{\frac{T}{M}} \tag{100}$$

where m(g) is the mass of vapour of molecular weight M which evaporates from an area $A(\mathrm{cm}^2)$ in time t(sec) if the Knudsen method is used, but the mass which evaporates when the Langmuir method is used is often *less* than the value given in eqn (100). This is because the evaporation of solids under the conditions required in the Langmuir method does not occur under equilibrium conditions and the fraction of the mass evaporated to that calculated from the corresponding Knudsen experiment is called the vaporisation coefficient.

In the effusion method of Knudsen the substance is enclosed in a gas-tight container into the lid of which a small orifice is drilled. The conditions for the application of eqn (100) to evaporation out of this container are that the orifice must be knife-edged and have a diameter less than one-tenth of the mean free path of the vapour inside the cell.

A channel may be used instead of a knife-edged orifice, but then it is important that the length-to-diameter ratio of the orifice is known. The correction factor to be applied to the equation above for this method was first deduced theoretically by Clausing, and has been calculated and tabulated by Searcy and Freeman (1954).

In the Langmuir method the sample is suspended freely in the vacuum system so that there is no impediment to evaporation from the whole of the specimen surface area. Although the true surface of a solid is never planar, the projected or apparent surface area is normally used in eqn (100).

The weight loss of a sample m_L in a Langmuir experiment can be calculated from a knowledge of the equilibrium vapour pressure p_K established by the Knudsen method, through the equation

$$m_L = 44tA\alpha p_K \sqrt{\frac{M}{T}} \tag{101}$$

where the symbols have been defined above, and α is the vaporisation coefficient $(0 < \alpha \leqq 1)$.

The weight loss during a Knudsen or Langmuir experiment was formerly obtained by weighing the cell and its contents at room temperature before and after an experiment. This should now be avoided since the outgassing error is difficult to allow for, and many satisfactory continuous weighing devices have been described in the literature following the early designs of Gulbransen and Andrew (1942c).

1. The Knudsen Effusion Method

In a modification of the Knudsen-effusion weight–loss method by Kubaschewski and Chart (1974f) a vacuum microbalance with a load-carrying capacity of 25 g and a sensitivity of 1 μg on its most sensitive range was employed. The cell was suspended from the microbalance in a vacuum furnace. By continuous weighing, with the cell at temperature, the progress of de-gassing and the onset of steady-state conditions could be determined. The method was designed to measure the activity of silicon in transition-metal silicides. Since direct dissociation would have required temperatures of well over 2000 K, the silicides were intimately mixed with silica, thus producing silicon monoxide vapour in the cell and reducing the temperature of measurement by 700 K or more compared with the temperature required for direct dissociation. One of the resulting advantages was that silica glass could be used to make the Knudsen cell.

For the highest sensitivity it is preferable to use radioactive isotopes and to condense a known fraction of the vapour on a target. The weight loss can then be obtained by radiochemical assay. Care must be taken in using this method to ensure that a significant fraction of the measured condensate has not been collected after scattering from the plates defining the solid angle of collection. The apparatus of Schadel and Birchenall (1950g, h) is a successful design for this method of vapour-pressure measurement.

The combination of effusion and tracer methods has also been applied by Kubaschewski, Heymer and Dench (1960g, h, i) to alloys of chromium. Higher temperatures (*ca.* 1400°C) were required, and resistance heating has been preferred to induction heating (Fig. 64). In addition to the main resistance heating element enclosed in radiation shields, heating elements at the top and bottom could be operated separately in order to establish uniform temperature within the molybdenum cell. The temperature was measured by means of two thermocouples. The effused chromium condensed on the molybdenum tube and on the target disc and was collected by solution in acid. The specimens were made from chromium which had been made radioactive, and the mass effused was determined by comparing the radioactivity of the solution with that of a standard solution of chromium from the same batch. The advantage of using the funnel is that all the escaping chromium could be collected, thus making the method even more sensitive. An improved method of determining the radioactivity has been devised by Feschotte and Kubaschewski (1964e): the molybdenum tube and target are burnt,

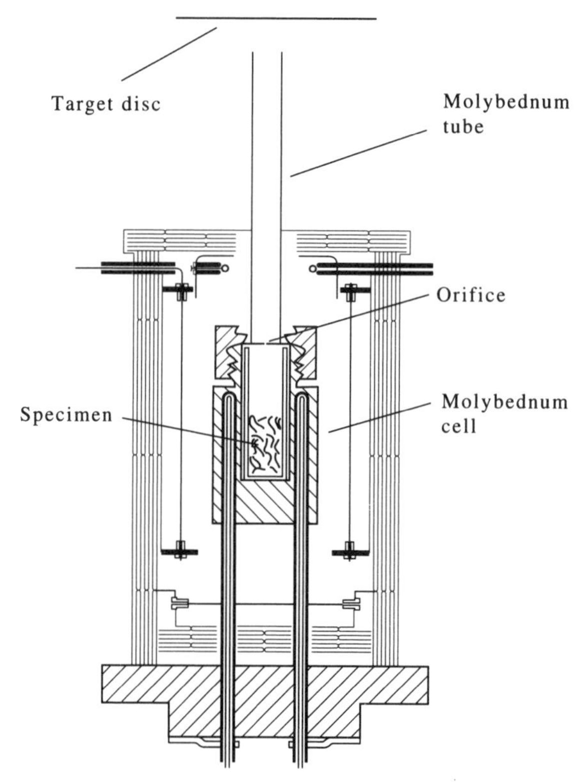

FIG. 64. Effusion cell and furnace for the determination of the vapour pressures of metals between 1200° and 1400°C.

together with the radioactive deposits, in oxygen, the molybdenum trioxide formed compacted into pellets and their radioactivity compared with that of a pellet obtained in the same way from an effusion experiment with pure chromium.

Two main ingredients for a successful measurement of the vapour pressure of a simple substance where the molecular weight of the effusing species is known, are the use of chemically inert materials for the crucible and lid of the assembly and the reduction to a minimum of temperature gradients in the cell. The effects of interaction between the sample and the container material may change the apparent vapour pressure either to a higher or a lower apparent value depending on the nature of the interaction.

Ward and co-workers (1967o, p) have published a valuable study of Knudsen effusion at high temperatures (1000–2000 K) which demonstrates the effects of container materials in reducing the reliability of results as a consequence of chemical interaction. The particles effusing through the orifice of a Knudsen cell should follow Lambert's cosine law which in two dimensions is given by

$$I_\theta = I_0 \cos\theta \tag{102}$$

where I_θ is the intensity in a direction at an angle θ to the axis of the orifice where the intensity is I_0. The actual distributions of particles which were directly measured by Ward *et al.* rarely conformed to this simple law, the departures usually being greater the more the sample reacted with the container. Gold and plutonium were used as metal samples, representing extreme cases of chemical reactivity and graphite and tantalum, and Y_2O_3, MgO and ThO_2 were among the container materials. The two metals are both liquid over the major part of the temperature range which was employed, and have approximately the same vapour pressure.

Gold placed in tantalum cells on ThO_2, Y_2O_3 and graphite cups yielded distribution curves relatively close to the ideal, but with some "peaking" of the distribution immediately along the line joining the orifice and a normal to the evaporating gold surface. This latter effect was found to change in magnitude depending on the curvature of the metal sample which changed as the surface tension altered with sample temperature. Some loss of effusate could be detected with the tantalum cell, probably due to alloy formation with the gold atoms, but in all-graphite assembly yielded the results most close to theoretical.

In the case of plutonium very variable results were obtained. With a Y_2O_3 cup placed inside a tantalum cell, the distribution approached the ideal as a number of successive experiments were carried out, but this was never achieved. With a ThO_2 cup, and more so with MgO, the plutonium vapour was severely depleted by interaction with the walls, and the great majority of the atoms leaving the cell came directly from the plutonium sample. In these containers approximately 80% of the anticipated effusate was lost by interaction. These results would therefore lead to an approximately five-fold error in measurements where the total weight loss was measured, or all of the effusate was collected, irrespective of distribution. For a spherical sample, under the conditions of these experiments, the material effusing directly from the sample was to be found within an angle of 10°. It can be concluded that the effusing beam from a Knudsen cell mirrors the relative emission rates from the various surfaces within the cell. Relatively minor losses of material at the cell walls can account for a drastic reduction of effusate from these areas. The importance of finding a chemically inert container, or as close an approximation to this as possible, emerges clearly from this study. It also demonstrates the value of repetition of vapour pressure measurements in a number of container materials where possible. Where container interactions cannot be sufficiently minimised, then a measurement of the total effusate coming *directly* from the sample surface will give the closest approximation to the correct value. The Knudsen experiment has now become virtually a Langmuir experiment since the effects of the vaporisation coefficient now become important.

Storms (1969l) has carried out a study of the effects of temperature gradients in the cell on the apparent vapour pressures of substances. The probability of there being significant temperature gradients in a Knudsen apparatus will vary very much with the average temperature of the study, and with the method of heating. Under relatively ideal circumstances, at moderate temperatures where resistance-heated furnaces having substantial even zones of temperature can be made, as in the case cited earlier, the effects will be minimal. If a cooled target is brought close to the sample in order to collect a defined beam of effusate, then the inevitable cooling of the Knudsen cell lid by radiative heat loss can cause condensation around the orifice. If vaporisation from these areas contributes significantly to the effusate which should be collected, the effects could be serious. If the effusate is collected over a narrow target angle so that direct effusion from a sample with unit vaporisation coefficient is the objective, this effect could be negligibly small except if the orifice becomes increasingly blocked during an experiment. At much higher temperatures, above 2000 K, an electron bombardment furnace is frequently used;

the cell is heated directly by energy transfer from accelerated electrons emitted by surrounding filaments. The assembly is contained in radiation shields to reduce radiative heat loss from the cell. Storms made use of such an arrangement with three separate electron sources allowing for independent irradiation of the top, centre and bottom of a Knudsen cell. The temperature profile of the cell could thus readily be modified and gradients of 200°C could be imposed. The temperature was normally measured either directly through the orifice and on the surface of the evaporating sample or in a "black-body" hole in the bottom of the cell. Using gold in a specially designed graphite crucible (Fig. 65) as the main object of study, it was shown that the error introduced into a

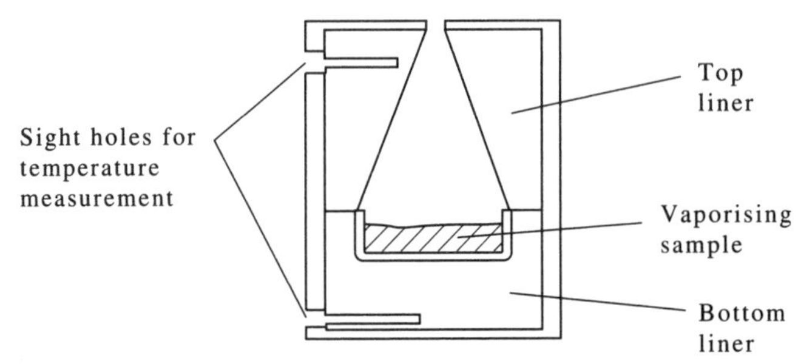

FIG. 65. Knudsen cell assembly containing a liner of good thermal conductivity in order to reduce thermal gradients.

measurement of the vapour pressure of gold was equivalent to an error of 5°C when the orifice above the sample was 177°C hotter than the bottom of the cell, but the effusion sample surface was observed for temperature measurement. The error was equivalent to 100°C when the black-body hole in the base of the cell was used for temperature measurement. It was suggested that since the vaporising surface controls the effusion rate, the effects of a significant temperature gradient causing errors in an effusion study should be looked for routinely. This could be done by comparing the apparent vapour pressure when the sample is placed on the bottom of the cell with that obtained when a disc is placed under the sample so that it now occupies a position nearer the top of the cell. Only when these two results agree is the absence of a significant temperature gradient established.

In the resistance furnace procedure at lower temperatures, it is also desirable to mount thermocouples at the top and bottom of the cell to check for the absence of a temperature gradient.

2. The Knudsen Cell–Mass Spectrometer Combination

It will have become apparent a number of times during this discussion of the measurement of the vapour pressure that the complexity of the vapour phase must be understood in any situation if a reasonably complete description is to be given of a vaporisation process. The traditional methods can be made to yield an answer in very many simple circumstances and still have a significant role to play, even when this is only to provide, for example, the total weight loss from a Knudsen measurement of a system which is not completely analysed for vapour complexity.

The mass spectrometer is now playing an increasingly dominant role in supplying information about the nature of this complexity because of the capability of separating, for mass determination, the components of complex vapour mixtures.

The separation of the vapour species can only be carried out in the mass spectrometer after a process of ionisation, and this is carried out by bombarding a molecular beam, effusing freely from a Knudsen cell orifice, with a beam of monoenergetic electrons. Since the number of ions produced per unit time is proportional to the flux of molecules through this ion source, and the flux from a Knudsen cell follows the inverse square law of intensity, the ion source must be placed as close as possible to the Knudsen cell. This is a very important aspect of the design of Knudsen cell–ion source systems and in most cases the entry to the ion source is placed directly above the Knudsen cell. This arrangement necessarily limits the length of the furnace, and hence the ease with which an even zone of temperature can be achieved. An alternative system in which the beam effuses from an orifice in the side of the Knudsen crucible, but which permits the use of a longer furnace, is shown in Fig. 66 compared with a "conventional" electron bombardment arrangement.

Having optimised the beam intensity into the ion source and produced the ions by electron bombardment, two methods of separating the ions into the various groups of

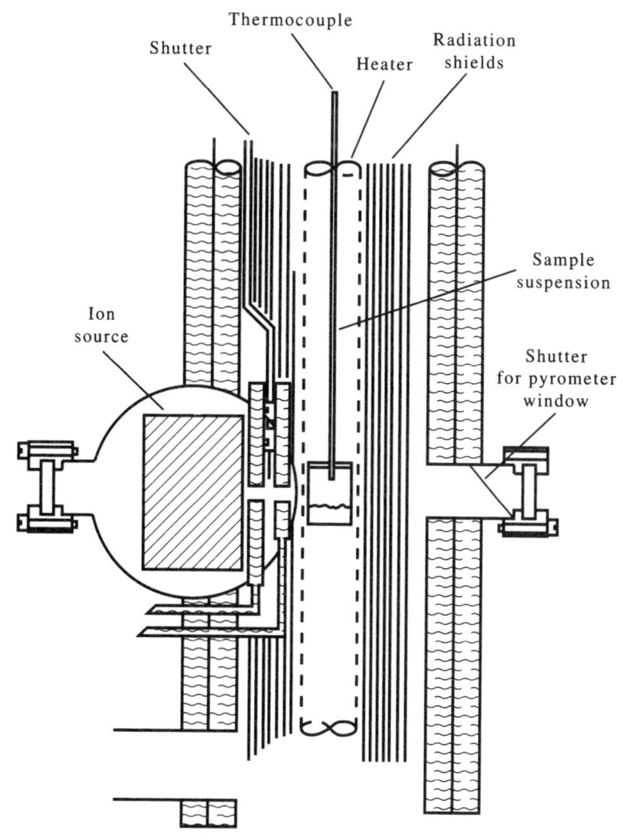

FIG. 66. Resistance-heated Knudsen cell-mass spectrometer source using a vertical furnace but a horizontal molecular beam from the Knudsen cell.

charged species are mainly used. In the more traditional of these, the ions are extracted continuously from the ion source by means of a fixed accelerating potential and are separated in a magnetic field into groups of constant charge to mass ratio. The alternative procedure is to accelerate the ions with a pulsed accelerating potential and allowing them to separate into groups of constant "time of flight" (TOF) down a straight vacuum chamber.

The immediately obvious advantage of magnetic resolution is that ions are extracted continuously instead of at repeated intervals as in the TOF spectrometer. The sensitivity of magnetic resolution machines can therefore normally be made higher than that of the TOF machine. An advantage of the latter is that high-speed electronic recording techniques may be used to display the arrival of a number of ion groups at the detector end of the "flight tube" apparently simultaneously. Although it is possible to scan through the masses in a magnetic resolution machine, this has not yet reached the speed of the TOF device. The latter has a decided advantage in the simultaneous recording of a number of ionic species. These are the major differences between the two alternatives which are mainly used in high-temperature studies.

These systems for mass analysis have a common requirement of a well-defined source of molecular beam and subsequently of ions, to function in a comprehensible manner. It is desirable to mount a movable slit or shutter between the Knudsen cell and the ion source so that ions which arise unambiguously from molecules leaving the Knudsen cell can be separated from those from background sources. If the shutter can be traversed across the molecular beam, then additional evidence on the beam angular distribution similar to that obtained by Ward *et al.* (1967o, p) can be used to reinforce this information. An alternative to a beam-scanning shutter is the capability to rotate the Knudsen cell so that the molecular beam can be removed from alignment with the entry port to the ion source.

The ion current which is produced from a given molecular beam intensity into the ion source is markedly dependent on the energy of the bombarding electron beam. A typical ionisation efficiency curve for a simple monatomic species is shown in Fig. 68. Here virtually no ions are produced until the electron energy passes a critical value, the "appearance" potential of the ion species, and then the ion current is a linear function of the electron energy over a wide range of electron energies until a maximum is reached. The slight curvature in the ionisation efficiency curve just above the appearance potential is the result of lack of precisely uniform conditions for ionisation, such as a small range of energies in the bombarding electron beam. The appearance potential is normally only a few electron volts, and a spectrometer is commonly operated at a little over 10 V for the bombarding electron energy.

In the more complex situations for which there is need of mass spectrometry, a number of molecular and atomic species are to be ionized. In this case non-linear ionisation efficiency curves can be obtained for a given mass number resulting from the fragmentation of complex molecules to smaller masses by electron bombardment (Fig. 67). This can complicate the interpretation of the resolved ion currents, but variations in the thermodynamic activities of the components in the Knudsen cell can be used to elucidate the true state of affairs. Thus the dimeric form of a metal vapour M_2 could contribute to the monomer ion beam M^+ as a result of the fragmentation process

$$M_2 + e^- \rightarrow M^+ + M + 2e^-$$

increasingly so as the electron beam energy is increased. However, the intensity of the

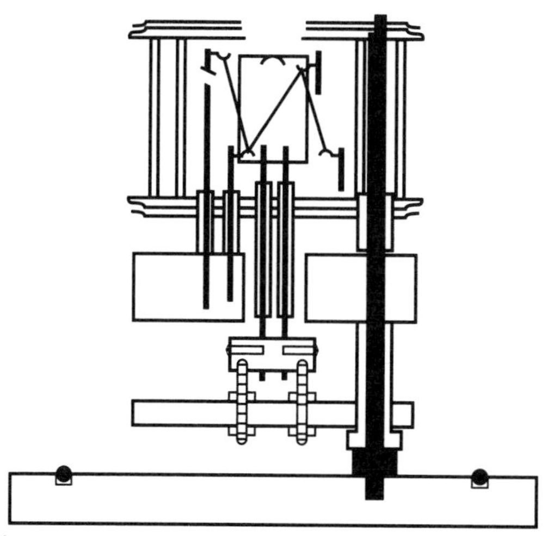

FIG. 67. Electron bombardment furnace with a coaxial molecular beam from a Knudsen cell.

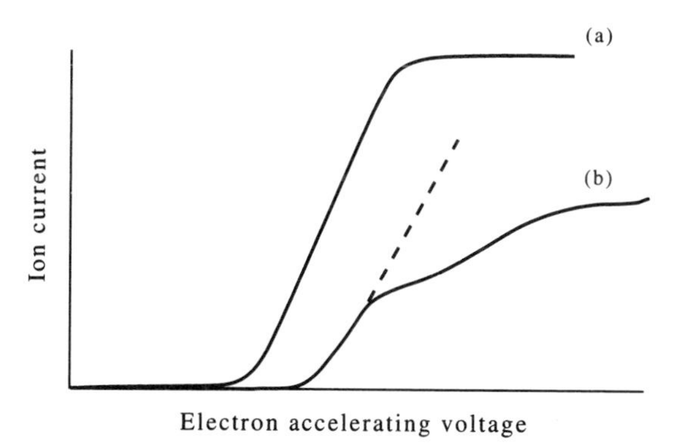

FIG. 68. Ion current–electron energy "appearance potential curves" for the ionisation of a simple monatomic species and (b) a molecular species showing the effects of fragmentation.

correct M^+ ion beam will vary directly as the activity of M, whereas that of the M_2 species will vary as the square of the activity of M, and hence the two effects can be separated. There are obviously numerous elaborations upon the effects of fragmentation and their recognition.

The ion beam is normally recorded by means of photomultiplier or other electron multiplying devices, and the response of these is a function of the mass of the ion which is detected. This instrumental factor is another contribution to the overall relationship between the intensity of a species in the mass spectrometer and the recorded intensity at the detector end of the spectrometer. The general relationship is

$$p = \frac{KI^+T}{\sigma D \Delta E} \tag{103}$$

where K is a geometric constant, I^+ the recorded ion current, T the absolute temperature of the Knudsen source, σ the ionisation cross-section, D the detector efficiency, and ΔE the electron beam energy minus the appearance potential.

Dimers of metallic elements have been detected in the vapour phase, and these may be used to advantage to reduce the effects of the instrumental factor on mass spectrometric studies. For example, if it is required to obtain the thermodynamics of the gaseous intermetallic species MN, one method is clearly to measure the heat change for the reaction

$$(\mathrm{M})+(\mathrm{N})\rightarrow(\mathrm{MN})$$

using the ion currents as a measure of the vapour pressures, and, applying the second law,

$$\frac{\partial}{\partial 1/T}\log\frac{I_{\mathrm{MN}}^{+}T}{I_{\mathrm{M}}^{+}I_{\mathrm{N}}^{+}}=\frac{-\Delta H^{0}}{\boldsymbol{R}} \tag{104}$$

and to estimate the entropy using the third law.

Alternatively, if the dimeric species M_2 or N_2 can be observed, the constants for the reactions

$$\mathrm{M}_2+\mathrm{N}\rightarrow\mathrm{MN}+\mathrm{M} \tag{a}$$

and

$$\mathrm{N}_2+\mathrm{M}\rightarrow\mathrm{MN}+\mathrm{N} \tag{b}$$

can be approximated very closely by the ion current ratios

$$K_a=\frac{I_{\mathrm{MN}}^{+}I_{M}^{+}}{I_{\mathrm{M}_2}^{+}I_{\mathrm{M}}^{+}};\quad K_b=\frac{I_{\mathrm{MN}}^{+}I_{\mathrm{M}}^{+}}{I_{\mathrm{N}_2}^{+}I_{\mathrm{M}}^{+}}. \tag{105}$$

This procedure has the advantage that the instrumental factors relating to the spectrometer, and the geometrical factors relating to the position of the sample and of the orifice of the Knudsen cell in the furnace, do not appear in the equilibrium constant because of cancellation. It follows that a value of the equilibrium constants K_a and K_b can be obtained satisfactorily from measurements of each of the four ion currents in any one experiment for a given temperature, and even though the instrumental and geometrical factors may alter before the experiment at the next temperature, a consistent value of the equilibrium constant will again be obtained. This is not so for the case where only the monomeric species M^+ and N^+ are used.

One disadvantage of the procedure involving the dimers is that the pressure of a dimeric species is usually at least an order of magnitude smaller than that of the corresponding monomer, and hence when the signal for the M_2^+ reaches an easily detectable level, the pressure of the monatomic gaseous species is frequently nearing the upper limit for the conditions of Knudsen cell operation.

Clearly ratios of pressures for a given molecule can readily be obtained providing the geometric factor, which reflects the position of the Knudsen orifice with respect to the ion source, can be accurately reproduced.

As this is not generally the case, then an internal calibrating substance of known vapour pressure placed in the Knudsen cell with the unknown material is useful if the equilibrium remains undisturbed within the cell. Silver is very commonly used as an internal calibrating agent because of its well-established vapour pressure and relatively low

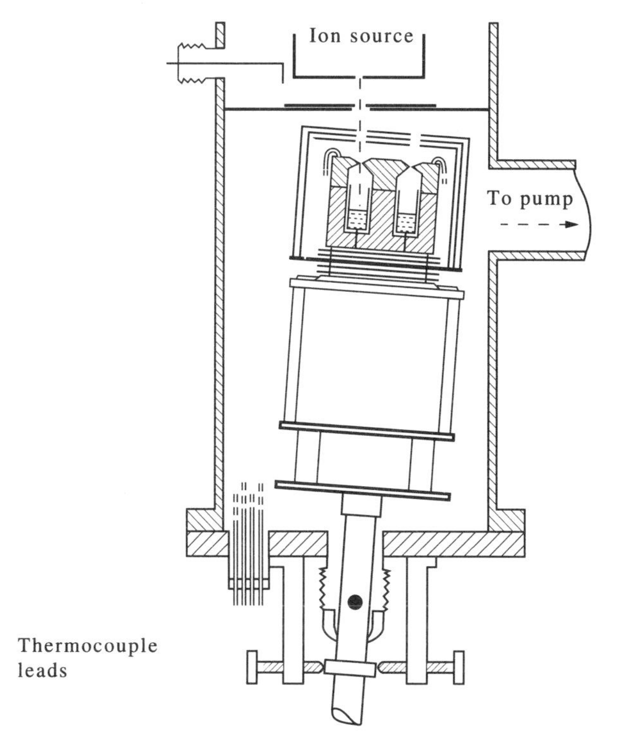

FIG. 69. Double-cell assembly for a Knudsen cell-mass spectrometer measurement with a calibrating standard material in one cell.

chemical reactivity. If interaction is unavoidable, a double-cell system can be employed, one cell of which contains the calibrating substance (Fig. 69).

When this has been considered, the product D is the remaining source of error in the mass spectrometric determination of absolute vapour pressures. This product can rarely be determined to better than a factor of two except in the trivial cases where the vapour pressure is already known in which case the product D is being measured. However, absolute pressure to within $\pm 100\%$ can be obtained in most circumstances when an unambiguous ionisation efficiency curve is at hand.

Finally, the limit of intensity is often set not by the detectability of the ion current but by the resolution of the required ion species from others of neighbouring mass. These latter can arise from a number of sources depending on the system, but one common source in high-temperature mass spectrometry is hydrocarbon fragments from vacuum-pump oil and other extraneous sources. The resolving power of the spectrometer can play an important part in deciding the lower limit of sensitivity, and the magnetic resolution machine normally has a considerably greater resolving power than the TOF machine.

A new high-temperature mass-spectrometer system (Fig. 70) incorporates all the advantages of modern components without the disadvantages associated with systems available on the market.

The vacuum chamber, which contains the Knudsen cell unit and the quadrupole mass-

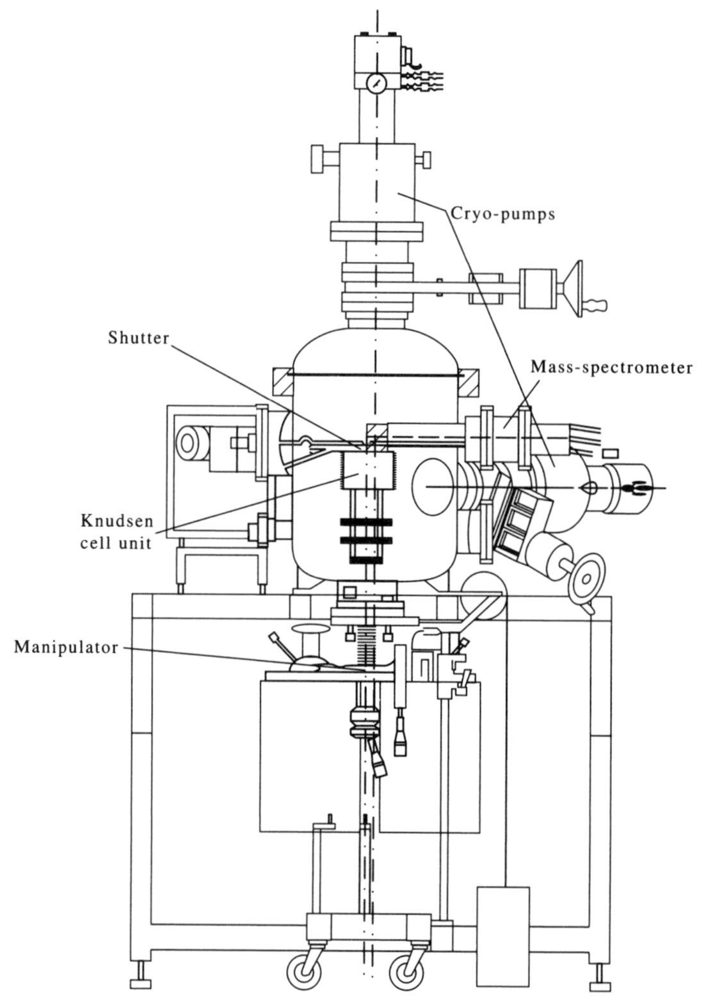

FIG. 70. Modern high-temperature Knudsen cell-mass-spectrometer.

spectrometer (mass range 1–1200), has a volume of about 100 l and is evacuated by means of adsorption and cryo-pumps (2 pumps, each with a capacity of 1500 l/min). The use of these pumps guarantees an oil-free residual gas in the apparatus. The working pressure is about 10^{-6} Pa. In choosing the mass-spectrometer, emphasis was placed on a high electron current density at low electron energy in order that a high ion current density at as low a fragmentation as possible be achieved. In addition, good resolution and excellent long-term stability are realized.

The Knudsen cell unit (Fig. 71) forms the heart of the apparatus. This can be heated inductively to about 2000°C. This method of heating largely obviates dissociation of gas molecules at hot wires as well as secondary reactions.

The Knudsen cell unit is mounted on a conventional manipulator which allows adjustment in an optimal position as well as rotation about the vertical axis. All necessary connectors for the temperature-measurement leads are also located on the manipulator.

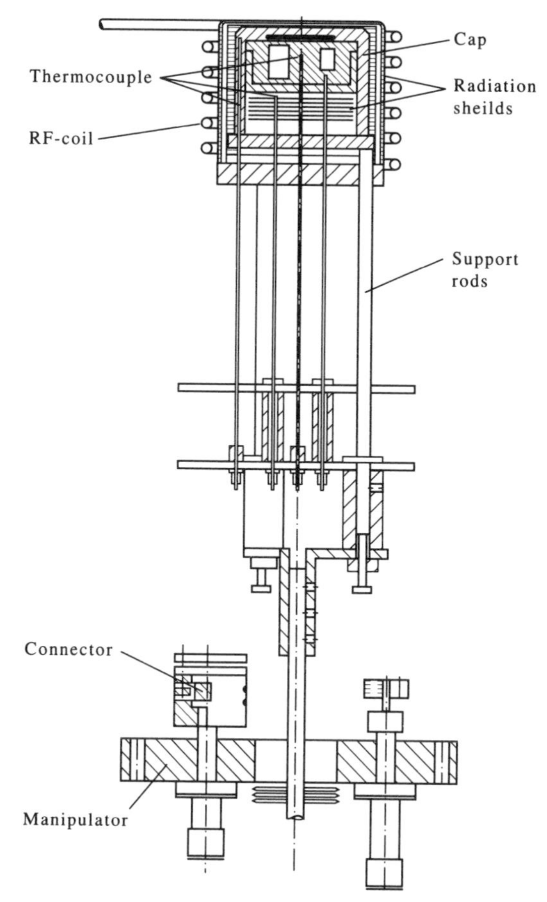

FIG. 71. Knudsen cell unit.

The Knudsen cell unit consists of a so-called "Revolver-holder" made of alumina, a molybdenum "cap" and various support and shield parts such as radiation shields, support rods, adjustment rings etc. The Knudsen cell assembly is provided with 8 thermocouples located in various positions, e.g. in the outer cap, on the base and on the lid of the cell. The revolver-holder enables various Knudsen cells to be used according to experimental requirements (Fig. 72a–d).

By use of the Revolver cells, it is possible to carry out direct quantitative measurements, since both the specimen and suitable reference materials can be placed in Knudsen cells and analysed mass-spectrometrically. The ratio of intensities of the species emanating from the reference and specimen materials corresponds to the ratio of the partial pressures.

Apart from its use in the elucidation of complex evaporation mechanisms and the approximate determination of absolute vapour pressures, one of the increasingly developing applications of the mass spectrometer in materials research is in the study of solution thermodynamics. There are two main routes along which this development is taking place, the common feature being the determination of activities from ion current

(a)

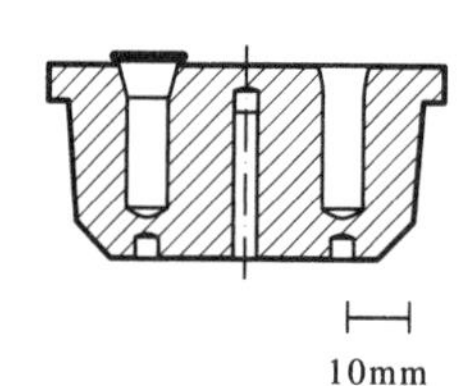

Simple 'Revolver' with 4 Knudsen cells constructed from a single material, e.g. copper, iron, graphite.

(b)

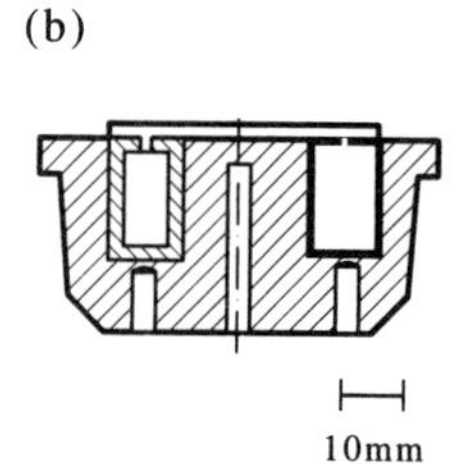

A 'Revolver' for use with Knudsen cells made of ceramic materials (e.g. Al_2O_3 or ZrO_2).

(c)

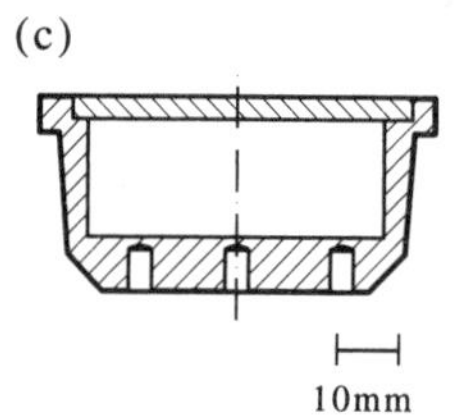

A 'Revolver' with a single large Knudsen cell and an orifice area: base area ratio of about 1 : 50,000.

(d)

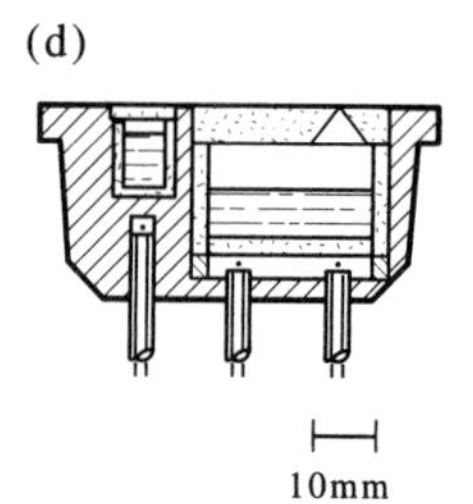

A 'Revolver' which contains one large measurement cell and one or more reference cells.

FIG. 72A, B, C and D.

ratios. The first technique makes use of the partial pressure ratios of gaseous polymers (1960b). For the metal M which yields the monatomic and diatomic species and the ions M^+ and M_2^+

$$a_M = \frac{p_M}{p_M^\circ} = \frac{I_M}{I_{M^+}^\circ}; \quad a_M^2 = \frac{P_{M_2}}{P_{M_2}^\circ} = \frac{I_{M_2^+}}{I_{M_2^-}^\circ}. \tag{106}$$

Hence

$$a_M = \frac{I_{M_2^+}}{I_{M_2^+}^\circ} \frac{I_M^\circ}{I_{M^+}}. \tag{107}$$

The ratio of the ion current I_M/I_{M_2} can normally be obtained independently of all instrumental and geometrical factors and hence if the ratio is measured over the pure element and then over the alloy, the activity is readily obtained. The complexity of the vapour phase is thus employed in the solution of thermodynamic problems instead of making them experimentally intractable.

The other technique which uses ion current ratios for the determination of activities was developed by Belton and Fruehan (1967b) and, independently, by Neckel and Wagner (1969j). The Gibbs–Duhem equation can be manipulated to produce a form which is of particular value in mass spectrometry thus; in an AB mixture

$$\log \gamma_B = -\int_1^{N_B} N_A \, d\left[\log \frac{a_A}{a_B} - \log \frac{N_A}{N_B}\right]$$

which, in the useful form for mass spectrometry, is

$$\log \gamma_B = -\int_1^{N_B} N_A \, d\left[\log \frac{I_{A^-}}{I_{B^-}} - \log \frac{N_A}{N_B}\right]. \quad (108)$$

This technique is most valuable for systems where the partial pressures of the two components over an alloy are approximately equal. The ion current ratio will then have a value around unity. Systems showing this kind of behaviour are those which are most difficult to resolve by traditional methods, and hence the mass spectrometer has its greatest potential for quantitative application in those very systems where its use is almost mandatory.

Very few systems have been studied by more than one group at the same temperature using what is ostensibly the same mass spectrometric method, but the liquid copper–germanium alloys have been measured by at least three groups (1970a, s; 1973b) and their results for the ion current ratios at one temperature and the resulting activities are shown in Figs 73a and b respectively. These graphs should give a representative picture of the state-of-the-art in alloy thermodynamics at the present time.

The copper–gallium system has been studied by this technique (1970a) and also by Laffitte and co-workers using a double cell assembly (1972b). Figure 74 shows the close agreement between activity measurements which can be achieved by such widely differing approaches. It should be clear from this discussion that mass spectrometric studies of activities in metallurgical solutions will probably become the method of preference in the future despite the many pitfalls to precise measurement which exist at present. An understanding of these in the particular context of high temperature studies is rapidly growing with experience. A review by Stafford (1971r) serves as a good guide.

3. The Langmuir Vaporisation Method

Where the lower pressure limit of application of the Knudsen method is usually about 10^{-4} mmHg, the method of Langmuir enables measurements to be made of pressures considerably lower than this. The Knudsen equation, multiplied by the vaporisation coefficient, can be used with samples with very much larger projected areas than the normal Knudsen orifice when this method is applied, but problems might be anticipated with the measurement of the temperature of the evaporating surface, which must be known. The Langmuir method is more often adopted to enlarge the rate of weight loss to a readily measurable quantity than to avoid difficulties associated with the availability of a

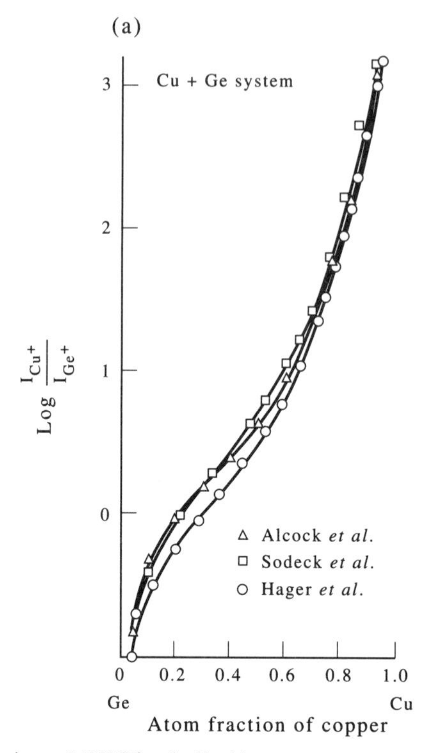

FIG. 73A. Ion current ratios at 1400°C for the liquid system copper–germanium from three separate mass spectrometric studies (1970a, s; 1973b).

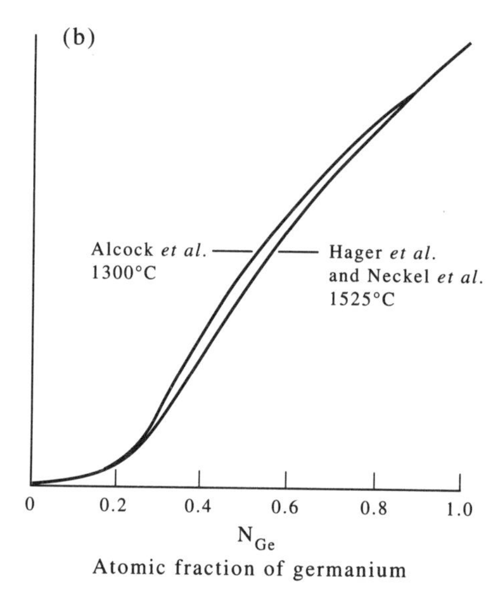

FIG. 73B. Activities for copper–germanium system at 1525°C obtained by Hager *et al.* (1973b) and Neckel *et al.* (1969j) and at 1300°C according to Alcock *et al.* (1970a).

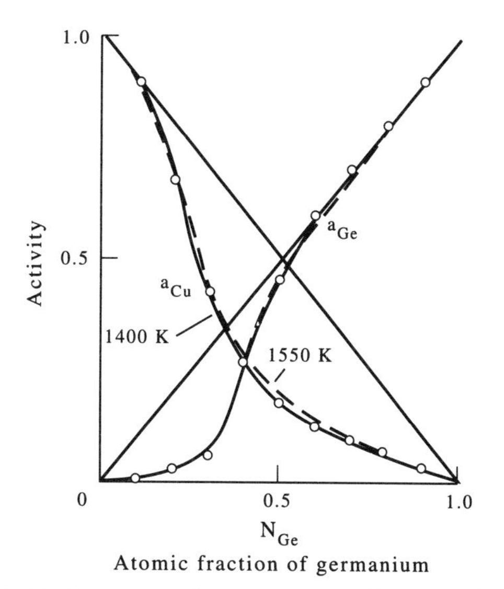

FIG. 74. Activities in the liquid copper–gallium system according to Carbonel *et al.* (1972b) and Alcock *et al.* (1970a).

suitable container for the Knudsen method. It therefore is most usefully applicable to substances having low vapour pressures at normal temperatures, and hence possessing high energies of sublimation.

Marshall *et al.* (1937d) obtained some of the first dependable results on metals using the apparatus shown in Fig. 75. They used metal samples in the form of an annular ring 22 mm o.d., 11 mm i.d., thickness 2.5–6 mm. These were heated by induction in an evacuated silica vessel, and the temperature was measured with an optical pyrometer. In measuring the vapour pressure of liquid copper, the metal was contained in a small annular trough of molybdenum.

A number of other studies have been made on pure metals by this technique or variants in which continuous weighing of the sample could be performed (1950c; 1960d) and the results have been in agreement with data from Knudsen measurements. This shows that the vaporisation coefficients of metals are unity to within experimental error. With oxide systems, in most cases a value of about 0.3 seems more appropriate for the solids. Burns (1966d) made a study of the vaporisation of Al_2O_3, Ga_2O_3 and In_2O_3 in both Langmuir and Knudsen conditions, using a mass spectrometer to analyse the vapours. He found that liquid Al_2O_3 and Ga_2O_3 showed unit vaporisation coefficients whilst the solids were all around 0.3. His analysis of the individual components of Al_2O_3 vaporisation, Al, O, O_2, Al_2O, and AlO being the vapour species, showed that all of these had about the same value for the vaporisation coefficients. Because the oxides are relatively poor thermal conductors it is tempting to follow Littlewood and Rideal's conclusions, made with low-temperature systems (1956e), that since metals are good thermal conductors and have a vaporisation coefficient of unity, then the apparent value of 0.3 for oxides which dissociate on evaporation, and hence have large enthalpies of vaporisation, is due to evaporative cooling of the surface. In order not to rely on temperature measurement on the freely evaporating surface, most workers employ a "black-body hole" in the sample for this

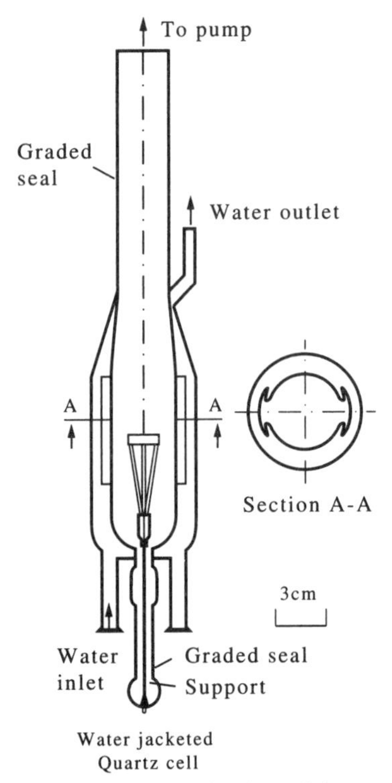

FIG. 75. Langmuir type apparatus for determining reaction pressures.

measurement. A 10–30° temperature drop accounts for a reduction of the vaporisation rate to a third when the enthalpy of sublimation is 400–800 kJ, and this probably exists near the surface of an evaporating oxide. The difference in thermal conductivities of these solids and metals at high temperatures is probably of the order of a hundred-fold.

Wolff and Alcock (19621) measured the free evaporation rates of alumina and thoria at temperatures around 2000°C. Their values for the vaporisation coefficient of Al_2O_3 were in agreement with that found by Burns, but the rate of evaporation of ThO_2 was quite in accordance with the Knudsen studies of Ackermann indicating α equal to one. The heats of vaporisation of Al_2O_3 to the predominant atomic species and of ThO_2 to ThO_2 gas, which is the principal species at 2000°C, are 3,033,400 and 661,072 J/mole respectively. It is probably this large difference which accounts for the difference of approximately three in the vaporisation coefficients of these two solids. The enthalpy of vaporisation of ThO_2 is not very much larger than those of some of the metals of established unit vaporisation coefficients, i.e. *ca*. 420,000 J/g-atom.

4. *Torsion Effusion*

The use of the Knudsen technique requires a precise knowledge of the vapour species under equilibrium conditions, but the pressure may be obtained directly from a measurement of the recoil momentum which is imparted to a Knudsen cell as a result of the effusion process. This measurement does not require a knowledge of the vapour species, and is valuable as a means for obtaining the total vapour pressure of a system in which a number of different species are present in the equilibrium vapour to a significant extent.

The method was first applied by Volmer (1931c) who used a two-orifice cell, the orifices in opposite directions on either side of a suspension fibre of known torsion constant. The size of orifice and the vapour-pressure range over which these may be used are subject to the same criteria as for Knudsen cells, and the vapour pressure may be calculated from the equation

$$p=2D\beta(q_1A_1+q_2A_2) \tag{109}$$

where q_1 and q_2 are the distances of the two orifices of area A_1 and A_2 from the axis of suspension. The torsion suspension fibre from which the double-orifice cell is suspended is turned through an angle β and has a torsion constant D. Useful materials for the suspension are quartz, tungsten and phosphor bronze, the last-named being preferred for strength and reproducibility.

Pratt and Aldred (1959e) constructed a torsion cell of graphite for the study of solid and liquid alloy systems. The suspension wire was of tungsten, and reliable measurements were obtained in the pressure range 10^{-8}–10^{-5} atm at temperatures up to 1000°C. Spencer and Pratt (1967n) constructed an apparatus for use at temperatures up to *ca.* 1800°C. Features of this equipment were the tantalum sleeve heater, boron nitride effusion cells, an electromagnetic control system for measuring the torque and a damping system for eliminating extraneous vibrations. Hildenbrand and Hall (1962c) calibrated their apparatus using gold as a vapour pressure standard, and this procedure is to be recommended. Searcy and Freeman (1955e) built an inductively heated apparatus having a graphite torsion cell which could be used up to 1900°C, but corrections for the torque effect of the radio frequency field on the cell had to be made. Peleg and Alcock (1966g) used an apparatus having a non-inductive doubly wound tungsten resistance heater for the study of the vaporisation of ceramic oxides up to 2500°C.

The torque which is produced by vaporisation can be measured directly, or in a null position by applying a measured current to solenoids mounted around a permanent magnet which is incorporated in the cold part of the suspension system. In their apparatus, Alcock and Peleg made use of a torsion head which could be smoothly rotated by means of a highly geared electric motor which enabled the whole suspension system to be rotated and returned to the null point. The angle of rotation was obtained by counting the number of revolutions required on the motor shaft with a revolution counter. The entire suspension system could be aligned in the vertical direction through a Wilson seal and horizontally by translation of the torsion head so as to place the orifice of the Knudsen cell in the centre of the furnace. The whole suspension was electromagnetically damped in the cold part of the apparatus (Fig. 76).

As long as the thermal requirements can be met to keep the torsion cell at a constant temperature, it is advantageous to keep the heating element and radiation shields as remote as possible from the Knudsen orifices. This is because atoms and molecules can rebound from the heater and radiation shields to the torsion cell, and depending on the precise geometric disposition of these furnace components with respect to the cell, an apparent torsion can be produced due to asymmetry in this rebound effect. The effect is more noticeable as the vapour pressure increases, and thus assumes its greatest significance at pressures where the technique might be considered in its most satisfactory range.

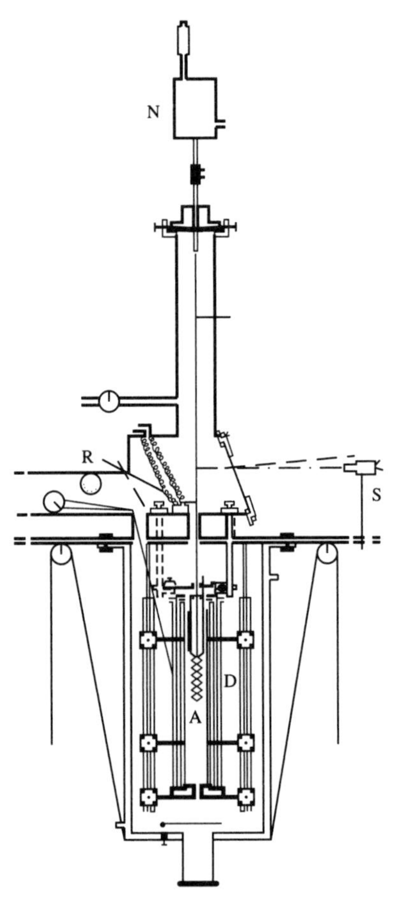

FIG. 76. Torque-effusion apparatus: A, non-inductively wound tungsten heater; D, radiation shields; S, rotation detection system; N, torsion head motor; R, electromagnetic damping.

5. Vapour Transpiration

The general limitation of the Knudsen method at the upper range of pressures results from the fact that the orifice diameter, which must be small compared with the mean free path in the vapour, becomes too small to manufacture. This practical limit corresponds in most systems to a vapour pressure of about 10^{-3} atm. The transportation method also has practical limitations at about this pressure, and there therefore exists a gap in the range 10^{-3}–10^{-2} atm between these two dynamic techniques and the direct pressure measurement techniques which were described earlier for use in the pressure range 10^{-2}–1 atm. A transpiration technique which was introduced by Gross *et al.* (1948e) could probably be very usefully applied in this intermediate range. The apparatus, which is shown in Fig. 77, consists of a closed end reaction tube in which the sample is held in the even zone of a furnace, a length of capillary tube which closes the open end of the reaction tube, and a condenser tube which receives the products escaping through the capillary from the reaction tube. The system is evacuated before being heated, and then the even zone of the reaction tube is brought up to the desired temperature. After a fixed time the

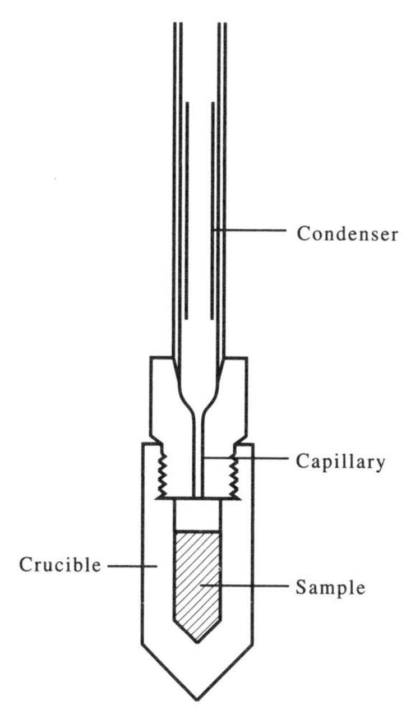

FIG. 77. Vapour transpiration apparatus of Gross *et al.* [268].

tube is cooled by removal of the furnace, and the contents of the condenser tube are analysed for the material which has evaporated from the sample. A number of experiments are made of varying duration, so that the appropriate allowance can be made for the amount of material which is lost by transpiration during the heating-up period. Since the capillary tube has a pressure drop from a few millimeters on one side where normal Poiseuille flow would occur, to a vacuum on the other side, where effusion would be the mode of gas motion, it is impossible to relate by means of a simple equation the mass of condensate which is collected per unit time to the vapour pressure of the substance within the reaction chamber. However, a number of standard substances are now available which have well-established vapour pressures over a wide pressure range. For example, the vapour pressure of lead has been established by the Knudsen and transportation techniques at pressures below 10^{-3} atm (1962a) and by static and boiling point methods from 10^{-2} to 1 atm (1925b; 1922b). Gross *et al.* were therefore able to use this element as a calibrating substance in the study of the equilibrium between aluminium and AlF_3 to form AlF. The calibration method can conveniently be applied with the standard substance incorporated with the system of unknown pressure provided no significant chemical interaction can take place.

Vapour transpiration has also been used in conjunction with levitation melting to measure vapour pressures of alloy systems at very high temperatures where the use of a container would be almost impossible because of severe chemical interaction. Mills and Kinoshita (1973k) have reported activity measurements on liquid titanium–vanadium alloys in the temperature range 1800–2000°C. The system, which showed slight positive departures from ideality, was the second to be studied by this technique, and no comparable results are available to assess the accuracy. The technique was established, however, in an earlier study of the iron–nickel liquid system in the temperature range

1900–2300°C (1972g). This system had already received considerable study by a variety of methods; Zellars *et al.* used the gas transportation method (1959g), Speiser *et al.* used the Knudsen method (1959f) and Belton and Fruehan (1967b) used the ion current ratio version of mass spectrometry. The system can therefore be said to be reasonably well documented. In the levitation–transpiration technique a sample of the alloy of known composition is levitated in helium gas contained in a water or air-cooled quartz tube. The temperature of the alloy bead can be controlled by the opposing heating effect of the electromagnetic field, and the cooling effect of the helium, which is a function of the gas flow rate past the sample. The principle of the method is that the flow of metal atoms from the hot sample to the cold walls, where they condense, occurs across a gas–metal boundary layer which is common to both species. Therefore, in the steady state

$$\frac{J_{Fe}}{J_{Ni}} = \frac{D_{Fe}}{D_{Ni}} \frac{P_{Fe}}{P_{Ni}}$$

where J is the flux of atoms, D the diffusion coefficient of the metal atoms in helium and P the equilibrium vapour pressure of the component at the sample temperature.

The flux ratio can be obtained by analysing the condensate on the quartz walls for iron and nickel. Although a number of theoretical treatments are in existence, it is probable that no serious error would be involved in assuming that the diffusivity ratio is equal to one. Then the activity ratio is obtained from the equation

$$\frac{N_{Fe}}{N_{Ni}} = \frac{a_{Fe}}{a_{Ni}} \frac{p^{\circ}_{Fe}}{p^{\circ}_{Ni}}$$

where N_{Fe}/N_{Ni} is the ratio of the atom fractions of iron to nickel in the condensate. Speiser *et al.* produced the form of the Gibbs–Duhem equation which should be used in conjunction with this technique to obtain the single activities in the liquid phase, viz.

$$\ln a_{Fe} = \int_{N_{Fe}=1}^{N_{Fe}} \frac{N^{1}_{Ni}}{N_{Ni}} \, d \ln N_{Fe} = \int_{N_{Fe}=1}^{N_{Fe}} \frac{N^{1}_{Ni}}{N_{Ni} N_{Fe}} \, d N_{Fe}$$

where N^{1}_{Ni} refers to the atom fraction of nickel in the liquid phase and the other atom fractions are for the condensate.

The results which were obtained in this study were quite consistent with the extrapolation of the earlier data to the higher temperature region. The high temperature which was used was the result of the necessity to balance the heating and cooling effects in the system. It is possible that improved design of the levitation system would make it possible to use this system in the more conventional temperature range if so required.

3. Electromotive Forces

The use of e.m.f. measurements to derive thermochemical data has been discussed on pp. 8, 18, 27–30, and 33. The chemical reaction to be investigated must be capable of being harnessed in a galvanic cell in such a way that its energy produces electromotive force. The principal problem which faces the experimentalist who wishes to apply the galvanic cell technique in order to measure thermochemical quantities is to find a suitable electrolyte for the cell which is envisaged. The second most important aspect of these measurements is the identification of a single electrode process which occurs reversibly at each electrode.

Consider a cell in which the oxygen dissociation pressure of a metal/metal oxide system is to be measured. This could consist of one electrode made of a platinum sheet attached to a pelletised mixture of the metal and its oxide in powder form, and the other electrode could be oxygen gas at 1 atm pressure in contact with platinum. The cell would be represented diagrammatically thus:

$$\mathrm{Pt}\langle \mathrm{M, MO}\rangle|\mathrm{electrolyte}|(\mathrm{O_2},\ 1\ \mathrm{atm})\mathrm{Pt}.$$

The electrode reactions will then be at the right-hand electrode

$$\tfrac{1}{2}\mathrm{O_2} + 2\mathrm{e}^- \rightarrow \mathrm{O}^{2-}$$

and at the left-hand electrode

$$\mathrm{M} + \mathrm{O}^{2-} \rightarrow \mathrm{MO} + 2\mathrm{e}^-$$

giving the overall cell reaction

$$\tfrac{1}{2}\mathrm{O_2} + \langle \mathrm{M}\rangle \rightarrow \langle \mathrm{MO}\rangle.$$

The cell

$$\langle \mathrm{Ag}\rangle|\langle \mathrm{AgBr}\rangle|(\mathrm{Br_2}),\ \mathrm{C}$$

has the electrode reactions

$$\mathrm{Ag} \rightarrow \mathrm{Ag}^- + \mathrm{e}^-$$

and

$$\tfrac{1}{2}\mathrm{Br_2} + \mathrm{e}^- \rightarrow \mathrm{Br}^-$$

with the overall cell reaction being

$$\mathrm{Ag} + \tfrac{1}{2}\mathrm{Br_2} \rightarrow \mathrm{AgBr}.$$

In these two typical examples there is one electrode at which electrons would be consumed—the cathode—and one electrode at which electrons are produced—the anode—when the cell reaction proceeds. If the electrolyte system conducts electricity by the migration of ions only, then the electrons must move from cathode to anode via an external electronic conductor. It is along this conductor that the electric potential which is generated by the cell reaction can be measured. When the resistance of this conductor is very high, the e.m.f. E is obtained, the so-called "open-circuit" potential. If the electrode components are all present at unit thermodynamic activity, the e.m.f. is the standard e.m.f. which is related to the standard Gibbs energy change for the cell reaction by the equation

$$-\Delta G^\circ = zFE^\circ. \tag{31}$$

Here, z is the number of electrons which are found in the separate electrode reactions which, added together, make the cell reaction. Thus for those cells shown above, z has the values 2 and 1 respectively.

In most practical high-temperature systems the so-called electrolyte usually has a small component of non-electrolytic conduction. The cell is never then in the open-circuit state because electrons can migrate through the electrolyte and permit the cell reaction to proceed at a speed which is not determined by the resistance of the external electronic conductor. A real electrochemical system therefore has a potential which is given by the equation

$$E = \frac{RT}{zF} \int_{a_i'}^{a_i''} t_i \, \mathrm{d} \ln a_i \tag{110}$$

where t_i is the transport number of the ionic species which is conducted through the electrolyte having the thermodynamic activity a_i' at one electrode and a_i'' at the other. When the "electrolyte" conducts significantly by electron migration, then the cell reaction can proceed spontaneously and one electrode is depleted of atoms whilst the other receives atoms.

It follows that the electrodes can only be maintained at a constant chemical potential of all components providing that equilibrium can be established more rapidly than the rate of arrival or loss of material at the electrode–electrolyte interface. Consider a cell which has two metal–metal oxide electrodes with a "leaky" electrolyte where t_{ion} is less than one. The electrode with the higher oxygen dissociation pressure can lose oxygen through the electrolyte by the arrival of electrons *through* the electrolyte to form oxygen ions, and hence the oxide is reduced at this electrode. At the other electrode oxygen ions are absorbed and the electrode is oxidised continuously. Providing the electrode reactions can accommodate this 'corrosion reaction" as a result of metal–oxygen reactions at the electrodes which restore the original mixture *at the electrode–electrolyte interface*, the cell can be used for thermodynamic studies. Failing this, the e.m.f. will drift continuously and the results are difficult to interpret. At the higher temperature, solid electrolyte systems can be permeable to gases, and interactions between the electrodes and their surrounding gaseous atmospheres can also play an important role in making experimentation difficult.

The porosity of a solid electrolyte depends very much upon the manufacturing method as well as on the temperature of the cell operation, and therefore no general description can be given of the conditions under which this effect could become significant. The effects of atmospheric reactions with electrode systems can be minimised, or eliminated to all practical extent, by providing samples of the electrode to the gas phase for pre-equilibration of the gas approaching the cell electrode and by separating the atmospheres surrounding the electrodes from one another.

The corresponding phenomena in molten salt electrolytes which can lead to drifting e.m.f.'s, are reactions which result from the solubility of metals in the molten electrolyte, which confers electron-conducting properties on the electrolyte, or the possibility of cations of more than one valency for a given cation in the melt. Both mechanisms can lead to the transport number of the ions being less than unity and to oxidation/reduction reactions at the electrodes.

In all electrochemical studies for the determination of thermodynamic properties it is therefore important to ascertain the conditions under which the electrolyte shows the closest approximation to the ideal system and to attempt to work in the region of best performance. It is in solid electrolyte systems, especially those in which oxide ions are the rapidly moving species, that the most complete study of the factors contributing to the ionic transport number have been carried out. By comparison, the study of molten salt systems is more complicated since this involves the possibility of the migration of a number of, usually, cationic species, and the results are specific for a given cation. A discussion of liquid and solid electrolytes of various types is presented in the following sections.

A. Liquid Electrolytes

1. Aqueous Solutions

The construction of cells used for e.m.f. measurements with aqueous solutions is relatively simple, and many have been described in the literature. A number of these measurements were carried out, for instance, by Japanese workers (Ishikawa, Shibata, Ueda, Watanabe) in the period 1930–5, most of their work being published in scientific reports of the Tôhoku Imperial University. The cell, usually of glass, consists of a simple H-shaped tube or of two beakers connected by a bridge containing the solution. In this work great attention must be given to the purity of the substances used (1940a, 1939a). Mercury should be distilled several times before use, amalgams should be melted under hydrogen, and salts should be recrystallised several times. Dakin and Ewing (1940a) have observed that equilibrium is attained much more readily when preparing electrodes (such as $\{Hg\}|\langle Hg_2Br_2\rangle|[KBr]_{aq}$) in the absence of air than otherwise. Bates (1939a) has pointed out the importance of ensuring that compounds are in their most stable state to avoid the occurrence of undefined transitions during experiment.

Richards and Conant (1922c) also devised a useful technique for investigating liquid alloys, such as sodium amalgams, which oxidise rapidly in contact with an aqueous electrolyte (in their case a sodium hydroxide solution). The two amalgams investigated drop continuously from two capillary tubes, thereby maintaining a practically fresh surface in contact with the electrolyte. (A similar device is used in the Heyrovsky type polarograph so extensively used nowadays for the routine analysis of alloys.)

For the measurement of the e.m.f.'s of *solid* alloys against mercury with an aqueous electrolyte. Ölander (1937e) used the cell shown in Fig. 78. It was 2.5 cm in diameter and was supplied with hydrogen to prevent oxidation. Four electrodes at a time were mounted in the cell.

2. Molten Salt Electrolytes

The electrolytes frequently used for high temperatures consist of alkali chlorides in which a small amount of a salt of the transported metal is dissolved. Eutectic mixtures of lithium and potassium chlorides (m.p. 359°C) and lithium and rubidium chlorides (m.p. 312°C) are commonly employed. For still lower temperatures the eutectic mixture of sodium and potassium acetates (m.p. 233°C) has been recommended. This however decomposes with charring at 300°C.

Careful attention must be paid to the choice and preparation of these electrolytes. When hygroscopic salts, such as lithium chloride, are used, they must be thoroughly dehydrated to avoid trouble arising from moisture of the formation of oxy-salts (e.g. (1927b). Elliott and Chipman (1954a) added 0.6% fused potassium hydroxide to a KCl–LiCl electrolyte to neutralise the hydrogen ions which might have been formed from hydrolysis by moisture in the salts. If the metal of the electrolyte is stable in more than one valency state, only one of these must be present. This may be ensured by the choice of a salt mixture which forms suitable complexes.

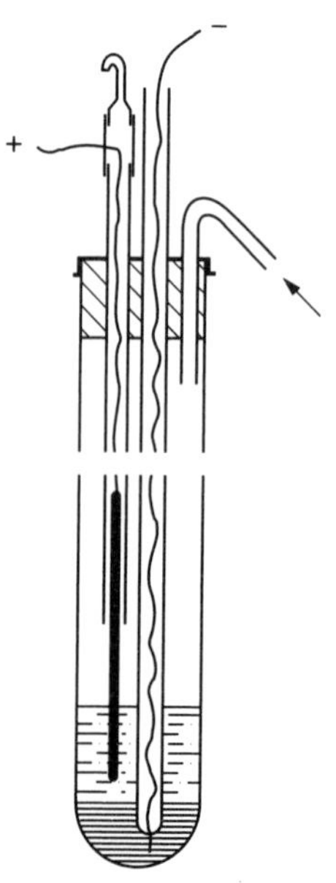

Fig. 78. Simple cell for e.m.f. measurements [37 Öla].

Weibke and v. Quadt (1939h) developed the apparatus shown in Fig. 79 to permit a carefully prepared molten salt mixture to be brought into contact with the electrodes without risk of contamination. The apparatus was of resistance glass ("Supremax"). The salts are purified and melted in the reservoir on the right, a vacuum being applied at the same time. The cell is then tilted and the vacuum released in order to drive the liquid into the cell proper, which is then sealed off at the constriction.

The molten salt electrolytes (halide or acetate mixtures) caused considerable trouble in the experiments of Ölander and Weibke, even when they were carefully prepared, i.e. dehydrated and degassed. The alkali chlorides were volatile and condensed on the electrodes and conductors and attacked other parts of the apparatus. Thermo-e.m.f.'s may be set up when there is a temperature gradient along the electrodes. These cause localised electrolysis resulting in a change in the composition of the alloy. Weibke and Matthes even observed in some experiments (1941e) a tapering of the alloy electrode with a deposition above the taper of a garland of crystals of the more electronegative metal. They obtained a more uniform temperature distribution along the electrodes by enclosing the cell inside their furnace in an iron or copper cylinder. The use of shorter electrodes also diminishes the thermoelectric effect.

The e.m.f. of halide formation at high temperatures was investigated by Lorenz and Velde (1929d) in a cell (Fig. 80) based on the scheme

$$\langle C\rangle\{M\}|\{MCl_x\}|(Cl_2)\langle C\rangle.$$

The electrolyte consisted of the fused chloride of the metal. The graphite electrodes were sealed into glass to prevent their combustion; one dipped into the molten metal and the other into the molten salt. To prevent diffusion of chlorine through the melt this latter electrode was surrounded by another wider glass tube. A special arrangement was used for

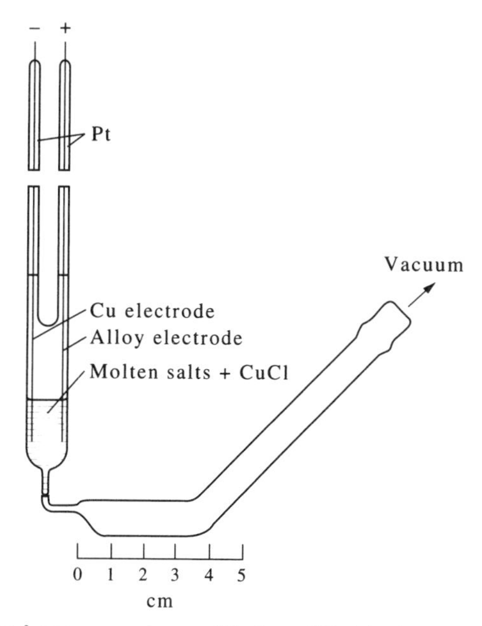

FIG. 79. Cell for e.m.f. measurements on solid alloys. Side tube for preparation of electrolyte.

magnesium cathodes. In order to saturate the anode with chlorine a polarizing current (0.2–0.3 A, 120 V) was applied, while chlorine gas was passed into the cell. Saturation took as much as 8 days and more. When saturation was complete, the polarising current was switched off and the e.m.f. measured. When all the necessary precautions had been taken the resulting thermodynamical data proved to be very accurate. The values obtained were somewhat different when the anode and cathode compartments were separated by a glass diaphragm, even though the electrolyte on either side was of the same composition.

According to Lorenz no junction potentials arise between the chlorides of different metals, so that the e.m.f.'s of cells (1934e, 1930e), of the type

$$\{Zn\}|\{ZnCl_2\}|\{PbCl_2\}|\{Pb\}$$

can be used additively in calculations. This does not, however, apply to the e.m.f.'s of cells containing mixtures of liquid salts, or to cells in which the liquid chlorides are separated by a glass diaphragm (1934e).

Rose *et al.* (1948g) reported a series of e.m.f. measurements on the reducibility of chlorides, oxides and sulphides of metallurgical interest. They pointed out that two serious sources of difficulty were hydrolysis of fused salts by atmospheric moisture, and the tendency of molten metals to disperse in a "fog" in the fused electrolyte. Both these troubles appeared to have been obviated to a large extent in the experiments on chlorides by passing a stream of dry hydrogen through the salt. The method of measuring e.m.f.'s consisted in taking current/voltage measurements and extrapolating to zero current. They were able to verify Lorenz's results for the cell

$$Ag|AgCl \vdots PbCl_2|Pb$$

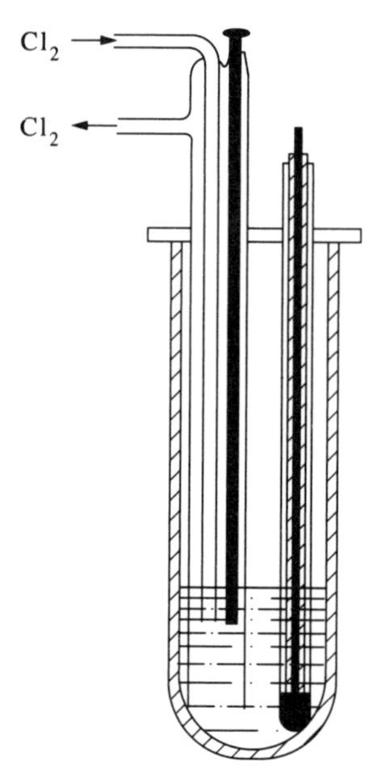

FIG. 80. Cell for e.m.f. measurements on molten chlorides (Lorenz).

and their results for the cell

$$Sn|SnCl_2 \vdots PbCl_2|Ag$$

gave a linear e.m.f./temperature relationship whereas Lorenz's results had been scattered.
The oxide cells investigated were:

$$O_2(Ag \text{ or } Pt)|NaOH, SnO_2|Sn$$

$$O_2(Ag)|NaOH \vdots NaOH, PbO|Pb.$$

The gas electrode was built to the same pattern as those used for measurements on aqueous solutions, but together with the containing vessel was made of non-siliceous sintered alumina. The oxygen electrode was surrounded only by molten caustic soda in the second cell in view of the tendency of PbO to become oxidised easily. Throughout these experiments allowance was made for thermo-e.m.f.'s set up at junctions between the electrode metals and other metals leading to the measuring instruments.

B. Solid Electrolytes

1. Glass Electrolyte

The first application of a solid electrolyte to studies of metallic systems is represented by the work of Hauffe (1940c), who used glass as a solid electrolyte in the electrochemical investigation of the molten Na–Hg and Na–Cd systems at 300–400°C. His apparatus is shown in Fig. 81. According to Warburg (1884) sodium ions in glass carry electric current. Faraday's law being obeyed. By using a glass of composition 70.2% SiO_2, 9.5% CaO, 14% K_2O, 6% Tl_2O, 0.3% As_2O_3 as the electrolyte, Vierk (1950l) was able to apply

Hauffe's apparatus to the investigation of liquid thallium alloys, e.g. thallium–tin. In this case, however, equilibrium was attained more slowly, owing obviously to the smaller diffusion rate of thallium ions in glass relative to that of sodium ions. Nevertheless the Hauffe–Vierk cell is probably the most suitable for the study of alloys containing sodium, potassium, thallium and possibly certain other metals as the less noble constituent.

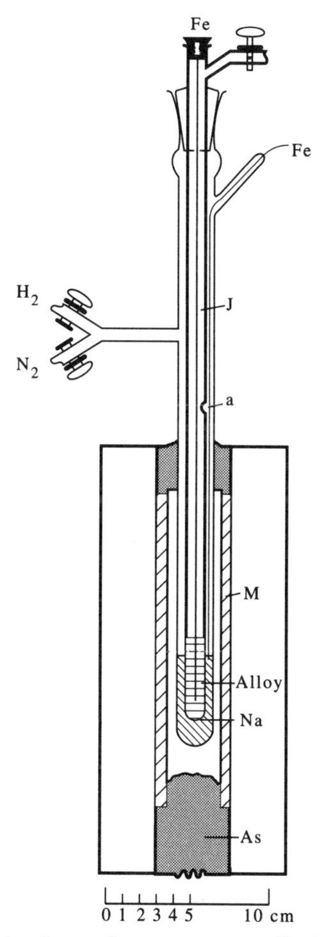

FIG. 81. Cell with glass electrolyte for e.m.f. measurements on liquid alloys. I = inner glass vessels, H to high vacuum, N to nitrogen supply, a = openaing to inner vessel, M = silumin tube.

Kubaschewski and Huchler (1948f) used solid glass as the solvent for silver ions in work on the system silver–gold. It was established that silver dissolves in glass in the ionic form (1936d). The type of apparatus used for such experiments is influenced, of course, by the low conductivity of the glass. That used by Kubaschewski and Huchler is shown in Fig. 82. A thin silver disc (3) is sealed on to the plate of glass (4) (0.12 mm thick) and is placed in contact with a more massive piece of silver (2) set in a ceramic cup (8) of 1.9 cm diameter. The gold–silver alloy, also in the form of a disc (5) is pressed on to the upper surface of the glass by an alumina tube (10) which also contains the thermocouple (11). The platinum conductors (1, 6) are welded to the alloy piece and to the lower piece of silver. The resistance of the cell described was 2500 ohms at 600°C. Great care was taken with this apparatus to exclude oxygen and to avoid small extraneous currents. The results were found to be quite reproducible and reliable.

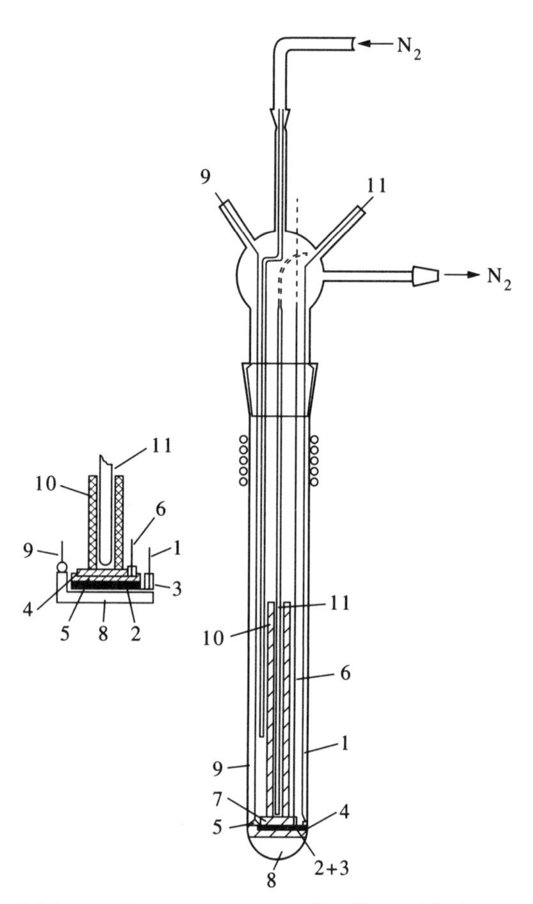

FIG. 82. Cell for e.m.f. measurements on solid alloys, with glass electrolyte.

2. Solid Oxide Electrolytes

Kiukkola and Wagner (1957c) aroused renewed interest in oxygen ion conducting solids with the fluorite structure when they reported results for simple cells of the type

$$\mathrm{Pt}\langle \mathrm{Ni, NiO}\rangle|\mathrm{CaO{-}ZrO_2}|\langle \mathrm{Fe, FeO}\rangle \mathrm{Pt}$$

in which the cell reaction is

$$\langle \mathrm{NiO}\rangle + \langle \mathrm{Fe}\rangle = \langle \mathrm{FeO}\rangle + \langle \mathrm{Ni}\rangle.$$

Since these studies a number of other solid oxide electrolytes have been studied in detail, the most important property being the nature and magnitude of the electrical conductivity as a function of both oxygen pressure in the surrounding atmosphere and temperature.

At a given temperature the general features of all systems show a region of semi-conduction by electron migration (n-type $\ominus$) at low oxygen pressures then follows a region of electrolytic conduction ($t_{ion} \simeq 1$), and finally at high oxygen pressures, another region of semiconduction develops with positive hole (p-type $\oplus$) conduction. All three mechanisms must be considered as operative at a given oxygen pressure, and we can merely discuss the regions of oxygen pressure where one mechanism may predominate with the other mechanisms playing a very minor role. Under any condition, the total conductivity is related to the number n_i, charge $|z_i e|$, and mobility μ_i of each species by the equation

$$\sigma = \sum_i n_i |z_i e| \mu_i = \sum_i \sigma_i$$

where σ_i is the partial conductivity due to species i. It is usually found that the semi-conduction components of the conductivities of oxides follow, approximately, a simpler power law relationship to the oxygen pressure. Thus, after Schmalzried (1962j).

$$\sigma_\ominus = p_\ominus p_{O_2}^{-1/n}; \quad \sigma_\oplus = p_\oplus p_{O_2}^{-1/m} \tag{111}$$

where $p_\ominus$, $p_\oplus$, n *and* m are constants for a given system. The ionic conductivity is independent of the oxygen pressure, and is related to the diffusion coefficient of oxygen in the solid by the Nernst–Einstein equation in the form

$$\frac{\sigma_{ion}}{D_{O_2-}} = \frac{n_{O^{2-}} 4e^2}{kT}. \tag{112}$$

It follows that the range of oxygen pressures over which electrolytic behaviour will extend should be higher when the diffusion coefficient of oxygen is high, and when the change of cationic valencies which can introduce n-type (reduction) and p-type (oxidation) is not readily brought about. Stable cationic valency in the principal cations is therefore an important contribution to the achievement of satisfactory electrolytes, and the presence of impurities which have variable valencies, e.g. Fe^{2+}, Fe^{3+} should be kept to a minimum. Because of the extreme difficulties associated with the preparation of samples, the control of atmospheres at high temperatures and the measurement of electrical conductivities, the data for the electrolyte systems show some scatter. It is quite well established (1965e) that the ionic conductivities of zirconia-based electrolytes are about an order of magnitude greater than those of the corresponding thoria-based electrolytes. The ionic transport number remains sensibly unity in the zirconia-based electrolytes up to 1 atm oxygen pressure, but is about 0.6 in the thoria-based electrolytes at 1000°C at 1 atm pressure of oxygen. At the lower limit, the zirconia-based electrolytes begin to show n-type semiconduction at p_{O_2} equal to 10^{-20} atm and 1000°C, whereas the thoria-based electrolytes function satisfactorily down to 10^{-30} atm. It is thus quite apparent that zirconia is the superior base material for oxygen potential measurements when the oxygen partial pressure is high in a given electrode, but thoria is to be preferred at the lower oxygen pressures. If a cell is made in which one electrode has a high oxygen partial pressure and the other side has a low oxygen partial pressure, it is possible to work with two electrolytes in series thus;

$$Pt(O_2, 1\ atm)|ZrO_2\text{–}CaO||ThO_2\text{–}CaO|Cr\text{–}Cr_2O_3|Pt$$

has been found to be a useful system up to 1200°C.

In these systems the activation energy of the ionic electrical conductivity is typically 105–125 kJ/mole and thus the change in conductivity in the typical temperature range of study, 600–1600°C, is found from the Arrhenius equation to be about 4 orders of magnitude with ZrO_2–CaO electrolytes having a specific conductivity of about 1 ohm^{-1} cm^{-1} at 1000°C.

Solid Electrodes in Oxygen Concentration Cells. A number of reference electrode systems have been established for oxygen concentration cells, and data for most of these have been collected in a review article by Rapp and Shores (1970p). This shows that a reference system is established at every decade of oxygen pressure from 1 to 10^{-18} atm at 1000°C, with three systems at 10^{-24} atm (Mn–MnO, NbO–NbO_2 and Ta–Ta_2O_5). It is therefore possible with our present fundamental knowledge to carry out a number of electrochemical studies in which oxide electrolyte galvanic cells can be used to obtain thermodynamic data.

The importance of having a number of reliable reference electrode systems available is that the correct one may be chosen which most nearly matches the oxygen potential of the unknown system under investigation. By minimising this difference, and hence the e.m.f. of the cell, the effects of any electronic short circuits within the electrolyte will also be minimised. Furthermore, it would be possible in many circumstances to reverse the polarity of the cell such that the unknown electrode is the cathode with one reference electrode and the anode with another reference. In this way the electrode can be checked for reversible behaviour. In this connection it has been found more recently that the imposition of a small a.c. perturbation is useful in equilibrating an oxide-electrolyte interface (1975b). This potential must be removed during the measurement of the e.m.f.

The assembly of equipment for making such measurements covers a wide range of possibilities, as might be expected, and only a few typical instances will be given here. The apparatus used by Steele and Alcock (1965e) to establish a number of reference electrodes, maintained a high vacuum around pellets of electrolytes and electrodes stack sequentially, and held under compression in an alumina assembly (Fig. 83).

Charette and Flengas (1968b, c) used an electrolyte in the form of a closed end tube. The tube was evacuated on the inside, which contained a metal–metal oxide pellet pressed against the bottom of the tube and a compacted mixture of metal and metal oxide powders or pure oxygen gas contacted the base on the outside of the tube (Fig. 84). The stated reproducibilities of e.m.f. in both studies were ± 1 mV for nearly all of the cells which were measured.

The pioneering study made by Kiukkola and Wagner, in which the stack of electrode and electrolyte pellets was exposed to a single flow of inert gas, and these results agreed well with the other studies.

This general state of affairs is by no means typical, and Rapp (1963f) for example, found in a study of the Mo–MoO_2 system using the "open-stack" system with Ni–NiO and Fe–FeO reference electrodes, that the e.m.f. appeared to be very gas-flow-rate dependent using helium as an "inert" atmosphere, and therefore he redesigned the cell to eliminate this effect. The electrolyte was formed into an H-shape (Fig. 85) so that electrode pellets could be recessed into the electrolyte and protected from parasitic reactions with the gas phase.

The most critical test of the satisfactory separation of electrodes from mutual interference is when one electrode is an alloy mixed with an oxide of one of the components, or a single phase of variable composition. Any transfer of oxygen to or from

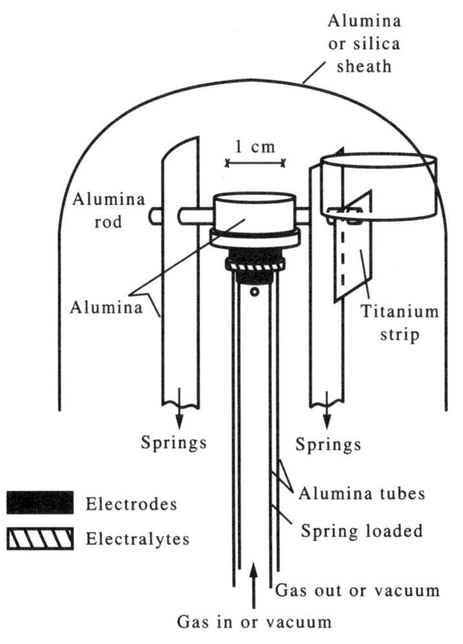

FIG. 83. Vacuum apparatus used by Steele and Alcock for metal–metal oxide electrodes.

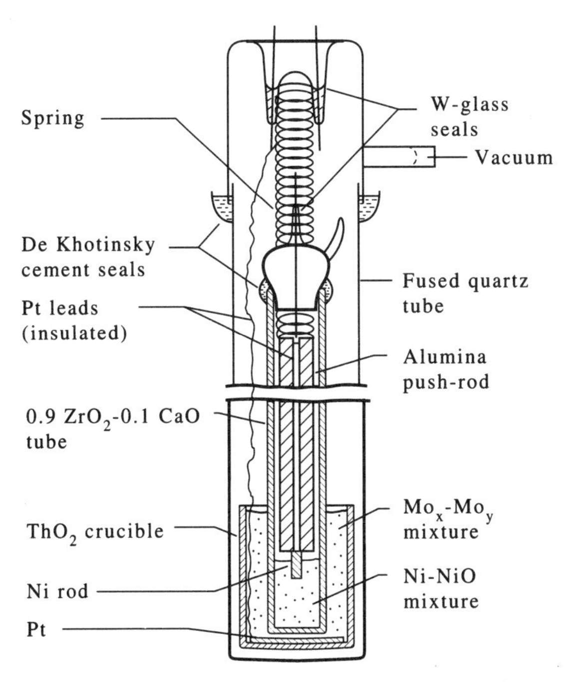

FIG. 84. Evacuated apparatus used by Charette and Flengas (1968a, b) incorporating a stabilised zirconia electrolyte tube for study of metal–metal oxide electrodes.

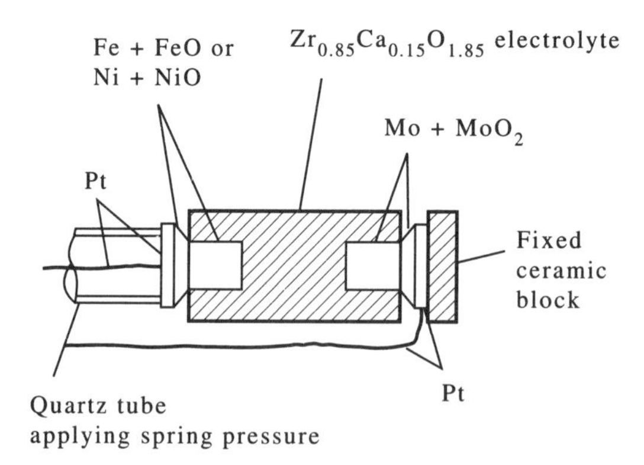

FIG. 85. Cell assembly used by Rapp (1963f) to minimize gas-electrode interactions.

an alloy electrode either via the gas phase or the electrolyte must necessarily involve a change in the composition of the alloy, and hence cause a drift in the e.m.f. of the cell. Dench and Kubaschewski (1969b) report a successful study of the cobalt–copper solid alloy system in which the first step was to eliminate the effects of thermal gradients by first placing the electrodes within the even zone to yield nearly zero for the e.m.f. of the symmetrical cell

$$\mathrm{Pt|Co,\ CoO|ZrO_2.CaO|Co,\ CoO|Pt.}$$

A precision of ± 0.2 mV could then be obtained from the cell

$$\mathrm{Pt|Co,\ CoO|ZrO_2\text{–}CaO|Co\text{–}Cu,\ CoO|Pt}$$

with the electrodes and electron leads disposed within the furnace even zone in the light of this knowledge.

The results for the cell

$$\mathrm{Pt|Cu,\ Cu_2O|ZrO_2\text{–}CaO|Ni,\ NiO|Pt}$$

agreed very well with those previously reported by Steele and Alcock (1965e) and by Rizzo *et al.* (1967m).

The effects of a temperature gradient within an electrochemical cell are to add another source of e.m.f. to that determined by the electrodes. The Seebeck coefficient which measures this effect for unit temperature gradient has been studied mainly with gas electrodes and with symmetrical cells. The indications are that a finite Seebeck coefficient must be anticipated when oxide electrolytes are used and that this coefficient will be a function of the absolute magnitudes of the oxygen potentials of the electrodes. Fischer (1967c) concluded from his study at relatively high pressures, 10^{-2}–1 atm of oxygen that the Seebeck coefficient for the ZrO_2–CaO electrolyte can be described by the equation

$$\frac{\Delta V}{\Delta T} = 0.46 + 0.0219 \ln \frac{159}{p} \text{ mV} \qquad (113)$$

where p is the oxygen pressure in Torr, and 159 is the partial pressure of oxygen in air.

Goto *et al.* (1969f) concluded that when air or oxygen was the gas in contact with the electrodes in a symmetrical cell, the coefficient α was 0.095 ± 0.005 mV/°C for ZrO_2–CaO and 0.050 ± 0.005 mV/°C for ThO_2–CaO if it can be assumed that the e.m.f. of a cell can be expressed as

$$E = \frac{R}{AF} \int_{T_1, p'_{O_2}}^{T_2, p_{O_2}} t_{O^{2-}} \, dT \ln pO_2 + \alpha(T_2 - T_1) \qquad (114)$$

(p'_{O_2} was the same value at each electrode). It should be remarked here that whereas t_{ion} was unity for the zirconia-based electrolyte in this study, the value for the thoria-based electrolyte was less than unity.

Ruka *et al.* (1968h) studied the e.m.f.'s which are generated across a ZrO_2–CaO electrolyte in the temperature range 700–1280°C, and oxygen pressures from 1 to 10^{-18} atm. They analysed the results using irreversible thermodynamics and considered the relationship

$$\frac{\Delta V}{\Delta T} = \frac{1}{F}\left[\frac{1}{2}\Delta\bar{\bar{S}}_{O^{2-}} - \Delta\bar{\bar{S}}_{(e^-,Pt)} - \frac{1}{4}\Delta\bar{S}_{O_2}\right] \qquad (115)$$

in which the first two terms on the right-hand side of the equation are the irreversible entropies of transfer for the oxygen ions and electrons across the temperature gradient and the third term is the reversible entropy change. They showed that the difference in $\Delta V/\Delta T$ between two different gas pressures but the same temperature gradient was constant as indicated by the equation above, and after application of the data of Cusack and Kendall (1968f, g) for $\Delta\bar{\bar{S}}_{(e^-,Pt)}$ obtained a constant value of $\Delta\bar{\bar{S}}_{O^{2-}}$ of about 45 J/deg mole.

These terms were not so clearly separated by Fischer and by Goto *et al.*, but the results cover a much wider range of oxygen pressures. Further studies are needed, however, before a complete analysis is at hand for the effects of thermal gradients in galvanic cells of this type.

A cell can be used to measure the activity of a component of a binary alloy system such as copper–nickel (1962h), or the Gibbs energy of formation of an intermetallic compound, such as the compound Co_3W which was obtained from the cell

$$\text{Pt}|\text{Fe}+\text{FeO}|\text{ThO}_2\text{–La}_2\text{O}_3|\text{Co, Co}_3\text{W, WO}_2|\text{Pt}$$

by Rezukhina and Proshina (1962i). The e.m.f. can only be interpreted to yield the activity of a metal in the electrode providing the oxide has a very narrow range of stoichiometry or there exists relatively complete information about the metal and oxygen potentials in the non-stoichiometric oxide. Thus in the cell for the study of copper–nickel alloys, Rapp and Maak could assume that nickel oxide was present at unit activity no matter what the nickel activity in the alloy. Hence the activity of nickel could always be obtained directly from a knowledge of the Gibbs energy of formation of the alloy and the oxygen potential of the electrode. Similarly in the cell for the study of Co_3W the oxide WO_2 could be assumed to have unit activity throughout.

However when the oxide phase is FeO_{1+x} which has a wide range of stoichiometry, the product $a_{Fe}p_{O_2}^{1/2}$ is not a constant since the *wüstite* activity, with respect to the phase in equilibrium with pure iron, does not remain at unity across the whole phase field. A correction had therefore to be applied in the study of the cell

$$\text{Pt}|\text{Fe, FeO}|\text{ThO}_2\text{–Y}_2\text{O}_3|\text{Fe–Au, FeO}|\text{Pt}$$

to take account of this before the correct activities of iron were obtained (1967g, h). It should be remarked that thoria-based electrolytes were used in this study because of the well-known interaction between iron oxides and zirconia-based electrolytes which is observed during prolonged contact at temperatures around 1000°C.

Finally, with solid electrodes which are single phase, it is almost mandatory to seal the electrode in a closed compartment one wall of which is the electrolyte in order to eliminate reactions with the gas phase. Several designs have been used to achieve this end, and two are shown in Figs 86a and 86b (1969m, 1970l). Because diffusion is slow in the oxide phase it is not advisable to use a sintered pellet of the oxide under investigation. Instead this can be placed loosely as a powder over the surface of the electrolyte, within the sealed chamber, and contacted with a platinum lead. The platinum then probably functions as the electron lead to an oxygen gas electrode in which the gas phase is the main instrument for distributing oxygen within the closed electrode compartment. If a controlled current is passed through the cell, the composition of the oxide can be altered, because the passage of current requires the removal of oxygen from one electrode and its transport to the other. The direction of transport depends upon the polarity of the impressed voltage which must be imposed on the cell to cause current to flow. We shall return to this topic later when discussing liquid electrode systems containing oxygen.

Gas Electrode in Oxygen Concentration Cells. It has been found that when a gas of fixed oxygen partial pressure is flowed over a solid oxide electrolyte which has a platinised surface, that the system behaves as a reversible oxygen electrode (1905). This system may be applied for measuring the oxygen potential of O_2-inert gas, H_2–H_2O, CO–CO_2 gas mixtures and others producing a defined oxygen potential over the temperature range 400–1700°C.

The gas electrode provides a very convenient reference electrode and has been used to establish the thermodynamics of metal–metal oxide solid electrodes, to monitor the gas phase in the equilibration of condensed non-stoichiometric phases with gases of fixed oxygen potential, and on the industrial scale to control annealing furnace atmospheres for alloy steels.

The most convenient assembly for the construction of a gas phase monitor involves the use of a closed-end ZrO_2–CaO tube which is platinised on the outside, and a platinum wire contacts this surface and acts as the electron lead in the e.m.f. measuring circuit. Inside the tube a solid or gas reference electrode may be used, but a gas must be kept flowing around the platinised electrode in order to eliminate possible thermal segregation in the mixture of gases which determine the oxygen potential. The high-temperature limit of a gas electrode depends on the oxygen availability of the gas, thus Alcock and Chan (1972a) found that the e.m.f. of the cell

$$\mathrm{Pt}|\mathrm{CO{-}CO_2}|\mathrm{ZrO_2{-}CaO\ tube}|\mathrm{air}|\mathrm{Pt}$$

was close to the theoretical value at 1000° and 1200°C but at 1400°C and 1600°C the results were 15 mV and 300 mV lower when the CO–CO_2 ratio was 100/1. The deviations were shown to be due to a flux of oxygen through the tube which increased to a significant value above 1200°C, but when Mo–MoO_2 was used as the reference electrode in place of air, the results were again close to the theoretical value. The gaseous reference electrode should therefore either consist of a gas with a low oxygen dissociation pressure or with a thick electrolyte to minimise this effect. A good indicator of a satisfactory gas electrode is that the measured e.m.f. will be found to be independent of gas flow rate over a wide range

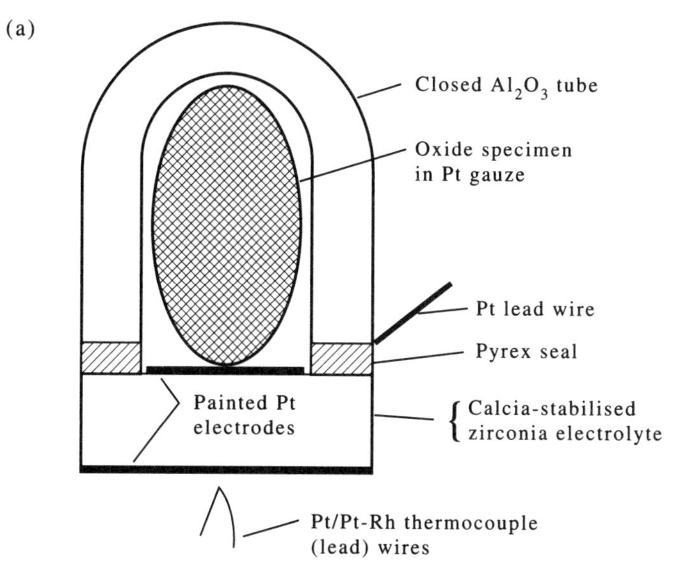

FIG. 86a. Apparatus used by Tretyakov and Rapp (1969m) for the study of the non-stoichiometry of NiO_{1+x}.

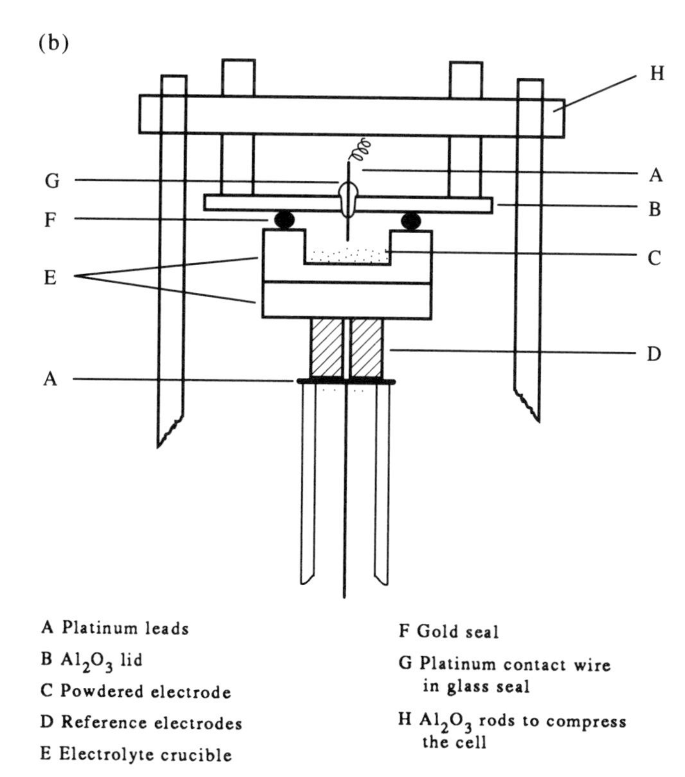

FIG. 86b. Apparatus used by Zador and Alcock (1970t) for the study of the non-stoichiometry of $MoO_{2\pm n}$.

of flow rates for a given gas mixture. Any leaks in the furnace tube which allow air to mix with the metered gas flow will, of course, vitiate the flow-rate test.

Smith, Meszaros and Amata have provided results for the permeability of ZrO_2–CaO, HfO_2–CaO and pure ThO_2 tubes (1965d) and the results show that these have approximately the same oxygen permeability in the temperature range 1100–2050°C. It would appear that the lower diffusion coefficient of oxygen ions in ThO_2- and HfO_2-based tubes is compensated for by the higher concentration of positive holes to provide equal fluxes of oxygen atoms in all of these systems.

The use of gas electrodes to monitor oxygen potentials is not only confined to metal–oxygen systems. Larson and Elliott used these electrodes for the study of sulphide systems (1967j) in such cells as

$$\mathrm{Pt}|\langle \mathrm{MnO}\rangle, \langle \mathrm{MnS}\rangle(\mathrm{SO_2}, 1\ \mathrm{atm})|\mathrm{ZrO_2{-}CaO}|(\mathrm{O_2}, 1\ \mathrm{atm})|\mathrm{Pt}.$$

Since the MnO and MnS have a common manganese activity, the oxygen and sulphur potentials are fixed by the requirement of equilibrium with these phases and with SO_2 at a fixed pressure. The electrode reaction at the left-hand side may be written thus:

$$\mathrm{MnS} + 3\mathrm{O}^{2-} \rightarrow \mathrm{SO_2} + \mathrm{MnO} + 6e^-.$$

This is a system with low oxygen and sulphur pressures, but at the other end of the scale in a study of PtS, the left-hand electrode consisted of a mixture of platinum metal and the sulphide. This is because platinum does not form a condensed phase oxide at high temperatures and with oxygen pressures less than 1 atm. Finally, if the oxide is very stable relative to the sulphide, as, for example, Ta_2O_5 compared with TaS_2, then the left-hand electrode is brought to equilibrium with a mixture of sulphur gas and SO_2.

Liquid Electrodes. The high diffusion coefficients which are found in liquid systems, when compared with typical solids, ensure that electrodes in the liquid state reach equilibrium very much more rapidly than is normally found with solids. There are a number of advantages in using liquid reference electrodes, not the least of which is their rapid equilibration. The ease with which contact can be made to a liquid electrode is a definite advantage in comparison with the pressure-assisted contact with a solid electrode. The major disadvantage appears to be the corrosion of metallic contact materials by the liquid electrode, and a great deal of attention should be directed to this aspect in designing cells with liquid electrodes.

Apart from this, the high diffusion coefficients are an advantage in the study of Henry's law behaviour of dilute solutions of oxygen in liquid metals. The possibility of coulometric oxygen transfer from one electrode to the other was first exploited by Alcock and Belford (1964a) in studies of dilute, unsaturated solutions of oxygen in liquid lead and tin. It was found that conformity to Henry's law could be demonstrated in a cell using, say, Ni/NiO as a reference electrode and titrating known amounts of oxygen in and out of the liquid metal. A known quantity of the metal held in a crucible made of electrolyte material serves as the electrode, and the e.m.f. of the cell at various oxygen levels up to saturation was measured. Figure 87 shows a typical titration curve for the cell

$$\mathrm{Pb[O]/ThO_2{-}Y_2O_3/Cu_2O,\ Cu}$$

at 700°C.

When a high temperature of operation is required, such as 1600°C for steelmaking studies, the long-term survival of a simple electrochemical cell requires that the electrolyte

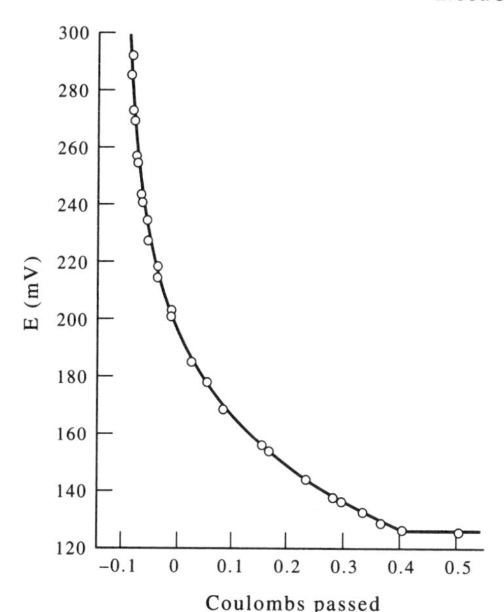

FIG. 87. Variation of the e.m.f. of the cell $Cu|Cu_2O|ThO_2\text{–}Y_2O_3|[O]_{Pb}$ as a function of the quantity of current passed through the cell. The curved portion corresponds to a solution of oxygen in lead, the flat portion is where PbO separates.

and reference electrode system are immersed slowly into the melt so as to avoid thermal shock to the electrolyte. The use of tubes, which are now commercially available, made of stabilized zirconia which contains a suitable reference electrode (Fig. 88), is mainly for laboratory studies of long duration and, again, the principal experimental problem consists of finding good contact materials which will remain in the liquid metal for a significant period without corrosion. Cermets of alumina or zirconia with refractory metals such as molybdenum have been found valuable for studies in liquid copper as well as liquid iron. For control devices which must be rapidly inserted into a liquid metal at temperatures above 1000°C, the "oxygen-probe" can be used for periods of immersion of 30–60 sec. Several commercial forms of this device have been proposed in most of which the tube made of electrolyte material is replaced by a pellet of electrolyte either flame-sealed or cemented into the end of an open quartz tube (Fig. 89). Reference electrodes for such oxygen probes can be made of molybdenum/MoO_2 (1968e), chromium/Cr_2O_3 (1967d), or gaseous systems O_2, CO_2 or air in contact with platinum.

Finally, there have been a number of studies in which oxide activities in a liquid slag have been obtained from studies using solid oxide electrolytes. Typical of these are the measurement of PbO activities in liquid lead silicates which were made by a number of groups of workers (1968b, c, 1971h, 1970l, 1964g) with the cell

$$\text{Pb, PbO–SiO}_3\text{/electrolyte/reference electrode.}$$

The results show a scatter in the measured activities which is large compared with the best available studies using this method. For example, at a mole fraction of PbO of 0.6, the range of measured activities at 1000°C is from 0.22 to 0.38 if all results are included. This corresponds to a range of partial free energies for PbO of ± 3000 J approximately, or a range of e.m.f. values of ± 15 mV.

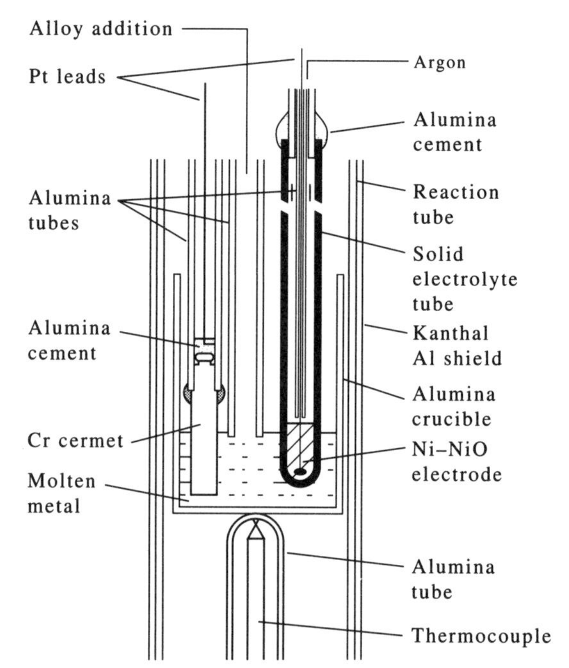

FIG. 88. Apparatus for the measurement of the dilute solution of oxygen in a liquid metal, incorporating a zirconia tube.

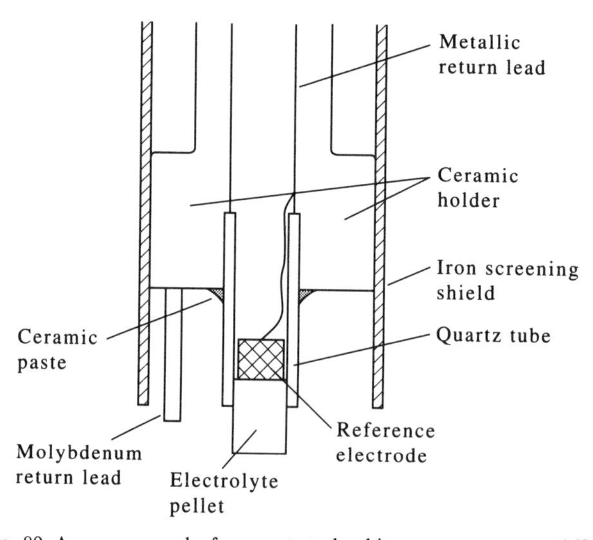

FIG. 89. An oxygen probe for use at steelmaking temperatures *ca.* 1600°C.

3. Other Cells

One of the main advantages of the zirconia and thoria electrolytes, apart from the valuable aspect of permitting measurement of oxygen activities, is that the solid electrolyte makes an effective barrier between the electrodes no matter what state is involved, and the electrolytes can be shaped into crucibles and tubes for electrode separation. This suggests that further developments of solid electrolytes would be very valuable in the study of high-temperature systems because of this constructional advantage. Unfortunately, very few additional electrolytes for use at high temperature have emerged during more than a decade of studies. Two of the most promising are calcium fluoride (CaF_2) which conducts electricity by fluoride ion migration and β-alumina which can be used for studies of sodium, potassium and silver activities. The original studies of conduction in CaF_2 made by Ure (1957f) show conductivities of the same order as the zirconia electrolytes, the moving ion being the interstitial fluoride ion. There appears to be little advantage in adding other fluorides in order to enhance the conductivity, and a single crystal of CaF_2 serves very well as the electrolyte. The material has been used up to 900°C with varying degrees of success, but it is necessary to pay close attention to the gaseous atmosphere which is allowed in contact with the electrolyte. The principal gas is argon or helium, which must be rigorously de-oxidised and dried before use, and even then it is useful to place some CaF_2 powder upstream of the electrolyte in order to equilibrate the gas phase with a sample of the solid other than the electrode surface. When this is successfully achieved, a number of reference electrodes can be used, e.g. Mg, MgF_2 (1974j), Cr, CrF_2 (1969h), U, Uf_3 (1974e) as a result of careful earlier studies.

The immediately apparent advantage of the fluoride electrolyte is the fact that studies may be made using electrodes containing metals which have a large solubility of oxygen, and hence would be unsuitable in oxide electrolyte studies. A disadvantage of the fluorides, in such a comparison, comes mainly from our lack of knowledge of metal–fluorine phase diagrams. The non-stoichiometry ranges of fluorides could well be less than those of oxides, but at present there is very little information one way or the other.

Another field where a considerable advantage is gained through the use of fluoride electrolytes is in the study of metallic carbides and borides. The pioneering work of Egan (1964d) using this electrolyte produced a value for the free energy of formation of thorium dicarbide (ThC_2) with the cell.

$$\mathrm{Th, ThF_4/CaF_2/ThF_4, ThC_2, C}$$

The cell obviously measures the activity of thorium in the carbide which is in equilibrium with carbon. A similar study by Aronson (1964b) produced values for the Gibbs energies of formation of ThC and the terminal carbon saturated α-thorium which is in equilibrium with the $ThC_{0.7}$. The solid solution contains about 10 at.% carbon, and therefore the metal activity is reduced to 0.84 in the α-Th-$ThC_{0.7}$ electrode. The cells which were operated between 800° and 1000°C are represented thus:

$$\mathrm{Th, ThF_4/CaF_2/ThF_4, ThC_2, ThC}$$

and

$$\mathrm{Th, ThF_4/CaF_2/ThF_4, ThC_{0.7}, \alpha\text{-}Th.}$$

Aronson and Auskern (1966c) used the same technique to study cells with thorium boride electrodes in the temperature range 800–950°C, thus opening another relatively unexplored field of solid state e.m.f. studies.

Because of the problems associated with hydrolysis of the electrolyte, Rezukhina and Pokarev (1971o) suggest that the cell be evacuated and preheated to 500–700 K before admitting the gas. Thorium, yttrium or lanthanum chips held in a tantalum boat were then put close to the cell in order to keep the argon atmosphere neutral. The cell was placed in double-wall silica vessels inside the furnace, and inert gas was passed continuously between the walls of the silica vessel.

Crucibles can easily be made from calcium fluoride by hot pressing techniques, but no e.m.f. application has been made of these materials to date.

β-alumina is another promising ceramic material which has been used in the measurement of sodium activities in alloy systems. This material, formula $NaAl_{11}O_{17}$ allows preferential, rapid, migration of sodium, potassium and silver ions along widely spaced planes, so that the material can be used satisfactorily from room temperature to about 800°C. Whittingham and Huggins (1971s) showed that silver β-alumina has an electron transference number less than 10^{-4} over the temperature range 550–790°C, and a wide range of oxygen pressure from 1 to 10^{-24} atm.

The material may be fabricated into pellets or tubes by pressing isostatically at about 20,000 psi and firing to about 1700°C (1971f). Joglekar *et al.* (1973f) succeeded in joining a β-alumina pellet to an alumina tube by the use of a Kovar glass-ethylene glycol paste which slowly heated to 300°C and then fired at 1000°C for a short time. This device could be used as a sodium probe, and was used to measure sodium in solution in liquid tin at 500°C. During the sintering of the β-alumina material there is a tendency for sodium loss to occur at high temperatures. There is therefore a considerable advantage to be obtained by fabricating components by hot pressing, where densification is normally achieved at lower temperatures than at atmospheric pressures. A novel way of minimising the vaporisation loss under atmospheric conditions is the zone-sintering technique described by Wynne-Jones and Miles (1971t) for the production of long tubes of β-alumina.

Fray and Savory (1975a) used β-alumina tubes in a study of sodium–lead alloys in the temperature range 350–500°C and $N_{Na}=0.044$–0.78. The cell can be represented thus:

$$\text{stainless steel, Na}/\beta\text{-alumina/Na–Pb, stainless steel.}$$

Some coulometric titrations were carried out from the sodium to the lead electrode in order to prepare the dilute sodium–lead alloys, and generally e.m.f.'s could be reproduced to ± 1 mV. Equilibrium was found to be very rapidly established during heating and cooling cycles. Very good agreement was found with the results of Hauffe and Vierk (1949e), who used a glass electrolyte and other electrochemical studies of this system.

Hydrogen Concentration Cells: A number of materials have been developed for use in electrochemical hydrogen cells but most of these suffer from dehydration at temperatures around 200°C, which destroys the electrochemical properties. Few electrolytes have been developed for temperatures higher than this, but Iwahara and Uchida (1981c) have demonstrated that ytterbium-doped strontium cerate $SrCe_{0.95}Yb_{0.5}O_{3-\delta}$ functions satisfactorily as the electrolyte in a hydrogen concentration cell operating at 800°C. The cell uses hydrogen at one atmosphere as the reference electrode, and responds to changes in the hydrogen concentration in a H_2/Ar mixture applied to the other electrode in about 30 sec. This material, which is of the perovskite family, has very promising properties for hydrogen cells, and suggests that other perovskites may be considered as candidate materials. The present upper limit of temperature of application is less than 1000°C, and the detailed mechanism of operation of this potential group of hydrogen sensors still awaits elucidation.

Electrolytes with a Dispersed Phase: It would appear that, in principle, the development of future electrochemical systems for use as concentration cells would depend upon the discovery of new electrolytes which conduct by, for example, sulphur or carbon ionic migration for sulphur and carbon cells respectively. The chances of discovering such materials seem very slim at present when an added requirement is that these should function at temperatures above 1000°C in pyrometallurgical operations.

A new possibility has been demonstrated at modest temperatures (up to 900°C) by recent studies of dispersed phase electrolytes. The solid electrolyte, made of a solid solution of strontium fluoride containing lanthanum fluoride, conducts electrolytically by fluoride ion migration. This may be shown by the achievement of the theoretical e.m.f. in the cell

$$\mathrm{Ni/NiF_2|SrF_2\text{–}LaF_3|Co/CoF_2}.$$

If a few volume percent of strontium oxide is dispersed as a second phase in this electrolyte, the cell now responds to oxygen potentials; thus the theoretical e.m.f. is obtained for the cell

$$\mathrm{Cu/Cu_2O|SrF_2\text{–}LaF_3/SrO|Ni/NiO}.$$

When strontium sulphide is the dispersed phase, the theoretical e.m.f. is obtained for the cell

$$\mathrm{Ag/Ag_2S|SrF_2\text{–}LaF_3/SrS|Cu/Cu_2S}.$$

This principle of using a dispersed phase to transmit another chemical potential gradient across a fluoride ion electrochemical cells depends upon the fact that in, for example, the oxygen cell, the oxygen potential gradient determines the strontium gradient since

$$K_{\mathrm{SrO}} = a_{\mathrm{Sr}} \cdot p^{\frac{1}{2}}\mathrm{O_2}$$

and this, in turn, controls the fluorine gradient

$$K_{\mathrm{SrF_2}} = a_{\mathrm{Sr}} p\mathrm{F_2}.$$

Hence the cell responds to a fluorine gradient (Alcock and Li, 1990i).

The two strontium activities above will be equal when the electrolyte mix can reach local equilibrium with the dispersed phase particles at the electrolyte surface.

This principle can now be extended to other non-metallic elements such as carbon and nitrogen using more refractory electrolytes such as stabilised zirconia as the matrix for a dispersed phase of carbide or nitride.

Electrolytes with a Sensing Electrode: An interesting development of the β-alumina electrolyte is its use by Jacob *et al.* to produce a sulphur probe. In this, β-alumina is used as the sodium-ion conducting electrolyte, and the electrodes are Na_2S in contact with two gas mixtures of differing sulphur potential

$$p\mathrm{H_2}s/p\mathrm{H_2}|\mathrm{Na_2S}|\text{Na-}\beta\text{ alumina}|\mathrm{Na_2S}|p'\mathrm{H_2S}/p'\mathrm{H_2}.$$

In this cell, the Na_2S sensing electrode delivers a sodium potential to the electrolyte which is inversely proportional to the sulphur activity in the gaseous phase with which it is in contact. The cell function is best in low oxygen potential gas atmospheres. At high oxygen potential, where there is a significant SO_2 partial pressure in a sulphur-bearing atmosphere, the sulphur potential may be ascertained from an oxygen potential electrochemical cell as shown by Larson and Elliott (1962e)

$$pSO_2/pS_2|\text{Zirconia electrolyte}|pO_2.$$

This is because the equilibrium SO_2/S_2 mixture establishes an oxygen pressure on the left-hand face of the zirconia, which is determined by the equilibrium constant for the reaction

$$\tfrac{1}{2}S_2+O_2\rightarrow SO_2;\ pO_2=K.pSO_2.p^{-1/2}S_2.$$

Worrell (1981d) has demonstrated that the cell with a fluoride solid electrolyte made from optical grade CaF_2 single crystals

$$\text{Cu, Cu}_2\text{S}|\text{CaS}|\text{CaF}_2|\text{CaS}|\text{Fe, FeS}$$

provides the correct e.m.f. corresponding to the sulphur pressure ratios of the Cu, Cu_2S and Fe, FeS systems in the temperature range 500–900°C.

Similarly, an electrochemical cell based on $CaF_2/CaSO_4$, as proposed by Jacob *et al.* (1987b) for the measurement of SO_2/SO_3 ratios, was found to take about 2 hr to reach a steady e.m.f. at 900°C.

CHAPTER 3

The Estimation of Thermochemical Data

No special skill or experience is required to learn the main thermodynamic formulae and to use them for calculating equilibria between pure stoichiometric substances from tables of heat capacities, enthalpies of formation and transition, and entropies. Moreover, the present-day availability of computer software, such as that referred to in Chapter 4, facilitates considerably these thermodynamic calculations of reaction equilibria. Unfortunately, published data are far from complete, especially in the ever-widening field of materials chemistry, where reliable thermodynamic data both for **stable** and **metastable** phases are becoming increasingly important. It is at this point, where present thermochemical data are lacking, that an important function of the experienced thermochemist is the estimation of missing data with reasonable accuracy. For stoichiometric inorganic compounds this task is not as difficult as it might appear, for many of the principles which are applied have been well established for a considerable time. For solution phases however, methods available are generally more complex, and sophisticated software incorporating various solution models has been written with the aim of providing missing data from the more limited information available.

Experience is required to enable the best choice of estimation method to be made in each particular case, and if necessary to develop new methods. A selection of current methods used to estimate thermodynamic values both for pure stoichiometric substances and for solution phases of different types, as well as examples of their application, are given in this chapter.

1. Heat Capacities

It is not the intention here to discuss the rules which apply at low temperature, such as Debye's equation, or the methods used to calculate electronic contributions to specific heat. These quantities are incorporated in room temperature entropy values, which will be discussed later. For most materials applications, it is sufficient to be able to estimate heat capacities at room temperature and above.

A. Gaseous Atoms and Molecules

Thermodynamic data for most monatomic and diatomic gaseous species important in practical materials problems are generally well-established. Estimation of heat capacity data is therefore very rarely required.

For polyatomic gaseous species it is preferable to calculate heat capacity data, as well as entropy values, from molecular constants. Where this information is incomplete, assumptions may be made concerning the molecular configurations and missing data for the fundamental vibrational frequencies, bond distances, bond angles and/or electronic levels estimated. Methods of evaluation of heat capacities and entropies from the molecular constants for a large number of gaseous molecules at various temperatures, including estimates of missing information are discussed in the JANAF Thermochemical Tables. For gaseous halide double molecules, the simple expression

$$\frac{C_p(\mathrm{DM})}{C_p(\mathrm{SM})} = 2.16 \pm 0.1$$

can be applied at all temperatures (DM = double molecule, SM = single molecule) (1984b).

B. Solids

Generally it will more often be necessary to estimate the heat capacities of solid substances, and here use can be made of Dulong and Petit's rule. This rule stated that the atomic heats of the elements generally approximate to the value 6.2 cal/deg at room temperature. The rule was also established from considerations of equipartition of energy. Since the atoms of a solid are fixed in a lattice, they are incapable of rotation or translation, but there exist three vibrational degrees of freedom (each of which must be counted twice). Hence $C_v = 6/2\boldsymbol{R} = 25.1$ J/K . mol when the solid lattice is well above the Debye temperature.

The relation between C_p and C_v for the solid state is more complicated than for the gaseous state and is given by a formula which includes the compressibility of the solid and its coefficient of thermal expansion. These quantities are generally not known very exactly, but this is not of such vital consequence since the difference between the two heat capacities amounts only to 0.84–2.09 J/K . mol at room temperature, giving a C_p value of about 25.9 to 31.5 J/K . mol.

Kellogg (1967f) has suggested an improved method for the estimation of the heat capacity of predominantly ionic, solid compounds at 298K. It is analogous to Latimer's method for the estimation of standard entropies (p. 173) and consists in adding together contributions from the cationic and anionic groups in the compound (here denoted Θ(cat) and Θ(an)). Working empirically, Kellogg derived from a compilation of experimental data, average values of Θ(cat) for the metallic ions independent of their valency and of Θ(an) for the anions—in the latter case, Latimer style, depending on valency. The estimated heat capacities at 298 K were then obtained by summing the various contributions, i.e.

$$C_p(298\ \mathrm{K}) = \sum \Theta.$$

For aluminium sulphate, $Al_2(SO_4)_3$, for instance:

$$C_p(298\ \mathrm{K}) = 2\Theta(\mathrm{Al}^{+++}) + 3\Theta(\mathrm{SO}_4^{--}).$$

Ünal (1977c) revised the tables of Kellogg with the aid of a computer program using experimental C_p(298 K) values for 90 compounds. It emerged that, within the scatter of the individual values, the Θ(an) values may be represented by a single number for each anion independent of valency. This is in contrast to Kellogg's findings but simplifies the procedure. The values finally arrived at are listed in Table IX and X. As an example, the heat capacity of one mole of aluminium sulphate at room temperature may be estimated:

$$C_p(298\ \text{K}) = 2\Theta(\text{Al}) + 3\Theta(\text{SO}_4) = 39.32 + 229.7 = 269.03\ \text{J/K.mol}$$

which compares with the measured value of 259.41 J/K.mol.

TABLE IX. *Cationic contributions to the heat capacity at 298 K*

Metal	Θ(cat) (J/K)	Metal	Θ(cat) (J/K)	Metal	Θ(cat) (J/K)
Ag	25.73	Hf	25.52	Rb	26.36
Al	19.66	Hg	25.10	Sb	23.85
As	25.10	Ho	23.01	Se	21.34
Ba	26.36	In	24.27	Si	—
Be	(9.62)	Ir	(23.85)	Sm	25.10
Bi	26.78	K	25.94	Sn	23.43
Ca	24.69	La	(25.52)	Sr	25.52
Cd	23.01	Li	19.66	Ta	23.01
Ce	23.43	Mg	19.66	Th	25.52
Co	28.03	Mn	23.43	Ti	21.76
Cr	23.01	Na	25.94	Tl	27.61
Cs	26.36	Nb	23.01	U	26.78
Cu	25.10	Nd	24.27	V	22.18
Fe	25.94	Ni	(27.61)	Y	(25.10)
Ga	(20.92)	P	14.23	Zn	21.76
Gd	23.43	Pb	26.78	Zr	23.85
Ge	20.08	Pr	24.27		

TABLE X. *Anionic contribution to the heat capacity at 298 K*

Anion	Θ(an) (J/K)	Anion	Θ(an) (J/K)
H	8.79	SO_4	76.57
F	22.80	NO_3	64.43
Cl	24.69	P	(23.43)
Br	25.94	CO_3	58.58
I	26.36	Si	(24.68)
O	18.41	CrO_4	90.79
S	24.48	MoO_4	90.37
Se	26.78	WO_4	97.49
Te	27.20	UO_4	107.11
OH	30.96		

The heat capacities of solids increase with temperature and at the melting point are roughly the same per ion or atom for all compounds. Kelley (1949f) took this value to be 29.3 J/K.mol. Ünal (1977c), with the more recent experimental evidence, increased this average to 30.3 ± 2.1 J/K.mol. Not included in the evaluation of the mean deviation are the data for those substances which undergo a solid–solid transformation below the melting point, nor for those that have a rather low melting point—say, below 420 K. If one includes the first transformation rather than considering the melting point alone, the average value turns out to be closer to Kelley's original value of 29.3 J/K . mol—a point that should be noted. These observations* taken together with the values in Tables IX and X may be used to recommend the estimation of the constants in a heat capacity equation of the type

$$C_p = a + b \times 10^{-3}\, T + c \times 10^5\, T^{-2} \tag{116}$$

Kelley (1949f) estimated quite a number of C_p vs T equations which have been widely employed in the thermochemical literature. To do so, he used a two-term expression. The introduction of a T^{-2} term, however, reflects the pronounced curvature at the lower temperatures above 298 K and around the Debye temperature. A list of values for c in eqn (116) evaluated by the investigators themselves or by Ünal (1977c) shows a considerable scatter due more to the inaccuracy of the differentiation of experimental enthalpy data than to the actual relations. A mean value of $c/n = -4.12$ J/K is obtained from the T^{-2} terms of 200 inorganic substances.

It is now possible to derive expressions for the constants in eqn (116). These are as follows:

$$a = \frac{T_m \cdot 10^{-3}(\sum\Theta + 1.125n) - 0.298n \cdot 10^5 T_m^{-2} - 2.16n}{T_m \cdot 10^{-3} - 0.298} \tag{117}$$

$$b = \frac{6.125n + 10^5 n \cdot T_m^{-2} - \sum\Theta}{T_m \cdot 10^{-3} - 0.298} \tag{118}$$

$$c = -4.12n. \tag{119}$$

Here, T_m is the absolute melting temperature of the compound and n the number of atoms in the molecule. If C_p(298 K) is known from low-temperature measurements, the experimental value is to be used instead of $\sum\Theta$, the estimated value. If the heat capacities of compounds similar in mass and chemical nature to the one under consideration are known, they should be used to adjust the estimated equation.

Hoch (1969g; 1970g; 1972e; 1974d) has shown that the high temperature heat capacity data of solids can be represented by the equation

$$C_p = 3\boldsymbol{R}F(\theta_D/T) + bT + dT^3 \tag{120}$$

where $F(\theta_D/T)$ is the Debye function, b is equivalent to the electronic heat capacity, and d reflects only the contribution of the anharmonic vibrations within the lattice. The equation is valid in the temperature range between θ_D and the melting point. By plotting the experimental heat capacity data for a number of metals and ceramics using eqn 120 in the rearranged form

$$\frac{C_p - 3\boldsymbol{R}F(\theta_D/T)}{T} = b + dT^3 \tag{121}$$

*In fact, the average value of C_p(m . pt)/n increases somewhat with molecular weight, as one might expect. However, in view of the scatter, this effect is not great enough to warrant the proposal of a two-term expression in terms of log M.

Hoch was able to derive values for the electronic heat capacity, b, and for the heat capacity due to the anharmonic vibrations, d, from the intercepts and slopes respectively of the linear plots. For insulating materials, $b=0$ the line passes through the origin. Using this approach, a very satisfactory description of the heat capacities of Nb, Ta, Cr, Re, Mo, W, Cu, Al, UO_2, UO_{2+x}, US, UN, UC, and Al_2O_3 was obtained. Equation (121) thus appears useful as a general equation for estimating heat capacity data for solids where values are scarce or rather unreliable.

If nothing better is known about the heat capacity of a compound, then ΔC_p may be assumed to be zero for **condensed reactions** without affecting most calculations too gravely. This postulate of the additivity of the heat capacities of the elements or the reactants to give the heat capacities of the compound or the products in a reaction is known as Neumann and Kopp's rule. It has been found to be approximately valid for a large number of reactions and frequent use has been made of it. Neumann-Kopp's rule holds good especially for alloy phases and can also be applied as a first approximation to compounds with a coordination lattice. In other words, for reactions involving these substances, the Gibbs–Helmholtz equation can be used in the simplified form

$$\Delta G(T)=\Delta H(298\text{ K})-T\Delta S(298\text{ K})$$

for any temperature T. At the transition points the enthalpies of transformation, fusion or evaporation must of course be taken into account. Their effects are much greater than that of variation in heat capacity and their estimation will be discussed later.

For accurate estimations, the Neumann–Kopp rule should not be applied to reactions in which gases are involved.

C. Liquids

The heat capacities of molten inorganic substances do not differ greatly from those of the corresponding solid materials. This indicates a close connection between the solid and the liquid state, and although it has been found difficult to envisage a simple model which will adequately describe the properties of liquids, it must be assumed that a form of potential energy, analogous in some way to the intermolecular binding in the solid state, is involved. Whatever the case may be, the heat capacity of an inorganic liquid amounts to 29.3 to 33.5 J/K.mol depending to some extent on the atomic weight of the substance concerned. The value 31.4 may be used if measurements are not available; alternatively, the atomic heats of the liquid constituent elements may be taken additively.

Hoch and Venardakis, in a series of publications [e.g. 1976f,g], have analysed experimental heat capacity data for a number of liquid metals, oxides and halides. They suggest that the experimental data indicate anharmonic contributions to the heat capacity of the liquids close to the melting point. As the temperature is raised, the anharmonic contributions decrease and the heat capacity at very high temperatures is made up of two parts—a term which corresponds to the Debye function, and a linear term which can be assigned to the electronic heat capacity.

The equation derived by Hoch and Venardakis to describe the specific heat of liquid inorganic substances of various types at high temperatures is

$$C_p=3\boldsymbol{R}F(\theta_D/T)+gT+hT^{-2} \tag{122}$$

where g is the electronic specific heat and h the anharmonic term.

Equation (122) was applied to a wide variety of materials (Li, K, Na, Pb, Hg, Fe, V, Ti, Al_2O_3, UO_2, NaF, $CaCl_2$, $MgCl_2$, $MnCl_2$, and MgF_2). The fit was very good in all cases, although scatter in the published enthalpy data results in values of the anharmonic contribution, h, ranging from 4% for Al_2O_3 to 18% for Fe.

Because of its apparent generality, eqn (122) seems suitable for estimating high temperature specific heat values for other liquid metals and inorganic compounds where the available data are scarce.

More recently, Hoch has shown (1990c) that, within experimental error, the term

$$d\,.\,h/m\,.\,\theta_D^2$$

is constant, where d and h correspond to the appropriate constants in eqns (120) and (122) for the heat capacity of solids and liquids respectively, and m is the mass per atom in a compound. Thus, if $C_p(L)$ is unknown for a particular compound, the constant value of $d\,.\,h/m\,.\,\theta_D^2$ can be used to estimate h, using the known value of d. The resulting error in $C_p(L)$ will be relatively small because only a correction term is calculated.

The general constancy of $d\,.\,h/m\,.\,\theta_D^2$ is illustrated by the values given in Table XI for a broad range of materials.

TABLE XI. *Relation between the anharmonic terms for substances in the solid and liquid state*

Substance	Debye T	d	h	mass/atom	$dh/m\,.\,\theta_D^2$
Li	344	4.60E−8	1.41E+6	7	7.81E−8
Pb	105	8.79E−9	1.81E+6	207	6.98E−9
Sn	212	2.35E−8	5.57E+5	119	2.44E−9
In	109	2.96E−8	3.30E+5	115	7.15E−9
Bi	117	8.28E−9	5.99E+5	209	1.73E−9
V	380	1.50E−9	9.14E+7	52	1.83E−8
Ti(α)	420	4.90E−9	6.11E+7	48	3.53E−8
Fe(**gam**)	470	1.61E−9	3.26E+7	56	4.25E−9
W	310	4.25E−10	6.30E+8	184	1.51E−8
Mo	380	7.66E−10	4.58E+7	96	2.53E−9
Ta	245	3.15E−10	8.80E+8	181	2.55E−8
Al_2O_3	895	5.02E−10	9.54E+7	20	2.93E−9
UO_2	590	1.39E−9	2.28E+8	90	1.01E−8
NaCl	1074	8.43E−9	1.28E+7	29	4.97E−8
NaF	1227	4.56E−9	2.07E+7	21	2.48E−8
$CaCl_2$	989	4.14E−9	1.59E+7	37	2.81E−8
$MgCl_2$	976	4.94E−9	8.57E+6	32	1.07E−8
$MnCl_2$	882	7.53E−9	7.30E+6	42	2.27E−8
MgF_2	1460	1.70E−9	1.85E+7	21	4.59E−9

D. Some Average Values

The following list gives estimated average values of the change in heat capacity for different types of reaction involving gases. They may be applied, when other information is lacking, over a wide temperature range. The values are, however, intended to be no more than a rough guide. With a little experience it is possible to make minor adjustments to these values for a particular chemical reaction. A and B represent two elements, x and y stoichiometric indices

Type of reaction	ΔC_p(J/K . mol)
$\langle A\rangle = (A)$	-7.5
$\{A\} = (A)$	-9.6
$\langle A_xB_y\rangle = (A_xB_y)$	$-9.6(x+y)$
$\{A_xB_y\} = (A_xB_y)$	$-11.3(x+y)$
$\langle A\rangle + x(B_2) = \langle AB_{2x}\rangle$	$+12.5x$
$\langle A\rangle + x(B_2) = \{AB_{2x}\}$	$+14.2x$
$\langle A\rangle + (B_2) = (AB_2)$	-9.2

2. Enthalpies and Entropies of Transformation, Fusion and Evaporation

As already mentioned, a knowledge of the various enthalpies of transition is more important than that of variation in heat capacity since the former have a relatively greater effect on the value of the Gibbs energy of a substance. Fortunately, the methods available for estimating their value, given a knowledge of the molecular structure of the substance, are fairly reliable.

A. Evaporation

The enthalpy of evaporation of any substance can be obtained using a rule first put forward by Pictet (1876) and later made more widely known by Trouton (1884). It states that the entropy of evaporation (i.e. the enthalpy divided by the absolute temperature of evaporation) at the normal boiling point is approximately the same for all substances.

$$\Delta S_e = L_e/T_e \approx 92.1 \text{ J/K . mol} \tag{123}$$

An analysis of available data reveals, however, that Trouton's constant increases somewhat with boiling temperature. It is also influenced by association or dissociation of the substance in the gaseous state. Since more recent mass spectrometric data reveal that virtually every compound species displays such behaviour in the gas phase, the value of Trouton's rule for estimating enthalpies of evaporation is somewhat diminished. It is nevertheless still useful for checking experimental results. Figure 90 shows a plot of L_e/T_e vs T_e for a number of pure elements and inorganic compounds, which vaporize congruently. The curve can be represented by the equation

$$L_e/T_e = 0.01037 \,.\, T_e + 75.96 \text{ kJ/mol . K.} \tag{124}$$

B. Fusion

The entropy of fusion is not as constant as that of evaporation. The change in state of order on melting is not as great as on evaporation, and the variation in the state of order of a solid, due to the variety of the possible binding forces (and to a lesser extent, preferred

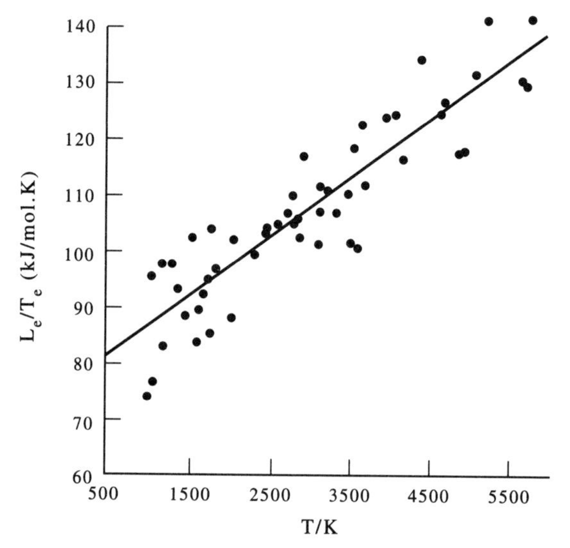

FIG. 90. L_e/T_e vs T_e for pure elements and inorganic compounds.

atomic or molecular configurations in certain composition ranges in liquid mixtures) result in a proportionately large effect on the entropy of fusion.

For pure metals, Crompton (1895) and Richards (1897) were the first to point out that the entropy of fusion should be nearly constant. An average value of 9.2 J/deg was suggested by Tamman (1913b). It has subsequently been found, however, that the entropy of fusion increases somewhat with temperature. For fcc and hcp metals, for example, the available data can approximately be represented by

$$\Delta S_m = 7.41 + 1.55 \times 10^{-3}\, T_m \text{ (J/K.mol).} \tag{125}$$

For bcc metals the values are, on average, a little lower

$$\Delta S_m = 6.78 + 0.71 \times 10^{-3}\, T_m \text{ (J/K.mol).} \tag{126}$$

The entropies of fusion of rare earth metals are lower still.

For metals which are strongly covalent in nature, e.g. Ga, Si, Ge, Sn, Sb, Bi, Se, Te, ΔS_m is substantially higher than 9.2. However, since the enthalpies of fusion of metals melting below 2000 K are virtually all known, a detailed analysis is of little point from the practical aspect.

It is rather more difficult to predict the entropies, and thus enthalpies of fusion of inorganic compounds, as the randomly chosen collection in Table XII shows. Since the entropy of fusion represents the difference in entropy between liquid and solid, its value for a particular substance depends on the nature of the atomic order and bond mechanism involved. To a certain extent, the crystallographic structure of the solid provides an indication of the bond mechanism, but AgCl and NaCl, or $CaCl_2$ and MgF_2, having the same structure, have rather different entropies of fusion. That the values for substances with molecular structures are lower than those for co-ordination lattices is understandable. With the former, only the intermolecular bonds are "broken" on melting, but the intra-molecular bonds much less so.

Fortunately for practical applications, enthalpies of fusion are comparatively small, and even an approximate estimate is adequate for calculating equilibria involving liquid species. For this purpose, entropies and enthalpies of fusion may be estimated by comparison with compounds similar to those listed in Table XII.

TABLE XII. *Entropies of fusion of inorganic compounds*

Compound	T(K)	ΔS_m (J/K) per mole	ΔS_m (J/K) per g-atom	Structure
NaF	1269	26.38	13.19	NaCl
NaCl	1074	26.22	13.11	NaCl
KF	1131	25.90	12.95	NaCl
KCl	1045	25.14	12.57	NaCl
MgO	3105	25.06	12.53	NaCl
AgCl	728	18.16	9.08	NaCl
BeO	2780	30.55	15.28	Wurtzite
FeS	1461	22.14	11.07	NiAs
TlCl	702	22.64	11.32	CsCl
TlBr	733	22.38	11.19	CsCl
$MgCl_2$	987	43.66	14.55	$CdCl_2$
$MnCl_2$	923	40.80	13.60	$CdCl_2$
$FeCl_2$	950	45.28	15.09	$CdCl_2$
$CaCl_2$	1045	27.03	9.01	Rutile
MgF_2	1536	38.14	12.71	Rutile
TiO_2	2143	31.24	10.41	Rutile
AsI_3	414	52.55	13.14	$FeCl_3$
$ScCl_3$	1240	54.32	13.58	$FeCl_3$
BiI_3	681	57.45	14.36	$FeCl_3$
$CeCl_3$	1095	49.29	12.32	UCl_3
UCl_3	1114	41.69	10.42	UCl_3
YCl_3	994	31.74	7.93	$AlCl_3$
$HoCl_3$	993	30.76	7.69	$AlCl_3$
MoO_3	1074	45.19	11.30	MoO_3
WO_3	1745	42.08	10.52	—
Al_2O_3	2325	46.02	9.20	Cr_2O_3
V_2O_5	952	70.32	10.05	Orthorh.
PbI_2	683	34.31	11.44	Layer latt
$PbCl_2$	774	28.27	9.42	Layer latt
$TeCl_4$	497	37.97	7.59	Mol. latt
$SnCl_4$	239	38.49	7.70	Mol. latt
UF_6	337	57.11	8.16	Mol. latt

3. Entropy and Entropy Changes

A. Standard Entropies

1. Solids

Elements: The standard entropies of nearly all elements in their stable structures are known more or less accurately. Methods for estimating them are therefore not required. Nevertheless, a knowledge of the entropies of metals in metastable or non-stable structures is needed, because in alloys a metal can be present in a phase displaying a structural form which the metal itself does not display. Calculation of phase equilibria in alloy systems could be refined significantly if information on the entropies (and enthalpies) of metals in different metastable structures were available. From an analysis of the phase equilibria in a large number of binary alloy systems, Kaufman (1970k) has derived approximate entropy (and enthalpy) differences between the stable and metastable structures of many metallic elements. In particular, the fcc, bcc and hcp structures have been considered. Figure 91 illustrates such differences for the hcp and bcc structures across the 2nd and 3rd Long Periods of the Periodic Table.

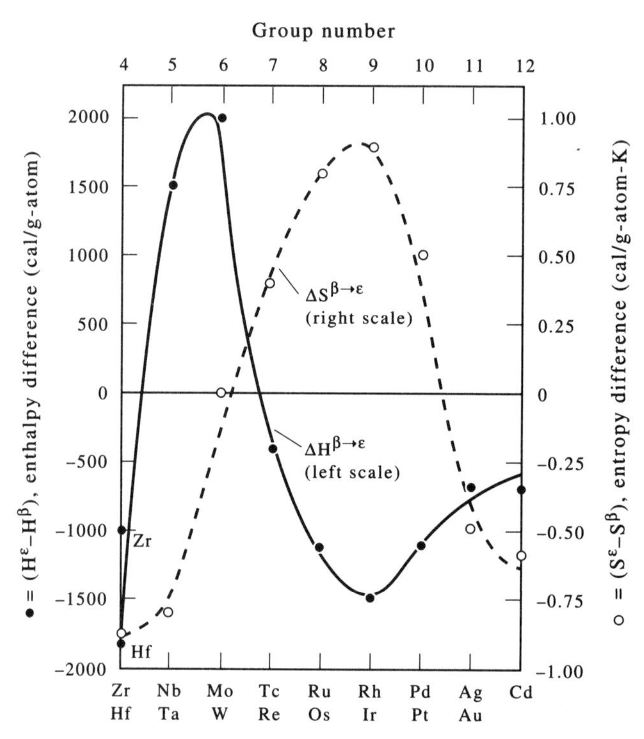

FIG. 91. Enthalpy and entropy differences between the hcp (ϵ) and bcc (β) forms of the transition metals.

With the help of curves such as these and available standard entropy values for the stable phases of the elements, standard entropy data for the elements in metastable structures can be derived.

Inorganic compounds: Using a method originally proposed by Latimer (1951e), in which the standard entropies of predominantly ionic compounds were obtained additively from values found empirically for the anionic and cationic constituents, Mills (1974i) has derived revised values of the anionic and cationic contributions from an analysis of experimental data for 308 compounds. The overall standard deviation of the resulting S(obs)-S(est) values, at 5.4 J/K . mol, is smaller than that associated with earlier values given by Latimer (1951e) and in earlier editions of this book. The data presented by Mills for the cationic and anionic contributions to the standard entropy of inorganic compounds of general formula MX_a were calculated for {M} and n{X} and are listed in Tables XIII and XIV.

TABLE XIII. "*Latimer*" *entropy contributions* $\{M\}$

M	{M} J/K . mol	M	{M} J/K . mol	M	{M} J/K . mol
Ag	57.6±2.5	Ho	56.0	Ru	53 ±8
Al	23.4±6.7	In	55.0±6.0	S	48
As	45.2±4.6	Ir	50.0	Sb	58.9±8.3
Au	58.5±2.0	K	46.4±0.8	Sc	36.0±1.3
B	23.5	La	62.3	Se	60.5
Ba	62.7±3.0	Li	14.6±3.8	Si	35.2
Be	12.6±4.2	Lu	51.5	Sm	60.2
Bi	65.0±9.2	Mg	23.4±4.2	Sn	58.2±7.6
Ca	39.1±2.9	Mn	43.8±6.7	Sr	48.7±2.5
Cd	50.7±3.4	Mo	35.9±5.2	Ta	53.8
Ce	61.9	Na	37.2±3.8	Tb	55.2
Co	34.1±3.3	Nb	48.1±2.5	Tc	(42)
Cr	32.9±5.9	Nd	60.7	Te	69
Cs	67.9±3.7	Ni	35.1±5.0	Th	59.9±0.8
Cu	44.0±5.0	Os	50 ±8	Ti	39.3±8.0
Dy	54.8	P	39.5	Tl	72.1±2.7
Er	54.8	Pb	72.2±5.0	Tm	52.3
Eu	60.2	Pd	45.6±2.2	U	64.0±5.2
Fe	35.0±7.8	Pm	(61)	V	36.8±6.3
Ga	40.0±2.5	Pr	61.1	W	40.9±3.3
Gd	56.0	Pt	39.3±1.5	Y	50.4±4.2
Ge	49.8±3.3	Rb	59.2±0.5	Yb	54.0
Hf	53.0	Re	42 ±6.5	Zn	42.8±6.3
Hg	59.4±5.4	Rh	(46)	Zr	37.2±9.6

TABLE XIV. "*Latimer*" *entropy contributions* n$\{X\}$ *as a function of the charge number*, n, *of the cations*

X	$n\{X\}$J/K . mol						
	$n=1$	$n=2$	$n=2.67$	$n=3$	$n=4$	$n=5$	$n=6$
O^{2-}	4.5	2.9	0.4	2.4	3.2	7.1	12.7
S^{2-}	20.6	18.4		20.1	17.0	22.4	
Se^{2-}	35.5	32.8		34.1	30.9		
Te^{2-}	38.3	41.9		44.1	40.1		
F^-	20.8	17.0		18.3	20.3	22.4	27.2
Cl^-	36.3	31.8		30.3	34.4		37.2
Br^-	50.3	45.7		44.7	50.8		
I^-	58.3	53.5		54.8	53.9	59.4	
CO_3^{2-}	62.4	46.6					
SO_4^{2-}	80.0	69.5		64.2			
NO_3^-	86.0	74.0					
NO_2^-	70.6	(61)					
SO_3^{2-}	42.9						

In order to obtain from these tables the standard entropy of a solid compound, the appropriate value for its cation in Table XIII is multiplied by the number of cations in the molecule and added to the value for the anion. The latter figure is obtained by multiplying the value given in Table XIV according to the charge on the cation by the number of anions in a formula weight. Thus the standard entropy of $Al_2(SO_4)_3$ is obtained as $S(298\ K) = (2 \times 23.4) + (3 \times 64.2) = 239.4$ J/K . mol.

Although Latimer originally devised the method described above for application to predominantly ionic compounds, Mills has nevertheless demonstrated its usefulness in estimating standard entropy values for non-ionic compounds also. He thus derived values of the {X} contributions to the standard entropy of metallic borides, carbides, silicides, nitrides, phosphides, arsenides and antimonides from the more limited data available for such substances and found that the uncertainties associated with the resulting values were not much larger than those found for ionic compounds. He attributed this consistency to the similarity of crystal structure and bonding characteristics for most compounds of an element X with a given stoichiometry.

Table XV presents the values obtained by Mills from an analysis of a number of compounds, which, because of the unusual valencies displayed, were categorized into different MX_a, MX_b types.

TABLE XV. *"Latimer" entropy contributions {X} for metallic borides, carbides, silicides, nitrides, phosphides, arsenides and antimonides*

X	{X}J/K . mol in $MX_{0.33}$	in $MX_{0.5}$	in $MX_{0.7\pm0.1}$	in MX	in MX_2	in MX_3
B					−2.2(±4.5)	
C		16(±10)	6.5(±4)	10.9(±5)	6.3(±4)	
S	−12(±8)		5.8(±3.5)	7.5(±3.5)	11.2(±2.5)	14
N				−4.6(±4)		20(±5)
P				10.5(±3)		
As				29(±10)		
Sb		40(±10)		40(±5)	36(±8)	

Richter and Vreuls (1979g) have estimated standard entropy values for solid and molten salts within a mean deviation of about 3.5% based on the linear dependence of the entropy on the radius of the cation constituent of the compound. The linear extrapolation of $r(c) \rightarrow 0$ ($r(c)$ = cation radius) leads to the anionic contribution of the molar entropy depending on the cationic charge. The method allows evaluation of cationic contributions to the entropy from experimental data and has the advantage that its application is not restricted to ionic compounds only.

Table XVI presents the anionic contribution to the standard entropy of solid salts at 298 K as a function of the cation charge and Table XVII presents the cationic contributions as evaluated by Richter and Vreuls.

2. Gases

The ease with which a solid lattice disrupts to form a liquid, and with which that liquid evaporates, depends broadly on the atomic size and type of bond. Thus one may expect to find a certain regularity in the normal entropy of gases in relation even to the molecular weight alone. This is because the translational partition function has a larger contribution than the vibrational and rotational contributions. That this is indeed so is illustrated in

TABLE XVI. *Anionic contributions to the entropy of solid salts at 298 K as a function of cation charge*

Anion	S(298 K) J/K . mol Monovalent cation	Divalent cation
F^-	8.9	18.6
Cl^-	37.8	30.5
Br^-	51.6	47.0
I^-	65.9	53.9
H^-	2.1	—
OH^-	25.1	18.3
NO_3^-	83.6	66.1
ClO_4^-	105.8	—
AlF_6^{3-}	114.2	—
$AlCl_6^{3-}$	229.0	—
$Al_2O_4^{2-}$	—	57.0
$B_2O_4^{2-}$	—	61.8
$B_4O_7^{2-}$	117.6	94.8
$B_6O_{10}^{2-}$	156.9	—
CO_3^{2-}	63.3	44.4
$Cr_2O_4^{2-}$	—	84.7
$Fe_2O_4^{2-}$	—	96.3
MoO_4^{2-}	—	84.3
O^{2-}	6.4	2.5
O_2^{2-}	21.9	—
O_2^-	76.0	—
PO_4^{3-}	—	60.3
S^{2-}	31.3	18.1
SiO_3^{2-}	43.4	39.9
SiO_4^{4-}	—	40.2
SO_4^{2-}	84.8	68.7
TiO_3^{2-}	—	51.6
TiO_4^{4-}	—	62.6
WO_4^{2-}	—	85.6

TABLE XVII. *Cationic contributions to the entropy of solid salts at 298 K*

Cation	S(298 K) J/K . mol	Cation	S(298 K) J/K . mol
Ag^+	55.3	Li^+	19.7
Ba^{2+}	59.3	Mg^{2+}	26.7
Be^{2+}	13.2	Mn^{2+}	48.1
Ca^{2+}	39.5	Na^+	34.0
Cd^{2+}	51.3	Ni^{2+}	33.3
Co^{2+}	44.8	Pb^{2+}	72.7
Cs^+	62.1	Rb^+	55.9
Cu^{2+}	41.5	Sn^{2+}	60.9
Fe^{2+}	52.2	Sr^{2+}	52.9
Hg^{2+}	67.1	Tl^+	68.3
K^+	46.4	Zn^{2+}	47.9

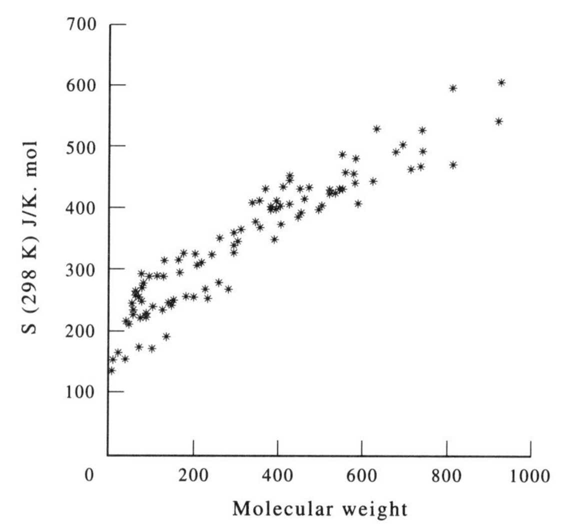

FIG. 92. Standard entropy of gases as a function of molecular weight.

Fig. 92, in which the normal entropies of a number of gaseous species are plotted against molecular weight.

Although the standard entropy of a gaseous species is best estimated by making use of information on the molecular parameters, the observed simple correlation with molecular weight (which is also a function of the number of atoms in the molecule) will allow a satisfactory value of the standard entropy of a particular gaseous species to be obtained very simply and rapidly. The following relations have been derived from available experimental information. (The numbers in brackets indicate the mean deviations from the average of the experimental values.)

No. of atoms in mol.	S(298 K) J/K . mol
1	$110.88 + 33.05 \log M$ ($\pm$ 6.7)
2	$101.25 + 68.20 \log M$ ($\pm$ 5.9)
3	$37.66 + 111.71 \log M$ ($\pm$ 7.5)
4	$-7.53 + 146.44 \log M$ ($\pm$ 6.7)
5	$-131.80 + 207.11 \log M$ ($\pm$11.3)

If there are more than five atoms in the molecule, the last equation may still be used, but it should be possible to derive further equations as more experimental information becomes available.

Standard entropy values for larger, double molecules, can be estimated with very reasonable accuracy if the standard entropy of the corresponding single molecule is known. A survey of available information for gaseous halide species (1984d), for example, has shown that the simple relation

$$\frac{S(298\ \mathrm{K})\,(\mathrm{DM})}{S(298\ \mathrm{K})\,(\mathrm{SM})}=1.52\pm 0.05$$

may be used to estimate missing values of $S(298\ \mathrm{K})$ for large gaseous halide molecules.

While very few data are available for gaseous oxide dimers, a similar, simple correlation appears to apply to these also. The quotient in this case is a little smaller with a value of about 1.45.

B. Entropies of Mixing of Non-metallic Solution Phases

The mixing of cations in, for example, double oxide compounds can be accompanied by additional entropy contributions. The spinels, MX_2O_4 or $MO.X_2O_3$, represent an important example. In this structure one third of the cations occupy tetrahedral sites in the close-packed oxide lattice. If all of the tetrahedral sites which are occupied contain only $M(2+)$ ions and the octahedral sites contain only $X(3+)$ ions, this structure, the normal spinel, does not have a cation-mixing contribution. If, however, some of the tetrahedral sites which are occupied contain $X(3+)$ ions, and some of the octahedral sites contain $M(2+)$ ions, there is such a contribution. In the general spinel of formula $(M_xX_{1-x})[M_{1-x}X_{1+x}]_2O_4$, where x is equal to zero in the **normal** spinel, and equal to unity in the **inverse** spinel, the value of x can be calculated from the equilibrium constant for the exchange reaction

$$(\mathrm{M})+[\mathrm{X}]=[\mathrm{M}]+(\mathrm{X})$$

$()$ = tetrahedral site, $[\,]$ = octahedral site

$$\Delta H(\mathrm{exchange})=-\boldsymbol{R}T\ln\frac{(1-x)^2}{x(1+x)}$$

ΔH(exchange) here is the exchange energy for M and X ions between tetrahedral and octahedral sites as indicated. Apart from considerations of Madelung energy, which normally cause the higher valency cations to seek the more highly coordinated octahedral sites, the cations of the transition metals have a contribution arising from the d electron population. This is the reasons for Co_3O_4 and Mn_3O_4 having the normal spinel structure, whereas Fe_3O_4 has the inverse structure. The entropy contribution due to cation mixing is given by the equation

$$S=-\boldsymbol{R}\left[x\ln x+(1-x)\ln(1-x)+(1-x)\ln\frac{(1-x)}{2}+(1+x)\ln\frac{(1+x)}{2}\right]\qquad(127)$$

In a compound such as Zn_2TiO_4 or $2ZnO.TiO_2$, the $Ti(4+)$ ion occupies half of the occupied octahedral sites because of the high cationic charge, and hence x has the value of 0.5. The entropy of mixing can be less than that given by the equation above due to ordering of the $Zn(2+)$ and $Ti(4+)$ ions on the octahedral sites. This effect cannot be easily predicted without detailed knowledge of the structure of the oxide and its deviations due to distortion from the ideal spinel structure. There will always be a finite contribution from cation mixing, however, which will yield a "residual" entropy for the compound, which cannot be calculated from the usual thermal data.

One of the classical rules which is applied to the calculation of activities in mixtures of non-metallic compounds is Temkin's Rule, which relates activities to the number of atoms in each molecular species. The rule is based upon the evaluation of the entropies of mixing cations and anions on their respective sub-lattices. Thus, in an A_2Y–B_2Y random solution, there are two g-atoms of cations mixing per mole of solution. Temkin's Rule states that

$$a_{A_2Y} = N^2_{A_2Y} \text{ and } a_{B_2Y} = N^2_{B_2Y}$$

In an ideal random mixture (where $\boldsymbol{R}T \ln a_{A_2Y} = -T.\Delta\bar{S}_{A_2Y}$)

$$\Delta\bar{S}_{A_2Y} = -2\boldsymbol{R} \ln N_{A_2Y} = -\boldsymbol{R} \ln N^2_{A_2Y}$$

and hence the rule.

Similarly in the aluminothermic reduction of Cr_2O_3:

$$2Al + Cr_2O_3 \rightarrow 2Cr + Al_2O_3$$

$$K = \left(\frac{a_{Cr}}{a_{Al}}\right)^2 \cdot \left(\frac{a_{Al_2O_3}}{a_{Cr_2O_3}}\right) \cong \left(\frac{a_{Cr}}{a_{Al}}\right)^2 \cdot \left(\frac{N_{Al_2O_3}}{N_{Cr_2O_3}}\right)^2$$

by the application of Temkin's Rule.

In mixed carbide phases, Richardson (1949l) applied Temkin's Rule to show that the activities in the mixed carbide phase $(Fe, Mn)_3C$ could be obtained if it is assumed that $a_{Fe_3C} = N^3_{Fe_3C}$.

Another example in which the equilibria involve ionic mixtures is treated is the work of Flood (1953a) who showed that the Gibbs energy change for the exchange reaction involving two cations

$$(Na, K)Cl + 1/2Br_2 \rightarrow (Na, K)Br + 1/2Cl_2 \qquad \Delta G^\circ_{III}$$

could be calculated from the separate equations involving single cations

$$NaCl + 1/2Br_2 \rightarrow NaBr + 1/2Cl_2 \qquad \Delta G^\circ_{I}$$

$$KCl + 1/2Br_2 \rightarrow KBr + 1/2Cl_2 \qquad \Delta G^\circ_{II}$$

where $\Delta G^\circ_{III} = N_{NaCl}\,\Delta G^\circ_{I} + N_{KCl}\,\Delta G^\circ_{II}$ and N_{NaCl} and N_{KCl} are the atom fractions of NaCl and KCl respectively in the two-cation exchange reaction. It can be shown that this will be true if the excess Gibbs energy of mixing of the NaCl–KCl mixture is equal to that in the NaBr–KBr mixture. This will be so if both solutions obey Temkin's Rule, or are regular solutions with the same enthalpy of mixing.

The application of this procedure to the calculation of phosphorus refining equilibria in steel refining was demonstrated by Flood and Grjotheim (1952e) who showed that the reaction

$$2[P] + 5[O] + 3\{O^{2-}\} \rightarrow 2\{PO_4^{3-}\}$$

where [P] and [O] represent phosphorus and oxygen dissolved in liquid iron, and $\{O^{2-}\}$ represents oxygen ions dissolved in the slag, can be represented by the equation

$$\log K = N_{Ca^{2+}} \log K_{Ca^{2+}} + N_{Fe^{2+}} \log K_{Fe^{2+}}$$

for the slag containing Ca^{2+} and Fe^{2+} ions in the ionic fractions $N_{Ca^{2+}}$ and $N_{Fe^{2+}}$. The two equilibrium constants involved are then for

$$2[P]+5[O]+3CaO \rightarrow Ca_3(PO_4)_2; \quad K_{Ca^{2+}}$$

and

$$2[P]+5[O]+3FeO \rightarrow Fe_3(PO_4)_2; \quad K_{Fe^{2+}}.$$

Since log $K_{Ca^{2+}}$ has the value 21 ± 1 at steelmaking temperatures, and log $K_{Fe^{2+}}$ has the value 11 ± 1, then the calcium equilibrium will predominate in steelmaking (1969c).

Finally, it should always be remembered that the entropy of formation of a mixture of cations contains not only the configurational entropy arising from the random distribution of cations on the cation sub-lattice, but also contains a thermal entropy term related to heat capacity changes. For example, in the formation of spinels from their constituent oxides, Jacob and Alcock (1975b) found that the configurational term described on p. 178 is always accompanied by a thermal entropy of formation which must be added to the configurational term to obtain the total entropy of formation. Such thermal terms arising from changes in the vibrational structure of the cations and their surrounding oxygen ions on formation of an inter-oxide compound should always be considered. In the spinel studies, it was found that the thermal entropy of formation of spinels such as Fe_3O_4, $FeAl_2O_4$, FeV_2O_4, and $FeCr_2O_4$ could be represented by the equation

$$\Delta S = -7.32 + \Delta S^M \text{ J/K.mol}$$

where -7.32 entropy units originate from the non-configurational source. Table XVIII shows the relation between the total entropy, the configurational entropy and the degree of inversion of some spinels.

TABLE XVIII. *Thermodynamics and cation distribution in spinels*

Spinel	ΔS J/mole	ΔS^M J/mole	Fe(2+) fraction on tetrahedral sites (x) at 1200 K
Fe_3O_4	6.15	14.31	0.11
FeV_2O_4	3.05	9.20	0.81
$FeAl_2O_4$	−3.26	4.35	0.94
$FeCr_2O_4$	−7.24	0.13	0.999

Here

$$\Delta S^M = -R\left[x \ln x + (1-x)\ln(1-x) + (1-x)\ln\frac{1-x}{2} + (1+x)\ln\frac{1+x}{2}\right]$$

in the compound $(Fe_xR_{1-x})[Fe_{1-x}R_{1+x}]O_4$ ($R = Fe^{3+}$, V^{3+}, Al^{3+} and Cr^{3+}).

4. Enthalpies of Formation

Without information on the enthalpies of formation of the substances taking part in a reaction, no reliable evaluation of Gibbs energy values, and hence of particular chemical equilibria of interest, can be carried out. Unfortunately, the methods available for estimating enthalpies of formation are often not very exact and apply to a relatively small group of compounds only. Consequently, as many methods as possible should be used to estimate a single value.

The enthalpies of the elements in their standard state at 298.15 K are taken to be zero by convention. From this basis, the enthalpies of formation of binary compounds only, will be considered. Since the variation of enthalpies of formation with temperature is generally small, they may often be assumed to be temperature-independent provided no change occurs in the state of aggregation.

A. General

If two elements form several compounds of different composition, there are certain fairly general relationships between their enthalpies of formation. To provide a consistent basis for comparison, all enthalpies of formation in the following discussion will relate to one mole of substance defined as A_xB_y with $x+y=1$.

If the melting points of the compounds in a given binary system are known, the one with the highest melting point may be expected to have the numerically largest enthalpy of formation per mol as defined above. If the melting points of the other compounds in the system are considerably lower, their enthalpies of formation may be approximated by the values, at the appropriate compositions, given by the straight lines joining the enthalpy of formation of the high-melting point compound to each of the pure elements. If their melting points are not much lower than that of the highest, their enthalpies of formation will have a value between the additive and the maximum. This can be illustrated by considering the enthalpy of formation curve for the Li–Sn system, shown in Fig. 93. Here, the compound with the highest melting point (Li_7Sn_2) displays the largest enthalpy of formation. The compositions Li_4Sn and LiSn, both of which show a maximum in the liquidus curve, are marked in the enthalpy of formation curve by pronounced angularities. The other compounds in the system are formed peritectically, and in accordance with this, there is practically no indication of them in the $\Delta H/N$ curve. This diagram is quite typical for such a system.

It is emphasised that a system must be considered as a whole when evaluating thermochemical data critically. Consideration of compounds individually, and not in relation to neighbouring phases, has often led to inconsistencies when the resulting data are used to calculate phase equilibria in a given system.

B. Homologous Series

There is some evidence that a certain relationship exists between the enthalpy of formation of metal compounds and the atomic number of the metal in compounds of the same stoichiometric proportion and the same common radical. Such a correlation has been demonstrated earlier, for example, by Roth (1940d) with the then available data for oxides and chlorides. Depending on the compounds concerned, the curves obtained from this correlation may show sharp maxima and/or minima, but missing values can nevertheless be predicted from the curves with a fair degree of reliability.

More recently, Pettifor (1986d) has reorganised the Periodic Table of the elements into a single string instead of the normal Periods and Groups—the relative ordering being given by what Pettifor terms the "Mendeleev number". The resulting sequence is illustrated in Fig. 94. The purpose of the re-ordering was to permit a better classification of the structures of binary compounds. Using this new arrangement of the Periodic Table, enthalpies of formation of particular types of compound with particular structures can be plotted as a

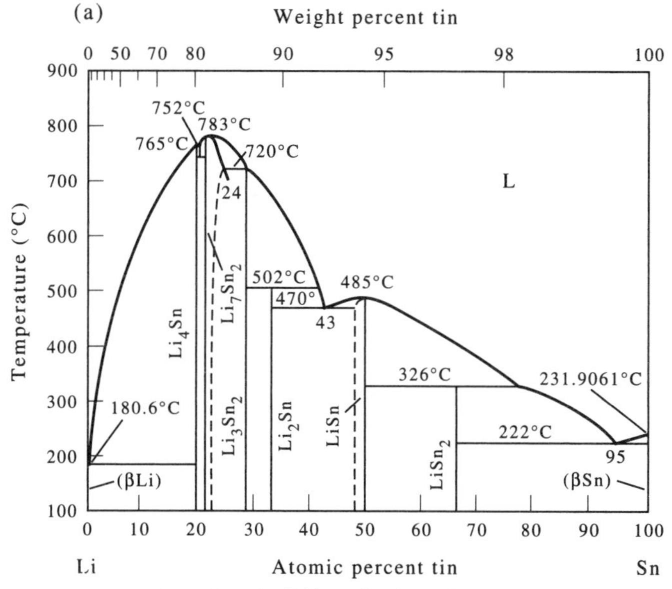

FIG. 93a. The lithium-tin phase diagram.

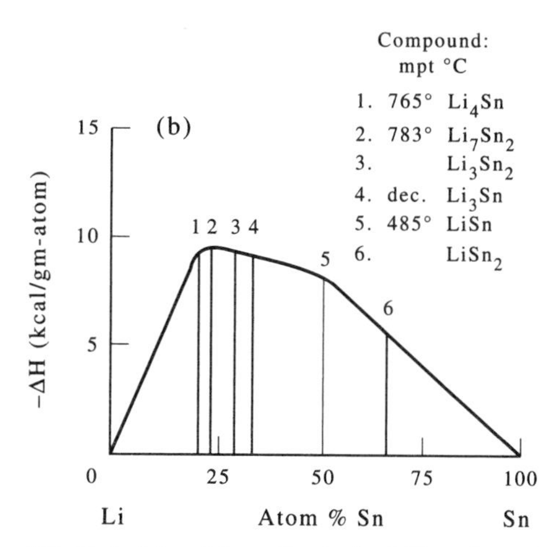

FIG. 93b. Enthalpies of formation in the lithium-tin system.

function of the Mendeleev number. Such a plot has been made by Stolten and is shown in Fig. 95 for metal nitrides and carbides with the cubic NaCl structure (1991d). The values lie on a smooth curve, rather than on the more irregularly-shaped curves resulting from the "conventional" Periodic arrangement. This enables not only missing values to be estimated for compounds which are known to exist, but also values which correspond to compounds in a metastable structure. Such information is particularly important in calculating the stability ranges of phases formed from mixtures of compounds.

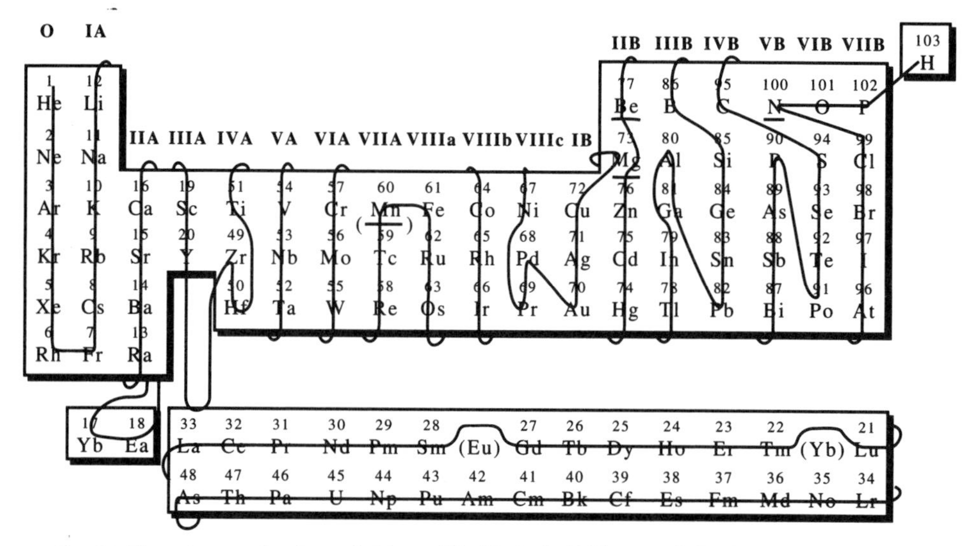

FIG. 94. The string running through this modified Periodic Table puts all the elements in sequential order, given by the Mendeleev number.

C. Empirical Relations

In a series of papers, Hisham and Benson (1985a, b; 1986b; 1987c–f; 1988c) have compiled information on the enthalpies of formation of a wide variety of inorganic compounds and derived empirical relations to enable known values to be calculated to within close limits, and missing values to be estimated, for particular groups of compounds. Some of the equations they have derived are presented below.

1. Polyvalent Metal Oxides (1985a, b)

For polyvalent oxides, MO_z, which have three or more well-defined stoichiometric valence states, the following relation holds:

$$-\Delta H(298\ \mathrm{K}) = az + bz^2.$$

The authors present values of a and b for 15 metals.

2. Metal Oxyhalide Compounds (1986b)

Examination of 35 solid metal oxyhalides, MO_xX_y, showed that their standard enthalpies of formation can be correlated quantitatively with the enthalpies of formation of the corresponding oxides and halides of the same oxidation states by the equation:

$$\Delta H(MO_xX_y) = a[(2x/z)\Delta H(MO_z/2) + (y/z)\Delta H(MX_z)] + C$$

where $z = 2x + y$ is the formal oxidation state of the metal and $MO_z/2$ and MX_z are the corresponding oxide and halide of the same oxidation state z. C is a correction factor in kcal/mol.

For main and first transition metal compounds, $a = 1$ and $C = 0$. For trivalent-state

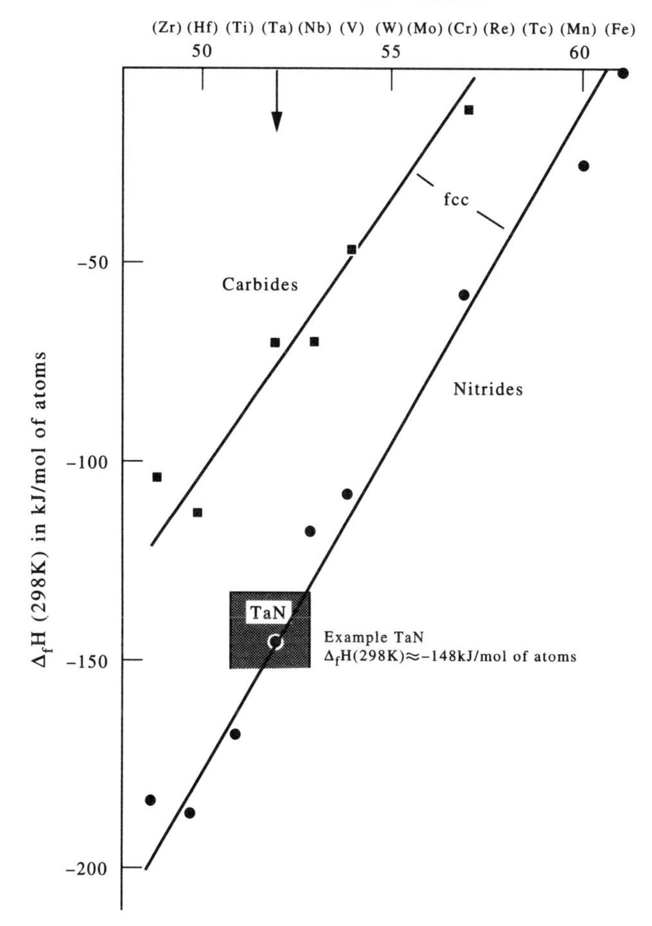

FIG. 95. Enthalpies of formation of carbides and nitrides with the cubic NaCl structure plotted using the Pettifor arrangement of the Periodic Table.

lanthanides, $a=2.155\pm0.12$ and $C=1078.6\pm5.4$ or 1047.7 ± 5.0 kJ/mol. For tetravalent oxychlorides, $a=1$ and $C=20.9$ kJ/mol. For penta- and hexavalent compounds, $a=1$ and $C=0$.

3. Double Salts with the Formula MX_aY_b (1987c–f)

The standard enthalpies of formation of double salts of the type MX_aY_b can be calculated additively from the enthalpies of formation of their binary salts MX_c and MY_d.

For divalent metals the relation takes the form:

$$\Delta H(\mathrm{MXY}) = 1/2\ \Delta H(\mathrm{MX_2}) + 1/2\ \Delta H(\mathrm{MY_2}) + C$$

where $C=-13.4$ or -17.6 kJ/mol.

From an analysis of the more limited amount of data available for trivalent and tetravalent metals, a simple additivity relation is again found

$$\Delta H(MX_aY_b) = (ax/z)\Delta H(MX_z/x) + (by/z)\Delta H(MY_z/y) + C$$

where x, y, and z are the formal valencies of X, Y, and M respectively, i.e. $z = ax + by$, and $C = 0$.

4. Oxides, Carbonates, Sulphates, Hydroxides and Nitrates (1987c–f)

The standard enthalpies of formation of any three compounds for a particular metal oxidation state can be correlated quantitatively by two-parameter linear equations.

For mono- and divalent compounds

$$\Delta H(SO_4) - \Delta H(O) = 1.36[\Delta H(CO_3) - \Delta H(O)] - 13.4 \text{ kJ/mol.}$$

For monovalent compounds

$$\Delta H(OH) - \Delta H(O) = 0.463[\Delta H(SO_4) - \Delta H(O)] - 9.6 \text{ kJ/mol}$$

and

$$\Delta H(NO_3) - \Delta H(O) = 1.02[\Delta H(SO_4) - \Delta H(O)] - 234.7 \text{ kJ/mol.}$$

For divalent compounds

$$\Delta H(OH) - \Delta H(O) = 0.318[\Delta H(SO_4) - \Delta H(O)] + 94.6 \text{ kJ/mol}$$

and

$$\Delta H(NO_3) - \Delta H(O) = 1.025[\Delta H(SO_4) - \Delta H(O)] - 500.4 \text{ kJ/mol}$$

where $\Delta H(O)$, $\Delta H(SO_4)$, $\Delta H(CO_3)$, $\Delta H(NO_3)$, and $\Delta H(OH)$ are the standard enthalpies of formation in kJ/mol of the oxide, sulphate, carbonate, nitrate, and hydroxide respectively of the metal. Figures 96 and 97 illustrate the observed linear correlations for the sulphates, nitrates and hydroxides of different metals.

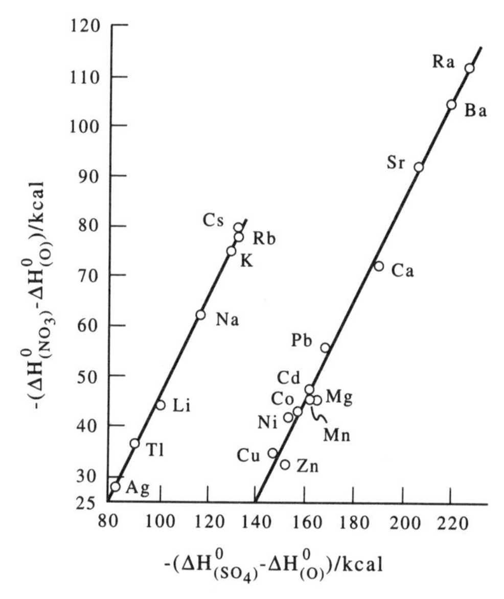

FIG. 96. Relationship between $(\Delta H^\circ_{(SO_4)} - \Delta H^\circ_{(O)})$ vs $(\Delta H^\circ_{(NO_3)} - \Delta H^\circ_{(O)})$.

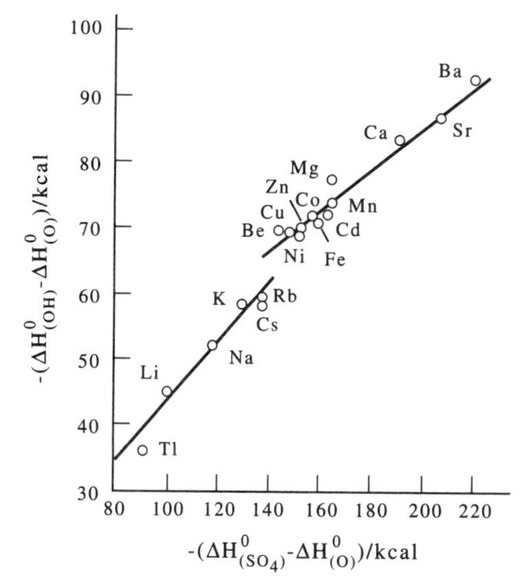

FIG. 97. Relationship between $(\Delta H^\circ_{(SO_4)} - \Delta H^\circ_{(O)})$ vs $(\Delta H^\circ_{(OH)}) - \Delta H^\circ_{(O)})$.

5. Halides (1987c–f)

The standard enthalpies of formation of any three solid halides, MX_n, MY_n, and MZ_n of any metal M with formal valence n (including cations such as NH^{4+}), can be correlated by the general equation

$$\Delta H(MX_n) - \Delta H(MY_n) = a[\Delta H(MX_n) - \Delta H(MZ_n)] + bn.$$

The coefficient a and b are the same for any particular main or subgroup of a given valence state. Values of b vary over a wide range, but a is always close to unity.

For any given group, maximum deviations are found to be no more than ± 12.5 kJ/mol. Figure 98 illustrates the observed correlation for subgroup metal halides.

D. Enthalpies of Formation of Double Oxides

Many methods are presented in the literature for estimating enthalpies of formation of double oxides. These are becoming increasingly important as new materials are developed. Some of the methods are specific to small groups of materials, others can be applied to a wider range of substances.

Slough (1973l) has made a comparison of methods available for the estimation of such values and has tabulated the results obtained. A summary of some of these methods is presented below.

1. Plots Involving the Ratio of Ionic Charge to Ionic Radius

Using the basic assumption that van der Waals and polarization forces are of major importance in determining the enthalpy change on reaction of two different oxides, Slough has developed a useful procedure for estimating this change (1972h). He found that for many double oxide combinations, good linear plots of C/R against $\Delta H(298\text{ K})$

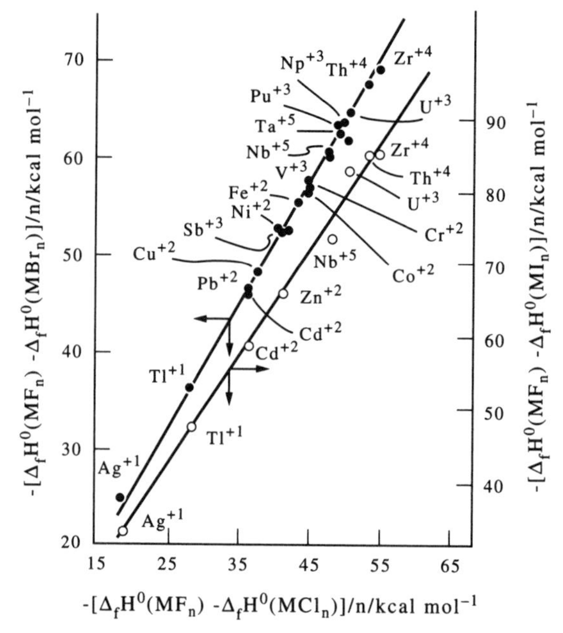

FIG. 98. Relationship among $\Delta_f H^\circ(MX_n)$ for subgroup metal halides.

from the component oxides were obtained. (C is the charge number and R the crystal ionic radius of the cation.) Ferrates, titanates, tungstates, vanadates, zirconates, silicates, selenites and borates were all analysed in this way. Deviations of individual points from the linear plots were usually not greater than 20 kJ/mol (see Fig. 99).

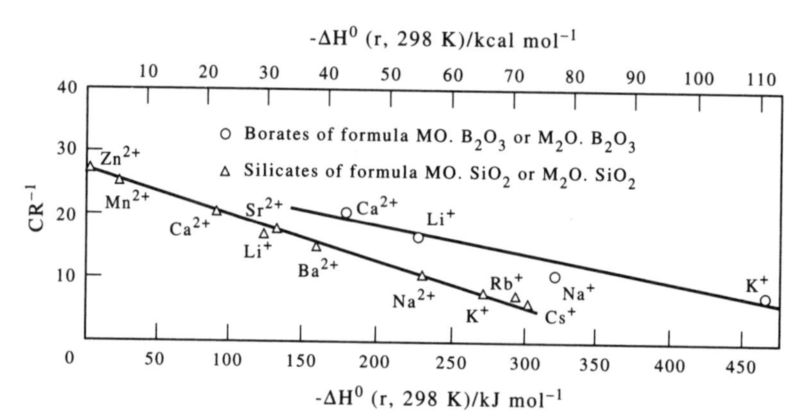

FIG. 99. O—O borates of formula $MO . B_2O_3$ or $M_2O . B_2O_3$. Δ—Δ silicates of formula $MO . SiO_2$ or $M_2O . SiO_2$.

Since the ionic charge of the cation is of major importance in applying this method, double oxides formed from metal oxides in which covalent bonding predominates (e.g. BeO, Ag_2O, or Cu_2O) do not fit into the plots concerned.

2. Statistical Analysis Methods

Schwitzgebel *et al.* (1971p) have produced a general relation for the estimation of enthalpies of formation of double oxides based on a statistical correlation of existing data. This relation takes the form

$$\Delta H(298\ \mathrm{K})\mathrm{kJ/mol} = -4.184\, b(K_\alpha - A_\beta)n$$

where $\Delta H(298\ \mathrm{K})$ is the enthalpy of formation from the component oxides, K_α represents the base strength of the oxide or alternatively, the stability of the cation in the double oxide combination, A_β represents the acid strength of oxide β. The exponent n_β is taken to be characteristic of the anion resulting from the double oxide combination.

Values of cation and anion parameters given by Schwitzgebel *et al.* are reproduced in Tables XIX and XX.

TABLE XIX. *Cation parameters*

α	K_α	α	K_α
Ag^+	8.97	Li^+	15.39
Al^{3+}	4.37	Mg^{2+}	8.68
Ba^{2+}	18.18	Mn^{2+}	9.10
Be^{2+}	4.93	Mn^{3+}	5.47
Bi^{3+}	7.75	Na^+	19.97
Ca^{2+}	13.10	Ni^{2+}	7.46
Cd^{2+}	8.53	Pb^{2+}	9.34
Ce^{4+}	9.05	Rb^+	24.60
Co^{2+}	8.95	Sb^{3+}	3.24
Cs^+	25.31	Sn^{4+}	1.41
Cu^+	4.08	Sr^{2+}	15.74
Cu^{2+}	4.53	Th^{4+}	7.73
Fe^{2+}	8.25	U^{4+}	6.79
Fe^{3+}	4.29	Zn^{2+}	6.56
K^+	23.73	Zr^{4+}	9.87

TABLE XX. *Anion parameters*

β	A_β	n_β
SO_4^{2-}	−9.77	1.45
CO_3^{2-}	0.00	1.43
SO_3^{2-}	−2.60	1.44
$Fe_2O_4^{2-}$	7.45	1.39
CrO_4^{2-}	0.13	1.43
$V_2O_6^{2-}$	1.55	1.47
TiO_3^{2-}	5.13	1.40
WO_4^{2-}	0.87	1.48
MoO_4^{2-}	−2.87	1.38
$Al_2O_4^{2-}$	5.24	1.38

3. Le Van's Method

Le Van (1972j) has described a method for estimating enthalpies of formation of oxyacids based essentially upon the assumption of additivity of bond energies. This allows

ΔH(298 K) for an oxyacid salt to be expressed in terms of two characteristic parameters, P and Q, as given by the relation

$$\Delta H(298\ \mathrm{K})\mathrm{kJ/mol} = [n(p)P + n(q)Q + 4.184(4(n(q))^2 + 4.184(n(p))^2]$$

where $n(p)$ represents the number of anions, $n(q)$ the number of cations, and the characteristic parameters P and Q refer to anion and cation respectively.

Values of the parameters P and Q are given in Tables XXI and XXII.

TABLE XXI. *Values of the parameter* Q

Cation	Q	Cation	Q	Cation	Q	Cation	Q
Ag^+	−92	Cs^+	−444	Li^+	−452	Sb^{3+}	−393
Al^{3+}	−916	Cu^+	−84	Mg^{2+}	−741	Sn^{2+}	−406
Ba^{2+}	−883	Cu^{2+}	−213	Mn^{2+}	−523	Sn^{4+}	−544
Be^{2+}	−653	Fe^{2+}	−372	Na^+	−448	Sr^{2+}	−862
Bi^{3+}	−469	Fe^{3+}	−423	NH_4^+	−326	Th^{4+}	−1435
Ca^{2+}	−858	Hg^+	−121	Ni^{2+}	−331	Ti^{2+}	−544
Cd^{2+}	−385	Hg^{2+}	−167	Pb^{2+}	−352	Tl^+	−205
Ce^{4+}	−1239	In^{3+}	−649	Pd^{2+}	−205	U^{4+}	−958
Co^{2+}	−343	K^+	−448	Ra^{2+}	−891	UO_2^{2+}	−1292
Cr^{2+}	−728	La^{3+}	−1213	Rb^+	−444	Zn^{2+}	−439

TABLE XXII. *Values of the parameter* P

Anion	P	Anion	P	Anion	P
$Al_2O_4^{2-}$	−1494	MoO_4^{2-}	−732	HSO_4^-	−724
AsO_4^{3-}	−460	NO_2^-	+29	$S_2O_3^{2-}$	−314
$HAsO_4^{2-}$	−607	NO_3^-	−67	$S_2O_5^{2-}$	−644
$H_2AsO_4^-$	−753	PbO_3^{2-}	−84	$S_2O_6^{2-}$	−908
BO_2^-	−653	PO_4^{3-}	−854	$S_2O_8^{2-}$	−1105
BO_3^-	−515	HPO_4^{2-}	−983	$S_4O_6^{2-}$	−950
$B_4O_7^{2-}$	−2853	$H_2PO_4^-$	−1146	SrO_3^{2-}	−335
BrO_3^-	+84	HPO_3^{2-}	−623	TiO_3^{2-}	−795
ClO^-	+50	$H_2PO_3^-$	−799	UO_4^{2-}	−1318
ClO_2^-	+63	$P_2O_7^{4-}$	−1766	VO_3^-	−762
ClO_3^-	+34	$H_2P_2O_7^{2-}$	−2004	VO_4^{2-}	−975
ClO_4^-	+21	$H_3P_2O_7^-$	−2121	ZnO_2^{2-}	−13
CNO^-	−13	ReO_4^-	−657	WO_4^{2-}	−845
CO_3^{2-}	−356	SeO_4^{2-}	−305		
HCO_3^-	−544	SiO_3^{2-}	−795		
CrO_4^-	−565	SnO_3^{2-}	−381	$HCOO^-$	−285
$Cr_2O_4^{2-}$	−1075	SO_3^{2-}	−314	CH_3COO^-	−326
IO_3^-	−109	HSO_3^-	−444	$CH_3CH_2COO^-$	−352
MnO_4^-	−393	SO_4^{2-}	−569		

4. Comparison of Data for Similar Compounds

Slobodin *et al.* (1975d) calculated enthalpies of formation for selected *ortho*-vanadates from a comparison of ΔH(298 K) values for various compounds with the same cation and the *ortho-*, $EO_4(n-)$, anion. Thus the sulphates, phosphates, molybdates, and orthovanadates of di- and trivalent elements were chosen for comparison, and calculations of the parameter A were made using the relation

$$A = \Delta H(298\ \mathrm{K})/r\,.\,m$$

where $\Delta H(298\ \mathrm{K})$ is the difference between the experimental and additive value for formation of the particular compound from the corresponding oxides (referred to one $EO_4(n-)$ group), r is the radius of the cation, and m is the number of cations per $EO_4(n-)$ group.

Slobodin *et al.* found for the four types of salt selected, and using Ba^{2+}, Ca^{2+}, Mg^{2+}, Sr^{2+}, Fe^{2+}, Ni^{2+}, Al^{3+}, Cr^{3+}, Fe^{3+} as the cations concerned, that the values of A for compounds with the same cation vary almost linearly and that the straight lines joining the values of A for compounds with the same cation are parallel (see Fig. 100). This enables missing values to be estimated with a fair degree of certainty. The method can presumably also be applied in a similar manner to other compounds.

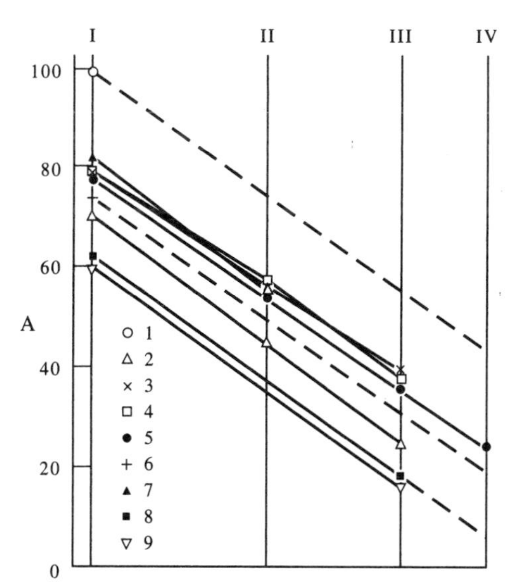

FIG. 100. Dependence of the parameter A for various compounds with the same cation on the anionic groups: (I) sulphates; (II) phosphates; (III) molybdates; (IV) vanadates; (*1*) Cr^{3+}; (*2*) Al^{3+}; (*3*) Sr^{2+}; (*4*) Ba^{2+}; (*5*) Ca^{2+}; (*6*) Fe^{3+}; (*7*) Mg^{2+}; (*8*) Ni^{2+}; (*9*) Fe^{2+}.

E. Volume Change and Enthalpy of Formation

If the enthalpy of formation is required of a compound whose density and structure is known, one can use a method of estimation which was first applied to intermetallic compounds (1941b). Its basis is that the deformation or polarisation of the atoms of the two metals which occurs on alloy formation depends upon their affinity. Attention was first brought to this relationship by Richards (1902b) who compared the enthalpies of formation of the halides with the difference between their molecular volume and the total atomic volumes of their constituents. From this striking connection he was led to the belief that enthalpy of formation was an enthalpy of compression.

In the formation of simple inorganic salt-like compounds the change in density is mainly due to the formation of ions, but surprisingly enough the relationship we have just mentioned still applies. A measure of the change in atomic size is given by the contraction in molecular volume, taking certain rules into observation. The contraction on formation of phases with the NaCl and CsCl structure is calculated in the following way. The atomic volumes of the metals are taken in their state of co-ordination number 12. The atomic volumes of nitrogen, oxygen and sulphur are not known for high co-ordination, but for this purpose they are assumed to be 5.0, 9.0 and 14.5, respectively. The atomic volumes are taken as those of the solid state. The molecular volume of the compound should be that of closest packing, in the face-centred cubic form. This requires that the measured molecular volume (molecular weight/density) should be multiplied by a factor e which has the values 0.95 and 0.825 for the CsCl and the NaCl types of lattice respectively. The percentage volume contraction is then given by

$$\Delta V = \frac{100(eMV - \sum AV)}{\sum AV} \tag{128}$$

where MV is the meaured molecular volume of the compound and $\sum AV$ the sum of the atomic volumes of the two components.

The values of ΔV for a number of compounds are plotted in Fig. 101 against their enthalpies of formation. The plots are seen to lie fairly close to a curve from which, with a few exceptions, they deviate no more than $\pm 25{,}000$ J/g-atom. The wide deviation of lithium thallide is explained by the large difference in the two ionic sizes which permits a much closer packing than is represented by the factor 0.95. The attempt to make an allowance for these ionic differences by introducing an atomic size ratio into the formula would make the relationship much more complicated without even then obtaining complete agreement. A similar relationship holds between compounds of other structures, such as the zinc sulphide or γ-brass type (1941b). It needs further detailed consideration, however, and at this stage it can only be taken as an empirical relationship which gives an approximate value for the enthalpy of formation. Its results should be used only in comparison with those of other methods of estimation, since large deviations can occur and since the densities are not always known with sufficient accuracy.

Instead of plotting the data of all binary compounds in one diagram it is preferable to consider the compounds of each structural type separately, using for the volume change term the expression

$$\Delta V = \frac{100(MV - AV)}{\sum AV}. \tag{129}$$

The unknown enthalpy of formation may then be estimated from the intersection of the curve by the corresponding ΔV value.

F. Enthalpy of Solution

It follows from the considerations in Section II.1.B.3 that the enthalpy of formation of a compound can be calculated from its enthalpy of solution in a solvent and those of the

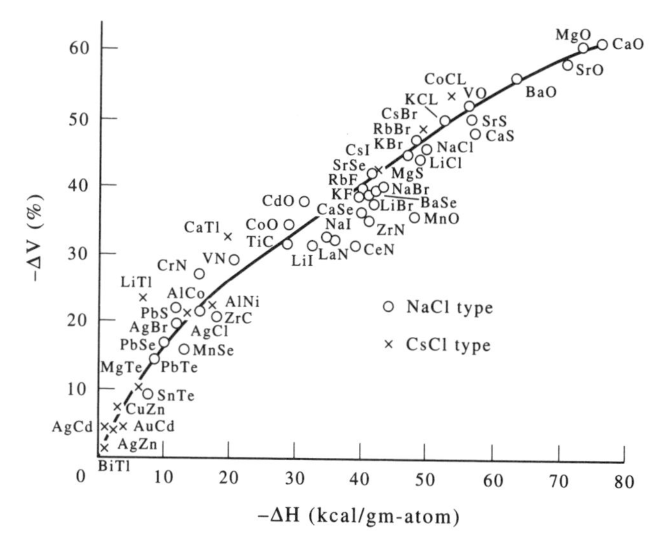

FIG. 101. Heat evolution and contraction on formation of compounds of simple structural type.

constituent elements in the same solvent provided that the final systems are identical. As a generalisation of the scheme given on p. 85 one may write for the solution of a compound of a divalent metal Me and a gaseous halogen:

$$\langle Me\rangle = [Me^{++}]_{aq\infty}; \qquad \Delta H_1$$
$$(X_2) = 2[X^-]_{aq\infty}; \qquad \Delta H_2$$
$$\langle MeX_2\rangle = [Me^{++}]_{aq\infty} + 2[X^-]_{aq\infty}; \qquad \Delta H_3$$

$$\langle Me\rangle + (X_2) = \langle MeX_2\rangle; \qquad \Delta H_4$$
$$\Delta H_4 = \Delta H_1 + \Delta H_2 - \Delta H_3.$$

It is conventional to take the standard states of ΔH_1 and ΔH_2 at infinite dilution at which any mutual interaction of the cations and anions can be neglected. These values are found by determining the enthalpy of solution at various concentrations and extrapolating to infinite dilution, denoted by the suffix "aq ∞".

The enthalpy of solution of ionic compounds in water (i.e. ΔH_3) are generally relatively small. On the other hand, the enthalpy of formation of a large number of anions and cations in water are known with quite a high accuracy, and it is possible to make use of these two facts for another method of estimating unknown enthalpies of formation. The hydrogen ion in water at infinite dilution is taken as the standard state in that the enthalpy of the reaction

$$\tfrac{1}{2}(H_2) = [H^+]_{aq\infty} \qquad (\Delta H = 0)$$

is taken to be zero and all other enthalpies of formation of ions are related to this value so that, for instance, the enthalpy of formation of hydrogen chloride in water at infinite dilution,

$$\tfrac{1}{2}(H_2) + \tfrac{1}{2}(Cl_2) = [HCl]_{aq\infty} = [H^+]_{aq\infty} + [Cl^-]_{aq\infty},$$

is the same as the standard enthalpy of formation of the chlorine ion:

$$\tfrac{1}{2}(Cl_2)=[Cl^-]_{aq\infty}.$$

In Table XXIII are given the standard enthalpies of formation of cations and anions according to the "selected values" of the Bureau of Standards,* supplemented or corrected by some more recent measurements. Further values for more concentrated solution of the hydrogen halides are obtainable from Table VIII.

TABLE XXIII. *Standard heats of formation of ions in aqueous solution at infinite dilution*

Ion	ΔH_{298} in $\frac{kJ}{g\text{-atom}}$	Ion	ΔH_{298} in $\frac{kJ}{g\text{-atom}}$	Ion	ΔH_{298} in $\frac{kJ}{g\text{-atom}}$	Ion	ΔH_{298} in $\frac{kJ}{g\text{-atom}}$
H^+	0.0	La^{3+}	−698.7†	Cu^+	+51.9	F^-	−328.9
Li^+	−278.2	Ce^{3+}	−699.6‡	Cu^{++}	+64.4	Cl^-	−167.4
Na^+	−241.4*	Pr^{3+}	−691.6†	Ag^+	−105.9	Br^-	−120.9
K^+	−252.7*	Nd^{3+}	−682.8‡	Au^+	+194.1	I^-	−56.1
Rb^+	−246.4	Th^{++}	−771.5§	Zn^{++}	−152.3	OH^-	−230.1
Cs^+	−261.9‡‡	U^{3+}	−514.6	Cd^{++}	−72.4	$S^=$	+32.6‖
Mg^{++}	−461.9	U^{4+}	−613.8	Hg^{++}	+174.1	SH^-	−17.2‖
Ca^{++}	−542.7	Mn^{++}	−218.8	Ga^{3+}	−210.9	$SO_4^=$	−907.5
Sr^{++}	−545.6	Fe^{3+}	−47.7	In^+	−52.3**	$Se^=$	(+96.2)
Ba^{++}	−538.1	Fe^{++}	−87.9	In^{3+}	−181.6**	$CO_3^=$	−676.1
Al^{3+}	−524.7	Co^{++}	−67.4	Tl^+	+5.4	PO_3^-	−985.8
		Co^{3+}	+88.7	Tl^{3+}	+115.9	PO_4^{3-}	−1284.1
		Ni^{++}	−64.0	Pb^{++}	+2.1	$CrO_4^=$	−870.3††

*E. E. KETCHEN and W. E. WALLACE, *J. Amer. Chem. Soc.*, **73**, 5810 (1951).
†H. R. LOHR and B. B. CUNNINGHAM, *J. Amer. Chem. Soc.*, **73**, 2025 (1951).
‡F. H. SPEDDING and C. F. MILLER, *J. Amer. Chem. Soc.*, **74**, 4195 (1952).
§L. EYRING and E. F. WESTRUM, *J. Amer. Chem. Soc.*, **72**, 5555 (1950).
‖J. W. KURY, A. J. ZIELEN and W. M. LATIMER, *J. Electrochem. Soc.*, **100**, 468 (1953).
W. KANGRO and T. WEINGARTEN, *Z. Elektrochem.*, **58, 505 (1954).
††C. N. MULDRON and L. G. HEPLER, *J. Amer. Chem. Soc.*, **79**, 4045 (1957).
‡‡H. L. FRIEDMANN and M. KAHLWEIT, *ibid.*, **78**, 4243 (1956).

If the enthalpies of solution are measured or can be estimated, Table XXIII supplies much information on enthalpies of formation of inorganic salts. Many values in Chapter 5 Table 1 are derived in this manner from known enthalpies of solution. Further values are best estimated individually by comparison with the known enthalpies of solution of chemically similar compounds. In order to extrapolate the enthalpy for infinite dilution a series of values at various concentrations must be determined, but if one value only has been measured in fairly dilute solution (molar ratio, H_2O:solute$>$400:1) this may be taken to represent approximately the enthalpy of solution at infinite dilution, particularly if the heat effect is small.

G. The Packing Effect

Kubaschewski (1958f, g) suggested that the stability of many multicomponent, truly metallic, phases stems mainly from a decrease in heat content due to an increase of co-ordination on formation from the component metals. To demonstrate this, it is necessary to re-define the co-ordination numbers of the crystallographers. Only *changes* in co-

*Now National Institute of Standards and Technology (NIST).

ordination during the formation of a phase from the component metals need to be considered, and a knowledge of the total energy of a lattice is not required.

Most metals crystallise in the f.c.c., b.c.c. and h.c.p. lattices. In the f.c.c. and h.c.p. structures every atom is surrounded by 12 nearest neighbours at a distance of $2r_A$, the gaps being sealed by 6 atoms at a distance $2r_A\sqrt{2}$. For the present purpose only the changes taking place on alloying within the distance up to $2r_A\sqrt{2}$ are being considered. The simplification is made that the metallic bond energy varies linearly with $1/d$ within the chosen distance d, which was found to be a reasonable approximation. When A forms a new phase with another metal B, the A atoms are surrounded by A and B atoms at various distances d_A and d_{AB}, respectively. If these distances are less than those corresponding to the atomic radii, i.e. $d_A < 2r_A$ and $d_{AB} < (r_A + r_B)$, the bond energy will be increased and the "effective" co-ordination number C_A^* higher than the actual co-ordination number, C_A, by a factor of either $2r_A/d_A$ or $(r_A + r_B)/d_{AB}$:

$$C_A^* = \sum \frac{2r_A}{d_A} C_A + \sum \frac{(r_A + r_B)}{d_{AB}} C_{AB}. \tag{130}$$

A corresponding relationship applies to the B atoms.

For distances $d_A > 2r_A$ and $d_{AB} > (r_A + r_B)$ a limit has been set above, the formal implication of which is that an A atom at the distance $2r_A\sqrt{2}$ just ceases to participate in the mutual attraction, the corresponding limit for A–B bonds being $(r_A + r_B)\sqrt{2}$. The corrections for obtaining the effective co-ordination numbers from the actual ones are then

$$C_A^* = \sum \frac{2(2r_A) - d_A}{(\sqrt{2} - 1)2r_A} C_A + \sum \frac{2(r_A + r_B) - d_{AB}}{(\sqrt{2} - 1)(r_A + r_B) C_{AB}} \tag{131}$$

and corresponding ones for the B atoms.

For the pure metals the atomic radii are taken to be equal to half the shortest atomic distances in their normal crystallographic structures. With the interpretation above, the f.c.c. and h.c.p. metals retain their co-ordination number of 12. For the b.c.c. metals the effective co-ordination number becomes 11.75. For hexagonal metals, in particular zinc and cadmium, the c/a ratios of which differ from the ideal one, the co-ordination number is less than 12. The accepted atomic radii and the corresponding effective co-ordination numbers of the metals are listed in Table XXIV. The designation of the structural types is that of *Strukturberichte* also used in *Metals Reference* Book [ed. C. J. Smithells] from which the details of most of the structures listed in Table XXIV have been obtained.

After having re-defined the co-ordination number, we now make the assumption that the energy of the A–B bonds may be obtained additively from those of the A–A and B–B bonds, and that the energy of attraction between the atoms of the pure metals, represented by the enthalpies of sublimation (L_A and L_B), is equally distributed over all the "links" up to a distance of $2r\sqrt{2}$, the energy being inversely proportional to the distance.

With these assumptions, ΔH_f becomes simply the negative product of the percentage increase in the effective co-ordination number and the enthalpies of sublimation of the components:

$$-\Delta H_f = \frac{N_A L_A [C^*(A)_{alloy} - C^*(A)_{metal}]}{C^*(A)_{metal}} + \frac{N_B L_B [C^*(B)_{alloy} - C^*(B)_{metal}]}{C^*(B)_{metal}}. \tag{132}$$

Using this equation, the enthalpies of formation of a number of intermetallic phases have been estimated and are listed in Table XXV and compared with the experimental values. Since the calculations are rather sensitive to small errors in the lattice dimensions, and since the experimental enthalpies of formation are often accurate to only ± 2000 J/g-atom, agreement between calculated and experimental values to within ± 4 kJ/g-atom may be regarded as an indication that the model is a reasonable one and that the simplifying assumptions are fair approximations.

There are notable exceptions to the quantitative application of the model. For instance, the experimental enthalpies of formation of the higher transition-metal aluminides are rather more exothermic than the estimated ones (1960g, h, i), whereas better agreement is found for the lower aluminides (Cr_2Al, Cu_3Al; Table XXV). Even so, comparisons between the experimental enthalpies and those estimated by means of the model contribute to the discussion of the bond mechanism in alloy phases. The bond mechanism of the B32 type (NaTl), originally discussed by Zintl (see Schneider and Heymer (1958j), derives from the interposition of two diamond lattices for the A and B atoms respectively, indicating the presence of covalent and polar as well as metallic bonds. It appears that the formation of the polar–covalent bonds absorbs some energy (see NaTl: Table XXV).

Further, where substantial differences between calculated and experimental values are found, the deviations tend to be of the same order and sign in homologous series—as has been pointed out by Alcock *et al.* (1966a).

The model implies that, for stability, it is essential that the larger atom should be the "stronger binder", i.e. have the higher enthalpies of sublimation, because it is the larger atom that will gain in co-ordination while the smaller atom will lose. As a result one may qualitatively predict the type of equilibrium diagram two metals, A and B, would form. For an empirical test, 350 binary phase diagrams have been ordered in groups and entered into a diagram of $(r_A - r_B)/\frac{1}{2}(r_A + r_B)$ vs $(L_A - L_B)/\frac{1}{2}(L_A + L_B)$, where A is always the larger atom. An also empirical, mostly small, correction term, $(\varepsilon_A - \varepsilon_B)^2$, has been added to the L term in order to account for the "electrochemical factor", ε_A and ε_B being the electro-negativities of the components, also listed in Table XXIV.

The results are shown in the simplified representation of Fig. 102. Systems forming large ranges of disordered solutions have been included in the survey although these are not covered by the model. A sharp distinction between simple eutectic systems and those showing liquid miscibility gaps cannot be made because it depends on the melting points whether the miscibility gap appears in the phase diagram or not.

Of the 350 systems examined of combinations between the metals of groups IA to VIIIA, IB and IIB (except Hg) less than 20 clearly do not fit into the pattern. The so-called "Hume-Rothery systems" also do not agree well with the present pattern, probably because of the relatively low co-ordination of pure zinc and cadmium.

H. Temperature Increase During Compound Formation

Descriptions of the preparation of alloys for metallographical work often contain statements of considerable increase in temperature on mixing the components in powder or liquid form. In suitable cases this information may be found quite useful. If the increase in temperature is rapid, little heat can be lost from the reacting materials to their surroundings, and in this event the increase in temperature multiplied by 30.33 (atomic heat) will give a useful minimum value for the enthalpy of formation per g-atom of alloy or compound. Allowance must of course be made for the heat of any melting (of components

TABLE XXIV. *Structure, atomic radius, effective co-ordination number, enthalpy of sublimation and electro-negativity of metals*

Metal	Structure	r in Å	Co-ordination number	L_{sb} (298 K) (kJ/g-atom)	ε
Li	A2	1.515	11.75	159.4	0.95
Na	A2	1.855	11.75	107.1	0.9
K	A2	2.31	11.75	89.2	0.80
Cs	A2	2.62	11.75	76.1	0.75
Be	A3	1.11	11.6	324.3	1.5
Mg	A3	1.60	12.0	146.4	1.2
Ca α	A1	1.98	12.0	178.2	1.0
β	A3	1.99	12.0	177.4	
Ba	A2	2.17	11.75	182.0	0.9
Y	A3	1.78	11.65	424.7	1.2
La	A3, A1	1.87	11.95	431.0	1.15
Ce	A3, A1	1.825	12.0	422.6	1.1
Th α	A1	1.80	12.0	575.3	IV:1.4
β	A2	1.72	11.75	572.8	
U γ	A2	1.50	11.75	525.1	IV: 1.4
Ti α	A3	1.45	11.8	469.9	1.6
β	A2	1.43	11.75	466.1	
Zr α	A3	1.585	11.7	608.8	1.5
β	A2	1.56	11.75	605.0	
Hf α	A3	1.57	11.67	619.2	1.4
V	A2	1.315	11.75	514.2	III: 1.5
Nb	A2	1.425	11.75	721.7	1.65
Ta	A2	1.43	11.75	781.6	III: 1.3; V: 1.7
Cr	A2	1.25	11.75	397.5	II: 1.4; III: 1.6
Mo	A2	1.36	11.75	658.1	IV: 1.6
W	A2	1.37	11.75	849.4	IV: 1.6
Mn α	A12	1.25	11.5	283.3	II: 1.4
β	A13	1.28	11.85		III: 1.5
γ	A1		12.0		
Re	A3	1.37	11.95	774.9	V: 1.8; VII: 2.2
Fe α	A2	1.25	11.75	415.5	II: 1.7
γ	A1	1.26	12.0	409.6	III: 1.8
Co	A1	1.25	12.0	428.4	1.7
Ni	A1	1.245	12.0	430.1	1.8
Ru	A3	1.325	11.72	651.4	2.0
Rh	A1	1.34	12.0	553.1	2.05
Pd	A1	1.375	12.0	376.6	2.0
Os	A3	1.34	11.7	788.3	2.0
Ir	A1	1.355	12.0	669.4	2.1
Pt	A1	1.38	12.0	564.8	2.1
Cu	A1	1.275	12.0	336.8	I: 1.8; II: 2.0
Ag	A1	1.44	12.0	284.1	1.8
Au	A1	1.435	12.0	368.2	2.25
Zn	A3	1.33	10.65	130.4	1.5
Cd	A3	1.49	10.5	111.8	1.5
Al	A1	1.43	12.0	329.3	1.5
In	A6	1.62	11.25	242.7	1.5
Tl	A3	1.71	11.90	181.0	I: 1.5; III: 1.9
Sn	A5	1.51	5.75	301.2	1.7
Pb	A1	1.745	12.0	195.0	1.6

or alloy) which occurs during the reaction. If the enthalpy of fusion of the alloy is unknown, the value 14.6 × (m.p., K) for ordered or 9.6 × (m.p., K) for disordered alloys may be used.

TABLE XXV. *Enthalpy of formation of intermetallic phases calculated from the increase in co-ordination*

Compound A_xB_y	Structure type	Lattice constant (Å)			Co-ordination number		$-\Delta H_f$(kJ/g-atom)	
		a	*b*	*cA*	B	Calculated	Experimental	
KNa_2	C14	7.48	—	12.27	14.45	11.20	3.43	1.0
LiTl	B2	3.42	—	—	12.85	14.7	28.9	26.8
NaTl	B32	7.47	—	—	14.25	13.85	26.4	18.8
$CaMg_2$	C14	6.22	—	10.10	15.75	11.96	18.2	19.7
$MgNi_2$	C36	4.81	—	15.77	16.45	12.2	22.8	28.2
$MgCu_2$	C15	7.05	—	—	15.65	11.9	12.6	11.7
MgAg	B2	3.32	—	—	13.9	12.3	14.9	18.8
MgCd	B19	3.22	5.27	5.00	12.55	11.4	8.2	8.2
Mg_3Cd	Do19	6.26	—	5.07	12.15	11.25	3.3	5.6
$Mg_{17}Al_{12}$	A12	10.54	—	—	13.4	11.3	2.4	(2.7)
$CaZn_5$	D2d	5.39	—	4.22	19.85	11.45	27.4	(23.0)
$CaZn_{13}$	D23	12.13	—	—	19.8	11.55	19.2	17.2
TiNi	B2	2.98	—	—	13.7	11.4	28.2	33.3
$TiNi_3$	Do24	5.09	—	8.29	15.3	11.85	30.3	35.0
$NbFe_2$	C14	4.835	—	7.88	13.85	11.25	31.0	20.5
Cr_2Al	C11	3.00	—	8.63	13.55	11.50	8.6	10.9
CoAl	B2	2.86	—	—	12.58	14.68	46.4	55.2
Cu_3Al	Orthorhombic	4.52	5.21	4.23	11.85	14.75	15.5	(16.3)
AuCu	L10	3.955	—	3.66	12.9	11.15	4.7	8.4
$AuCu_3$	L12	3.74	—	—	14.0	11.35	6.7	6.7

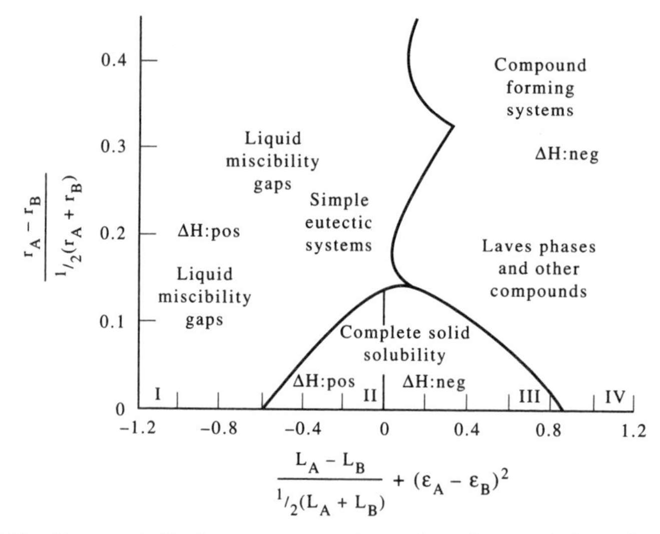

FIG. 102. Diagram indicating, in terms of atomic radius, enthalpy of sublimation, and electro-negativity of the component metals A and B, regions of preference for the formation of certain types of binary equilibrium diagrams.

The values of $-\Delta H$ calculated in this manner are low as some of the enthalpy of reaction is imparted to the surroundings from the moment reaction begins. They are only slightly lower when reaction is rapid and when contact with the crucible is poor. If the powder mixture is packed closely in a metallic crucible surrounded by thermal lagging material, the change in heat content of the crucible may be added. A knowledge of the relative weight of charge and crucible is of course necessary.

The figures in Table XXVI give an indication of how the minimum values for enthalpies of formation obtained by this means differ from the actual values. It may again be emphasised that the extent of the departure depends upon the rate of reaction. The measure of agreement found in the table shows that careful assessment of apparently casual information to be found in the literature can be of great value to the desperate thermochemist.

TABLE XXVI. *Minimum enthalpy of formation from temperature increase during compound formation* (all Joule values: per g-atom)

System	Composition	Temperature of reaction (°C)	$\Delta\theta$	State after reaction	Reference	L_f (J)	$\Delta\theta \times 30.33$	Minimum $-\Delta H$	$-\Delta H$ (meas.)
Fe–Si	80% Fe	1250	560	molten	(1921)	(16,736)	16,985	>33,700	34,300
Pb–Tl	34% Tl	330	52	solid	(1909b)	—	1577	>1600	2175
Mg–Sn	Mg_2Sn	450	250	molten	(1909b)	15,900	7585	>23,500	26,150
Mg–Sn	Mg_2Sn	590	220	molten	*	15,900	6675	>22,600	26,150
Mg–Bi	Mg_3Bi_2	480	450	molten	*	21,085	13,650	>34,700	34,100

*Kubaschewski, unpublished work.

This method cannot be expected to be reliable for determining exothermic enthalpies of formation much in excess of 60,000 J/g-atom, and so it can only be applied to intermetallic compounds or to other suitable reactions.

5. Thermodynamic Properties of Alloys

A number of methods of approximating the thermodynamic properties of alloy phases are described in Chapter 1 (see e.g. pp. 44–63) and in Chapter 4 (see e.g. the references to Section IV.8). The reader is recommended to refer to the cited literature for more comprehensive information. Here, two more recently published methods for calculating thermodynamic values of alloys from available physical property information are presented.

A. Calculations Using the Wigner–Seitz Model of Metals

In the Wigner–Seitz model of the alkali metals each atom is segregated into a volume within which all of the electrons on the atom are treated as being localized and in normal atomic states except one electron, which is in an s state and is treated as the electron giving rise to the electrical conductivity of the metal. This electron has a wave function which is different from that in the isolated atom and the s wave function at the periphery of the cell occupied by each atom joins smoothly with the s wave function in neighbouring cells, thus permitting easy movement of s electrons among the cells. This gives rise to metallic conduction.

When an alloy is formed between two elements, the cells containing the metallic ions will differ from one metallic species to the other. The nature of these species and their relative attraction for valence electrons, which is reflected in the Pauling electronegativities, makes an important contribution in a negative sense to the enthalpy of formation of alloys according to Miedema (1973j). Counteracting this effect is the mismatch of conduction electron wave functions and electron densities at the joining plane between two cells containing different metallic ions. This density term (Δm) leads to a positive contribution to the enthalpy of mixing. The enthalpy of formation of an intermetallic compound can thus be expressed in simplified form by the equation

$$\Delta H = -A(E_A - E_B)^2 + B(\Delta m) \tag{133}$$

Miedema identifies the first term, following the model of metallic contact potentials in free electron theory, with the difference in work function for the two metals, and shows that the electron density term can be related to the respective compressibilities of the metals forming the compound

$$\Delta m = B/V \text{ where } V \text{ is the molar volume.}$$

Miedema *et al.* (1973j; 1975c; 1976b, c, h) have published a series of papers in which they have demonstrated that experimental enthalpies of formation for solid and liquid binary alloys of transition metals can be accounted for within reasonable limits.

The expression they derive for the enthalpy of formation is

$$\Delta H \approx [-Pe(\Delta\phi^*)^2 + Q_0(\Delta n_{ws}^{1/3})^2] \tag{134}$$

where P and Q_0 are constants having nearly the same values for widely different alloy systems (e.g. intermetallic compounds of two transition metals or liquid alloys of two non-transition metals), ϕ^* is obtained by adjusting the experimental work functions, and n is obtained from estimates of the charge density at the Wigner–Seitz cell boundary.

For regular liquid or solid solutions, the concentration dependence of ΔH contains the product $c_A^s c_B^s$. For ordered compounds the area of contact between dissimilar cells is larger than the statistical value. Near the equiatomic composition, experimental results show that the ordering energy of alloys ($f(c_A^s, c_B^s)$) is, quite generally, of the order of 1/3 of the total enthalpy of formation.

Consideration of the physical origin of the two terms in eqn (134) suggests that in addition to the ordering function, there is another factor, $g(c_A, c_B)$, that varies somewhat with the relative concentration of the two metals. Concentration dependent values of ΔH can thus be derived using the more general expression

$$\Delta H/N_0 = f(c_A^s, c_B^s) \cdot g(c_A, c_B)[-Pe(\Delta\phi^*)^2 + Q_0(\Delta n_{WS}^{1/3})^2]$$

N_0 is Avogadro's number, ΔH is expressed per g-atom of alloy, and

$$g(c_A, c_B) = 2(c_A V_A^{2/3} + c_B V_B^{2/3})/(V_A^{2/3} + V_B^{2/3}).$$

Table XXVII presents the values of ϕ^*, $n_{WS}^{1/3}$ and $V_m^{2/3}$ selected by Miedema *et al.* (1975c) and used in calculating enthalpies of formation.

The method due to Miedema described above has the great advantage that enthalpies of formation can be calculated for very many alloy systems where no experimental information whatsoever is available.

TABLE XXVII. *Parameters of eqn (134) to be used in calculating enthalpies of formation of alloys*

Metal	ϕ^* Volt	$n_{WS}^{1/3}$	$V_m^{2/3}$ cm^2	Metal	ϕ^* Volt	$n_{WS}^{1/3}$	$V_m^{2/3}$ cm^2
Sc	3.25	1.27	6.1	Li	2.85	0.98	5.5
Ti	3.65	1.47	4.8	Na	2.70	0.82	8.3
V	4.25	1.64	4.1	K	2.25	0.65	12.8
Cr	4.65	1.73	3.7	Rb	2.10	0.60	14.6
Mn	4.45	1.61	3.8	Cs	1.95	0.55	16.8
Fe	4.93	1.77	3.7	Cu	4.55	1.47	3.7
Co	5.10	1.75	3.5	Ag	4.45	1.39	4.7
Ni	5.20	1.75	3.5	Au	5.15	1.57	4.7
Y	3.20	1.21	7.3	Ca	2.55	0.91	8.8
Zr	3.40	1.39	5.8	Sr	2.40	0.84	10.2
Nb	4.00	1.62	4.9	Ba	2.32	0.81	11.3
Mo	4.65	1.77	4.4	Be	4.20	1.60	2.9
Tc	5.30	1.81	4.2	Mg	3.45	1.17	5.8
Ru	5.40	1.83	4.1	Zn	4.10	1.32	4.4
Rh	5.40	1.76	4.1	Cd	4.05	1.24	5.5
Pd	5.45	1.67	4.3	Hg	4.20	1.24	5.8
La	3.05	1.09	8.0	Al	4.20	1.39	4.6
Hf	3.55	1.43	5.6	Ga	4.10	1.31	5.2
Ta	4.05	1.63	4.9	In	3.90	1.17	6.3
W	4.80	1.81	4.5	Tl	3.90	1.12	6.6
Re	5.40	1.86	4.3	Sn	4.15	1.24	6.4
Os	5.40	1.85	4.2	Pb	4.10	1.15	6.9
Ir	5.55	1.83	4.2	Sb	4.40	1.26	6.6
Pt	5.65	1.78	4.4	Bi	4.15	1.16	7.2
Th	3.30	1.28	7.3	Si	4.70	1.50	4.2
U	4.05	1.56	5.6	Ge	4.55	1.37	4.6
Pu	3.80	1.44	5.2	As	4.80	1.44	5.2

B. Properties of Mixing from "Free Volume" Theory

In a recent series of papers (1990g, h; 1992c), Tanaka, Gokcen, Morita *et al.* have described how the thermodynamic properties of mixing in liquid binary alloys can be derived from physical properties using the "free volume" theory proposed by Shimoji and Niwa (1957e), the first approximation of the regular solution model as described by Gokcen (1986a), and a consideration of the configuration and vibration of the atoms in the alloys.

Assuming that an atom vibrates harmonically in its cell surrounded by its nearest-neighbours, the following equations may be used to calculate the excess properties of mixing:

$$\Delta G_m^E = \Delta H_m - T\,\Delta S_m^E \tag{135}$$

$$\Delta H_m = N_{AB}\Omega_{AB}/Z \tag{136}$$

$$\text{with}\quad N_{AB} = ZN_0X_AX_B(1 - X_AX_B\Omega_{AB}/kT) \tag{137}$$

$$\Delta S^E = \Delta S_{conf}^E + \Delta S_{nonconf}^E \tag{138}$$

$$\Delta S^{E}_{conf} = -X_A^2 X_B^2 \Omega_{AB}^2 / 2kT^2 \tag{139}$$

$$\begin{aligned}\Delta S^{E}_{nonconf} &= 3/2kN_0\{X_A \ln(v_A/v_{AA}) + X_B \ln(v_B/v_{BB})\} \\ &= 3/2kN_0\{2X_A \ln(L_A/L_{AA}) + 2X_B \ln(L_B/L_{BB}) \\ &\quad + X_A \ln(U_{AA}/U_A) + X_B \ln(U_{BB}/U_B)\}\end{aligned} \tag{140}$$

where N_{AB} is the number of A–B pairs; Z is the co-ordination number; Ω_{AB} is the exchange energy; k is Boltzmann's constant; N_0 is the Avogadro number; X_A, X_B are mole fractions; v_A, v_B, v_{AA} and v_{BB} are free volumes; L_A, L_B, L_{AA} and L_{BB} are the distances which the interatomic potential extends in a cell; U_A, U_B, U_{AA} and U_{BB} are the potential energy depths in a cell. In these equations the suffices AA and BB denote pure elements and A and B the states of A and B atoms in an A–B alloy.

In eqn (140), if the free volumes of A and B in an A–B alloy are larger than those in the pure states, i.e. $v_A > v_{AA}$ and $v_B > v_{BB}$, the region in which an atom moves randomly in its cell surrounded by its nearest-neighbours increases, $\Delta S^{E}_{nonconf}$ becomes positive. With $v_A < v_{AA}$ and $v_B < v_{BB}$ a negative $\Delta S^{E}_{nonconf}$ results.

By differentiation and rearrangement, the above equations can also be used to derive partial thermodynamic properties of mixing:

$$\Delta \bar{H}_B = \Omega_{AB} \tag{141}$$

$$\begin{aligned}\Delta \bar{S}^{E}_B &= \Delta \bar{S}^{E} \\ &= 3/2kN_0[(L_{AA} - L_{BB})^2/L_{AA}L_{BB} \\ &\quad + \{4U_{AA}U_{BB} - 2\Omega_{AB}(U_{AA} + U_{BB}) - (U_{AA} + U_{BB})^2\}/2U_{AA}U_{BB}\}\end{aligned} \tag{142}$$

where U_{AA} and U_{BB} in eqn (142) can be obtained from:

$$U_{ii} = -2\pi^2 L_{ii}^2 M_{ii} \nu_{ii}^2 / N_0 \qquad (i = A \text{ or } B) \tag{143}$$

with M_{ii} being the atomic weight, and L_{ii} half the nearest-neighbour distance, given by

$$L_{ii} = 1/2(2^{1/2} v_{ii}/N_0)^{1/3} \qquad (i = A \text{ or } B) \tag{144}$$

where v_{ii} is the molar volume. ν_{ii} in eqn (143) is the frequency of an atom, which can be evaluated using the following equation proposed by Iida and Guthrie (1988d):

$$\nu_{ii} = 2.8\ 10^{12}\ \beta_{ii}(T_{m'ii}/M_{ii} v_{ii}^{2/3})^{1/2} \qquad (i = A \text{ or } B) \tag{145}$$

where $T_{m'ii}$ is the melting point and β_{ii} is the coefficient required to transform the solid state frequency to that in the liquid state at the melting point. Values of β_{ii} can be obtained from experimental data for the surface tension of the pure elements in the liquid state.

It can be seen from eqns (141)–(145), that if the partial enthalpy of mixing is known, both the partial excess entropy and partial excess Gibbs energy of mixing can be calculated.

Using the above equations, Tanaka *et al.* have demonstrated the relation between enthalpy and excess entropy of mixing in liquid binary alloys. The necessary enthalpy of mixing data for the calculations were obtained both from published experimental values and also by use of the Miedema method described above. An equation allowing for the influence of temperature on the enthalpy–entropy relation has been derived. Figure 103 illustrates the relation in the case of the partial enthalpy and excess entropy of mixing.

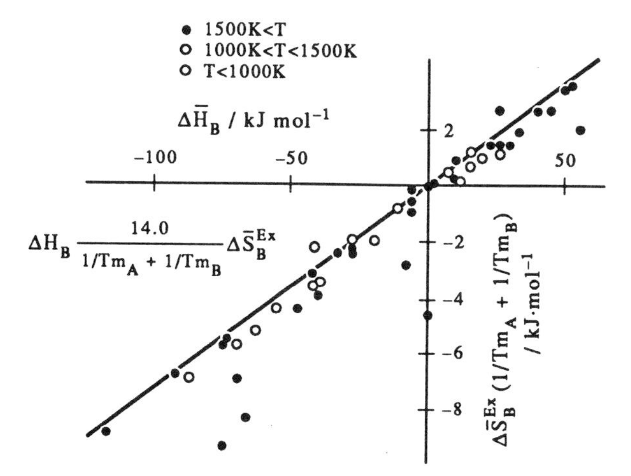

FIG. 103. Calculated results for the relationship between $\Delta\bar{H}_B$ and $\Delta\bar{S}_B^{Ex}/(1/Tm,_A + 1/Tm,_B)$ in infinite dilution of liquid binary alloys.

In addition, Tanaka *et al.* have used the free volume theory to calculate successfully activity coefficients of solutes at infinite dilution in liquid iron-base binary alloys (1992c). Some results of these calculations, expressed in the form of partial excess Gibbs energy values, are illustrated in Fig. 104.

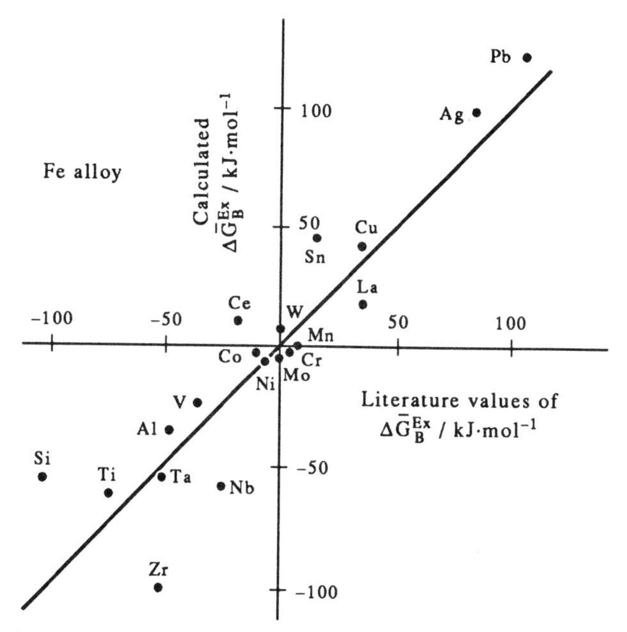

FIG. 104. Calculated partial excess Gibbs energy values for solutes in dilute solution in liquid Fe-base binary alloys [1992c].

CHAPTER 4

Examples of Thermochemical Treatment of Materials Problems

The real raison d-être for the continuation of extensive experimental research in materials thermochemistry is the potential application of its principles and data to practical, in particular industrial, problems. For this purpose the gathering of raw experimental data is obviously not enough. Missing numerical information must be supplemented by estimates to which the preceding Chapter 3 has therefore been devoted. Raw data must be sifted and critically evaluated to provide for every chemical system a consistent set of thermochemical properties. One example of such an evaluation is elaborated in the present chapter, and a large store of critically evaluated data is presented in the final Chapter 5.

In addition, however, materials scientists in general must increasingly be made aware of the help that can be obtained from the thermochemists. This is why the present chapter deals with a number of practical applications pertaining to chemical equilibrium. The number of examples is admittedly small compared with the virtually infinite number of possibilities, but the perceptive reader will soon see that every chosen example can very similarly be employed in many other cases.

Some of the examples can be worked through by the reader using the tabulated data; others involve solution phases containing many components and these must be treated using suitable models with the aid of software for complex equilibrium calculations. Such software is now available commercially (e.g. 1988a, b, g; 1985d; 1980a; 1990b) and some of the examples presented in this chapter have been treated using such computer programs in combination with evaluated data for complex alloy, slag or other types of system.

In practice, it is true, a knowledge of reaction rates is as important as that of equilibrium, sometimes even more so. Nevertheless, the kinetic aspects can only be tackled when the thermodynamic equilibria have been settled, because it is the equilibria which determine the limits of a process, no matter how rapidly these are achieved.

With respect to the accuracy of the results of such calculations, one should distinguish two types of question that may be asked. Often only a cursory answer is needed to the question whether the equilibrium conditions of a particular chemical process are such that the process is feasible. To this type of question an answer is almost always possible, since the accuracy of the pertinent data need not be better than several kilojoules. In other cases the inquirer may be far more demanding and may wish to know if an impurity of, say, a few hundredths of a per cent can still be reduced under given conditions. In such cases the thermochemist may often find that the available data are just not accurate enough, and this is why the improvement of his experimental methods and measurements must continue.

1. Iron and Steelmaking

A. Deoxidation of Steel

Dissolved oxygen may be removed from a molten steel bath by the addition of an element the oxide of which is more stable than that of iron. The extent of deoxidation depends primarily on the Gibbs energy of formation of the oxide and on the concentration of the deoxidising element remaining in the liquid steel when equilibrium has been attained. The relationship between residual oxygen and oxidiser is of great importance to the steelmaker, and it is closely described by the mass action product of the reaction which is known as the "deoxidation constant".

In even the simplest systems, this quantity is not necessarily a true constant for any one temperature. It may change with the residual concentration of the oxidising element due to variation in the activity coefficients of the deoxidiser and the oxygen. (The activity coefficient of oxygen in pure iron is practically constant, but is modified by the presence of a third element.) In practical steelmaking, the "deoxidation constant" is influenced by numerous other factors, among which are the presence of various alloying elements and the assimilation of the primary product of deoxidation in a slag of complex composition. For this reason steelmakers control their deoxidation as well as other refining reactions from largely empirical relationships known to be valid for the production of certain types of steel composition under certain types of slag. Much information is being gained by laboratory-scale investigation and thermodynamic calculation of the simpler equilibria.

1. Deoxidation with Silicon

There have been several determinations of this equilibrium, most of which have been based on the addition of silicon to oxygen-containing iron melted in a silica crucible. A determination by Gokcen and Chipman (1952f) involved further measurements, namely those of equilibria between melts of silicon and oxygen in iron and controlled mixtures of hydrogen and water vapour. The mass-action product of the reaction

$$\langle SiO_2 \rangle = [Si]_{Fe} + 2[O]_{Fe} \qquad (\alpha)$$

(i.e. $K_c = c_{Si} c_O^2$) was found to have a constant value of 2.8×10^{-5} at 1600°C, the concentrations being expressed as weight percentages.

The individual Gibbs energy equations from which the mass action constant of eqn (α) may be approximately derived are the following:

$$\langle SiO_2 \rangle = \{Si\} + (O_2); \qquad \Delta G^\circ = 952{,}697 - 203.8\ T\ \text{J}$$

$$(O_2) = 2[O]_{Fe}; \qquad \Delta\bar{G} = -233{,}676 + 50.84\ T + 38.28\ T \log N_O\ \text{J}$$

$$\{Si\} = [Si]_{Fe}; \qquad \Delta\bar{G} = -131{,}378 + 15.02\ T + 19.14\ T \log N_{Si}\ \text{J.}$$

Upon summation, the Gibbs energy of reaction (α) is obtained:

$$\Delta G = 587{,}643 - 137.94T + 38.28T \log N_O + 19.14\ T \log N_{Si}\ \text{J.} \qquad (\varepsilon)$$

At equilibrium $\Delta G = 0$, and consequently,

$$2 \log N_O + \log N_{Si} = -30{,}700\ T^{-1} + 7.20. \qquad (\beta)$$

At 1600°C

$$2 \log N_O + \log N_{Si} = -9.19.$$

This result can be transformed into the weight percentage basis c_O and c_{Si} by the use of the general equation for a very dilute solute i, in the solvent j,

$$\log N_i = \log \frac{\text{wt\%}}{M_i} \frac{M_j}{100}.$$

Therefore

$$2 \log c_O + \log c_{Si} = -4.58$$

$$c_O^2 c_{Si} = 2.7 \times 10^{-5}.$$

This result is approximate because the interaction coefficients of oxygen and silicon $\varepsilon_O^{Si} = \varepsilon_{Si}^O$ in solution in iron have been ignored. However, the concentrations are quite small and during deoxidation the minimum amount of silicon would be added; hence the effect can be ignored for this application.

2. *Deoxidation with Aluminium*

The equilibrium constant for the corresponding reaction with aluminium

$$\langle Al_2O_3 \rangle = 2[Al]_{Fe} + 3[O]_{Fe} \qquad (\gamma)$$

can also be calculated from the thermochemical data which are available. The result

$$c_O^3 c_{Al}^2 = 10^{-13}$$

disagrees, however, with the bulk of the deoxidation constants observed in practice, as may be seen from Table XXVIII which is reproduced from a paper by Fitterer (1957a). The possible reason for this discrepancy has also been given by Fitterer.

TABLE XXVIII. *Deoxidation constants for aluminium in iron (1600°C)*

Investigators	$(\%Al)^2(\%O)^3$
Hessenbruch, 1929	6.2×10^{-8}
Herty, Fitterer and Byrns (exptl), 1930	6.3×10^{-9}
Herty, Fitterer and Byrns (calcd), 1930	0.19×10^{-9}
Wentrup and Hieber, 1939	0.11×10^{-9}
Geller and Dickie, 1943	4×10^{-13}
Hilty and Crafts, 1950	3.0×10^{-9}
Present calculation	1×10^{-13}

One contributory factor probably is: when aluminium "shot" is added to undeoxidised liquid steel, the reaction in the immediate vicinity of the aluminium "particle" proceeds to a very high degree because the reaction with oxygen proceeds at a much higher rate than that at which aluminium dissolves and diffuses through the iron. However, other areas of the steel are essentially unaffected. The second contributory factor appears to be that when a resultant Al_2O_3 particle in a deoxidised area is transmitted to an undeoxidised area by turbulence, etc, oxygen in the undeoxidised area comes out of solution to form FeO which in turn combines with the Al_2O_3 particle to form a spinel, $FeO.Al_2O_3$. This would result in a high FeO content as well as a larger deoxidation constant. Both of these factors would tend to explain the discrepancy between the calculated and the experimental "equilibrium" constants. From the thermochemical data, one can estimate that the spinel is, indeed, stable when the oxygen content exceeds approximately 0.01 wt% at the melting temperature of iron.

However, the discrepancy has been resolved to some extent by Repetylo, Olette and Kozakevitch (1967l) who made careful experiments over longer periods under purified argon at 1600°C and demonstrated that the deoxidation constant varies with time approaching a "final" figure of 5×10^{-13}. The phenomenon shown to be responsible for this behaviour is the elimination of the alumina suspension, the coarser particles being eliminated in 7–10 min, whereas a finer suspension ($<20\ \mu$) remains in the melt even 20 min after the addition of aluminium. Further, Foerster and Richter (1969d) have observed the actual concentration of oxygen in killed steel by an e.m.f. method and found that supersaturation can reach a very high value in relatively good agreement with nucleation theories by Volmer.

B. The Decarburisation of Iron–Chromium–Carbon and Iron–Silicon–Carbon Liquid Alloys

Liquid alloys of iron and chromium are produced by reduction of the oxides in a submerged electric arc furnace with carbon. The temperature of the operation normally reaches about 1700°C, and in order to maintain the integrity of the graphite electrodes an excess of coke is added to the furnace charge. The product always has a high carbon content as a result of this procedure. In refining the alloy it is desirable to reduce the carbon content to about 0.01 wt%. Typical ferro-chromium alloys might contain 10–25 wt% Cr, and the refining process is required to reduce the carbon content with the minimum loss of chromium from the alloy. Since the oxidation products are substantially carbon monoxide and Cr_2O_3, the equilibrium constant for the reaction

$$\tfrac{2}{3}\mathrm{Cr} + (\mathrm{CO}) \rightarrow \tfrac{1}{3}\mathrm{Cr_2O_3} + \mathrm{C} \qquad K = \frac{a_{\mathrm{Cr_2O_3}}^{1/3} a_{\mathrm{C}}}{a_{\mathrm{Cr}}^{2/3} p_{\mathrm{CO}}}$$

can be used to calculate the carbon monoxide pressure which would be in equilibrium with a given iron–chromium–carbon alloy and pure solid Cr_2O_3 at the refining temperature, and providing the carbon monoxide pressure can be kept below this level, oxidation of the chromium could not take place unless Cr_2O_3 could be formed in solution in some other oxide ($a_{Cr_2O_3} < 1$).

The equilibrium constant for this reaction varies with temperature according to the equation

$$\log_{10} K = 12{,}580\ T^{-1} - 9.10$$

and has the values 3.32×10^{-3} at 1900 K and 1.55×10^{-3} at 2000 K, which straddle the operating temperature range. It follows from the equilibrium constant that when the chromium activity in an alloy is 0.1, the carbon activity is 3×10^{-4} when the system is in equilibrium with pure Cr_2O_3 and 1 atm pressure of carbon monoxide, and one-hundredth of this when the carbon monoxide pressure is 10^{-2} atm at 2000 K. The activity of carbon in iron–chromium alloys at these temperatures has been reported by Richardson and Dennis (1953e). Chromium forms stable carbides whereas those of iron have very low stability. It might therefore be expected that the chromium–carbon bond strength would be very much stronger than the iron–carbon bond and since the iron–chromium system is almost Raoultian, then chromium should lower the activity of carbon in solution in iron. The interaction coefficient ε_C^{Cr} in iron as solvent has the value -4.3 and e_C^{Cr}, which is $\partial \log \gamma_C / \partial$ wt% Cr, is correspondingly -0.020 at these temperatures. The measurements show that when the atom fraction of chromium has the values 0.1 and 0.23 the activity coefficient of carbon with respect to graphite as standard state is lowered from the value in iron, 0.57, to 0.35 and 0.21 respectively at 1660°C when the carbon content is less than 1 at.%. The presence of chromium in the alloy will therefore approximately double the atom fraction of carbon for a given activity in solution in liquid iron for typical alloy compositions.

If it is possible to reduce the carbon activity only to 3×10^{-4} by oxidation of carbon before Cr_2O_3 is formed, then the best alloy that could be made would contain about 0.04 wt% C. By lowering the oxygen partial pressure, values much lower than this are theoretically possible. Krivsky (1973h) has described the use of argon–oxygen mixtures in the removal of carbon in an industrial furnace, and states that by finishing the refining period with practically pure argon it is possible to reach a carbon level of 0.01 wt% in a typical alloy containing 18.5 wt% Cr.

The iron–silicon–carbon system has two significant differences from the iron–chromium–carbon system with respect to the oxidation behaviour and the first of these is that silicon has a very strong affinity for iron. Therefore, despite the fact that silicon forms a very stable carbide, the interaction parameter ε_C^{Si} has a value of approximately $+10$ and silicon *raises* the activity coefficient of carbon in solution in liquid iron.

If we now evaluate the oxidation equilibrium constant for iron–silicon–carbon in the same way as was done for chromium,

$$[Si]_{Fe} + 2(CO) \rightarrow SiO_2 + 2[C]_{Fe}; \qquad K = \frac{a_{SiO_2} a_C^2}{a_{Si} p_{CO}^2}$$

$$\log K = 38{,}100\, T - 19.8.$$

The constant has the value of unity at 1924 K which is in the range of steelmaking temperatures, and the activity of carbon is equal to the square root of the silicon activity when p_{CO} is 1 atm and one-tenth of the square root of a_{Si} when p_{CO} is 10^{-1} atm.

The partial Gibbs energy of solution of silicon in liquid iron at a mole fraction of 0.2 is given by the equation

$$\Delta \bar{G}_{Si} = -109{,}620 + 6.44\, T \text{ J}$$

and hence the activity of silicon is 2.1×10^{-3} at 1900 K and 3.0×10^{-3} at 2000 K. The activity of carbon has a value equal to the square roots of these numbers, *ca.* 5×10^{-2} when SiO_2, carbon monoxide at 1 atm pressure and an iron–silicon–carbon liquid alloy containing silicon at an atom fraction of 0.2 are in equilibrium in the temperature range

1900–2000 K. This carbon activity corresponds to a carbon content of 1.3 wt% in the iron–carbon system; when p_{CO} is held at 10^{-2} atm this content falls to about 0.02%. However, this reaction is not the only one to be considered in the oxidation behaviour of iron–silicon–carbon because the formation of gaseous SiO, a product having no parallel in the iron–chromium–carbon system, becomes important at higher temperatures above 1700°C. The reaction would be described by the equilibrium

$$[\mathrm{Si}]_{\mathrm{Fe}}+(\mathrm{CO})\rightarrow(\mathrm{SiO})+[\mathrm{C}]_{\mathrm{Fe}}; \qquad K=\frac{a_{\mathrm{C}}\,p_{\mathrm{SiO}}}{a_{\mathrm{Si}}\,p_{\mathrm{CO}}}$$

$$\log K=1990\,T-1.94.$$

This constant has a value of approximately 0.1 at steelmaking temperatures and hence the two gases will be formed with the same partial pressure when the activity of carbon is one-tenth that of silicon; carbon and silicon will be evolved together at the same rate. No dilution of the oxidising gas with argon would permit further decarburisation without the accompanying silicon loss in this case.

Carbon and silicon have greatly differing activity coefficients in solution in liquid iron, and neglecting the $\varepsilon_{\mathrm{C}}^{\mathrm{Si}}$ and conjugate parameter $\varepsilon_{\mathrm{Si}}^{\mathrm{C}}$ for the present purposes, the atom fraction ratio for a given activity ratio is inversely related to the activity coefficients. The carbon content of an alloy in which CO and SiO are being formed at the same rate will be given approximately by

$$10a_{\mathrm{C}}=a_{\mathrm{Si}}; \qquad 10N_{\mathrm{C}}=\frac{\gamma_{\mathrm{Si}}}{\gamma_{\mathrm{C}}}N_{\mathrm{Si}}.$$

Table XXIX shows the values at 1873 K for both solutes in iron and in the range of compositions which are of interest in ferro-silicon. The silicon content is usually large compared with the carbon content because of the activity coefficient effect and when N_{Si} is 0.2 (11 wt%), N_{C} is only 0.0003 (0.008 wt%) when carbon and silicon are oxidised at the same rate. As the silicon content increases, the activity coefficient of silicon rises rapidly, and when N_{Si} is 0.5, γ_{Si} is now 0.446 and the activity of carbon must therefore be 0.0223. The effect of silicon on the activity coefficient of carbon is no longer negligible and should be taken into account when attempting to calculate N_{C}.

There is reasonable experimental evidence that the interaction coefficient $\varepsilon_{\mathrm{C}}^{\mathrm{Si}}$ is relatively constant over the range $N_{\mathrm{Si}}=0$–0.25 (1962f) and therefore the activity coefficient of

TABLE XXIX. *Activity coefficients of carbon and silicon dissolved in liquid iron at 1873 K*

N_{C}	γ_{C}	N_{Si}	γ_{Si}
0.00	0.57	0.00	0.0013
0.05	0.85	0.1	0.0030
0.10	1.37	0.2	0.0090
0.15	2.30	0.3	0.040
0.20	4.12	0.4	0.178
		0.5	0.446

carbon in the dilute solution range will be given by the equation

$$\ln[\gamma_C]_{Fe-Si} = \ln[\gamma_C]_{Fe} + N_{Si}\varepsilon_C^{Si}$$

$$= -0.56 + (0.2 \times 10)$$

$$= 1.44$$

$$[\gamma_C]_{Fe-Si} = 4.21.$$

Therefore when N_{Si} is equal to 0.2 the limiting carbon content, when $p_{CO} = p_{SiO}$, is

$$N_C = \frac{0.0223}{4.21} = 0.01 = 0.22 \text{ wt\% C}.$$

Bringing the results for SiO_2 and SiO formation together it can now be concluded that lowering the partial pressure of carbon monoxide in the equilibrium gas by using $Ar–O_2$ mixtures could avoid the formation of SiO_2 as a separate phase, and the carbon level could be reduced below the value predicted for this particular equilibrium. As the carbon content was decreased, however, the partial pressure of SiO would increase until at about 0.2 wt% C the rate of loss of silicon from the melt would equal that of carbon on an atomic basis.

C. Chill Factors

The knowledge of the change in temperature of a bath of liquid iron caused by the addition of a master alloy, is of considerable interest to the steelmaker—especially when the addition is made to the ladle, isolated from an external heat source, because of the necessity to arrive at a predetermined teeming temperature.

The temperature change of a bath of molten metal due to the addition of a master alloy is given by the total heat effect caused by adding this alloy divided by the heat capacity of the alloy produced. The total heat effect is composed of the heat required to raise the master alloy from room temperature to the bath temperature and the enthalpy of solution of the alloy in liquid iron. It is convenient to consider the total heat effect as taking place in three stages: dissociation of the master alloy into its component elements, followed by raising the individual elements from room- to bath-temperature, and their subsequent dissolution in liquid iron, and then to invoke the First Law of Thermodynamics.

The procedure is best illustrated with a practical example, say, the dissolution of a 50 at.% Si–Fe alloy in iron to give a 1 at.% Si solution.

$$\langle Fe_{0.5}Si_{0.5}\rangle_{298\text{ K}} = \tfrac{1}{2}\langle Fe\rangle_{298\text{ K}} + \tfrac{1}{2}\langle Si\rangle_{298\text{ K}} \qquad \Delta H_1 = +38{,}493 \text{ J}$$

$$\tfrac{1}{2}\langle Fe\rangle_{298\text{ K}} = \tfrac{1}{2}\{Fe\}_{1873\text{ K}} \qquad \Delta H_2 = +37{,}698 \text{ J}$$

$$\tfrac{1}{2}\langle Si\rangle_{298\text{ K}} = \tfrac{1}{2}\{Si\}_{1873\text{ K}} \qquad \Delta H_3 = +45{,}647 \text{ J}$$

$$\tfrac{1}{2}\{Fe\}_{1873\text{ K}} = \tfrac{1}{2}[Fe]_{\{Fe_{0.99}Si_{0.01}\}1873\text{ K}} \qquad \Delta H_4 = -1 \text{ J}$$

$$\tfrac{1}{2}\{Si\}_{1873\text{ K}} = \tfrac{1}{2}[Si]_{\{Fe_{0.99}Si_{0.01}\}1873\text{ K}} \qquad \Delta H_5 - 65{,}584 \text{ J}$$

Summing up,

$$\langle Fe_{0.5}Si_{0.5}\rangle_{298\ K}=\tfrac{1}{2}[Fe]_{\{Fe_{0.99}Si_{0.01}\}1873\ K}+\tfrac{1}{2}[Si]_{\{Fe_{0.99}Si_{0.01}\}1873\ K}$$

$$\Delta H_1+\Delta H_2+\Delta H_3+\Delta H_4+\Delta H_5=+56{,}253\ J.$$

For the heat capacity of the final solution one may take that of pure liquid iron, i.e., $C_p=46$ J/deg mole. Since a positive sign indicates an endothermic heat effect, the temperature of the bath is lowered. Hence

$$\Delta\theta=-0.02\times 56{,}253/46=-24.4°C$$

independent of total mass.

Thus, if the pertinent enthalpy data are known or can be reliably estimated, the calculation of the temperature changes is fairly simple. "Chill factors", defined as the temperature change accompanying the formation of a 1 wt% master alloy solution from hypothetically pure liquid iron at 1600°C and solid master alloy at 25°C, have occasionally been published, for example, for 53 commercial master alloys by Chart (1971d) on behalf of the Metallurg Group of Companies. The number of possibilities is, of course, legion.

These chill factors have been calculated to minimise the departure from the desired casting temperatures of steels which occurs in practice and is an important factor affecting the reject rate of steel ingots. A secondary advantage of the use of chill factors is increased furnace life. This is because in many instances the melt will not have to be preheated in the furnace to such an extent, prior to the addition of the ferro-alloys, when the chill factors are applied.

D. "Window" for Liquid Calcium Aluminates in Continuous Casting

A general problem encountered in the continuous casting of steel is the gradual clogging of the submerged nozzles which occurs as a result of the deposition of finely dispersed oxide or sulphide particles which are produced during the prior deoxidation and desulphurisation treatments with aluminium and calcium. The clogging necessitates replacement of the nozzles after a certain time and prevents the casting process being truly continuous.

With the help of thermodynamic data and calculations, it is possible to investigate whether ranges of dissolved aluminium, calcium and oxygen exist for particular steels, such that no solid oxide or sulphide phases are formed, but that a liquid slag only is in equilibrium with the liquid steel. The undesirable effects resulting from deposition of solid particles in the submerged nozzles as well as oxide or sulphide inclusions in the cast steel, can thereby be avoided.

Figure 105 shows calculated ranges of solid sulphide and calcium aluminate stability in a plot of dissolved oxygen concentration against added calcium concentration for a liquid steel of given aluminium and sulphur content at 1600°C. The calculations indicate that there is indeed a range of calcium and oxygen concentrations in which no solid phases are precipitated but only a liquid slag is formed. The calculations show very satisfactory consistency with the experimental results of Pellicani (1986c), which are also presented in the figure.

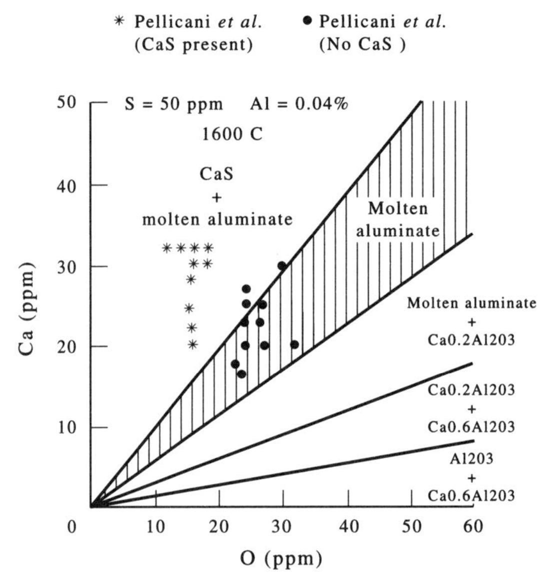

FIG. 105. Calculated liquid aluminate window avoiding solid sulphide or oxide precipitation.

The calculated information is of particular help in choosing appropriate deoxidation and desulphurisation conditions for given steels prior to carrying out the casting operation.

E. Precipitation of Carbide and Nitride Phases from Dilute Solution in Alloy Steels

The presence of carbide and nitride precipitates in alloy steels can have a beneficial effect on the mechanical properties of the steels concerned. However, the amounts, morphology and distribution of the precipitated phases must be carefully controlled in order to achieve the properties required. Because the presence of hard precipitates in a steel during hot-rolling operations can result in damage both to the rolls and to the steel, it is important that information be available on the ranges of temperature and composition in which precipitated phases are stable. For this reason, and also to achieve desired precipitation characteristics using the minimum amount of expensive precipitating elements such as niobium, titanium, vanadium etc, it is helpful to carry out prior calculations of the stability of precipitates in steels of different compositions.

Table XXX presents the results of an equilibrium calculation relating to the precipitation of a number of phases, including the niobium carbonitride phase, from dilute solution in an austenitic steel at 1223 K. The calculation shows that at this temperature, the Nb(N, C) phase itself is stable, together with AlN, BN and MnS as other precipitated phases.

By carrying out a series of such calculations for a number of temperatures for the given steel, the changes in the amounts of the precipitated phases can be determined as well as the composition of the carbonitride phase. This is illustrated for a chosen steel in Fig. 106.

TABLE XXX. *Calculated equilibrium in the Fe–B–C–Mn–S–N–Al–O–Nb system at 1223 K, including the niobium carbonitride phase*

T = 1223.00 K			
P = 1.0000E + 00 BAR			
V = 0.0000E + 00 DM3			
REACTANTS	WEIGHT/GRAM		
Mn/Solid–Fe/	1.2000E + 00		
S/Solid–Fe/	2.0000E – 02		
B/Solid–Fe/	5.0000E – 03		
Al/Solid–Fe/	4.0000E – 02		
Nb/Solid–Fe/	3.5000E – 02		
O/Solid–Fe/	1.0000E – 03		
C/Solid–Fe/	2.5000E – 01		
N/Solid–Fe/	2.0000E – 02		
Fe/Solid–Fe/	9.8429E + 01		
	EQUIL AMOUNT	PRESSURE	FUGACITY
PHASE: Gas	MOL	BAR	BAR
N2	0.0000E + 00	6.0288E – 04	6.0288E – 04
CO	0.0000E + 00	3.6891E – 06	3.6891E – 06
Mn	0.0000E + 00	1.3277E – 07	1.3277E – 07
CS	0.0000E + 00	1.7697E – 10	1.7697E – 10
Fe	0.0000E + 00	1.0575E – 10	1.0575E – 10
TOTAL:	0.0000E + 00	6.0670E – 04	
PHASE: Nb–Carbnit	GRAM	WEIGHT FRACTION	ACTIVITY
NbC	3.5953E – 02	9.1284E – 01	9.1432E – 01
NbN	3.4330E – 03	8.7164E – 02	8.5676E – 02
TOTAL:	3.9386E – 02	1.0000E + 00	1.0000E + 00
PHASE: Solid–Fe	GRAM	WEIGHT FRACTION	ACTIVITY
Fe	9.8429E + 01	9.8571E – 01	9.7657E – 01
Al	1.4774E – 02	1.4795E – 04	3.0339E – 04
B	6.4861E – 05	6.4954E – 07	3.3243E – 06
C	2.4588E – 01	2.4624E – 03	1.1343E – 02
Mn	1.1657E + 00	1.1674E – 02	1.1757E – 02
N	6.4446E – 04	6.4539E – 06	2.5494E – 05
Nb	1.8014E – 04	1.8040E – 06	1.0744E – 06
O	3.6926E – 11	3.6979E – 13	1.2788E – 12
S	4.8712E – 06	4.8782E – 08	8.4178E – 08
TOTAL:	9.9856E + 01	1.0000E + 00	1.0000E + 00
	GRAM		ACTIVITY
MnS	5.4254E – 02		1.0000E + 00
AlN	3.6613E – 02		1.0000E + 00
BN	1.1329E – 02		1.0000E + 00
Al_2O_3	2.1243E – 03		1.0000E + 00
C	0.0000E + 00		1.9258E – 01
FeB	0.0000E + 00		1.9259E – 02
Fe_2B	0.0000E + 00		1.6090E – 02
Mn	0.0000E + 00		1.4936E – 02
Nb_2C	0.0000E + 00		3.9978E – 04
MnB	0.0000E + 00		2.8785E – 04
$NbFe_2$	0.0000E + 00		2.2720E – 04
NbB_2	0.0000E + 00		9.1932E – 05
B	0.0000E + 00		2.9421E – 05
Nb	0.0000E + 00		6.7450E – 06
$FeO.Al_2O_3$	0.0000E + 00		2.9835E – 06
Nb_2N	0.0000E + 00		2.3191E – 06
Al	0.0000E + 00		4.9043E – 07
Mn_4N	0.0000E + 00		2.5902E – 07
MnB_2	0.0000E + 00		3.4273E – 08
S	0.0000E + 00		2.9168E – 08

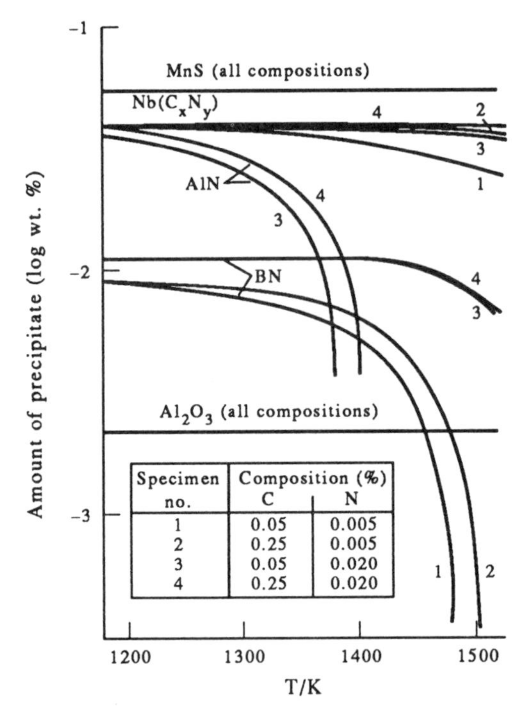

Specimen no.	Composition (%) C	N
1	0.05	0.005
2	0.25	0.005
3	0.05	0.020
4	0.25	0.020

FIG. 106. Calculated precipitation behaviour in a steel containing 1.2% Mn, 0.02% S, 0.005% B, 0.04% Al, 0.035% Nb, 0.001% O, 0.05 to 0.25% C and 0.005 to 0.02% N.

2. Non-ferrous Metallurgy

A. Aluminothermic Type Reactions

1. The Production of Uranium by Reduction of its Fluoride with Calcium and Magnesium

A number of metals are produced by reduction of a suitable compound with a very reactive metal. In such a process it is desirable to arrange it so that the heat generated by the chemical reaction will be sufficient to raise at least one of the products of the reaction to the liquid state. This has the practical advantage of allowing a liquid/liquid or liquid/solid separation to occur. The process is used for the reduction of transition metal chlorides in the so-called Kroll process or of their oxides by aluminothermic reaction. All these processes have strongly negative Gibbs energies at all practical temperatures. The example selected here has been taken from the monograph of Rand and Kubaschewski (1963e).

The reaction of uranium tetrafluoride with calcium,

$$\frac{\langle UF_4\rangle + 2\langle Ca\rangle}{\text{at 298 K}} \rightarrow \frac{2\{CaF_2\} + \{U\}}{\text{at temp. } T}$$

provides sufficient heat to raise the temperature after ignition above the melting points of calcium fluoride and uranium. The maximum temperature increase may be estimated in the following way. The standard enthalpy of formation of two moles of calcium fluoride is $-2{,}443{,}456$ J, that of 1 mole of uranium tetrafluoride $-1{,}897{,}444$ J, the difference being $-546{,}012$ J in favour of the formation of calcium fluoride. The change in enthalpy of reaction with temperature may be neglected for such an approximate estimate, but the enthalpies of fusion of the products must be added if these are to be heated above their melting point. The combined heat effect of the transformation and fusion for $2CaF_2$ is 69,036 J that for uranium 18,410 J. The heat available for heating the mixture is thus

$$\Delta H = -2{,}443{,}456 + 1{,}897{,}444 + 69{,}036 + 18{,}410 = -458{,}566 \text{ J}.$$

The reacting mixture may now be considered as a calorimeter, the water equivalent of which may be obtained from the molar heats of the products. If these are not known, it is sufficient to ascribe 29.3 J/deg to every g-atom or g-ion present, which in this case amounts to 205 J/deg since there are 7 g-atoms + g-ions in ($2CaF_2 + 1U$). (From the actual values of the heat capacities, one finds an average value of 196.7 J/deg over the temperature range 300–2000 K.) The maximum temperature increase is then obtained by dividing the heat produced by the water equivalent, i.e.

$$\Delta\theta_{max} = 458{,}566/205 = 2237°\text{C}.$$

In fact the temperature can never increase by this amount because heat is being conducted away as soon as the reaction starts. To what extent it can be approached depends on the rate of reaction and on the insulation. In the present case, with good insulation, the boiling point of calcium will certainly be attained, and the evaporation of excess calcium would consume more heat. It is clear, however, that the enthalpy of reaction suffices to melt the reaction products so that a dense billet of uranium metal separated cleanly from the slag is likely to be obtained.

This is, however, not true when magnesium is used in a similar manner as the reducing agent. We may repeat the above calculation for the reaction

$$UF_4 + 2Mg = 2MgF_2 + U.$$

The enthalpy of formation of magnesium fluoride is $-1{,}112{,}944$ J/mole, its melting point 1263°C and its enthalpy of fusion 58,158 J/mole. The total heat available to heat and melt the reaction products is therefore $\Delta H = -2{,}225{,}888 + 1{,}897{,}444 + 116{,}316 + 18{,}410 = -193{,}718$ J. The average molar heat of the products is 200.8 J/deg between 500 and 1750 K, whereas the estimated figure, which we shall use here, is again 205 J/deg. The maximum temperature increase thus amounts to

$$\Delta\theta_{max} = 193{,}718/205 = 945°\text{C}.$$

This is clearly not enough to heat the reaction mixture above the melting points of MgF_2 and uranium. Since it is desirable to attain a temperature of, say, 1500°C, a preheating of the reactants to about 600°C is required, the alternative of using a "booster" being ruled out because of the danger of contamination. Harper and Williams (1957b) have indeed devised a satisfactory method for the reduction of uranium tetrafluoride by magnesium based on a preheating procedure.

In view of the heat losses during reaction owing to radiation and conduction, more elaborate calculations of the temperature increase in this type of reaction are not worthwhile.

2. The Production of Manganese and Chromium by the Aluminothermic Process

In the aluminothermic reduction of manganese, one of the variables under the control of the operator is the initial state of oxidation of the metal. In the ore pyrolusite, manganese is quadrivalent, but MnO, Mn_3O_4 and Mn_2O_3 are also well-defined lower oxides. The enthalpies of formation of the oxides are

$$\begin{aligned} Mn+\tfrac{1}{2}O_2 &= MnO; & -\Delta H^\circ &= 384{,}928 \text{ J/mole} \\ 3Mn+2O_2 &= Mn_3O_4; & -\Delta H^\circ &= 1{,}386{,}578 \text{ J/mole} \\ 2Mn+1\tfrac{1}{2}O_2 &= Mn_2O_3; & -\Delta H^\circ &= 956{,}881 \text{ J/mole} \\ Mn+O_2 &= MnO_2; & -\Delta H^\circ &= 520{,}071 \text{ J/mole.} \end{aligned}$$

Using the enthalpy of formation of alumina,

$$2Al+1\tfrac{1}{2}O_2 = Al_2O_3; \qquad -\Delta H^\circ = 1{,}673{,}600 \text{ J/mole}$$

we may calculate the heat evolved in the reduction of each oxide by aluminium.

In the process, the products must reach a temperature of about 2000°C, and hence we should calculate the heat required to raise manganese metal and alumina to this temperature from room temperature. The results of this calculation using data in the tables are as follows:

$$\text{Mn, } H^\circ_{2300} - H^\circ_{298} = 100{,}416 \text{ J/g-atom}$$

$$Al_2O_3\text{, } H^\circ_{2300} - H^\circ_{298} = 355{,}640 \text{ J/mole.}$$

The alumina is here in the liquid state at the upper temperature. A table is now compiled showing the heat required and that available to heat the products to 2300 K for three of the oxides, for 1 mole of alumina produced. Also shown is the Mn/Al ratio since the lower this ratio becomes, the more expensive is the reduction.

Oxide	*Heat required*	Heat available/mole Al_2O_3	*Mn/Al*
MnO	656.9 kJ	518.8 kJ	1.5
Mn_3O_4	581.6 kJ	636.0 kJ	1.125
MnO_2	506.3 kJ	895.4 kJ	0.75

It can thus be seen that the process should become self-sustaining if the Mn/O ratio in the oxide charged into the reactor is between that of MnO and Mn_3O_4. It should be noted that the reduction of MnO_2 would probably be explosively fast once initiated owing to the great excess of heat available.

In practice there will be heat losses from the reduction assembly which have not been accounted for in the calculation so far. An approximate value for these losses may be obtained, purely by chance, from a consideration of the aluminothermic reduction of chromium sesquioxide.

The standard heat change for the reaction

$$Cr_2O_3 + 2Al = 2Cr + Al_2O_3$$

is $\Delta H^\circ = -543.9$ kJ. The heat required to raise the products of this reaction from room temperature to 2300 K is found to be 539.7 kJ. Thus the heat available just balances the heat produced, and if there were no heat losses the reduction could be made satisfactorily. However, it is industrial practice to pre-heat the mixture of chromium sesquioxide and aluminium to about 550°C before charging to the aluminothermic crucible. The heat required to do this amounts to 104.6 kJ per mole of Cr_2O_3, and hence this value must be close to the heat lost by radiation and convection, during the reaction, to the surroundings.

B. The Chlorination of Metal Oxides

An important first step in the extraction of a number of the nuclear metals, such as zirconium, vanadium and titanium, is the chlorination of the oxides which are the naturally occurring form. From the type of the equation for chlorination to form a gaseous chloride,

$$\langle MeO\rangle + (Cl_2) = (MeCl_2) + \tfrac{1}{2}(O_2)$$

and the corresponding equilibrium constant

$$K = \frac{p_{MeCl_2} p_{O_2}^{1/2}}{a_{MeO} p_{Cl_2}}$$

it can be seen that chlorination efficiency can be increased by lowering the oxygen potential (1950e). This is usually achieved in industry by briquetting the oxide with carbon. The oxygen pressure which can be placed in the equilibrium constant above will now be that for the C–CO equilibrium above about 800°C, and close to that for C–CO_2 below 600°C. This is the oxygen potential suggested by Goodeve (1948d) as being appropriate to the situation in the iron blast furnace.

Using the available Gibbs energy data it can be shown that the chlorination of all the important metal oxides is close to complete under normal industrial operating conditions where the temperature is between 500° and 1000°C in most cases.

What is a more interesting problem to resolve is that of the selection of an unreactive lining material for the chlorination reactor. Silica can be used successfully in most cases, but in the chlorination of beryllia to form gaseous $BeCl_2$, graphite must be used. The attack of silica by $BeCl_2$ can be demonstrated by the following argument.

For the chlorination of cristobalite,

$$\langle SiO_2\rangle + 2(Cl_2) = (SiCl_4) + (O_2); \qquad \Delta G^\circ = 259{,}408 - 43.93\ T$$

and for beryllia,

$$2\langle BeO\rangle + 2(Cl_2) = 2(BeCl_2) + (O_2); \qquad \Delta G^\circ = 515{,}469 - 200\ T.$$

On combining these two equations we obtain the one for the attack of silica by $BeCl_2$ vapour:

$$\langle SiO_2\rangle + 2(BeCl_2) = (SiCl_4) + 2\langle BeO\rangle; \qquad \Delta G^\circ = -256{,}061 + 156.07\ T$$

which has a negative standard Gibbs energy change in the temperature range of industrial operation.

On the other hand, for titania,

$$\langle TiO_2\rangle + 2(Cl_2) = (TiCl_4) + (O_2); \qquad \Delta G^\circ = 161{,}084 - 56.48\ T$$

and for the attack of silica by $TiCl_4$ vapour,

$$\langle SiO_2\rangle + (TiCl_4) = (SiCl_4) + \langle TiO_2\rangle; \qquad \Delta G^\circ = 98{,}324 + 11.72\ T$$

which is positive in the same temperature range.

It may thus be concluded that the attack of a silica lining in a TiO_2 chlorination plant will be virtually absent except as a result of the direct access of chlorine to the lining, whereas in the BeO operation, a silica lining cannot be employed.

C. Refining of Lead

In the removal of impurities from blast-furnace-produced lead, there are two steps concerned with the removal of silver and gold which are interesting to analyse thermodynamically. In the first step, zinc is added to the liquid lead, and the temperature is reduced to just above the melting point of lead. A crust separates which contains alloys of zinc and silver or gold. After separation of these metals, the zinc must be removed from the lead, and this is sometimes accomplished by bubbling chlorine through the liquid metal to form, principally, zinc chloride.

To begin the analysis we consider the zinc–silver system. Satisfactory phase diagram and thermodynamic studies have been made of this sytem so that chemical potentials of silver, as a function of composition across the solid region and in dilute solution in liquid zinc, can be fairly accurately described.

The liquid silver–zinc system can be considered as approximately regular, and the partial enthalpy of solution of silver in dilute solution in zinc is then found to be −14,225 J. The distribution of silver between liquid lead and zinc has been studied by a radioactive technique by Peterson and Kontrimas (1960j). The results for the distribution coefficient K_d show that it is virtually independent of the concentration of silver in either metal from 5×10^{-5} to 5×10^{-2} for the molefraction of silver. Hence Henry's law may be used for silver in dilute solution in both lead and zinc.

Values of K_d as a function of temperature

T(K)	K_d	$RT \ln K_d = \Delta\bar{G}^{Ag}_{Zn\rightarrow Pb}$
711	55.7 ± 2.1	23,765 J
781	27.5 ± 0.5	21,422 J
821	20.1 ± 1.1	20,502 J

These results were used to calculate the two-term Gibbs energy equation for the transfer, at constant composition of 1 mole of silver, from a dilute solution in liquid zinc to one in liquid lead:

$$\Delta\bar{G}^{Ag}_{Zn\rightarrow Pb} = 45{,}605 - 29.3\ T \quad \text{J/g-atom.} \qquad (\alpha)$$

Now as the silver–zinc system can be considered as a regular solution, we may calculate from the results given above that for the dilute solution of silver in zinc, with solid silver as standard state,

$$\Delta \bar{G}_{Zn}^{Ag} = -14{,}225 + \boldsymbol{R}T \ln N_{Ag} \text{ J/g-atom.} \qquad (\beta)$$

Hence, combining (α) and (β) we derive a Gibbs energy equation for the dilute solution of silver in liquid lead:

$$\Delta \bar{G}_{Pb}^{Ag} = 31{,}380 - 29.3\ T + \boldsymbol{R}T \ln N_{Ag} \text{ J/g-atom.} \qquad (\gamma)$$

This equation allows the calculation of the chemical potential of silver at various concentrations in liquid lead, and as an example, these have been made for 610 K and for 0.02, 0.2, 2, 20 and 200 oz silver per ton of lead, which represents the range of concentration covered in the desilvering process.

Now, clearly, the solid silver–zinc alloy which is in equilibrium with these solutions at this temperature will have the same silver potential and this is demonstrated in Fig. 107. This shows the integral Gibbs energy of formation of the silver–zinc system at 610 K obtained from the data given above together with tangents which intercept the pure silver ordinate at the potentials for the given silver contents of liquid lead at the same temperature. The points of tangency on the integral silver–zinc curve represents the compositions of the solids in equilibrium with these liquids. Thus, at 200 oz/ton the liquid is in equilibrium with a $\beta+\gamma$ mixture of the silver–zinc phases, whilst at and below 0.2 oz/ton the liquid is in equilibrium with the η solid solution.

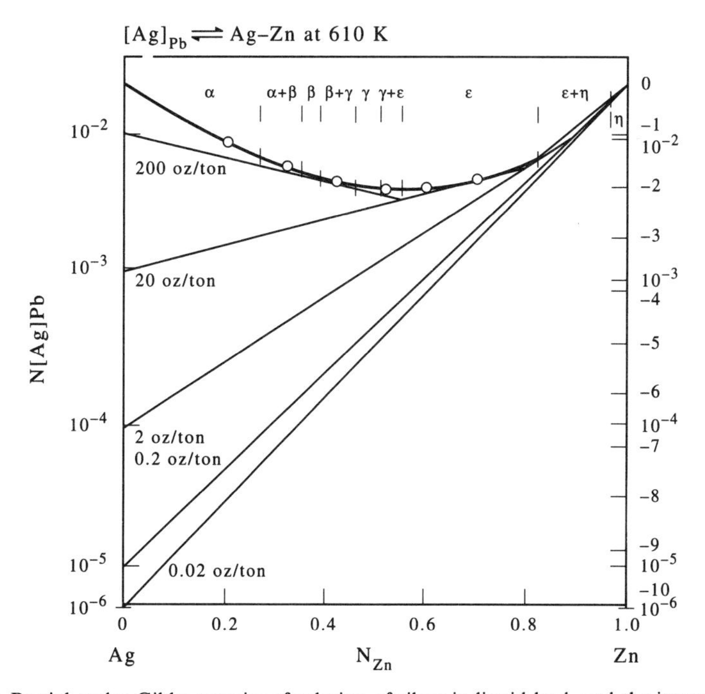

FIG. 107. Partial molar Gibbs energies of solution of silver in liquid lead, and the integral Gibbs energies of formation of silver–zinc alloys at 610 K.

1. The Removal of Zinc from Lead

After the silver and gold have been removed, the lead bullion is left with a zinc content which must be reduced. This can be done by bubbling chlorine through the metal when the zinc content is considerably reduced by preferential chloride formation. The chloride phase which is formed will consist, assuming the process is carried out at or near equilibrium, of a liquid chloride slag containing both zinc and lead, the proportions being determined by the metal activities in the metal phase. The amounts of lead and zinc in the slag can be calculated by considering the equilibrium

$$PbCl_2 + Zn = ZnCl_2 + Pb$$

$$\Delta G^\circ = -63{,}137 \text{ J at } 400°\text{C}$$

$$K = \frac{a_{ZnCl_2} a_{Pb}}{a_{PbCl_2} a_{Zn}} = 7.96 \times 10^4 \text{ at } 400°\text{C}. \qquad (\delta)$$

Because of the large difference in the standard Gibbs energies of formation of these two halides, the reaction will be displaced to the right and the lead activity can be taken as unity. Lacking quantitative information, we will assume that the halide slag is an ideal liquid mixture of the two component chlorides; then

$$K = \frac{N_{ZnCl_2}}{N_{PbCl_2}} \frac{1}{a[Zn]_{Pb}} \qquad (\varepsilon)$$

$$\Delta G^\circ = -\boldsymbol{R}T \ln K = \boldsymbol{R}T \ln a[Zn]_{Pb} - \boldsymbol{R}T \ln \frac{N_{ZnCl_2}}{N_{PbCl_2}}. \qquad (\zeta)$$

The dilute solution of zinc in lead shows a positive partial heat of solution of zinc, and an excess partial entropy of +8.4 J/deg. We may therefore write for the partial Gibbs energy of solution of zinc in lead, as a function of zinc content,

$$\boldsymbol{R}T \ln a[Zn]_{Pb} = \Delta\bar{G}[Zn]_{Pb} = 20{,}920 - 8.4\,T + \boldsymbol{R}T \ln N[Zn]_{Pb}. \qquad (\eta)$$

Hence at 400°C, when the slag contains a mole fraction of lead chloride of 0.1, the metal phase has a mole fraction of zinc of 4×10^{-5}. This can be shown by substitution into eqns (ζ) and (η).

3. Stability and Production of Ceramics

A. The Free Evaporation of an Oxide Ceramic in Vacuo

When ceramic materials are used as structural components in high-temperature vacuum furnaces it is important to consider the loss in weight and size which will occur as a result of vaporisation at high temperature. The calculation of this rate may be made by assuming that the equilibrium vapour pressure is exerted under conditions of free evaporation. This means that the vaporisation coefficient is assumed to be unity for the material in question. We will consider as an example of this calculation the weight loss of magnesia at a temperature of 2000 K in a vacuum where the residual gases offer no resistance to evaporation of the solid.

The vaporisation mechanism of magnesia is principally

$$\langle \mathrm{MgO} \rangle = (\mathrm{Mg}) + (\mathrm{O})$$

and the Gibbs energy change for this process may be obtained from data for the Gibbs energy of formation of MgO at 298 K, thermal data for the oxide up to 2000 K and the Gibbs energy of vaporisation of magnesium and that of dissociation of oxygen. This yields a simplified equation:

$$\Delta G^\circ = 1{,}007{,}507 - 282\ T \text{ J/mole of oxide}$$

where

$$K = \frac{p_{\mathrm{Mg}} p_{\mathrm{O}}}{a_{\mathrm{MgO}}}.$$

Now concerning ourselves only with pure magnesia ($a_{\mathrm{MgO}} = 1$), there is a relation betwen p_{Mg} and p_{O} resulting from the fact that equal numbers of magnesium atoms n_{Mg} and oxygen atoms n_{O} must leave the surface in unit time in order to retain the stoichiometric ratio in the solid. Thus the weights of the two elements leaving the surface must be in proportion to their atomic weights:

$$\frac{n_{\mathrm{Mg}}}{n_{\mathrm{O}}} = \left(\frac{m_{\mathrm{Mg}}}{M_{\mathrm{Mg}}} \frac{M_{\mathrm{O}}}{m_{\mathrm{O}}} \right) = 1.$$

The weight of each constituent leaving unit area in unit time is given by Knudsen's equation

$$m_{\mathrm{Mg}} = 44.4\ p_{\mathrm{Mg}} \sqrt{\frac{M_{\mathrm{Mg}}}{T}}.$$

Hence

$$\frac{n_{\mathrm{Mg}}}{n_{\mathrm{O}}} = \frac{p_{\mathrm{Mg}}}{p_{\mathrm{O}}} \sqrt{\frac{M_{\mathrm{O}}}{M_{\mathrm{Mg}}}} = 1; \quad p_{\mathrm{O}} = p_{\mathrm{Mg}} \sqrt{\frac{M_{\mathrm{O}}}{M_{\mathrm{Mg}}}}.$$

Substituting in the equilibrium constant, then,

$$K = 0.81\ p_{\mathrm{Mg}}^2 = 3.98 \times 10^{-13} \text{ at } 2000 \text{ K}$$

$$p_{\mathrm{Mg}} = 7 \times 10^{-7} \text{ atm.}$$

The weight loss per unit area of the solid is then

$$\Sigma m = m_{\mathrm{Mg}} + m_{\mathrm{O}} = 44.4 \frac{p_{\mathrm{Mg}}}{\sqrt{T}} [\sqrt{M_{\mathrm{Mg}}} + 0.81 \sqrt{M_{\mathrm{O}}}]$$

$$= 5.67 \times 10^{-6} \text{ g/cm}^2 \text{ sec.}$$

The alternative mode of vaporisation would involve the evaporation of oxygen as O_2 molecules. The Gibbs energy change for this process is

$$\langle \mathrm{MgO} \rangle \rightarrow (\mathrm{Mg}) + \tfrac{1}{2}(\mathrm{O_2})$$

$$\Delta G^\circ = 759{,}814 + 30.84\ T \log T - 316.7\ T \text{ (J)}$$

which yields at 2000 K

$$\Delta G^\circ = 329{,}908 = -38{,}284\ (\log p_{Mg} + \tfrac{1}{2}\log p_{O_2}).$$

As can be concluded from the previous discussion, the relationship between p_{Mg} and p_{O_2} is such that the fluxes are in the ratio 2:1 so that the stoichiometry of the oxide remains at 1:1. Hence

$$p_{Mg} = 2p_{O_2}\sqrt{\frac{M_{Mg}}{M_{O_2}}} = 1.74\ p_{O_2}.$$

Hence solving for p_{Mg} and p_{O_2}, we find $p_{Mg} = 2.14 \times 10^{-6}$ atm; $p_{O_2} = 1.23 \times 10^{-6}$ atm at 2000 K.

Under equilibrium conditions the evaporation of MgO to give magnesium vapour and O_2 molecules is slightly favoured over the evolution of oxygen atoms. In a Knudsen experiment the correct ratio p_O/p_{O_2} should be obtained for the equilibrium condition when at 2000 K

$$\frac{p_O}{p_{O_2}} = \frac{0.568}{1.23} = 0.46$$

but under free evaporation conditions the less-complicated desorption of oxygen atoms would seem to be the more probable event.

B. Metal–refractory Interaction

The extent of reaction of metals with oxides is important for the problem of finding suitable refractories for the melting of metals and alloys. The problem is of growing interest as more and more high-melting-point metals are coming to the fore.

As has frequently been mentioned in this monograph, the higher the temperature the less likely are kinetic checks and the more significant are the positions of the thermochemical equilibria. For the high-affinity metals of high melting point, such as titanium, satisfactory refractories in which the molten metal can be kept have not yet been found. But even when the affinity of the metal for non-metals is relatively low, interactions with the refractory are likely to occur and should be carefully assessed.

Let us consider as an example the amount of oxide pure molten iron is likely to dissolve when held in an alumina crucible. The appropriate chemical equation, the Gibbs energy of which must be found, is as follows:

$$\langle Al_2O_3 \rangle = 2[Al]_{Fe} + 3[O]_{Fe}. \qquad \text{(a)}$$

In order to obtain the corresponding Gibbs energy equation, the chemical equation should be broken up as follows:

$$\langle Al_2O_3 \rangle = 2\{Al\} + 1\tfrac{1}{2}(O_2); \qquad \Delta G_b \qquad \text{(b)}$$

$$1\tfrac{1}{2}(O_2) = 3[O]_{Fe}; \qquad 3\Delta\bar{G}_c \qquad \text{(c)}$$

$$2\{Al\} = 2[Al]_{Fe}; \qquad 2\Delta\bar{G}_d. \qquad \text{(d)}$$

The corresponding Gibbs energy equations are as follows:

$$\Delta G_b^\circ = 1{,}677{,}366 - 318.8\ T \qquad (\alpha)$$

$$3\Delta G^\circ_c = -350{,}619 + 57.32\ T \log N_O + 76.36\ T \qquad (\beta)$$

$$2\Delta G^\circ_d = -217{,}568 + 38.28\ T \log N_{Al} + 46.86\ T \qquad (\gamma)$$

where N_O and N_{Al} are the atomic fractions of oxygen and aluminium respectively in liquid iron.

When equilibrium is reached in reaction (a), the sum $\Delta G^\circ_b + 3\Delta G^\circ_c + 2\Delta G^\circ_d$ becomes zero, so that

$$0 = 1{,}109{,}179 - 195.58\ T + 57.32\ T \log N_O + 38.28\ T \log N_{Al}.$$

Since Al_2O_3 dissolves with the atomic proportion $1.5\ N_{Al} = N_O$, the equilibrium concentrations of aluminium and oxygen can be evaluated. For $T = 1873$ K, for instance,

$$\log N_{Al} = -4.25 \quad \text{and} \quad \log N_O = -4.07$$

corresponding to the weight percentages: $c_{Al} = 0.0027\%$ Al and $c_O = 0.0024\%$ O.

The concentrations of aluminium and oxygen in pure iron after melting 11 kg ingots in alumina crucibles observed by Hopkins *et al.* (1951b) at the National Physical Laboratory were about 0.001 to 0.002% each. The difference is probably not due to errors in the calculation but to the fact that equilibrium had not been reached in the large melts. The kinetic delay in the metal–refractory interactions is, indeed, the smelter's best chance for keeping down the impurities, an advantage that is progressively lost with increasing temperature. A particularly smooth surface of the refractory also assists to a considerable extent to cut down the reaction rate at the interface: a fact that is well known to the practical smelter.

Let us consider next the interaction of an alloy containing a baser metal, such as a 10% chromium–nickel alloy, with alumina. Again the alumina will be dissolved to a certain extent in the alloy, according to the equation

$$\langle Al_2O_3 \rangle = 2[Al]_{Cr\text{–}Ni} + 3[O]_{Cr\text{–}Ni}. \qquad (e)$$

The calculation corresponds to the one above for iron. Since the solubility of oxygen in liquid chromium–nickel alloys is unknown, the Gibbs energies may be taken to be additive, according to

$$\tfrac{1}{2}(O_2) = [O]_{Ni}; \qquad \Delta\bar{G}_f \qquad (f)$$

$$\tfrac{1}{2}(O_2) = [O]_{Cr}; \qquad \Delta\bar{G}_g \qquad (g)$$

and

$$\tfrac{1}{2}(O_2) = [O]_{Ni\text{–}Cr}; \qquad \Delta G_h = 0.9\ \Delta G_f + 0.1\ \Delta G_g \qquad (h)$$

for a 10 at.% alloy. This simplification admittedly introduces an additional error, and the total error in the finally calculated concentrations may amount to as much as half a logarithmic unit. In view of the analytical uncertainties at these low concentrations, however, the calculated values still represent useful estimates.

In the present case the apparent concentrations of aluminium and oxygen in the chromium–nickel alloys are obtained as 0.0019% Al and 0.0017% O by weight at 1550°C.

The stability of Cr_2O_3 is much greater than those of Fe_3O_4 and NiO, and although it is still much lower than that of Al_2O_3 it approaches it more closely. This is why the thermochemist must consider another possible equilibrium, namely

$$\langle Al_2O_3\rangle + 2[Cr]_{Cr-Ni} = 2[Al]_{Cr-Ni} + \langle Cr_2O_3\rangle \qquad \text{(i)}$$

that is, chromic oxide may be precipitated on or even dissolved in the refractory while the equivalent amount of aluminium goes into solution.

Again, eqn (i) may be broken up into individual equations, the Gibbs energies of which can be assessed.

$$\langle Al_2O_3\rangle = 2\{Al\} + 1\tfrac{1}{2}(O_2); \qquad \Delta G_b = 1{,}677{,}366 - 318.8\ T \qquad \text{(b)}$$

$$2\{Cr\} + 1\tfrac{1}{2}(O_2) = \langle Cr_2O_3\rangle; \qquad \Delta G_j = -1{,}158{,}550 + 277.8\ T \qquad \text{(j)}$$

$$2\{Al\} = 2[Al]_{Ni}; \qquad 2\Delta\bar{G}_k = -317{,}984 + 38.28\ T \log N_{Al} \qquad \text{(k)}$$

$$2[Cr]_{Cr-Ni} = 2\{Cr\}; \qquad 2\Delta\bar{G}_l = -38.28\ T \log N_{Cr} \qquad \text{(l)}$$

The Gibbs energy equation for reaction (l) implies that chromium and nickel form ideal solutions. This is only an approximation since recent measurements have shown a deviation from ideality.

When these four chemical equations are added, eqn (i) results, and

$$\Delta G_b + \Delta G_j + 2\Delta\bar{G}_k + 2\Delta\bar{G}_l = 200{,}832 - 41\ T + 38.28\ T \log N_{Al} - 38.28\ T \log N_{Cr}$$

represents the Gibbs energy of eqn (i). At equilibrium, ΔG_i is zero, and for 1550°C and $N_{Cr} = 0.1$, for instance, $\log N_{Al} = -2.81$. This corresponds to 0.07% Al by weight in the liquid alloy.

The equilibrium constant of reaction (e) in terms of weight percentages according to the calculation further above is

$$\log K_p = \log[c_{Al}^2 c_O^3] = \log[0.0019^2 \times 0.0017^3] = -13.8.$$

In fact, however, another equilibrium, namely reaction (i), must be taken into account. This was shown to produce an aluminium concentration of $c_{Al} = 0.07$, and, using the equilibrium constant, we find that

$$c_O = \sqrt[3]{1.6 \times 10^{-14}/c_{Al}^2} = 0.00015\%$$

so that the oxygen percentage in the chromium–nickel alloy melted in an alumina pot should actually be very low.

The evaluation of reaction (i) shows that the assumption that $1.5N_{Al} = N_O$ made to obtain c_{Al} and c_O for reaction (e) was not justified, and this example shows the importance of finding the correct chemical equations representing a particular metal-refractory interaction. In the case of pure iron or nickel as the metal phase the reaction of the metal with the aluminium in the refractory corresponding to eqn (i) need not be considered because of the lower stabilities of iron and nickel oxides.

The above calculation with chromium–nickel alloys would also apply to more complex chromium alloys provided that the alloying metals have a lower affinity for oxygen than chromium—to Ni–Co–Cr–Mo alloys for instance. In this way, the thermochemist can, in fact, handle fairly complex systems with some confidence provided that an error of, say, half a logarithmic unit in the final concentrations can be tolerated.

If such alloys contain carbon, however, considerable interaction with the refractory oxides must be anticipated, owing to the stability and volatility of the carbon oxides. Carbon in iron would react with an alumina pot in the following manner:

$$\tfrac{1}{3}\langle Al_2O_3\rangle + [C]_{Fe} = (CO) + \tfrac{2}{3}[Al]_{Fe}. \tag{m}$$

Using the Gibbs energies of formation of $\langle Al_2O_3\rangle$ and (CO) and the Gibbs energies of solution of carbon and aluminium in liquid iron, one obtains at 1600°C for the equilibrium constant

$$p_{CO}N_{Al}^{2/3}/N_C = 0.05$$

where p_{CO} is the pressure of carbon monoxide in atmospheres. In a vacuum, and even in an inert gas, carbon monoxide will continuously be removed from the system and equivalent amounts of aluminium dissolved. Eventually the melt will become virtually carbon-free and contain aluminium instead. Adcock (1937a) therefore, when making pure carbon-containing iron, used thoria linings or, alternatively, the carbon can be added just before casting. In the second case, use is made of the kinetic delay in reaction (m), in the first of the much lower equilibrium constant for the reaction:

$$\tfrac{1}{2}\langle ThO_2\rangle + [C]_{Fe} = (CO) + \tfrac{1}{2}[Th]_{Fe} \tag{n}$$

$$p_{CO}N_{Th}^{1/2}/N_C = 0.0002.$$

Because of the continuous removal of carbon monoxide in a good vacuum, even here gradual dissolution of thorium is expected, only it takes longer owing to the low equilibrium constant.

Silicon in the iron would act similarly to carbon, but, owing to the low equilibrium constant,

$$p_{SiO}N_{Al}/N_{Si} = 6.5 \times 10^{-5}$$

the danger is not serious.

When the oxygen potential is increased over a sample of iron which is held in an alumina crucible, a reaction ensues between metal and container at the point where the spinel phase $FeAl_2O_4$, hercynite, can be formed. This phase has the normal spinel structure in which the ferrous ions occupy tetrahedral and the aluminium ions octahedral holes in a f.c.c. oxide structure. Because of the similarity in structure between the spinel phase and γ-Al_2O_3 which is the same as $FeAl_2O_4$ but with vacancies where some of the Fe^{2+} ions are present in hercynite, there is a measurable range of solid solutions between these phases at temperatures exceeding 1500°C.

The spinel which is formed when iron is oxidised in contact with alumina is therefore that which is saturated with Al_2O_3 and to predict the behaviour of the iron–oxygen–Al_2O_3 system it is necessary to have thermodynamic information about the Al_2O_3-saturated phase. Results for this system were obtained by Chan *et al.* (1973a) over the temperature range 700–1600°C using a combination of galvanic cells with solid oxide electrolytes. For the low-temperature region Fe–FeO was used as the reference electrode, and for the high-temperature system Mo–MoO_2 was employed. The electrode containing hercynite was a mixture of Fe, Al_2O_3, and $FeAl_2O_4$. The Gibbs energy equation which was obtained from the study for the reaction

$$2Fe + O_2 + Al_2O_3 \rightarrow 2FeO\,.\,Al_2O_3(Al_2O_3\text{ saturated})$$

is of special interest in the present context for liquid iron since a knowledge of this makes it possible to calculate the oxygen content of the liquid iron when the oxidation just begins. For iron in the liquid state ($T \geq 1536$°C) the equation is

$$\Delta G^\circ = -612{,}496 + 152.63\ T.$$

Combining this equation with that for oxygen dissolved in liquid iron,

$$(O_2) = 2[O]_{Fe}; \qquad \Delta G^\circ = -233{,}676 + 50.84\ T + 38.28\ T \log N_O$$

we obtain the equation for the Gibbs energy change of the reaction

$$2Fe + 2[O] + Al_2O_3 \rightarrow 2FeO\,.\,Al_2O_3(Al_2O_3 \text{ saturated})$$

$$\Delta G^\circ = -378{,}820 + 101.79\ T + 38.28\ T \log N_O.$$

At equilibrium this is equal to zero since iron, alumina and hercynite are all present at essentially unit activity. At 1873 K the solution of this equation, after conversion from atom fraction to weight percentage, yields an oxygen content close to 0.07 wt%. The saturation value for iron in contact with pure FeO at this temperature is 0.22 wt%. The FeO activity in the spinel phase in contact with iron and alumina is therefore 0.07/0.22 or about one-third.

Another popular refractory is magnesia. Here the chemical reactions with molten metals are somewhat different. The solubility of magnesium in iron is virtually nil. On the other hand, the vapour pressure of magnesium is high, so that we have to consider the following chemical reaction:

$$\langle MgO \rangle = (Mg) + [O]_{Fe} \tag{p}$$

At 1600°C, the individual Gibbs energies are as follows:

$$\langle MgO \rangle = (Mg) + \tfrac{1}{2}(O_2); \qquad \Delta G_q = 355{,}640 \tag{q}$$

$$\tfrac{1}{2}(O_2) = [O]_{Fe}; \qquad \Delta G_c = -69{,}036 + 35{,}815 \log N_O. \tag{c}$$

Hence

$$-\log p_{Mg} = \log N_O + 8.0 \quad \text{at } 1600^\circ\text{C}.$$

There is no defined pressure of magnesium. This would evaporate from the reaction zone and condense in the cooler parts of the system. Thus the iron would gradually pick up oxygen up to the saturation point, simultaneously dissolving the magnesia pot. From the above equation one obtains:

at	$p_{Mg} = 10^{-3}$	10^{-2}	1	mmHg of magnesium
	$c_O = 0.2$	0.02	0.0002	wt% O.

Fischer and Hoffmann (1958d), who determined the oxygen concentrations in iron after holding the metal for various periods of time in crucibles made of magnesia, found that after 22 hr 0.15% O had accumulated in the iron. For rougher magnesia surfaces the pick-up of oxygen was faster.

Thus although the affinity of magnesium for oxygen is higher than that of aluminium, and the solubility of magnesium in iron less, alumina should be preferred as a container because of the more favourable mechanism of reaction.

Refractories other than oxides can be treated in a similar manner. Roughly, the enthalpy of formation of a refractory compound gives an indication of its suitability, but individual combinations should be considered more fully. Among the sulphides, cerium monosulphide appears to be particularly stable. Among the carbides and silicides, those of zirconium and titanium have highly negative heats of formation. Borides seem to gain some importance as refractory materials, but their consideration by the thermochemist is more difficult because thermochemical data are few and less reliable.

It is, however, seen that in quite a large number of practical cases the thermochemist may be of assistance in assessing the maximum degree of interaction between a refractory and a molten alloy, although the scarcity of accurate data may affect the accuracy of his computations. Knowing the chemical mechanism of an interaction may often help to reduce the pick-up of impurities; hopeless combinations may be eliminated at an early stage, thus avoiding costly trial runs; and current methods may be checked. If, for instance, a higher pick-up of impurity were found in a particular case than estimated thermochemically one would conclude that side-reactions, with impurities in the refractory for example, take place which could possibly be eliminated.

C. Electrochemical Cells and the Stabilities of Ceramics

One of the earliest applications of e.m.f. measurements to ceramic oxide systems was made by Benz and Wagner (1961a) who determined the Gibbs energies of formation of calcium silicates by means of the following cells:

$$\text{Pt, } O_2|CaO|CaF_2|Ca_2SiO_4,\ SiO_2|O_2, \text{ Pt.}$$

The electrolyte, CaF_2, conducts electricity principally by fluoride ion migration, and the cell measures the difference in fluorine potentials which are established at the two electrodes. These potentials are the result of the calcium activities in the two electrodes, which are established between the condensed phases and the gaseous oxygen atmosphere. The left hand electrode fixes the calcium activity in equilibrium with pure CaO and oxygen gas, and the right hand electrode fixes the calcium activity in equilibrium with oxygen gas and calcium orthosilicate saturated with silica.

Left hand electrode

$$a_{Ca} = K_{CaO}\frac{1(a_{CaO}=1)}{p_{O_2}^{1/2}}$$

$$p_{F_2} = K_{CaF_2}\frac{1(a_{CaF_2}=1)}{a_{Ca}}$$

Right hand electrode

$$a'_{Ca} = K_{CaO}\frac{a_{CaO}(Ca_2SiO_4,\ SiO_2)}{p_{O_2}^{1/2}}$$

$$p'_{F_2} = K_{CaF_2}\frac{1(a_{CaF_2}=1)}{a'_{Ca}}$$

$$E = \frac{RT}{2F}\ln\frac{p'_{F_2}}{p_{F_2}}.$$

When the same electrolyte is used to measure sulphur potentials, the same considerations apply. Thus in the cell

$$\text{Pt, } H_2/H_2S/Ar|CaS|CaF_2|CaS|H'_2/H'_2S, \text{ Ar, Pt}$$

the sulphur potentials established at each electrode by the H_2/H_2S mixtures determine the calcium activities in the calcium sulphide layers and hence determine the fluorine potential at the two faces of the CaF_2 electrolyte.

In a cell used by Levitskii and Scolis (1974g) the more complicated arrangement of electrodes

$$Pt, O_2|Sr_3Al_2O_6, SrAl_2O_4, SrF_2|CaF_2|Sr_2WO_5, SrWO_4, SrF_2|O_2Pt$$

has an e.m.f. yielding the Gibbs energy change for the reaction

$$Sr_3Al_2O_6 + 2SrWO_4 \rightarrow Sr_2Al_2O_4 + 2Sr_2WO_5.$$

This is because the presence of two coexisting aluminates $Sr_3Al_2O_6$ and $SrAl_2O_4$ fixes both the alumina and SrO activities and these, together with the oxygen pressure, in turn determine the strontium and fluorine activities in the SrF_2 phase present in the electrode. An analogous argument applies to the right hand side electrode, where the $Sr_2WO_5/SrWO_3$ mixture fixes the WO_3 and SrO activities.

When stabilised zirconia electrolytes are used for the study of oxide systems, these are limited to those ceramics based on oxides of lower stability than those of the alkaline earths, and the oxygen partial pressure which must be applied when fluoride electrolytes are used, as in the examples above, can be omitted. An example of this is the cell which was used to measure the stability of the compound $SrCu_2O_2$ (or $SrO.Cu_2O$)

$$Ar, Pt|Cu, Cu_2O|ZrO_2 + Y_2O_3|Cu, SrCu_2O_2, SrO|Pt, Ar.$$

In this cell, the activity of Cu_2O in the right hand electrode is controlled by the presence of pure copper and pure SrO. The cell thus measures the difference in Cu_2O activity between the two electrodes through the oxygen pressures exerted by the electrodes.

$$E = \frac{RT}{2F} \ln \frac{a_{Cu_2O}(\text{in Cu, } SrCu_2O_2\text{, SrO})}{1 a_{Cu_2O}(\text{in Cu, } Cu_2O)}$$

Finally, the cell

$$Pt|Ag\text{–}Zn \text{ alloy, } ZnO|CaO\text{–}ZrO_2|ThO_2\text{–}Y_2O_3|CaO\text{–}ZrO_2|Ag_2Zn \text{ alloy, } ZnTiO_3, TiO_2|Pt$$

was used to measure the Gibbs energy of formation of zinc titanate through the oxygen potentials exerted by the Ag–Zn alloy and ZnO phases in each electrode.

$$E = \frac{RT}{2F} \ln \frac{a_{ZnO}(ZnTiO_3, TiO_2)}{1(a_{ZnO} = 1)}$$

D. Equilibrium Phase Relations Relevant to the Production of Oxide Superconductor Materials

The discovery of a new range of superconducting materials with very high critical temperatures has led to an enormous research effort worldwide to find mixtures of substances with still-higher values of T_c and to establish reliable and reproducible methods for their large-scale production. Much of this research has proceeded in a rather haphazard way, in that a variety of oxides, for example, have been mixed together in different proportions in the hope that a superconducting phase with the required characteristics will result. Very few attempts have been made to carry out systematic studies of the phase constitution of the systems concerned, or by performing thermodynamic measurements, to determine the dependence of constitution, including the stability of the superconducting phases, on values of oxygen partial pressure. If such information were available, the appropriate, optimum conditions for the production of the superconducting phases in particular systems could be calculated. At the same time, conditions leading to stability of other, undesired phases, could be avoided.

One method of production of superconducting oxide materials is by oxidation of the metallic components of the systems concerned. In the case of the Y–Ba–Cu–O system, difficulties in production using this method can be traced back to the presence of a miscibility gap in the liquid phase of the ternary Y–Ba–Cu system. The extent of this miscibility gap was unknown, but with the help of thermodynamic calculations based on evaluated data for the binary systems Y–Ba, Y–Cu and Ba–Cu (1990d, e), phase equilibria in the ternary Y–Ba–Cu system have been calculated (1990d). Figure 108 shows an isothermal section of the ternary at 1273 K, calculated using the principles described on pp. 253.

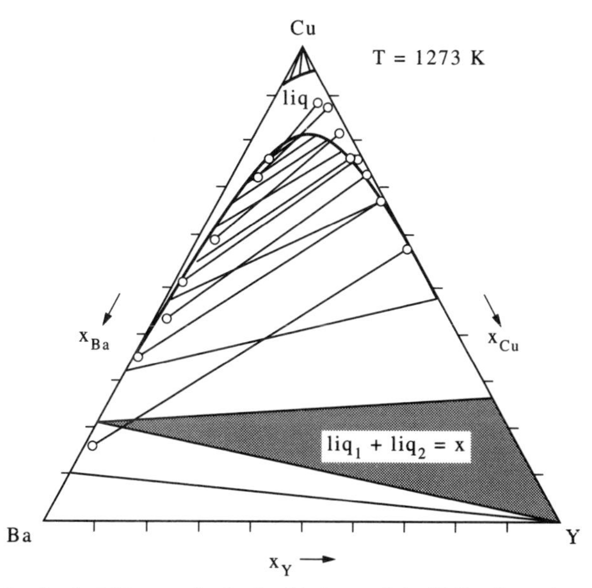

FIG. 108. Calculated miscibility gap in the liquid phase of the Y–Ba–Cu system at 1273 K. The experimental points are from unpublished work by Itagaki (see ref. (1990d)).

The thermodynamic data produced in this evaluation work can be used in association with appropriate software for calculating complex phase equilibria (e.g. (1990b)), to investigate aspects of Y–Ba–Cu–O superconductor production based on the metallic components of the system.

4. Chemical Vapour Deposition (CVD) and Physical Vapour Deposition (PVD) Processes

A. CVD Production of Ultra-pure Silicon

A simple, but representative, example of the application of thermochemical calculations to the investigation of chemical vapour deposition (CVD) processes relates to the production of ultra-pure silicon by the thermal decomposition of $SiHCl_3$ gas. In this process, it is of technological and economic importance to establish the optimum temperatures for maximum yield of pure silicon from the decomposition reaction.

Figure 109 illustrates the calculated silicon yield (mols) from 1 mol of $SiHCl_3$ gas as a function of temperature at 1 atm pressure. The plot shows that an optimum yield can be expected at temperatures around 1100 K. This is in good agreement with experimental observations. The corresponding composition of the gas phase as a function of temperature is illustrated in Fig. 110.

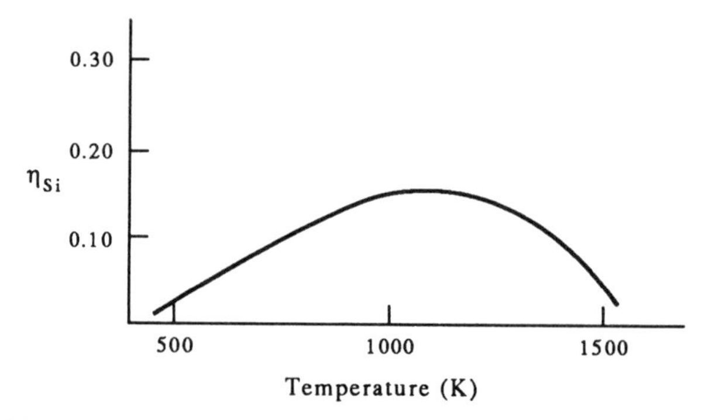

FIG. 109. Si yield (mols) produced by the dissociation of 1 mol $SiHCl_3$ gas, as a function of temperature.

B. Vapour Phase Transport of Silicon Carbide

Transport reactions involving volatile halides are of considerable use in the purification of metals such as żirconium by the van Arkel process, and in vapour phase coating and single crystal growth. What these processes have in common is the formation and subsequent decomposition of gaseous halides in a system incorporating a temperature gradient. In some instances the gaseous compound decomposes usefully in going to a higher temperature, and in others the decomposition occurs at lower temperatures.

The van Arkel process mentioned above uses the formation of ZrI_4 at low temperatures, around 600°C, between iodine vapour and impure zirconium, and decomposition of the tetraiodide at higher temperatures, around 1400°C, to yield zirconium of a high purity.

In the Gross process an aluminium alloy is reacted with $AlCl_3$ gas to form the monohalide AlCl at a high temperature, around 1200°C, and this then disproportionates at a lower temperature, about 700°C, to produce the pure metal and to regenerate $AlCl_3$ which is used again.

One criterion of a successful vapour transport reaction is obviously that the equilibrium constant should change appreciably over the temperature gradient. Secondly, it is necessary that the partial pressure of the transporting species is reasonably large so that a satisfactory rate of transport may be achieved. It is because the rate of vapour transport of a metal can be increased by the formation of a volatile halide, above the rate of vaporisation of the pure metal, that this process is sometimes referred to as "catalytic distillation".

The change in the value of the equilibrium constant for the formation of a volatile halide with temperature depends mainly on the entropy change for a reaction, and this in turn depends mainly on the change in the number of gaseous molecules in the vapour transport reaction. It follows that the greatest possible valency changes are most desirable in the metal being transported when the gaseous species is formed or is decomposed. The value

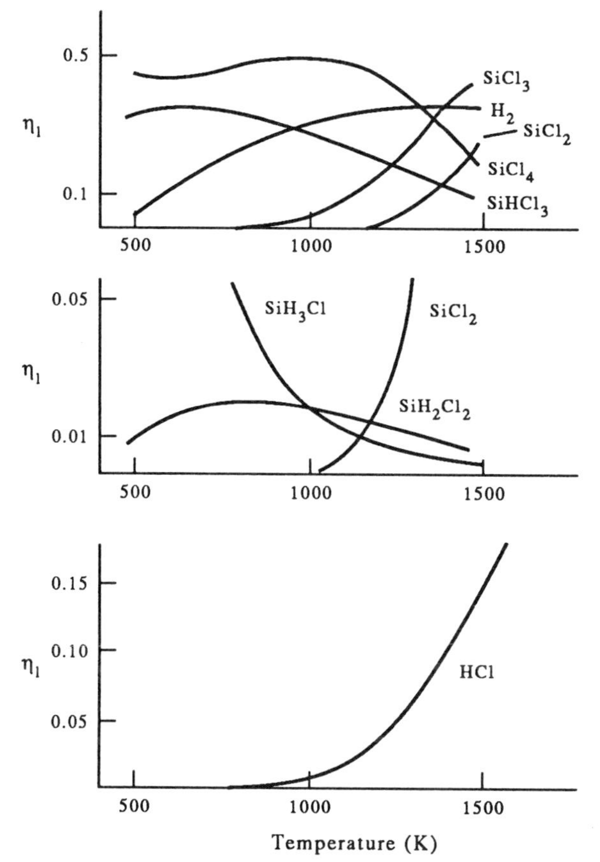

FIG. 110. Gaseous species resulting from the dissociation of 1 mol $SiHCl_3$ gas, as a function of temperature.

of the equilibrium constant for the formation of a compound should be relatively close to unity in order that the entropy change can have the largest effect. This is shown in Table XXXI which was prepared by Alcock and Jeffes (1967a) in a survey of the thermodynamics of vapour phase transport reactions. For a reaction $\langle X \rangle + (Y_2) \rightarrow (XY_2)$ the equilibrium constant will change over a range of temperatures according to the van't Hoff equation

$$\frac{\partial \ln K}{\partial(1/T)} = \frac{-\Delta H^\circ}{R}$$

(clearly a reaction where ΔH° is equal to zero is of no interest in vapour phase transport).

The table shows the values of p_{XY_2} for a total pressure of 1 atm $(p_{Y_2} + p_{XY_2})$ when K changes as a result of a change in temperature from 10^{-5} to 10^5, and how each value of p_{XY_2} differs from the value $p^\circ_{XY_2}$ which is the equilibrium pressure when K is equal to unity.

It is apparent from these calculated values that the largest change in p_{XY_2} occurs around $K = 1$, and at the extremes a change of K by one order of magnitude changes p_{XY_2} very little. Thus, although a large value of ΔH° will produce a large temperature coefficient of K, this must be balanced by a sufficiently large entropy change of the same sign, so that, ideally, K should pass through unity in some experimentally useful temperature range.

TABLE XXXI. *Pressures of the species XY_2 formed by the reaction $\langle X\rangle + (Y_2) \rightarrow (XY_2)$ as a function of the value of the equilibrium constant* K

K	p_{XY_2} (atm)	$p_{XY_2} - p^{\circ}_{XY_2}$ (atm)
10^{-5}	9.1×10^{-6}	0.49999
10^{-4}	9.1×10^{-5}	0.4999
10^{-3}	9.1×10^{-4}	0.499
10^{-2}	9.1×10^{-3}	0.49
10^{-1}	9.1×10^{-2}	0.41
1	0.50 ($p^{\circ}_{XY_2}$)	0.00
10	0.91	0.41
10^2	0.99	0.49
10^3	0.999	0.499
10^4	0.9999	0.4999
10^5	0.99999	0.49999

In the same paper it was also pointed out that the large collection of thermodynamic data on gaseous metal halides indicated that in the general reaction

$$\langle M\rangle + \frac{n}{2}(X_2) \rightarrow (MX_n)$$

the entropy changes are, as would be expected, approximately constant for a given value of n but that the heat change for the formation of a metal bromide was about 96.2 kJ per mole Br_2 and the iodide about 230 kJ per mole I_2 less negative than the heat change for the corresponding chloride per mole Cl_2. Thus if a given chloride reaction was found to have too large a negative heat change for the equilibrium constant to approach unit in some useful temperature range, the bromide and more so the iodide would lower this temperature range providing ΔH° did not then approach too closely to zero for reasons given above.

Turning to the particular example of the transport not of an element, but of the compound silicon carbide, the same general guiding principles must be applied. Now, however, a vapour phase transport reaction must be used which will transport, effectively, not only silicon, as one of the tetrahalides, but carbon as well. Clearly no pure halogen is of much use in this respect because of the very low affinity of carbon when compared with silicon for the halogens at temperatures where equilibrium is achieved (1968f).

If HCl is used as the source of chlorine, the partial pressure of $SiCl_4$ which could be generated would be considerably reduced because hydrogen has a comparable affinity for chlorine as has silicon. The hydrogen which would be displaced by $SiCl_4$ formation would now be available to react with carbon forming gaseous methane. Methane is a suitable candidate for the transport of carbon and the low stability of silane SiH_4 in comparison with CH_4 and with silicon halides suggests that a probable transport reaction for SiC would be

$$\langle SiC\rangle + 4(HCl) \rightarrow (SiCl_4) + (CH_4)$$

$$\Delta G^{\circ} = -301{,}248 + 251.5\,T; \qquad \log K = \frac{15{,}738}{T} - 13.14.$$

The considerable reduction in the chlorine potential which is brought about by the introduction of hydrogen into the system lowers the partial pressure of $SiCl_4$ gas and it is thus made comparable with the partial pressure of the carbon-carrying species CH_4. The ratio of CH_4/H_2 which is achieved at equilibrium is unity at 500°C and 0.03 at 950°C and the equilibrium constant for the vapour transport reaction changes from approximately 10^{-7} to 0.5 in the same temperature range. It follows that this reaction is satisfactory in terms of the criteria for vapour phase transport which were elaborated above. The equilibrium constant achieves the value of unity at 925°C.

C. Prediction of Metastable Phase Formation during PVD of Mixed Coating Materials

A major application of hard metal carbides and nitrides is as coatings for tool materials. Such coatings have frequently been produced by CVD (Chemical Vapour Deposition) methods, but there is an ever-increasing interest in physical vapour deposition (PVD) techniques, whereby material of known compositions is transferred from a target (e.g. by sputtering) to a substrate maintained at a lower temperature. PVD processes are of particular importance because they enable homogeneous, multi-component phases with improved properties to be deposited easily. The stoichiometry of these phases can be controlled by adjusting the process parameters, while by controlling the substrate temperature, some materials can be deposited in different crystalline structures or as an amorphous phase.

Saunders and Miodownik (1986e) have shown for metallic alloys that low substrate temperatures result in diffusion processes being too slow to produce the normal equilibrium multiphase structures in the deposited materials. As a result, a single phase coating is formed from the phase with the lowest Gibbs energy at the over-all coating composition and substrate temperature concerned. Using thermodynamic calculation methods, Saunders and Miodownik were able to predict with success the ranges of stability (both stable and metastable) of different phases found in codeposited alloys at low substrate temperatures.

Using a 'complex equilibrium program' (1990b), together with evaluated thermodynamic data for the solution phases in the ternary Ti–B–C system, prediction has been made of the likely ranges of existence of metastable phases across different sections of the ternary (1990f).

Figure 111 shows the calculated Gibbs energy curves at 500 K for the hexagonal, cubic-NaCl, and liquid phases in the section TiB_2–TiC. From consideration of the fact that it is the **single phase** with the lowest Gibbs energy that is observed during PVD of the coating compositions concerned, the metastable ranges for the different phases can be predicted. First experimental results for the structures observed at a number of compositions are indicated along the composition axis.

5. Corrosion

A. The Oxidation of Iron–Chromium Alloys

The b.c.c. solid solutions of chromium in iron, the α solid solutions, exist over the whole composition range at temperatures above 1392°C, and below this temperature a closed

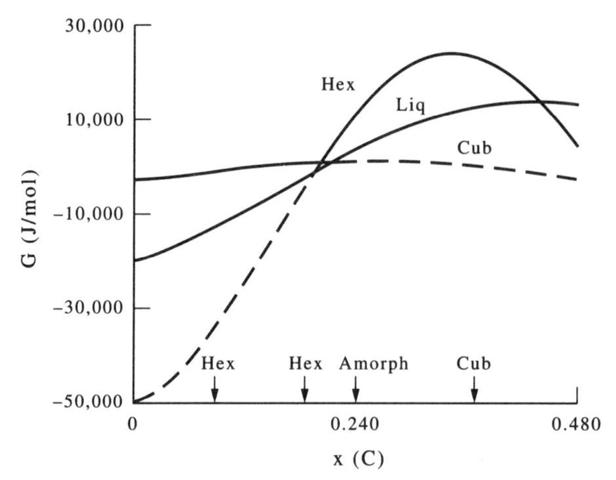

FIG. 111. Calculated Gibbs energy curves for the cubic, hexagonal and liquid phases of the Ti–B–C system in the section TiB_2–⟨TiC⟩ at 500 K. The predicted metastable ranges for the three phases are indicated by the dashed portions of the curves. Preliminary results from experimental observations of the structures observed for four coatings prepared by PVD are indicated on the abscissa.

range of compositions exist which are iron-rich and have the f.c.c. structure (γ solid solutions). The thermodynamic properties of the α solid solutions have been reviewed by Müller and Kubaschewski who obtained the results shown in Table XXXII. These are only quoted for the compositions up to $N_{Cr}=0.25$ because those are all of the results which are needed for the present discussion. More data are available in the original paper (1969i).

TABLE XXXII. *Partial thermodynamic properties in αFe–Cr alloys at high temperatures*

N_{Cr}	$\Delta\bar{H}_{Cr}$	$\Delta\bar{S}_{Cr}$	$\Delta\bar{H}_{Fe}$	$\Delta\bar{S}_{Fe}$
0.05	+5415	8.162	15	0.109
0.10	4860	6.584	60	0.233
0.15	4335	5.609	135	0.371
0.20	3840	4.878	240	0.524
0.25	3375	4.274	375	0.700

The other information which is required to complete the thermodynamic properties of this system with respect to oxidation is that for the formation of the oxides. The Gibbs energy equations are as follows:

$$2\langle Fe\rangle + (O_2) \rightarrow 2\langle FeO\rangle \qquad \Delta G^\circ = -519{,}234 + 125.1\ T\ \text{J}$$

$$\tfrac{4}{3}\langle Cr\rangle + (O_2) \rightarrow \tfrac{2}{3}\langle Cr_2O_3\rangle \qquad \Delta G^\circ = -746{,}844 + 173.2\ T\ \text{J}$$

$$2\langle Fe\rangle + (O_2) + 2\langle Cr_2O_3\rangle \rightarrow 2\langle FeCr_2O_4\rangle \qquad \Delta G^\circ = -633{,}458 + 145.2\ T\ \text{J}.$$

These data lead to the ready conclusion that the lowest oxygen potential in most of the potential alloy–oxide systems in the ternary Fe–Cr–O system is that of the Cr–Cr_2O_3

system, $FeCr_2O_4$ being the next and the oxides of iron only being formed at much higher oxygen pressures. The results for the oxides are shown in the Ellingham diagram (Fig. 112).

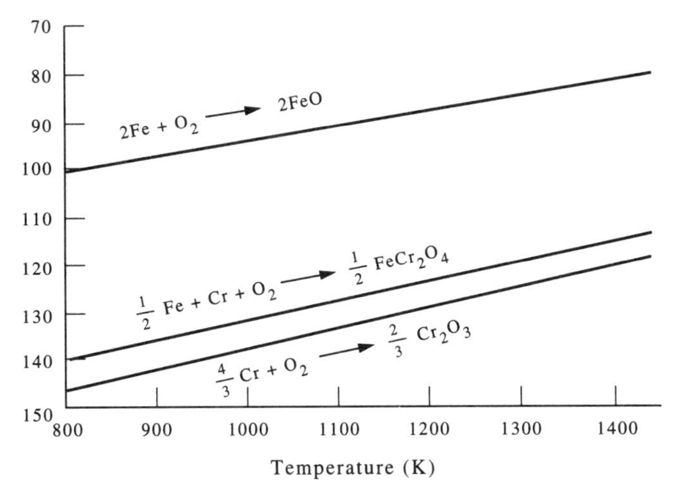

FIG. 112. Ellingham diagram for the oxides FeO, $FeCr_2O_4$ and Cr_2O_3.

The experimental evidence for the oxidation mechanism by Moreau (1953d) and Rahmel (1962g) at 1000°C shows that alloys containing greater than 15% Cr are protected from corrosion by formation of $FeCr_2O_4$. Those with less than this chromium content oxidise to form a multilayer oxide comprising a mixture of FeO and $FeCr_2O_4$ in contact with the metal, then Fe_3O_4 containing Cr^{3+} in the octahedral sites and finally Fe_2O_3. There appears to be no role for Cr_2O_3, the most stable oxide, in the reaction mechanism. However, there are a number of points which emerge when a thermodynamic analysis is made and these largely centre around the relative stabilities of Cr_2O_3 and $FeCr_2O_4$. The equilibrium constant for the displacement reaction

$$Fe+\tfrac{4}{3}Cr_2O_3 \rightarrow FeCr_2O_4+\tfrac{2}{3}Cr$$

provides information concerning the alloy composition ranges in which this reaction goes from left to right or vice versa. When the oxide phases are present at unit activity, the constant simplifies to

$$K=a_{Cr}^{2/3}/a_{Fe}.$$

It is soon found by insertion of typical values (the activities of iron and chromium are linked by the fact that N_{Cr} equals $1-N_{Fe}$) that the reaction only proceeds to the right when chromium is present in the coexisting alloy phase at very high dilution. From the data which have been given for the formation of the oxides, we obtain for the displacement reaction

$$\Delta G^\circ = 56{,}693-14.02\,T \qquad K=\frac{a_{FeCr_2O_4}a_{Cr}^{2/3}}{a_{Cr_2O_3}^{4/3}a_{Fe}}.$$

Since iron is present at nearly unit activity when the reaction proceeds from left to right, we may write, as a good approximation for this Gibbs energy,

$$\Delta G^\circ = -RT \ln K = -\tfrac{2}{3}\Delta\mu_{Cr}.$$

In order to convert these chemical potentials into the corresponding chromium concentrations in the alloys it is necessary to obtain an equation between the chromium potentials and the chromium atom fraction over a range of temperatures. Using the results quoted above from Müller and Kubaschewski, we calculate the excess partial entropy for chromium at each of the stated compositions. These are then plotted against composition and a reasonable extrapolation produced for the low values of N_{Cr} (Fig. 113). The results fit the equation

$$\Delta \bar{S}^E_{Cr} = 10.04 - 15.06\, N_{Cr}$$

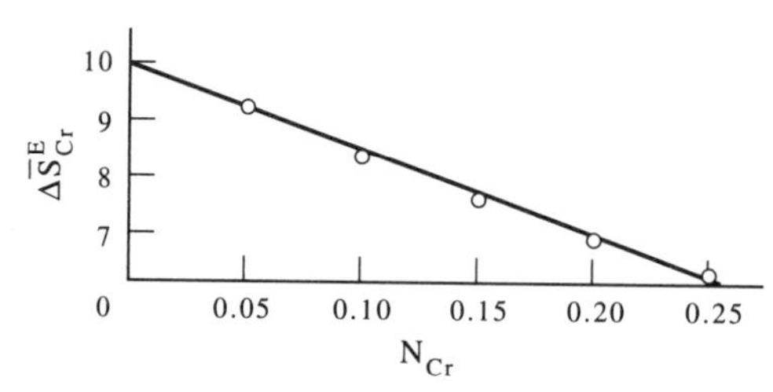

FIG. 113. Excess partial entropy of chromium in dilute Fe–Cr alloys.

but for the region below 1 at.% Cr we shall assume the constant value 2.40. The integral heat of mixing curve shows a strictly regular solution behaviour and can be described by the equation

$$\Delta H^M = 25{,}104\, N_{FE} N_{Cr}$$

from the results of Dench (1963b). At N_{Cr} equal to 0.01 this yields the partial enthalpy solution for chromium

$$\Delta \bar{H}_{Cr} = 25{,}104\, N^2_{Fe} = 24{,}602.$$

Hence the equation for the partial Gibbs energy of chromium in solution in iron is

$$\Delta \bar{G}_{Cr} = 24{,}602 - 10.04\, T + 19.142\, T \log N_{Cr} \qquad (N_{Cr} < 0.01).$$

From the value for the Gibbs energy change of the displacement reaction, we may now calculate the values of N_{Cr} at equilibrium with iron, Cr_2O_3, and $FeCr_2O_4$ at any temperature. The results are given in Table XXXIII and it must be observed that when the chromium content is higher than these values then chromium displaces iron from $FeCr_2O_4$ to produce Cr_2O_3. These results show that Cr_2O_3 will always be the first product of oxidation of iron–chromium alloys in this temperature range excepting for those containing less than 1.5 wt% and which are oxidised above 1400 K. Below this temperature the manner of formation of $FeCr_2O_4$ must be the result of a secondary reaction between FeO and Cr_2O_3. It is probably because of this that this compound is formed below the original surface during oxidation at 1000°C. This is demonstrated by the observation that inert markers which were originally placed on the surface of the alloy are found after oxidation between the FeO layer and the FeO–$FeCr_2O_4$ dispersion.

TABLE XXXIII. *Chromium contents of Fe–Cr alloys in equilibrium with* Cr_2O_3 *and iron chromite*

T(K)	$\Delta G°$ (displacement)	N_{Cr} (equilibrium)
800	45,480	3×10^{-6}
1000	42,675	7.9×10^{-5}
1200	39,875	7.1×10^{-4}
1400	37,070	3.4×10^{-3}

6. Environmental and Energy Problems

A. Calculation of Hazardous Emissions During Sintering of Ores

A knowledge of the gas species resulting from the sintering of ores and other materials is particularly important from the environmental point of view. Above all, information on the conditions necessary to avoid the formation of toxic species containing, for example, halogens or As, Cd, Pb, etc., are required, as are data for the likely amounts (or partial pressures) of such substances formed. Here again, thermochemical calculations can be used to provide the relevant information.

A series of calculations carried out to investigate the sintering behaviour of a representative iron ore has provided information on the likely gas species formed, their amounts, and the phases expected to condense from this gas phase on cooling under oxidising or reducing conditions.

The composition of the ore (wt%) is given below:

Fe	53.015	BaS	1.281	S	0.024
O_2	23.011	Mn	0.505	Zn	0.009
CaO	7.887	F	0.244	Pb	0.005
H_2O	5.410	K_2O	0.081	As	0.002
C	4.494	Cl	0.079	Cd	0.0002
FeO	3.925	Na_2O	0.027		

Figure 114 illustrates the calculated partial pressures of gas species formed during sintering, assuming cooling of the gas phase at atmospheric oxygen pressure.

Figure 115 illustrates the amounts of condensed phases forming from this gas phase and the ranges of temperature in which the phases form.

Examination of these two figures shows that arsenic is calculated to remain in gaseous form as AsF_3 at all temperatures between 400 and 1200°C. Lead is present in the gas phase as PbO, but on cooling below about 800°C condenses to form $PbSO_4$. Cadmium gas also

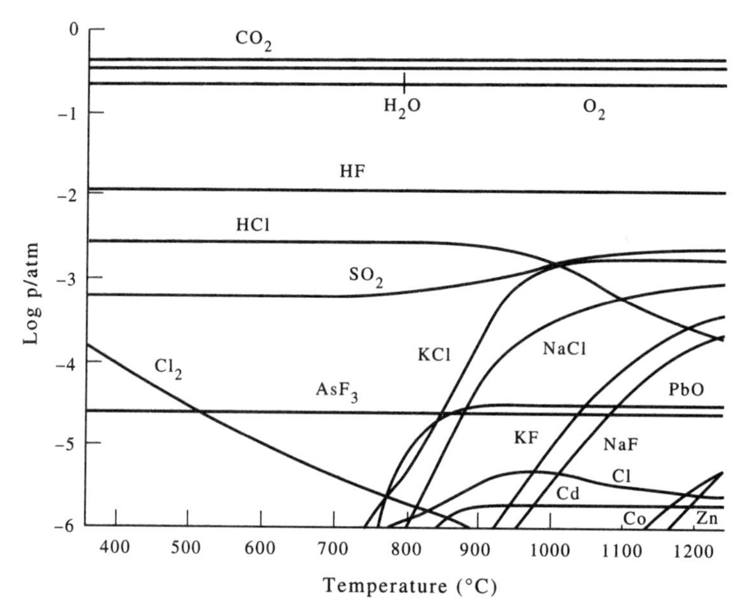

FIG. 114. Partial pressures of gaseous species resulting from cooling in air of the gas phase produced by sintering at 1200°C.

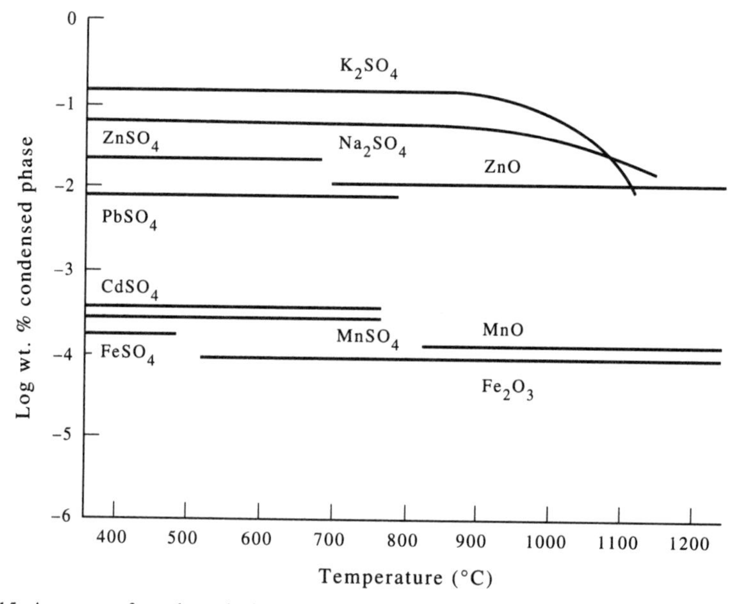

FIG. 115. Amounts of condensed phases formed on cooling in air of the gas phase produced by sintering at 1200°C.

condenses to form cadmium sulphate at temperatures close to 800°C. CO_2, H_2O and O_2 are the major gas species if the process is carried out at atmospheric oxygen pressure, but but Fig. 114 shows that HF and HCl are also present in measurable quantities.

If the gas phase is cooled under reducing conditions (no interaction with atmospheric oxygen), AsF_3(gas) is still the stable form in which arsenic is present at lower temperatures, but lead and cadmium condense as sulphides at temperatures around 600°C.

Once again, calculated information of this nature enables appropriate process parameters to be selected thereby enabling undesirable emissions to be kept within regulation limits. At the same time, a knowledge of the form in which toxic species are present enables appropriate treatment of the emissions to be chosen.

B. Incineration of Waste in a Molten Iron Bath

A more recently utilized method for the disposal of hazardous waste, such as waste oil with high chlorine content, or heavy-metal containing sewage, is by feeding into a bath of molten iron. During this process, most of the waste is incinerated and released as gaseous species, but heavy metals present in the waste in very small quantities may also be dissolved in the melt or in a slag phase covering the melt.

To investigate the composition of the gas phase resulting from the incineration, as well as the quantities and distribution of toxic elements such as lead or cadmium between the gas phase and the melt (or slag), calculations have been carried out for a variety of possible waste oils and sewage compositions.

It should be pointed out, that to perform the calculations, the thermodynamic properties of all gaseous species which could possibly be formed must be known, and the thermodynamic properties of the dissolved elements in iron, as well as their interactions with each other, must be described. This applies too to the components of the slag phase.

Figure 116 illustrates the calculated equilibrium composition of the gas phase resulting from addition of waste oil containing an **arbitrarily selected** 10% hexafluorobenzine to an iron bath maintained at different temperatures. The major products of the incineration are hydrogen and carbon monoxide, but HF gas is also produced in significant quantities. The partial pressures of other gaseous species present can be read from the plot.

For the incineration of sewage, a representative composition containing the following amounts of metallic elements was chosen:

Cd	0.0028%	Pb	0.067%	Cu	0.12%
Zn	0.35%	Cr	0.08%.		

The distribution of these elements between the gas phase and the molten iron bath was calculated for several bath temperatures. The results are presented in Table XXXIV.

From calculated information of the type shown in Fig. 116 and Table XXXIV it is possible to adjust process parameters such as input waste composition or bath temperature such that the amounts of hazardous species in the gaseous emissions are kept within the regulation limits. The calculations can, of course, be performed for a variety of wastes, temperatures and total gas pressures.

Additional calculations show, that due to the presence of sulphur in the product gas, for example as H_2S, condensed sulphide species such as ZnS and CdS will form as the gas is

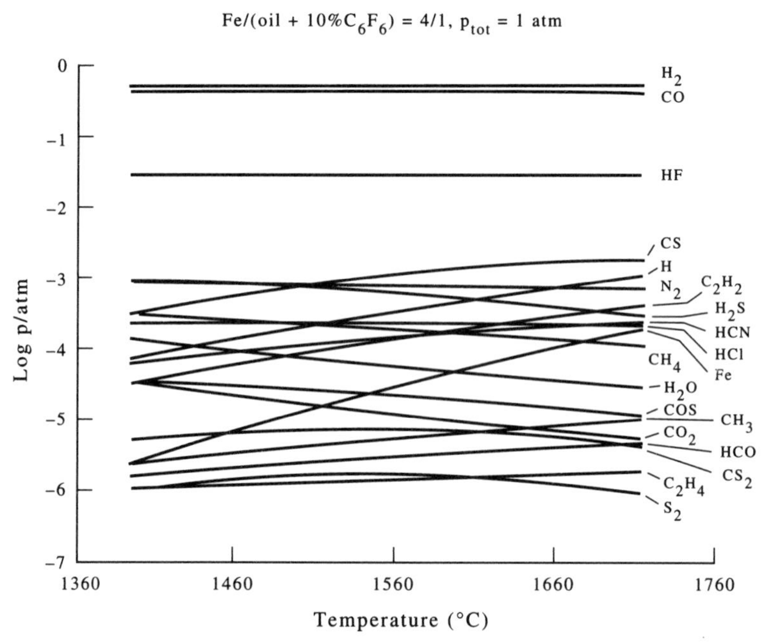

FIG. 116. Equilibrium gas phase composition produced by combustion of waste oil containing 10% hexafluorobenzine.

TABLE XXXIV. *Distribution of metallic elements between the gas phase and the melt during incineration of sewage in a molten iron bath*

Element	Bath temp °C	% in Fe-melt	% in gas phase
Cd	1200	40.51	59.49
	1400	21.05	79.95
	1600	12.18	87.82
Pb	1200	3.34	96.66
	1400	4.12	95.88
	1600	5.25	94.75
Cu	1200	100.0	—
	1400	99.997	0.003
	1600	99.98	0.02
Zn	1200	65.4	34.6
	1400	38.69	61.31
	1600	22.49	77.51
Cr	1200	100.0	—
	1400	100.0	—
	1600	100.0	—

cooled. These species have a low solubility in water and can be recovered by appropriate filtering operations.

C. Thermodynamic Conditions for Formation of Dioxin in Waste Incineration

The formation of small amounts of the highly toxic species 2, 3, 7, 8-TCDD (tetrachlor-dibenzo-dioxin) has repeatedly been confirmed during the operation of certain industrial processes, in particular processes used for the incineration of waste. Although practical procedures to avoid formation of Dioxin have, in certain cases, been established empirically, prior calculation of suitable operating parameters for specified processes, such that conditions for Dioxin formation can be avoided, would be invaluable. Spencer and Neuschütz (1992b) have recently carried out a study of the thermodynamic conditions leading to the formation of Dioxin. The process variables used in the calculations were

—the chlorine content (between 0.001 and 20 mol%): corresponding to the chlorine content in organic wastes.
—the C/H mol ratio (between 0.1/1 and 1000/1): corresponding to different fuels and organic wastes.
—the oxygen partial pressure (between 10^{-50} and 0.21 bar: corresponding to different possible gas atmospheres.
—the temperature of the system (between RT and 1500°C): corresponding to temperatures encountered in incineration processes and during cooling of the gas phase.

Some of the findings of this study are summarised below:

—In the whole range of conditions considered, Dioxin is thermodynamically metastable. If the predicted free carbon formation is not suppressed in the calculations by removing C from consideration as a possible product phase, Dioxin is calculated to dissociate completely into C, CO_2 and HCl, even at room temperature.

All subsequent discussion relates to calculations made with the formation of free carbon suppressed.

—For Cl-contents between 0.001 and 10 mol%, and for all C/H ratios, Dioxin partial pressures are insignificant ($<10^{-15}$ bar) if the oxygen partial pressure $p(O_2)>10^{-20}$ bar.
—Significant Dioxin partial pressures are calculated for Cl-contents between 0.01 and 10 mol%, when C/H > 1/1 and $p(O_2)<10^{-20}$ bar.
—For p(TCDD) values $>10^{-15}$ bar and temperatures between 500 and 1500°C, very low oxygen partial pressures (around 10^{-40} bar) give rise to higher p(TCDD) values at lower temperatures and higher $p(O_2)$ values to higher p(TCDD) values at higher temperatures.
—For a given C/H ratio, a higher Cl-content leads to higher p(TCDD) values at all $p(O_2)$ values and temperatures.

Some of these findings are illustrated in Fig. 117 where log p(TCDD) values are plotted against temperature for a selected chlorine content and C/H ratio and for varying $p(O_2)$ values.

Of particular relevance for the safe operation of the process are plots such as that illustrated in Fig. 118 where for a particular C/H-ratio, the ranges in which the gaseous

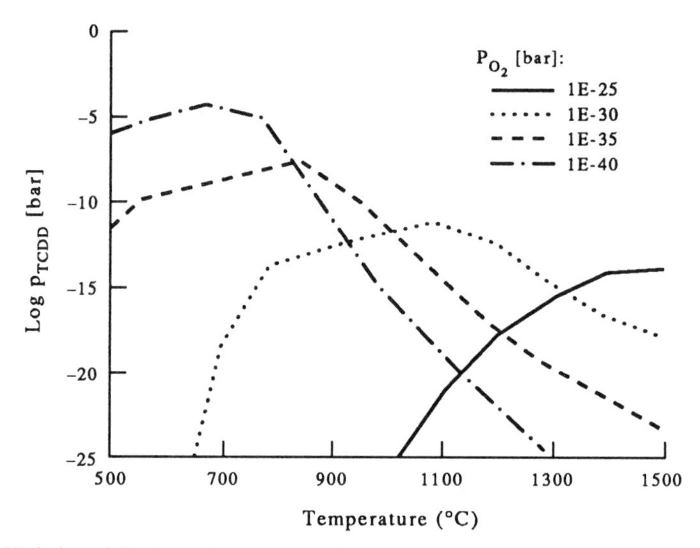

FIG. 117. Variation of dioxin partial pressure with temperature for given chlorine content and C/H ratio and varying oxygen partial pressure.

Dioxin level reaches 0.1 ng/m^3 are plotted as a function of oxygen partial pressure and temperature for various chlorine contents.

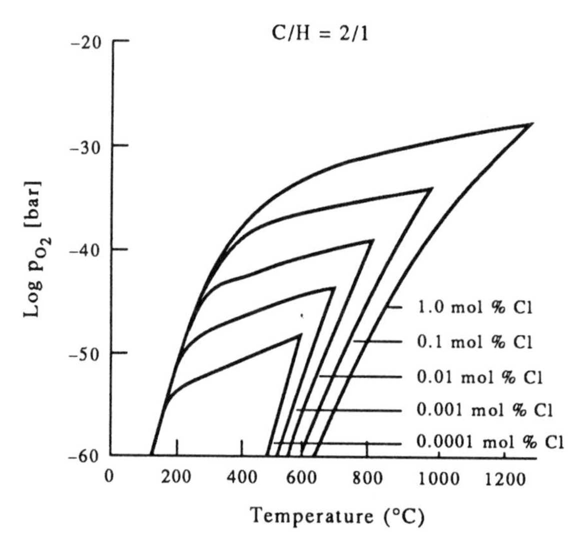

FIG. 118. Envelope of curves within which the dioxin level exceeds 0.1 ng/m^3.

It is clearly difficult in waste incineration operations to guarantee a homogeneous composition of the waste throughout the furnace, or to maintain a uniform partial pressure of oxygen. Kinetic factors are also important in determining the rate and location at which Dioxin forms. Nevertheless, the background information provided by calculations of the type described above is of great value in helping to select safe conditions for carrying out incineration of particular types of waste.

Since Dioxin formation is calculated to occur under conditions of low oxygen partial pressures, the use of oxygen-rich atmospheres for waste incineration appears to be appropriate. The amounts of gas of different oxygen content needed for complete incineration, as well as the enthalpy effects accompanying the incineration in each case, can also be determined using thermochemical calculations.

D. Energy Conservation in Waste Incineration

Waste incineration can be achieved using a variety of fuels. The efficiency of the incineration and the amount and composition of the resulting gas phase in each case may vary considerably, however.

Thermochemical calculations can be used to provide information not only on the product species (gaseous or condensed) resulting from a particular incineration process, but also on the accompanying enthalpy change. This is particularly useful in determining the costs of the operation and in choosing optimum process parameters to allow safe incineration with minimum wastage of energy.

Figure 119a and b presents a simple comparison of the incineration of C_6H_5Cl, in the one case in air and in the other case in pure oxygen. Apart from the partial pressures of the resulting gas species in the two cases, the diagrams present information on the amounts of air or oxygen needed to achieve complete incineration, and the accompanying enthalpy change at three temperatures. It is clear that incineration in air requires a considerably greater input of gas than is the case with pure oxygen. At the same time, because the nitrogen in the air plays a very minor role in forming incineration products, much of the energy produced during incineration in air goes towards heating up the nitrogen. Incineration in oxygen is a considerably more exothermic, and hence more efficient, operation. However, the greater costs associated with the use of pure oxygen (or other possible gas mixtures) must be set against the improved incineration efficiency.

7. Assessment of Standard Values

*A. A Pure Stoichiometric Substance—Silicon Monoxide**

It is often found that several independent measurements involving a particular substance are available which when considered as a whole will permit a more or less accurate assessment of its thermochemical standard values. Unfortunately, individual authors are sometimes inclined to see only their own results and to disregard, apart from an honorable mention, those of their fellow observers. Their evaluations thus require reassessment in the light of the total information available. This is particularly true of equilibrium measurements. Authors usually overlook the fact that the temperature coefficients of such results, on which heat and entropy values depend, are rarely reliable and occasionally quite inaccurate despite the care with which the measurements may have

*The more recent results on the equilibrium $\langle Si \rangle + \langle SiO_2 \rangle = 2(SiO)$ by Chart have not been incorporated in this evaluation but were found to agree well with the assessed values.

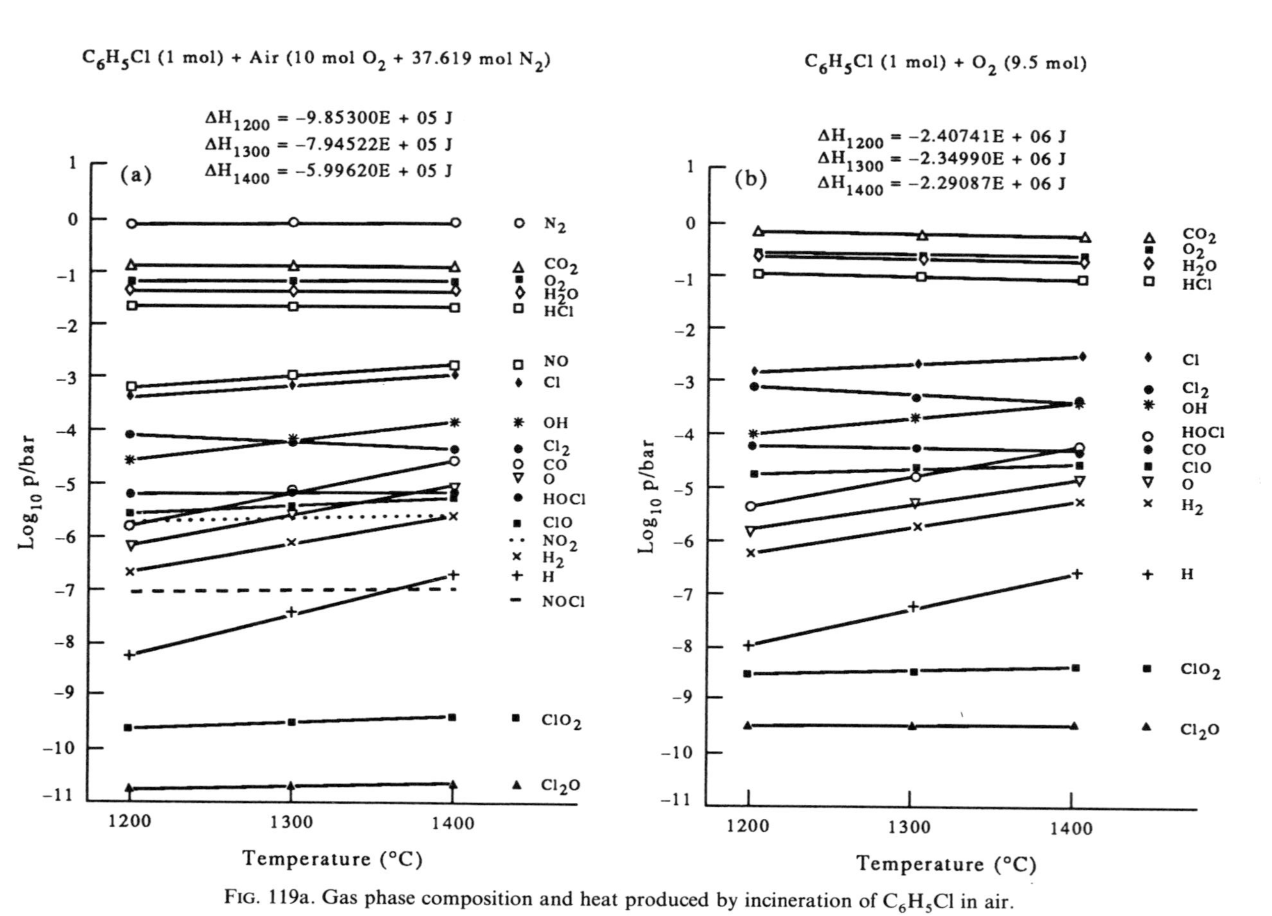

FIG. 119a. Gas phase composition and heat produced by incineration of C_6H_5Cl in air.

FIG. 119b. Gas phase composition and heat produced by incineration of C_6H_5Cl in oxygen.

been made. It is generally safer to combine the observed Gibbs energies with estimated entropies rather than to derive the entropies from the temperature coefficients.

In this section the thermochemical properties of silicon monoxide may be assessed as a demonstration of how available information is to be digested. This information is first summarised, equilibrium pressures being converted initially into Gibbs energy equations by means of eqn (29a).

The equilibrium

$$\langle SiO_2 \rangle + (H_2) = (SiO) + (H_2O) \tag{a}$$

was studied by means of a transportation method by Tombs and Welch (1952i) (1228–1653°C), Grube and Speidel (1949d) (1200–1500°C) and Ramstad and Richardson (1961c) (1425–1600°C), the results being represented by

$$\text{T \& W (1710 K):} \quad \Delta G_a^\circ = 327{,}189 - 87.86\ T\ \text{J} \tag{α}$$

$$\text{G \& S (1620 K):} \quad \Delta G_a^\circ = 468{,}608 - 148.53\ T\ \text{J} \tag{β}$$

$$\text{R \& R (1800 K):} \quad \Delta G_a^\circ = 531{,}786 - 188.57\ T\ \text{J}. \tag{γ}$$

The average absolute temperatures are indicated.

The SiO pressure over mixtures of silicon and silica (presumably in the form of cristobalite) was measured by Tombs and Welch (1952i) (1300–1647°C) and Ramstad and Richardson (1961c) (1237–1485°C) who used an entrainment technique, and by Schäfer and Hörnle (1950i) (1050–1250°C) who used the Knudsen method. Tombs and Welch's results were presented by

$$\Delta G^\circ = 244{,}973 - 106.48\ T \tag{δ}$$

per mole SiO. Since silicon is liquid over most of the temperature range studied by the last mentioned observers, the Gibbs energy of fusion of half a g-atom of silicon, $\Delta G = 25{,}313 - 15.06\ T$, should be added to reaction (δ) in order to get the corresponding Gibbs energy equation for solid silicon, according to

$$\tfrac{1}{2}\langle SiO_2 \rangle_{cr} + \tfrac{1}{2}\langle Si \rangle = (SiO). \tag{b}$$

By using the Gibbs energies of formation of water vapour and cristobalite obtained from Table I in Chapter 5, the experimental Gibbs energies of reaction (a) may be converted to yield the corresponding Gibbs energies of reaction (b). We thus obtain six experimental two-term expressions for the Gibbs energy of reaction (b), as follows:

$$\text{T \& W (1750 K), from } \delta: \quad \Delta G_b^\circ = 270{,}286 - 121.55\ T \tag{ε}$$

$$\text{S \& H (1420 K), directly:} \quad \Delta G_b^\circ = 322{,}168 - 155.23\ T \tag{ζ}$$

$$\text{R \& R (1600 K), directly:} \quad \Delta G_b^\circ = 340{,}829 - 161.59\ T \tag{η}$$

$$\text{T \& W (1710 K), from } \alpha: \quad \Delta G_b^\circ = 122{,}591 - 55.86\ T \tag{θ}$$

$$\text{G \& S (1620 K), from } \beta: \quad \Delta G_b^\circ = 264{,}010 - 116.52\ T \tag{κ}$$

$$\text{R \& R (1800 K), from } \gamma: \quad \Delta G_b^\circ = 327{,}189 - 156.57\ T. \tag{λ}$$

Gel'd and Kochnev (1948c) claim to have prepared amorphous silicon monoxide by heating an intimate mixture of silica and silicon to 1250° to 1350°C. This is somewhat puzzling in view of the observations of the other authors, but for the time being their claim to have measured the vapour pressure of the monoxide in this temperature range may be accepted. They give their results in the form of the equation

$$\text{G \& K (1300 K):}\quad \Delta G^\circ_c = 318{,}068 - 140.16\,T \qquad (\mu)$$

for the reaction

$$\{SiO\}_{am} = (SiO). \qquad (c)$$

There are a few more thermochemical values available for SiO. Wartenberg (1949n) determined the enthalpies of solution of silicon and amorphous silicon monoxide, made by sublimation of $\langle SiO_2\rangle + \langle Si\rangle$ mixtures and condensation at 900°C, in hydrofluoric acid. His results have been recalculated, and the value

$$\Delta H^\circ_d = -445{,}596 \pm 21{,}000$$

has been obtained for the heat of formation at 298 K of amorphous silicon monoxide, according to the equation

$$\langle Si\rangle + \tfrac{1}{2}(O_2) = \{SiO\}_{am}. \qquad (d)$$

Spectroscopic data were used by Kelley to calculate the standard entropy of gaseous silicon monoxide as 211.5 J/K, and by Zintl (1940f) to calculate the heat of the reaction

$$(Si) + (O) = (SiO) \qquad (e)$$

as

$$\Delta H \simeq -711.3 \text{ kJ}.$$

All this information will now be used for the assessment.

First it is of interest to compare the entropy terms of eqns (ε) to (λ) by converting them to a common temperature of 25°C and calculating the "standard" entropy of (SiO). For this the change in molar heat ΔC_p must be estimated for reaction (b). The molar heat of $(SiO)_{gas}$ is assumed to be about half-way between that of (CO) and that of (S_2). The molar heats of $\langle Si\rangle$ and $\langle SiO_2\rangle$ are known from Table I, and an average $\Delta C_p = -11.72$ is obtained. Its inaccuracy is not likely to introduce an appreciable error in the present calculations. Then the standard entropy of (SiO) may be obtained from the relationship

$$S_{(SiO)} = \tfrac{1}{2}S_{\langle SiO_2\rangle} + \tfrac{1}{2}S_{\langle Si\rangle} + \Delta S_{298}$$

where

$$\Delta S_{298} = \Delta S_b - \Delta C_p \ln T + \Delta C_p \ln 298.$$

Hence, with $\Delta C_p = -11.72$

$$S_{(SiO)} = \Delta S_b + 26.99 \log T - 66.77 + \tfrac{1}{2}(41.46) + \tfrac{1}{2}(18.83)$$

where ΔS_b is the negative factor of T in eqns (ε) to (λ). The following values for $S_{(SiO)}$ at 25°C are thus derived:

for (ε), 172.45 (ζ), 203.69 (η), 211.44 (θ), 106.49 (κ), 166.52 (λ), 207.80.

This comparison clearly demonstrates how little trust can be placed on the evaluation of the temperature coefficients of equilibrium measurements. The most reliable value for the standard entropy is probably the one given by Kelley: 211.5. The estimate by means of the pertinent equation on p. 176 yields $S_{(SiO)}=213.4$ in good agreement with the spectroscopic value. The present authors are quite generally inclined to use estimated entropies rather than those obtained from equilibrium measurements when calorimetric or spectroscopic values are not available, and observers are urged to use estimates, at least as a check on their experimental entropy terms; it will save them and the later assessors much unnecessary work.

It may be noted that the fair agreement between the estimated value and the experimental values of Kelley, Ramstad and Richardson, and Schäfer and Hörnle also indicates that SiO is a molecular gas.

The entropy value of 211.5 may now be used to calculate the entropies of reaction (b) at the various average temperatures of measurement, by reversing the procedure employed above: in fact, this would have been the course taken right in the first place if the discussion had not been extended for the sake of demonstration. In other words, a third-law evaluation should generally be preferred to a second-law evaluation. By combination of the ΔS_b value thus obtained with the experimental ΔG_b value at the same temperature, values for ΔH_b may be calculated, and the standard heat of reaction is derived from the relationship

$$\Delta H_{298}=\Delta H_b-\Delta C_pT+\Delta C_p\,.\,298=\Delta H_b+11.72\;(T-298).$$

For the individual ΔG_b equations the results are as follows:

	*T*K	ΔS_b°	ΔG_b°	ΔH_b°	ΔH_{298}°
T & W (ε)	1750	160.60	57,582	338,632	355,649
S & H (ζ)	1420	163.04	101,747	333,264	346,414
R & R (η)	1600	161.65	82,291	340,931	356,190
T & W (θ)	1710	160.87	27,077	302,165	318,714
G & S (κ)	1620	161.50	75,241	336,871	352,365
R & R (λ)	1800	160.27	45,371	333,857	351,460

By comparing the ΔH_{298} values it is now seen that the disagreement between the various sets of measurements is not so bad as it might have appeared in a comparison of the original equations. The average value, disregarding (θ), is $\Delta H_{298}=352{,}416$ J.

With the selected enthalpy of reaction (b) and the enthalpy of formation of cristobalite, the standard enthalpy of formation of gaseous monoxide becomes

$$\langle Si\rangle+\tfrac{1}{2}(O_2)=(SiO) \qquad \text{(f)}$$

$$\Delta H_{298}=352{,}416-\frac{905{,}418}{2}=-100{,}293\pm8400\;J.$$

From Zintl's spectroscopic enthalpy of reaction (e), the enthalpy of dissociation of O_2, and the enthalpy of sublimation of silicon, an approximate value of $\Delta H_f=-98.3$ kJ is obtained but, since the probable error is rather large, the excellent agreement with the third-law value is obviously fortuitous. However, it tends to support the above assessment to some extent.

We may now turn to the evaporation reaction (c). The enthalpy of evaporation follows from the enthalpies of formation of the gaseous and amorphous monoxide,

$$L_{298}(c) = 445{,}596 - 100{,}293 = 345{,}303 \pm 29{,}500.$$

The entropy of amorphous $\{SiO\}$ must be estimated. For the hypothetically crystalline substance the value $S_{\langle SiO \rangle} = 36$ follows from Latimer's tables (pp. 173); for the entropy of fusion one would expect about 20.9 J/deg from Table XII but empirical experience suggests halving the entropies of fusion for solid-amorphous transformations. Then the entropy of evaporation at 25°C becomes

$$\sigma_{298}(c) = 211.5 - 36 - 10.45 = 165.05 \pm 8.5 \text{ J/K . mol.}$$

Hence

$$\Delta G_c^\circ = 345{,}303 - 165.05\ T \quad (298 \text{ K}).$$

Since

$$\Delta G_b^\circ = 352{,}416 - 181.35\ T \quad (298 \text{ K})$$

$$\Delta G_c^\circ - \Delta G_b^\circ = -7113 + 16.3\ T$$

for the reaction

$$\{SiO\}_{am} = \tfrac{1}{2}\langle SiO_2 \rangle + \tfrac{1}{2}\langle Si \rangle. \tag{g}$$

From various sources it is found that the transformation of cristobalite to silica glass is accompanied by a Gibbs energy change represented by

$$\tfrac{1}{2}\langle SiO_2 \rangle_{cr} = \tfrac{1}{2}\{SiO_2\}_{gl}; \quad \Delta G_h = 2930 - 2.3\ T \tag{h}$$

at room temperature. Then

$$\Delta G_c - \Delta G_b + \Delta G_h = -4183 + 14.0\ T = \Delta G_i.$$

For this reaction, namely

$$\{SiO\}_{am} = \tfrac{1}{2}\{SiO_2\} + \tfrac{1}{2}\{\langle Si \rangle\} \tag{i}$$

ΔC_p is probably negligibly small, and the ΔG_i equation may be used at any temperature. It follows, if $\Delta G_i = -4183 + 14\ T$, that amorphous silicon monoxide should be stable in relation to silicon and silica at any temperature above 250 K. This contradicts the observation of Grube and Speidel, Schäfer and Hörnle, and others, that reaction (i) proceeds to the right of the equation at 1000–1150°C. On the other hand, the entropy term of ΔG_i appears to have the correct sign and to be of the order to be expected (± 6.28 J/deg). The heat term should therefore be more negative. It is possible, though it is unsupported by X-ray analysis, that Gel'd and Kochnev's claim is correct that eqn (i) moves to the left at 1250–1350°C. If ΔG_i were zero at about 1150–1200°C, then, with the entropy term left unchanged, ΔG_i becomes about

$$\Delta G_i = -20{,}270 + 14\ T.$$

This may be taken to imply that the enthalpy of formation of $\{SiO\}_{am}$ is $-429{,}509$ rather than $-445{,}596$, the alteration being within the limits of error of v. Wartenberg's experimental determination. With this alteration the Gibbs energy of evaporation of $\{SiO\}_{am}$ becomes

$$\Delta G_{298}(c) = 329{,}216 - 165.05\ T.$$

The change in molar heat of this reaction may be estimated to be $\Delta C_p = -23$ on an average, and the complete Gibbs energy equation may then be obtained in the manner described previously in this monograph.

$$\Delta G_c = 340{,}715 + 53.0\ T \log T - 319\ T$$

corresponding to the vapour pressure equation

$$\log p_{SiO}\ (\text{mmHg}) = -17{,}800\ T^{-1} - 277 \log T + 19.55$$

ΔG_c becomes zero at about 2430 K, that is, at the normal boiling point, and the enthalpy of evaporation at this temperature is

$$L_e = 340{,}715 - 23\ T = 284{,}825.$$

The entropy of evaporation is therefore 117.2 J/deg which is a little higher than Trouton's constant but may be accepted as reasonable.

At the average temperature of Gel'd and Kochnev's measurements (1300 K) ΔG_c is found to be 140,566 J whilst the Russian authors quote 135.980 J. The corresponding value for ΔG_b is 116,660 J. It thus appears to be probable that Gel'd and Kochnev did in fact have silicon monoxide in the condensed state, but the differences in free energies for the two reactions are too small to allow definite conclusions to be drawn, particularly in view of the observations of Schäfer and Hörnle, Grube and Speidel, and Wartenberg that $\{SiO\}$ is unstable at about 1100°C.

We have attempted to show that a full assessment of all the available information renders the resulting standard values for a particular substance far more reliable than the evaluation of a single series of measurements is likely to accomplish. It follows necessarily, however, that the evaluator must preserve an open mind and must be prepared to review his assessment whenever new measurements are forthcoming.

From the present example, the values derived from all the results available to date may now be summarised.

Thermochemical properties of silicon monoxide:

Gas:

$$\Delta H^\circ_{298} = -100{,}300 \pm 8000\ \text{J/mole}$$

$$S_{298} = 211.5 \pm 0.4\ \text{J/mole K.}$$

Amorph.:

$$\Delta H_{298} = -429{,}510 \pm 21{,}000\ \text{J/mole}$$

$$S_{298} = 46.5 \pm 6.3\ \text{J/mole K.}$$

b.p. 2160°C:

$$L_e(298\ \text{K}) = 329{,}215 \pm 25{,}000\ \text{J/mole}$$

$$L_e(\text{b.p.}) = 284{,}825 \pm 25{,}000\ \text{J/mole.}$$

B. A System Exhibiting Wide Solution Ranges—Chromium–Nickel

Although computer programs are now available which greatly facilitate the consistent evaluation of thermodynamic and phase diagram data for systems exhibiting wide solution ranges, the principles involved should be well understood and are therefore discussed in detail below using the system chromium–nickel as example (1976e). The equilibrium diagram as eventually accepted is shown in Fig. 120.

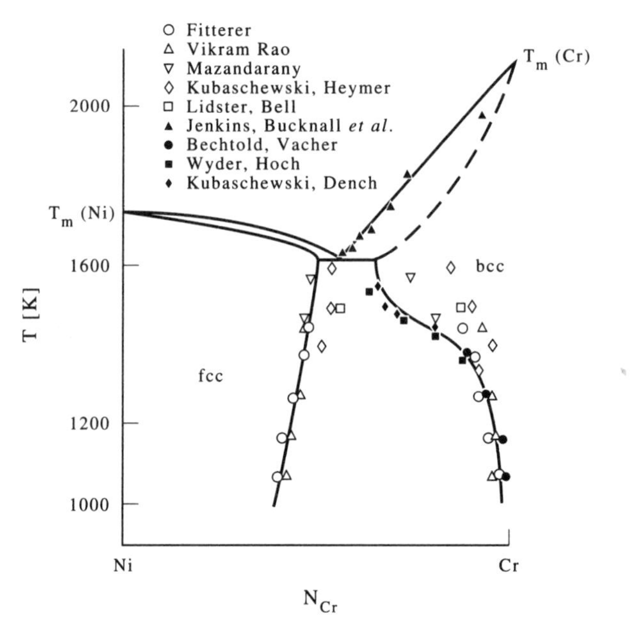

FIG. 120. Experimental and calculated equilibrium diagram of the system chromium–nickel.

Although such an evaluation may vary from system to system, there are certain rules of general significance to be obeyed. Since truly accurate experimental information is hardly ever available, the emphasis of the evaluation should be on consistency rather than ideal accuracy. Whenever possible, the evaluation should begin with good calorimetric enthalpies of mixing in either the solid or the liquid state. In the present case, the enthalpies of mixing of solid nickel–chromium have been determined by adiabatic calorimetry at 1514–1571 K by Dench (1963b) (Table XXXV). The results may be taken as independent of temperature, as has been found for some selected compositions (1973c, d; 1967g).

Enthalpies of mixing are usually obtained as integral values whereas experimental Gibbs energies are obtained as partial values. Thus for combination the former must be differentiated or the latter integrated in order to evaluate the entropies of reaction. Differentiation of enthalpy vs concentration curves may be carried out graphically or by expressing them by convenient polynomials (p. 38) with subsequent conversion. The former course leads at first to inaccurate values which have to be adjusted during a following integration. In Table XXXV integral heats of Dench and the partial enthalpies of Hack are listed for readers who wish to repeat the procedure(s) as an exercise.

The next step is to combine the partial enthalpies of solution with the measured Gibbs energies of solution. Several investigators who employed different experimental methods have provided such Gibbs energies for the solid solution of chromium in nickel–

TABLE XXXV. *Thermochemical values for the formation of solid alloys from solid metals*

X_{Cr}	ΔH	$\Delta\bar{H}_{Cr}$	$\Delta\bar{H}_{Ni}$	ΔS	$\Delta\bar{S}_{Cr}$	$\Delta\bar{S}_{Ni}$	ΔS^E	$\Delta\bar{S}^E_{Cr}$	$\Delta\bar{S}^E_{Ni}$
0	0	−17,573	0	0	∞	0	0	2.824	0
0.05	−661	−9008	−222	1.757	25.062	0.715	0.285	0.159	0.289
0.10	−879	−879	−879	3.067	22.238	0.937	0.364	3.096	0.059
0.15	−720	4586	−1657	4.092	21.046	1.100	0.577	5.272	−0.251
0.20	−268	9113	−2615	5.075	20.301	1.268	0.916	6.920	−0.582
0.25	481	12,774	−3615	5.991	19.364	1.536	1.318	7.837	−0.854
0.30	1406	15,648	−4699	6.874	18.828	1.753	1.795	8.820	−1.218
0.35	2489	17,811	−5761	7.719	17.991	2.188	2.339	9.263	−1.389
0.40	3577	19,338	−6929	8.477	17.364	2.552	2.883	9.745	−1.690
0.45	4874	20,305	−7753	9.205	16.548	3.197	3.485	9.908	−1.770
0.50†	6360	20,790	−8071	9.807	15.962	3.653	4.046	10.201	−2.109
0.55		14,799	−2079	9.975	11.414	8.217	4.255	6.443	1.582
0.60		14,799	−2079	10.138	11.414	8.217	4.556	7.167	0.640
0.645‡	8807*	11,799*	3372*	10.263	8.732	13.046	4.853	5.088	4.427
0.65	8845*	11,468*	3975*	10.242	8.598	13.046	4.862	5.017	4.573
0.70	8987	8427	10,297	9.945	7.176	16.410	4.866	4.209	6.397
0.75	8657	5849	17,079	9.301	5.418	20.949	4.628	3.025	9.435
0.80	7862	3745	24,330	8.364	3.954	26.004	4.205	2.100	12.623
0.85	6598	2105	32,062	7.058	2.866	30.815	3.544	1.515	14.205
0.90	4866	937	40,225	5.473	2.238	34.581	2.770	1.364	15.422
0.95	2665	234	48,852	3.418	1.142	46.664	1.770	0.715	21.803
0.98	1121	38	54,220	1.791	0.456	67.191	0.975	0.289	34.346
1.00	0	0	57,852	0	0	∞	0	0	—

*Extrapolated.
†Phase boundary Ni f.c.c. at T_{eut}.
‡Phase boundary Cr b.c.c. at T_{eut}.

The unit for ΔH is $\left[\dfrac{\text{J}}{\text{g atom}}\right]$ and for $\Delta S\left[\dfrac{\text{J}}{\text{g atom K}}\right]$.

chromium. Electromotive force measurements have been carried out by Panish *et al.* (1958i) who used a molten NaCl–RbCl mixture for the electrolyte at 1023 and 1238 K, by Fitterer and Pugliese (1970e) who employed stabilised zirconia at 1073–1448 K and by Mazandarany and Pehlke (1973i) who used stabilised thoria at 1273–1573 K. For the assessment the suitability of the experimental method, in this case particularly the nature of the electrolyte, should be considered. Molten chlorides should only be employed below, say, 800°C. At such temperatures, however, diffusional delays in the electrodes may produce errors in systems melting above, say, 1500°C; zirconia electrolytes are on the borderline of usefulness for cells having solid Cr, Cr_2O_3 as the one electrode, because of significant electronic conductivity, so that one would rely most on the results obtained with thoria electrolytes. Rao and Flores-Magon (1971e) suggested a modified evaluation of the results of Fitterer and Pugliese.

In addition, Dench and Heymer (1960i) have measured dissociation pressures, that is the partial pressures of chromium, at 1399–1600 K by means of the Knudsen effusion method combined with tracer analysis of ^{51}Cr—a method that has been found to give reasonably good results. The first to investigate the present system, Grube and Flad (1942b) equilibrated the alloys with H_2O/H_2 mixtures and Cr_2O_3. The method is prone to certain errors (p. 114), but can give quite reasonable results.

All the individual partial Gibbs energies obtained by these investigators have been converted with the accepted partial enthalpies of solution to yield partial excess entropies of solution which are plotted against the atomic fraction of chromium for the solutions in nickel in Fig. 121. The scatter of the data which stem from the various Gibbs energy measurements is not unusual and should make the compilers as well as the investigators think about the precision that can be achieved. Obviously, a consistent set of thermochemical properties cannot be assessed on the basis of these data alone. Additional information must be sought.

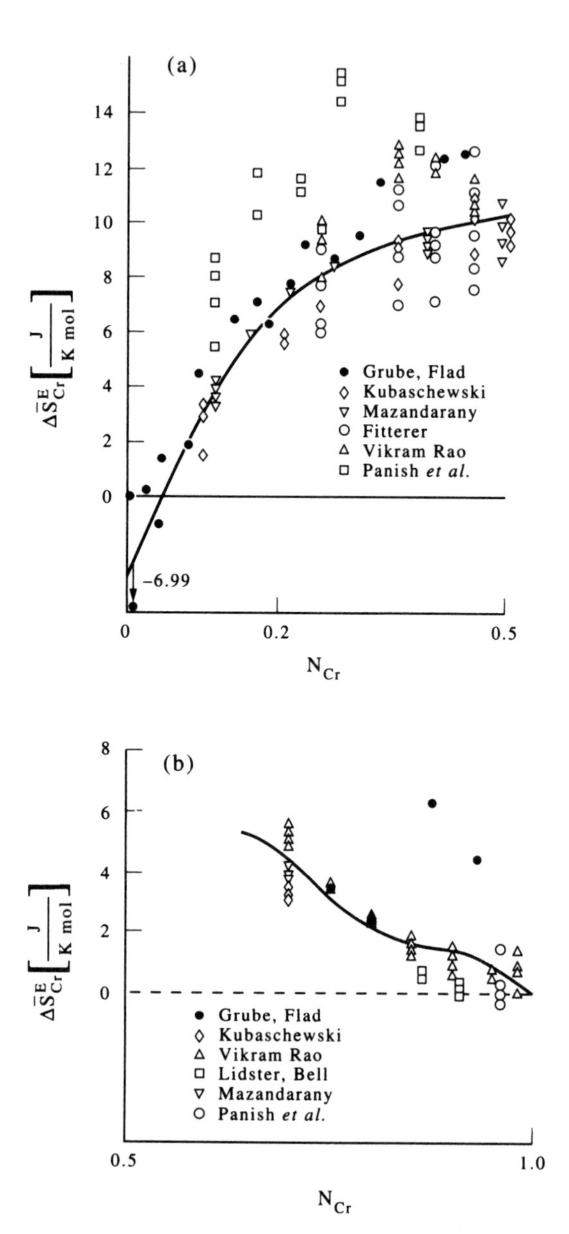

FIG. 121. Partial excess entropies of chromium (a) in the (Ni)-f.c.c. phase (b) in the (Cr)-b.c.c. phase.

The curve shown in Fig. 121 is somewhat arbitrary, but is drawn with due regard to the relative reliability of the experimental methods. The next step is to accept the phase boundaries of the heterogeneous region above 1200 K as presented by Hultgren *et al.* (1973c, d) and to calculate the partial Gibbs energies of chromium at the compositions of its saturated solutions in nickel from its established partial enthalpies and entropies of solution. These values are identical with those at the chromium-rich phase boundary for a given temperature. When combined with the partial enthalpies of solution of chromium derived from the results of Dench (Table XXXV) the excess entropies of solution shown as full curve in Fig. 121 are obtained. Again, it may be seen that the corresponding values derived from the experimental Gibbs energies are rather scattered around the calculated curve.

Finally, for the solid system the partial Gibbs energies are integrated at, say, 1600 K after Duhem–Margules (pp. 34, 35) and combined with the integral enthalpies of mixing to yield integral entropies of mixing. These are tabulated together with the excess functions in Table XXXV which provides a consistent set of thermochemical data for solid nickel–chromium alloys.

For a check, the phase boundaries of the heterogeneity region are evaluated by means of the common-tangent method (p. 37) from plots of the integral Gibbs energies derived, at various temperatures, from the tabulated enthalpies and entropies of mixing. Figure 122 is an example of such an evaluation at a temperature of 1500 K. During this last stage it is usually found that some slight adjustments to the assessed values well within the experimental accuracy must be made. This is because of the high sensitivity of the position of the calculated phase boundaries to minor deviations in the thermochemical values—as a glance at Fig. 122 shows.

To complete the assessment, consistent data for the liquid solutions must be derived. However, relatively little experimental evidence on liquid nickel–chromium is available. Fruehan (1968d) measured the e.m.f. of cells of the type $\langle Cr\rangle\langle Cr_2O_3\rangle/\langle ZrO_2\text{–}CaO\rangle/\langle Cr_2O_3\rangle\{Cr\text{–}Ni\}$ at 1873 K. Here, again, doubts exist as to the suitability of the zirconia electrolyte. Gilby and St. Pierre (1969e) determined the activities of chromium in $\{Cr\text{–}Ni\}$ solutions with $N_{Cr}=0.01\text{–}0.5$ at 1873 K via the vapour pressures, using a mass spectrometer.

Since enthalpies of mixing for the liquid are not available nor can be derived, various simplifications may be introduced. One may accept the enthalpies *or* the entropies of mixing of the solid for the liquid solutions, for instance. Most commonly, however, one assumes the liquid solutions to be regular and, hence, takes the excess Gibbs energies as equal to the enthalpies, the entropies being then, of course, the ideal ones.

The liquidus curves, where they have been reliably determined, are used as additional information: in the present case in the range 0–60 at.% Cr (1973c, d), including the composition and temperature of the eutectic. There is no easy mathematical way for this incorporation. The integral Gibbs energies of formation of the solid solutions are therefore plotted at the eutectic temperature and other temperatures above: various curves for the integral Gibbs energies of formation of the liquid solutions are also plotted at the respective temperature using the ideal entropies until, by the common tangent method, the curve is found that reproduces the liquidus curve exactly. (Naturally, the ΔG curves for both the solid and liquid solutions must be related to the pure components in their state of aggregation at that temperature by incorporating the Gibbs energies of fusion for the supercooled or superheated metal, as the case may be.) Thus the mode of

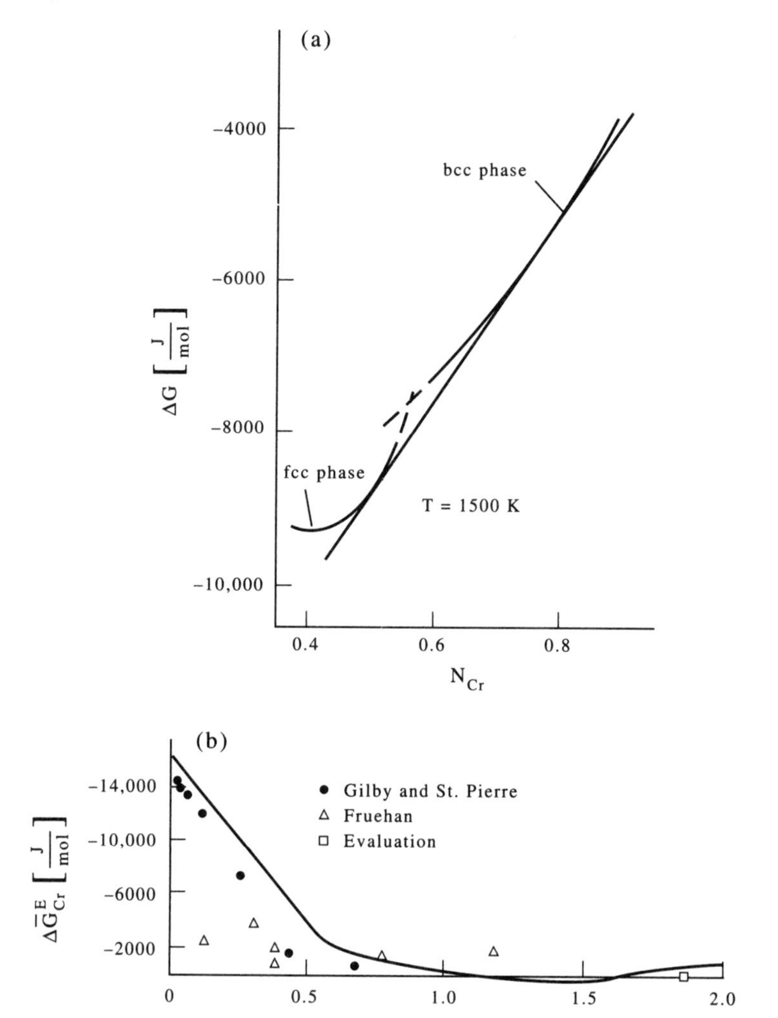

FIG. 122. (a) Integral Gibbs energies of the f.c.c. and b.c.c. phases in the system chromium–nickel indicating the graphical evaluation of phase boundaries; (b) experimental and assessed partial excess Gibbs energies of chromium in liquid chromium–nickel alloys.

evaluation is one of trial and error. It must produce a single ΔG^E vs N curve which, in the present case, may then be extrapolated to concentrations above 60 at.% Cr.

The differentiation of this curve yields the $\Delta \bar{G}^E_{Cr}$ vs N_{Cr}/N_{Ni} curve in Fig. 122b which also shows the experimental results of the investigators mentioned above. It may be emphasised that, once the curve in Fig. 121 has been selected, the full curves in Figs 122(a) and 122(b) follow without any reference to the experimental Gibbs energy results. The only additional experimental evidence that has been incorporated are the phase boundaries at temperatures high enough to warrant complete equilibration—provided the investigators were not impatient.

This does not apply in general and in this particular case to the solidus curves which, for reasons pointed out elsewhere, are rarely equilibrium data. Thus the solidus curves in Fig. 120 are to be preferred to those obtained by conventional methods of phase-boundary determination. Here the calculated solidus–liquidus gaps are narrower than those found experimentally—a common observation.

8. Calculation of Metallurgical Equilibrium Diagrams

In recent years there has been a concerted international effort directed to the acquisition, evaluation and dissemination of phase diagram information. As a result of workshops directed to this subject, user needs for phase diagram information in such diverse areas as primary metal production, iron and steel-making, superalloys, non-ferrous alloys, high-temperature structural ceramics, glass processing, production of superconductors, semiconductors, hard-metal coatings, the electric power industry, thermal energy storage technology, magnetohydrodynamics, etc, have been defined by specialists in the topics concerned. Emphasis has been laid on the fact that industrial requirements for phase diagram information cannot be met by the presently available binary phase diagram compilations alone. Very many materials finding practical application consist of several constituents and for such systems phase diagram information is very scarce, although compilations such as *Ternary Alloys* (*Ternary Alloys*, ed. G. Petzow and G. Effenberg, VCH Verlag, 1988 onward), *Phase Diagrams for Ceramists* (E. M. Levin, C. R. Robbins, H. F. McMurdie, *Phase Diagrams for Ceramists*, American Ceramic Society, 1964 onward), *Slag Atlas* (*Slag Atlas*, Verlag Stahleisen, Düsseldorf, 1981), etc., represent important continuing contributions.

There is, of course, a simple explanation for this dearth of phase diagram information for more complex systems. A realistic estimate of the time required to determine reliably, using conventional experimental techniques, just **one ternary** phase diagram, shows that experimental investigation of the vast number of possible ternary metal, oxide, salt, etc., systems is a completely unrealistic proposition. Experimental studies of systems with still more components are too time-consuming and costly to pursue, other than for a very restricted range of compositions and temperatures.

It has therefore been apparent for some time that there exists an important practical need to develop methods for calculating multicomponent phase equilibria and for utilising modern computer techniques to automate and facilitate the calculation process as much as possible.

The thermodynamic basis for the calculation of phase equilibria was already explicitly established by van Laar (1908c) at the beginning of this century. Important later contributions in which experimental results were applied to phase diagram calculations have been made by Kubaschewski (1949h), by Wagner (1952j) and by Kubaschewski and Chart (1965b). In particular, Meijering (1957d) and Kubaschewski (1967g) demonstrated how ternary diagrams could be calculated from the basis of thermodynamic data for the three bordering binary systems using simple solution models.

The basis for computer calculation of a wide range of metallurgical phase diagrams was laid by Kaufman (1970k), who in a series of publications, reported values for the "lattice stabilities" of pure metals in metastable structures. This permitted the phase equilibria in alloy systems containing a number of different phases and components to be calculated. To perform the calculations, models allowing the thermodynamic description of different types of phases were required, as were methods of calculating the thermodynamic

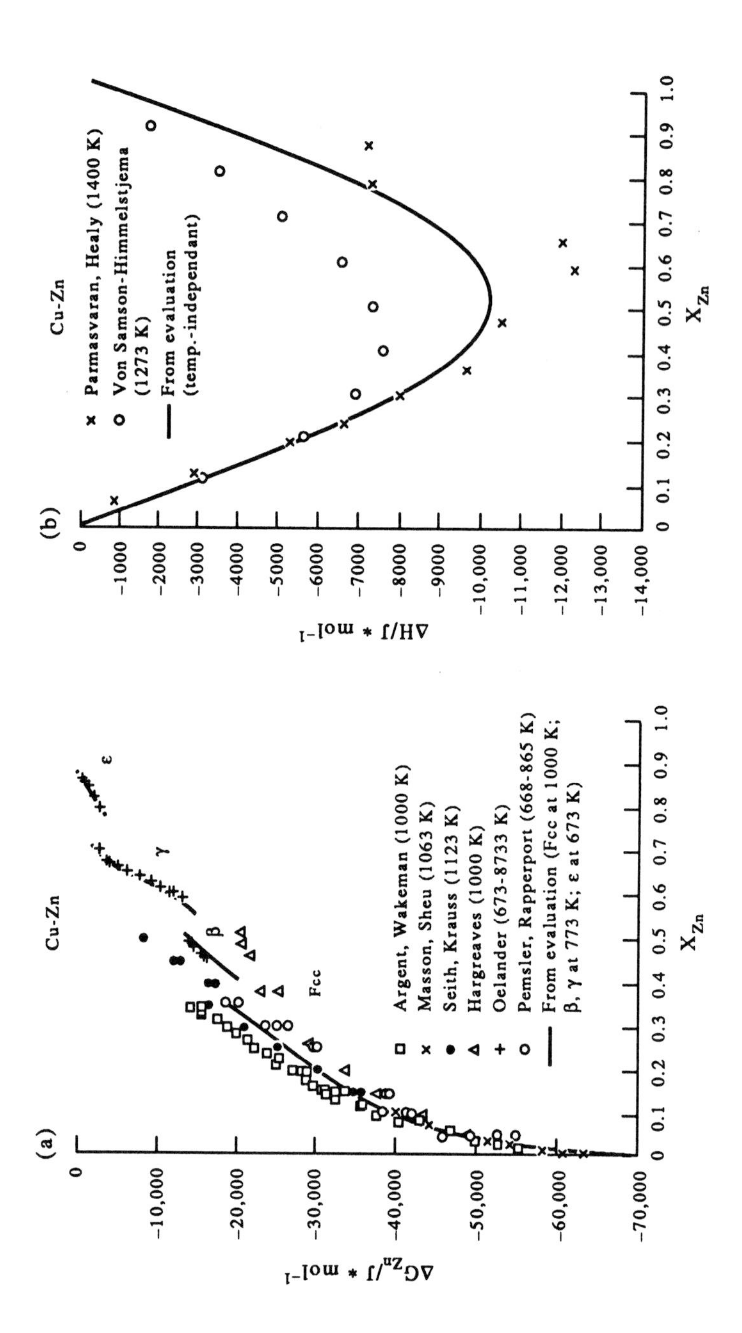

FIG. 123a. Partial Gibbs energy of Zn in solid Cu–Zn alloys.

Fig. 123b. Enthalpy of mixing of liquid Cu–Zn alloys.

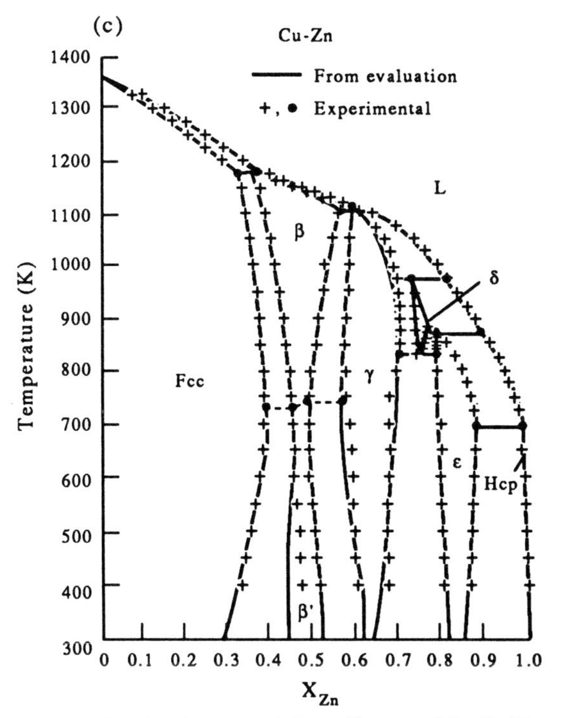

FIG. 123c. Calculated and measured phase diagram of the Cu–Zn system.

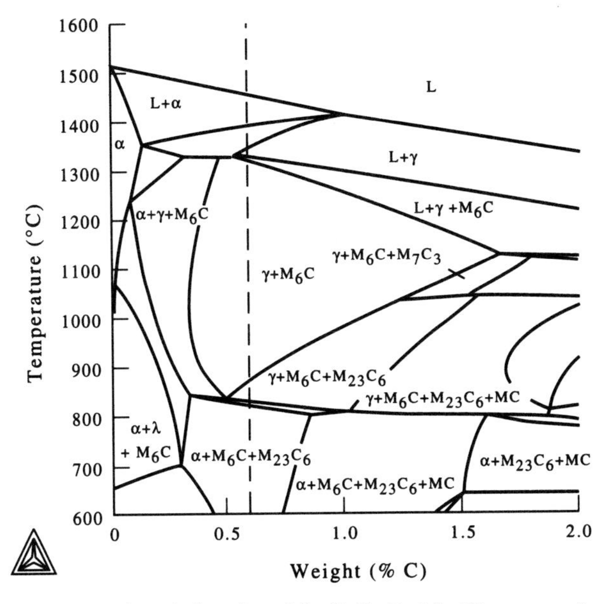

FIG. 124. Computer-generated vertical section of the C–Cr–Fe–Mo–W system at 4 wt% Cr, 6 wt% Mo and 6 wt% W (1988f).

properties of multicomponent alloys from available data for the relevant binary systems. Review papers covering this field have been written by Ansara (1971a), by Spencer and Barin (1979h), by Hillert (1970h), by Pelton (1991c) and by others.

In evaluating the binary or ternary thermodynamic properties required for the phase equilibrium calculations, computer programs such as those due to Lukas (1977d) and to Jansson (1984c) are also of particular importance. These programs provide optimised coefficients with which the experimental thermodynamic properties and equilibrium phase boundaries of a system can be reproduced in a self-consistent mathematical description. For example, Figs 123a, b, c present results of an evaluation of the Cu–Zn system carried out using the Lukas program. In Fig. 123a, calculated relative partial Gibbs energies of zinc resulting from the "optimisation" process are compared with the published experimental information. Figure 123b, presents a comparison of evaluated and experimental enthalpies of formation of liquid alloys, and Fig. 123c shows the final calculated equilibrium phase diagram.

It should be pointed out that these calculations include a suitable treatment of the ordering in the β-phase of the system and that the properties of the γ-phase are described using a sub-lattice model (1970h). An example of calculations relating to a more-complex alloy system is shown in Fig. 124. The system involved is C–Cr–Fe–Mo–W, for which prior assessment of experimental information for all relevant binary, ternary and quaternary sub-systems was necessary to enable reliable calculations in the 5-component system to be performed. It is clear that only with the aid of suitable models and software, as well as the availability of fewer, carefully chosen experimental measurements, can complex systems of this type be investigated.

The CALPHAD journal (CALPHAD, Computer Coupling of Phase Diagrams and Thermochemistry, Pergamon, 1977 onward) contains very many examples of calculations relating both to metallic and non-metallic systems. Different models and calculation methods are also published in this journal.

A very useful review of metallurgical thermochemical databases and their application to phase diagram calculations of many different types has been written by Bale and Eriksson (1990a).

CHAPTER 5

Thermochemical Data

Tables I and II

The format for the presentation of the tables of data for the thermodynamic properties of inorganic compounds (Table I) and for metallic solutions (Table II) has been changed in this edition to facilitate their use. Table I presents all of the data, with some additions, which were displayed in Tables A to C2 in the 5th edition. The data for a given element or compound are now assembled together, and can be applied with simple computer programs to calculate the vapour pressures previously given in Table D and the Gibbs energy change in Table E. These latter tables have therefore been omitted from this edition.

The Table II has not been altered from the 5th edition content for metallic alloys except for the conversion to S.I. units.

We wish to express our gratitude to Dr Rajiv Doshi of the University of Notre Dame who has carried out the programming, conversion and reprinting of data in Table I and II in the form in which they are now presented.

TABLE I

Substance	$S_{298} \pm \delta S$ (J/deg mol)	$-\Delta H_{298} \pm \delta H$ (kJ/mol)	Phase	T K	$C_p = A + BT + C/T^2 + DT^2$ (J/deg mol)				H_t (kJ/mol)
					A	$B \times 10^3$	$C \times 10^{-5}$	$D \times 10^6$	
Ag	42.6	0.0	fcc	298	21.30	8.54	1.51	—	—
			liq	1235	33.47	—	—	—	11.3
Ag(g)	172.9	−284.9	gas	298	20.79	—	—	—	—
AgBr	107.1 ± 0.4	100.7 ± 0.8	cubic	298	33.18	64.43	—	—	—
			liq	700	62.34	—	—	—	13.0
AgBr(g)	257.0	−96.4	gas	298	37.49	—	−0.79	—	—
$AgBrO_3$	154.0	26.4	tetrag	298	66.94	125.52	—	—	—
AgCl	96.2	127.0	cubic	298	62.26	4.18	−11.30	—	—
			liq	728	66.94	—	—	—	13.2
AgCl(g)	246.1 ± 0.2	−92.6	gas	298	37.24	—	−1.42	—	—
AgF	83.7 ± 5.0	205.9 ± 3.8	cubic	298	53.05	16.15	−8.79	—	—
AgF(g)	235.8	− 7.1	gas	298	35.56	—	−1.42	—	—
AgI	115.5 ± 1.3	61.9 ± 0.4	α	298	35.77	71.13	—	—	—
			β	420	56.48	—	—	—	6.3
			liq	830	58.58	—	—	—	9.4
AgI(g)	264.2	−159.3	gas	298	37.40	—	−0.54	—	—
Ag_2O	120.9 ± 0.4	31.1 ± 0.6	cubic	298	59.33	40.79	−4.60	—	—
Ag_2CO_3	167.4 ± 4.2	505.8 ± 2.5	cryst	298	79.37	108.16	—	—	—
Ag_2CrO_4	218.0 ± 1.3	721.3	orth	298	132.21	66.94	−8.91	—	—
$AgNO_3$	140.9 ± 0.4	124.5 ± 1.0	orth	298	36.65	189.12	—	—	—
			hex	433	106.69	—	—	—	2.5
			liq	483	128.03	—	—	—	11.7
Ag_2S	143.5 ± 1.3	31.8 ± 0.8	monocl	298	64.60	39.96	—	—	—
			cubic	449	81.34	2.93	—	—	3.9
			cubic	859	82.72	—	—	—	0.5
			liq	1103	93.09	—	—	—	7.9
Ag_2SO_4	199.8 ± 0.4	717.1 ± 1.3	orth	298	96.65	116.73	—	—	—
			hex	700	96.65	116.73	—	—	15.7
			liq	933	205.02	—	—	—	18.0
Ag_2Se	150.7 ± 0.8	43.5 ± 0.8	orth	298	65.14	54.89	—	—	—
			cubic	406	80.50	9.50	—	—	7.1
Ag_2Te	153.6 ± 1.7	36.0 ± 0.6	orth	298	49.20	109.62	2.76	—	—
			cubic	421	78.87	3.85	10.42	—	6.6
			cubic	620	84.01	—	—	—	—

Ag_2WO_4	205.0	925.5 ± 22.6	cryst	298	132.47	69.20	—	—	—
Al	28.3	0.0	fcc	298	31.38	−16.40	−3.60	20.75	—
			liq	934	31.76	—	—	—	10.7
$Al(g)$	164.4	−330.0	gas	298	20.75	—	0.54	—	—
$AlAs$	60.2 ± 4.2	116.3 ± 6.3	cubic	298	43.93	6.28	—	—	—
$AlBr(g)$	239.5 ± 0.2	−15.9 ± 14.6	gas	298	37.26	0.59	−1.63	—	—
$AlBr_3$	180.2 ± 0.4	511.3	monocl	298	49.96	169.58	—	—	—
			liq	371	124.98	—	—	—	11.3
$AlBr_3(g)$	349.1 ± 2.9	410.9	gas	298	82.55	0.31	−6.44	—	—
$Al_2Br_6(g)$	547.3	937.2	gas	298	182.00	0.46	−13.77	—	—
Al_4C_3	88.9 ± 0.8	209.2 ± 6.3	hex	298	154.68	28.74	−41.84	—	—
$AlCl(g)$	227.9 ± 0.2	51.5 ± 9.2	gas	298	36.59	1.26	−2.06	—	—
			gas	800	37.38	0.46	−3.05	—	—
$AlCl_2(g)$	287.9	288.7	gas	298	57.78	0.22	−4.95	—	—
$AlCl_3$	109.3 ± 0.8	705.6 ± 1.7	monocl	298	64.94	87.86	—	—	—
$AlCl_3(g)$	314.3 ± 1.3	584.5 ± 2.5	gas	298	81.96	0.63	−9.92	—	—
$Al_2Cl_6(g)$	475.5	1295.4 ± 3.3	gas	298	180.92	1.05	−20.42	—	—
$AlF(g)$	215.1 ± 0.2	265.7 ± 3.3	gas	298	33.56	4.08	−2.53	—	—
			gas	800	37.28	0.44	−7.66	—	—
$AlF_2(g)$	263.2	732.2	gas	298	51.76	6.13	−6.92	—	—
			gas	800	57.93	0.10	−16.16	—	—
AlF_3	66.5 ± 0.4	1510.4 ± 1.3	hex	298	90.96	17.66	−18.79	—	—
			hex	500	3.79	126.82	62.68	—	—
			β	728	92.68	9.06	−8.94	—	0.6
$AlF_3(g)$	277.8 ± 1.3	1207.9 ± 3.3	gas	298	79.16	2.26	−15.44	—	—
$Al_2F_6(g)$	387.0	2633.6	gas	298	162.90	22.09	−31.67	−6.49	—
AlH_3	30.0 ± 0.4	11.4 ± 0.8	hex	298	45.19	—	—	—	—
AlI_3	189.5 ± 8.4	309.2 ± 5.9	hex	298	70.63	94.81	—	—	—
			liq	462	121.34	—	—	—	15.9
$Al_2I_6(g)$	582.8	514.2	gas	298	182.42	0.26	−10.21	—	—
AlN	20.2 ± 0.2	318.4 ± 2.1	hex	298	32.26	22.68	−7.91	—	—
			hex	600	50.21	1.17	−26.07	—	—
			hex	1000	50.12	0.39	−17.41	—	—
Al_2O_3	50.9 ± 0.2	1675.7 ± 1.3	hex	298	117.49	10.38	−37.11	—	—
			liq	2325	184.10	—	—	—	107.0
$Al_2O(g)$	256.9 ± 6.7	130.5 ± 16.7	gas	298	50.84	3.97	−8.70	−2.38	—
$Al_4B_2O_9$	155.6 ± 5.0	4690.7 ± 6.7	orth	298	270.29	108.57	−71.13	—	—
$9Al_2O_3.2B_2O_3$	654.0	17727.6 ± 30.1	orth	298	980.31	613.37	−288.28	−176.56	—
$AlOCl$	54.4 ± 4.2	793.3 ± 1.3	orth	298	55.35	34.35	−7.78	—	—
$AlOCl(g)$	248.8 ± 2.9	353.1 ± 10.5	gas	298	58.74	3.47	−8.70	−0.84	—
$AlOOH$	35.2 ± 0.4	1002.1 ± 4.2	diasp	298	52.76	—	—	—	—

TABLE I—*continued*

Substance	$S_{298} \pm \delta S$ (J/deg mol)	$-\Delta H_{298} \pm \delta H$ (kJ/mol)	Phase	T K	$C_p = A + BT + C/T^2 + DT^2$ (J/deg mol)				H_t (kJ/mol)
					A	$B \times 10^3$	$C \times 10^{-5}$	$D \times 10^6$	
AlOOH(b)	48.4	985.3	boch	298	60.42	17.57	—	—	—
$Al_2O_3.3H_2O$	140.2 ± 1.3	2586.5 ± 4.6	monocl	298	72.38	381.58	—	—	—
$AlCl_3.6H_2O$	318.0 ± 2.5	2691.6 ± 1.7	hex	298	297.06	—	—	—	—
AlOF(g)	237.2 ± 16.7	581.6 ± 29.3	gas	298	58.66	2.05	−11.13	—	—
AlP	47.3 ± 1.3	164.4 ± 2.5	cubic	298	40.17	6.28	—	—	—
$AlPO_4$	90.8 ± 0.4	1733.0 ± 2.1	hex	298	51.46	139.33	—	—	—
				853	167.36	—	—	—	1.3
				978	163.18	—	—	—	1.1
AlS(g)	230.5 ± 0.4	−238.9	gas	298	36.84	0.69	−3.26	—	—
Al_2S_3	123.4	723.4	hex	298	102.17	36.07	—	—	—
			liq	1370	156.90	—	—	—	56.5
$Al_2(SO_4)_3$	239.2 ± 1.3	3441.3 ± 2.1	hex	298	366.31	62.59	−111.63	—	—
AlSb	64.9 ± 0.8	50.2 ± 1.0	cubic	298	43.51	9.62	—	—	—
			liq	1333	58.99	—	—	—	82.0
AlSe(g)	243.1	−221.3 ± 12.6	gas	298	107.74	34.31	—	—	—
Al_2Se_3	154.8 ± 25.1	566.9 ± 25.1	hex	298	107.74	34.31	—	—	—
$Al_6Si_2O_{13}$	274.9 ± 0.4	6820.8 ± 8.4	mullit	298	233.59	633.88	−55.86	−385.77	—
				600	503.46	35.10	−230.12	−2.51	—
Al_2SiO_5	83.8 ± 0.4	2589.5 ± 2.1	kyanit	298	177.99	23.89	−57.95	—	—
Al_2SiO_5(a)	93.2 ± 0.2	2587.0 ± 2.1	andal	298	173.64	26.11	−52.63	−0.63	—
Al_2SiO_5(s)	96.1 ± 0.4	2584.0 ± 2.1	sillim	298	169.74	28.74	−50.84	—	—
Al_2Te_3	188.3 ± 18.8	318.8 ± 9.2	hex	298	110.88	34.73	—	—	—
			liq	1163	176.56	—	—	—	50.0
Al_2TiO_5	109.6 ± 2.1	2607.1 ± 16.7	rhomb	298	182.55	22.18	−46.90	—	—
As	35.7 ± 0.4	0.0	rhomb	298	23.03	5.74	—	—	—
As_2(g)	239.3 ± 1.3	−190.8 ± 6.3	gas	298	37.20	0.15	−2.02	—	—
As_4(g)	328.4 ± 5.0	−156.1 ± 2.5	gas	298	82.94	0.13	−5.13	—	—
$AsBr_3$(g)	363.8 ± 0.4	132.1 ± 4.2	gas	298	83.26	—	−4.23	—	—
$AsCl_3$	212.5	315.5	liq	298	133.89	—	—	—	—
$AsCl_3$(g)	326.2 ± 0.8	271.1	gas	298	82.09	1.00	−5.94	—	—
AsF_3	181.2 ± 0.4	956.9 ± 3.3	liq	298	126.78	—	—	—	—
AsF_3(g)	288.9 ± 0.4	921.3 ± 3.3	gas	298	75.98	6.90	−11.21	—	—
AsH_3(g)	223.0 ± 0.6	−66.4 ± 1.3	gas	298	42.01	22.80	−9.08	—	—
AsI_3	213.0	64.9	hex	298	71.21	116.32	—	—	—
			liq	414	133.89	—	—	—	21.8

As_2O_3	122.6	665.8 ± 2.1	monocl	298	59.83	171.96	—	—	—
			liq	587	152.72	—	—	—	22.6
As_2O_5	105.4 ± 2.5	920.1 ± 5.0	cryst	298	42.47	246.86	—	—	—
As_2S_3	163.6 ± 5.0	167.4 ± 16.7	monocl	298	105.65	36.44	—	—	—
			liq	585	177.86	16.86	—	—	28.7
AsS	63.4 ± 1.7	71.1 ± 10.5	monocl	298	41.46	18.66	—	—	—
			liq	580	73.22	—	—	—	6.1
AsS(g)	232.2 ± 4.2	−202.9	gas	298	34.94	1.63	—	—	—
As_2Se_3	194.6 ± 5.0	102.5 ± 16.7	rhomb	298	95.81	85.77	—	—	—
			liq	650	195.39	—	—	—	40.8
As_2Te_3	226.4 ± 4.2	37.7 ± 8.4	monocl	298	135.19	44.35	−18.58	—	—
			liq	648	167.36	—	—	—	46.9
Au	47.5	0.0	fcc	298	31.46	−13.47	−2.89	10.96	—
			liq	1338	50.33	−12.69	—	—	12.6
			liq	1500	30.96	—	—	—	—
Au(g)	180.4	−368.2	gas	298	20.79	—	—	—	—
			gas	1000	22.26	−3.01	—	1.51	—
AuBr	98.3	14.2 ± 2.9	cryst	298	49.37	5.44	−0.84	—	—
AuCl	85.8 ± 4.2	36.4 ± 2.1	rhomb	298	48.53	5.44	−1.26	—	—
$AuCl_3$	164.4 ± 6.3	118.4 ± 4.2	monocl	298	97.91	5.44	−4.18	—	—
AuH(g)	211.1 ± 0.2	−273.4	gas	298	29.29	—	—	—	—
AuI	111.1	− 1.3 ± 2.1	cryst	298	50.21	5.44	—	—	—
Au_2O_3	130.3	3.4	cubic	298	107.53	21.76	—	—	—
Au_2P_3	150.6 ± 12.6	97.5 ± 12.6	monocl	298	108.37	37.66	—	—	—
$AuPb_2$	175.3	6.3 ± 1.7	tetrag	298	61.92	74.48	—	—	—
			liq	527	96.65	—	—	—	24.0
AuS(g)	267.6 ± 4.2	−230.5 ± 25.1	gas	298	37.32	0.05	−1.62	—	—
$AuSb_2$	119.2 ± 1.3	19.5 ± 1.3	tetrag	298	71.63	19.41	—	—	—
AuSe	80.8 ± 6.3	7.9 ± 6.3	monocl	298	41.84	27.91	—	—	—
AuSn	98.1 ± 2.9	30.5 ± 1.7	hex	298	46.57	15.90	—	—	—
			liq	692	60.67	—	—	—	24.7
$AuTe_2$	141.7 ± 0.4	18.6 ± 4.6	monocl	298	63.60	37.40	1.72	—	—
B	5.9	0.0	hex	298	17.91	9.64	−8.58	−1.80	—
			liq	2350	30.54	—	—	—	50.2
B(a)	6.5	− 4.2 ± 1.3	amorph	298	16.05	10.00	−6.28	—	—
B(g)	153.3 ± 0.2	−560.7 ± 12.6	gas	298	20.79	—	—	—	—
BBr_3	228.7	238.9 ± 1.3	liq	298	128.03	—	—	—	—
BBr_3(g)	324.5 ± 1.7	204.4 ± 1.3	gas	298	80.37	1.44	−11.92	—	—
B_4C	27.1 ± 0.3	71.5 ± 11.7	hex	298	96.52	21.92	−44.98	—	—
			liq	2734	135.98	—	—	—	104.6
BCl_3(g)	290.1 ± 0.8	403.3 ± 0.8	gas	298	78.41	2.41	−15.48	—	—

TABLE I—*continued*

Substance	$S_{298} \pm \delta S$ (J/deg mol)	$-\Delta H_{298}$ $\pm \delta H$ (kJ/mol)	Phase	T K	$C_p = A + BT + C/T^2 + DT^2$				H_t (kJ/mol)
					A	$B \times 10^3$ (J/deg mol)	$C \times 10^{-5}$	$D \times 10^6$	
BF_3(g)	254.3 ± 0.4	1136.0 ± 1.0	gas	298	56.82	27.49	−12.76	−7.74	—
BH(g)	171.8	−442.7	gas	298	30.46	2.97	−2.47	—	—
B_2H_6(g)	231.8 ± 0.8	−38.5 ± 3.8	gas	298	11.30	180.08	−3.46	−55.31	—
BI_3(g)	348.6	−71.1	gas	298	82.01	−0.42	−10.46	—	—
BN	14.8 ± 0.2	252.3 ± 1.3	hex	298	41.21	9.41	−21.76	—	—
B_2O_3	54.0 ± 0.4	1273.5 ± 1.7	hex	298	57.03	73.01	−14.06	—	—
			liq	723	129.70	—	—	—	24.1
B_2O_3(a)	78.4 ± 0.8	1253.5 ± 2.5	amorph	298	23.72	131.29	—	—	—
			amorph	500	845.59	−741.82	−963.99	—	—
B_2O_3(g)	283.7 ± 4.2	836.4 ± 4.2	gas	298	78.78	22.18	−16.78	−4.44	—
BOCl(g)	237.2 ± 1.7	315.9 ± 8.4	gas	298	47.66	14.23	−5.90	−3.82	—
$B(OH)_2$(g)	243.1	477.0	gas	298	64.18	31.51	−19.58	−6.07	—
$B(OH)_3$	88.7 ± 0.6	1094.1 ± 1.3	tricl	298	81.34	—	—	—	—
BP	26.8 ± 0.4	115.5 ± 5.9	cubic	298	21.97	28.03	—	—	—
B_2S_3	92.0 ± 20.9	252.3 ± 8.4	monocl	298	98.03	64.02	—	—	—
			liq	836	151.67	—	—	—	48.1
Ba	62.4	0.0	bcc	298	−473.21	1586.99	128.16	—	−1306.04
			bcc	648	0.67	−975.71	−710.02	−428.44	—
			liq	1003	22.84	7.82	126.78	—	7.7
			liq	1300	40.58	—	—	—	—
Ba(g)	170.1	−179.1	gas	298	20.79	—	—	—	—
			gas	900	48.66	−36.32	−49.04	13.43	—
$BaBr_2$	148.5 ± 4.2	757.7 ± 1.7	orth	298	70.54	21.67	—	—	—
			liq	1130	104.85	—	—	—	32.0
$BaBr_2$(g)	341.8 ± 8.4	424.7 ± 12.6	gas	298	58.45	0.65	—	—	—
BaC_2	88.7 ± 8.4	74.1 ± 12.6	tetrag	298	73.64	3.77	−9.67	—	—
$BaCl_2$	123.7 ± 0.2	858.6 ± 12.6	cubic	298	69.45	19.16	—	—	—
			β	1198	123.85	—	—	—	16.9
			liq	1245	108.78	—	—	—	16.0
$BaCl_2$(g)	325.6 ± 5.0	498.7 ± 16.7	gas	298	58.20	—	−1.80	—	—
BaF_2	96.4 ± 0.4	1208.8 ± 4.2	cryst	298	70.21	18.37	−2.83	—	—
			1250	70.21	18.37	18.37	−2.83	—	17.1
			liq	1641	99.83	—	—	—	23.3
BaF_2(g)	301.2 ± 2.1	803.7 ± 6.3	gas	298	57.99	0.15	−3.77	—	—

BaH_2	64.4 ± 6.3	178.7 ± 4.2	cryst	298	37.24	17.15	—	—	—
BaI_2	165.1 ± 0.4	605.4 ± 3.3	cubic	298	71.55	20.00	—	—	—
			liq	984	112.97	—	—	—	26.5
$BaI_2(g)$	348.1 ± 8.4	302.9 ± 16.7	gas	298	57.82	—	—	—	—
Ba_3N_2	152.3 ± 8.4	341.0 ± 31.4	ps.hex	298	87.86	98.32	—	—	—
BaO	72.1 ± 0.4	548.1 ± 2.1	cubic	298	50.29	7.20	−4.60	—	—
			liq	2286	66.94	—	—	—	58.6
$BaO(g)$	235.4 ± 0.8	123.4 ± 8.4	gas	298	39.54	−3.85	−5.27	1.61	—
BaO_2	93.1 ± 9.6	634.3 ± 11.7	cubic	298	62.34	28.03	—	—	—
$Ba_3Al_2O_6$	301.2	3508.3 ± 12.6	cryst	298	266.52	30.84	−53.43	—	—
$BaAl_2O_4$	148.5 ± 9.2	2324.2 ± 7.9	hex	298	143.30	73.89	−45.90	—	—
			β	600	138.20	31.63	—	—	—
$BaCO_3$	107.5 ± 1.3	1217.5 ± 12.6	orth	298	86.90	48.95	−11.97	—	—
			tetrag	1079	154.81	—	—	—	18.8
			cubic	1241	163.18	—	—	—	2.9
$BaHfO_3$	121.3 ± 12.6	1789.9 ± 12.6	monocl	298	123.85	13.39	−17.36	—	—
$BaMoO_4$	146.9 ± 4.6	1516.3 ± 6.7	tetrag	298	135.27	28.45	−25.10	—	—
$Ba(NO_3)_2$	213.8 ± 1.0	992.4 ± 2.1	nit.ba	298	125.73	149.37	−16.74	—	—
BaS	78.2 ± 1.3	463.6 ± 3.3	cubic	298	51.17	7.87	−3.68	—	—
$BaSO_4$	132.2 ± 1.3	1481.1 ± 12.6	baryte	298	141.42	—	−35.27	—	—
$BaSiO_3$	104.6 ± 8.4	1618.0 ± 7.1	cryst	298	122.01	7.11	−31.21	—	—
			1300	62.76	51.04	—	—	—	
Ba_2SiO_4	177.8 ± 8.4	2272.3 ± 3.8	orth	298	175.35	11.46	−39.71	—	—
Ba_2Sn	126.8 ± 10.5	376.6 ± 33.5	cryst	298	60.67	41.84	—	—	—
Ba_2TiO_4	196.6	2233.4 ± 12.6	monocl	298	179.91	6.69	−29.12	—	—
$BaTiO_3$	107.9	1647.7 ± 11.7	cubic	298	121.46	8.54	−19.16	—	—
$BaUO_4$	178.1 ± 0.4	1988.2 ± 2.1		298	140.58	20.92	—	—	—
BaV_2O_6	193.7 ± 10.0	2282.0 ± 6.7	cryst	298	181.59	81.17	−29.29	—	—
$BaZrO_3$	124.7	1769.0 ± 5.9	cubic	298	127.90	5.94	−21.21	—	—
Be	9.5	0.0	hcp	298	21.21	5.69	−5.86	0.96	—
			bcc	1543	32.22	—	—	—	2.5
			liq	1562	29.46	—	—	—	12.6
$Be(g)$	136.1	−324.0	gas	298	20.92	—	—	—	—
			gas	1900	18.95	0.59	27.24	—	—
$BeBr_2$	100.4 ± 4.6	355.6 ± 12.6	ortho	298	60.21	29.50	—	—	—
			liq	781	112.97	—	—	—	10.0
Be_2C	16.3 ± 3.8	117.0 ± 1.0		298	38.37	45.02	−8.41	−9.58	—
$BeCl_2$	82.7 ± 0.4	490.8 ± 3.3	orth	298	76.61	12.34	−13.77	—	—
			liq	688	121.42	—	—	—	8.7
$BeCl_2(b)$	75.8 ± 0.2	496.2 ± 3.3	orth.β	298	65.65	20.88	−8.41	—	—
$BeCl_2(g)$	251.0 ± 4.2	360.2 ± 10.5	gas	298	58.12	3.47	−0.75	−6.99	—

TABLE I—*continued*

Substance	$S_{298} \pm \delta S$ (J/deg mol)	$-\Delta H_{298}$ $\pm \delta H$ (kJ/mol)	Phase	T K	$C_p = A + BT + C/T^2 + DT^2$ A	$B \times 10^3$ (J/deg mol)	$C \times 10^{-5}$	$D \times 10^6$	H_t (kJ/mol)
$Be_2Cl_4(g)$	387.0 ± 8.4	821.3	gas	298	131.38	0.63	−14.98	—	—
$BeF(g)$	205.6	174.9 ± 10.5	gas	298	30.04	7.91	−2.18	−2.13	—
BeF_2	53.3 ± 0.4	1026.8 ± 4.2	β−qtz	298	20.63	104.60	−0.03	—	—
			hex	500	47.36	33.47	—	—	0.2
			liq	1447	52.55	32.89	—	—	4.8
$BeF_2(g)$	227.4	796.0	gas	298	51.17	8.45	−6.65	—	—
			gas	1000	61.84	0.17	−32.76	—	—
BeH_2	17.6 ± 4.2	19.0 ± 2.1	cryst	298	20.50	29.29	—	—	—
$BeH_2(g)$	173.2	−125.5	gas	298	31.38	23.43	−8.16	−4.77	—
BeI_2	120.5 ± 4.2	188.7 ± 20.9	tetrag	298	79.50	13.93	—	—	—
			liq	753	112.97	—	—	—	20.9
Be_3N_2	34.3 ± 0.2	589.5 ± 2.5	cubic	298	53.93	103.55	−17.66	—	—
			hex	430	114.52	15.06	−59.54	—	17.8
			liq	2470	158.99	—	—	—	129.3
BeO	13.8 ± 0.4	609.4 ± 3.3	hex	298	41.59	10.21	−17.24	−1.34	—
			cubic	2370	46.86	4.60	—	—	6.7
			liq	2780	64.85	—	—	—	84.9
$BeAl_2O_4$	66.3 ± 0.4	2301.2 ± 5.9	orth	298	368.78	66.44	−101.67	−1.97	—
$BeAl_6O_{10}$	175.6 ± 0.4	5624.1 ± 10.0	cryst	298	389.32	47.07	−123.22	—	—
$Be_3B_2O_6$	92.5 ± 12.6	3133.8 ± 4.2	cryst	298	142.26	211.12	−54.81	−66.02	—
$Be(OH)_2$	50.2 ± 5.0	905.8 ± 2.1	β	298	19.54	153.76	—	—	—
BeS	33.5 ± 6.3	233.5 ± 4.2	cubic	298	41.71	8.12	−9.62	—	—
$BeSO_4$	78.0 ± 0.8	1200.8 ± 3.3	tetrag	298	112.80	−9.25	−27.70	76.02	—
			β	863	162.05	—	—	—	1.1
			gamma	908	290.03	−52.55	−717.56	8.49	19.5
$BeSO_4.2H_2O$	163.2	1820.9		298	71.55	269.87	—	—	—
Be_2SiO_4	64.4 ± 0.4	2146.0 ± 8.4	trigon	298	124.06	78.53	−45.35	−33.47	—
$BeWO_4$	88.4	1513.4	cryst	298	112.38	42.89	−25.10	—	—
Bi	56.7	0.0	rhomb	298	11.84	30.46	4.10	—	—
			liq	545	19.04	10.38	20.75	−3.97	11.3
			liq	1200	27.20	—	—	—	—
$Bi(g)$	186.9	−209.6	gas	298	20.92	—	—	—	—
$Bi_2(g)$	273.2	−220.1	gas	298	37.24	—	—	—	—
$BiBr_3$	181.6 ± 6.3	276.1 ± 8.4	cryst	298	97.49	25.10	—	—	—
			liq	492	157.74	—	—	—	21.8

$BiCl(g)$	255.0 ± 0.4	−25.1 ± 8.4	gas	298	37.32	0.84	−1.17	—	—
$BiCl_3$	171.5 ± 8.4	378.7 ± 5.0	cubic	298	68.83	133.89	—	—	—
			liq	507	127.61	—	—	—	23.6
$BiF(g)$	244.0 ± 0.4	29.3 ± 20.9	gas	298	37.03	0.84	−2.55	—	—
BiF_3	122.6 ± 5.9	910.4 ± 14.6	cubic	298	104.56	99.75	—	180.50	—
			liq	922	184.60	—	—	—	21.6
$BiH(g)$	214.7 ± 0.2	−179.9 ± 29.3	gas	298	37.78	—	−9.37	—	—
BiI_3	224.7 ± 6.3	150.6 ± 6.3	hex	298	39.96	110.04	2.97	—	—
			liq	681	157.74	—	—	—	39.1
Bi_2O_3	151.5 ± 2.9	570.7 ± 3.3	monocl	298	103.51	33.47	—	—	—
			cubic	978	146.44	—	—	—	56.9
			liq	1097	152.72	—	—	—	59.8
$BiOCl$	102.5 ± 12.6	371.1 ± 4.2	tetrag	298	57.32	43.10	—	—	—
Bi_2S_3	200.4 ± 6.3	201.7	orth	298	114.47	27.70	—	—	—
			liq	1036	188.28	—	—	—	79.5
$Bi_2(SO_4)_3$	312.5 ± 18.8	2543.9 ± 25.1	cryst	298	228.45	169.03	—	—	—
Bi_2Se_3	239.7 ± 8.4	140.2 ± 3.3	rhomb	298	118.53	19.25	—	—	—
			liq	995	188.28	—	—	—	86.6
Bi_2Te_3	261.1 ± 8.4	78.7 ± 2.1	hex	298	107.99	55.23	—	—	—
			liq	850	167.36	—	—	—	119.7
Br_2	152.2	0.0	liq	298	75.73	—	—	—	—
$Br_2(g)$	245.3	−30.9	gas	298	37.36	0.46	−1.30	—	—
$Br(g)$	174.9	−111.9	gas	298	19.87	1.49	0.42	—	—
C	5.7	0.0	graph	298	0.11	38.94	−1.48	−17.38	—
				1100	24.43	0.44	−31.63	—	—
$C(d)$	2.4 ± 0.1	− 1.5 ± 0.1	diam	298	9.12	13.22	−6.19	—	—
$C(g)$	158.0	−716.7	gas	298	20.77	0.05	—	—	—
			gas	2000	19.49	0.72	—	—	—
$C_2(g)$	199.3 ± 0.1	−832.6 ± 5.9	gas	298	30.69	4.75	10.18	−0.43	—
$C_3(g)$	237.2	−820.1 ± 12.6	gas	298	31.84	14.98	1.21	−2.62	—
$CBr_4(g)$	358.0 ± 0.8	−50.2	gas	298	105.90	1.14	−13.68	—	—
CCl_4	214.4 ± 1.3	135.4 ± 2.1	liq	298	133.89	—	—	—	—
$CCl_4(g)$	309.8 ± 0.4	102.9 ± 2.1	gas	298	104.18	2.01	−19.82	—	—
$CF_2(g)$	240.7 ± 1.3	171.5 ± 18.8	gas	298	43.01	16.32	−7.87	−4.69	—
$CF_3(g)$	260.9 ± 3.8	484.1 ± 13.8	gas	298	58.16	26.32	−13.31	−7.45	—
$CF_4(g)$	261.3 ± 0.4	933.2 ± 10.5	gas	298	85.75	18.37	−27.51	−3.87	—
$C_2F_6(g)$	332.0 ± 1.7	1325.5 ± 25.1	gas	298	136.23	46.82	−38.66	−13.64	—
$CF_2Cl_2(g)$	300.7	475.3	gas	298	107.65	—	−31.38	—	—
$CFCl_3(g)$	309.8	277.0	gas	298	108.07	—	−26.78	—	—
$CH_2(g)$	181.2 ± 1.7	−397.5 ± 20.9	gas	298	25.26	27.37	−1.44	−6.00	—
$CH_3(g)$	192.9	−133.6	gas	298	22.97	49.16	−0.03	−11.66	—

TABLE I—*continued*

Substance	$S_{298} \pm \delta S$ (J/deg mol)	$-\Delta H_{298}$ $\pm \delta H$ (kJ/mol)	Phase	T K	$C_p = A + BT + C/T^2 + DT^2$ A	$B \times 10^3$ (J/deg mol)	$C \times 10^{-5}$	$D \times 10^6$	H_t (kJ/mol)
$CH_4(g)$	186.3 ± 0.2	74.8 ± 0.4	gas	298	12.45	76.69	1.45	−17.99	—
$C_2H_2(g)$	200.8 ± 0.2	−226.8 ± 1.3	gas	298	43.63	31.65	−7.51	−6.32	—
$C_2H_4(g)$	219.2 ± 0.4	−52.5 ± 0.4	gas	298	32.64	59.83	—	—	—
$C_2H_6(g)$	229.5 ± 0.4	84.7 ± 0.8	gas	298	28.19	122.63	−9.12	−27.82	—
$CH_3Cl(g)$	234.3 ± 0.4	83.7 ± 2.1	gas	298	42.51	44.89	−13.74	−8.47	—
$CN(g)$	202.5 ± 0.4	−435.1	gas	298	27.82	5.69	−0.46	−0.33	—
$C_2N_2(g)$	241.5 ± 0.6	−309.1 ± 2.1	gas	298	56.07	27.45	−6.23	−6.86	—
$CO(g)$	197.5	110.5	gas	298	28.41	4.10	−0.46	—	—
$CO_2(g)$	213.7	393.5	gas	298	44.14	9.04	−8.54	—	—
$COCl_2(g)$	283.7 ± 0.8	220.1 ± 5.0	gas	298	65.02	18.16	−11.13	−4.98	—
$COF(g)$	248.1 ± 1.7	171.5	gas	298	39.37	18.66	−5.19	−5.19	—
$COF_2(g)$	258.8 ± 0.4	633.9 ± 5.4	gas	298	53.22	30.42	−12.97	−8.37	—
$CH_2O(g)$	218.7 ± 0.2	115.9 ± 0.8	gas	298	21.07	53.87	0.78	−13.4	—
CH_3OH	126.6 ± 0.8	239.5 ± 0.4	liq	298	81.59	—	—	—	—
$CH_3OH(g)$	239.7 ± 0.8	202.0 ± 0.4	gas	298	4.31	128.72	4.54	−44.10	—
C_2H_5OH	161.0 ± 0.4	277.0 ± 0.8	liq	298	112.13	—	—	—	—
$C_2H_5OH(g)$	282.4 ± 2.1	234.6 ± 0.8	gas	298	31.38	112.97	—	—	—
$CS(g)$	210.5 ± 0.2	−280.3 ± 25.1	gas	298	29.37	8.66	−1.86	−2.38	—
CS_2	151.3 ± 0.8	−89.1 ± 1.0	liq	298	76.99	—	—	—	—
$CS_2(g)$	237.9 ± 0.2	−116.9 ± 1.3	gas	298	49.58	13.68	−6.99	−3.77	—
$COS(g)$	231.5 ± 0.4	138.5 ± 1.0	gas	298	47.40	9.12	−7.66	—	—
Ca	41.6	0.0	fcc	298	16.02	21.51	2.55	—	—
			bcc	716	−0.45	41.35	—	—	0.9
			liq	1115	33.47	—	—	—	8.5
$Ca(g)$	154.8	−177.8	gas	298	20.79	—	—	—	—
$CaAl_2$	85.4 ± 10.0	216.7 ± 16.7	cubic	298	82.84	7.95	−11.21	—	—
			liq	1353	94.14	—	—	—	63.2
$CaBr_2$	129.7 ± 4.2	682.8 ± 4.2	tetrag	298	77.24	9.62	−5.48	—	—
			liq	1015	112.97	—	—	—	29.1
$CaBr_2(g)$	420.9 ± 8.4	384.9 ± 8.4	gas	298	62.34	—	−1.72	—	—
CaC_2	70.3 ± 1.7	59.4 ± 5.0	tetrag	298	68.62	11.88	−8.66	—	—
				720	64.43	8.37	—	—	5.6
$CaCl_2$	108.4	795.0 ± 3.3	tetrag	298	69.83	15.40	−1.59	—	—
			liq	1045	122.26	−14.90	0.71	—	28.2

$CaCl_2(g)$	287.4	481.6	gas	298	62.13	0.14	−2.55	—	—
CaF_2	68.9 ± 0.4	1229.3	cubic	298	41.05	55.48	8.49	—	—
			cubic	1430	154.10	—	−729.69	—	—
			liq	1691	99.16	—	—	—	31.2
$CaF_2(g)$	273.7	800.0	gas	298	57.24	0.61	−5.44	—	—
$CaH(g)$	201.7 ± 0.4	−228.9	gas	298	32.13	3.10	−2.93	—	—
CaH_2	41.8 ± 6.3	181.6 ± 8.4	orth	298	29.92	37.24	—	—	—
			β	1053	69.04	—	—	—	6.7
			liq	1273	75.31	—	—	—	22.0
CaI_2	145.3 ± 0.4	533.0 ± 2.1	hex	298	71.34	19.75	—	—	—
			liq	1052	103.34	—	—	—	41.8
$CaI_2(g)$	327.4 ± 8.4	258.2 ± 16.7	gas	298	62.38	—	−1.51	—	—
$CaMg_2$	102.6 ± 1.0	39.3 ± 1.7	hex	298	66.94	26.07	−0.86	—	—
Ca_3N_2	107.9 ± 11.7	439.3 ± 12.6	cubic	298	138.78	15.48	−26.36	—	—
CaO	38.1 ± 0.3	634.9 ± 0.8	cubic	298	50.42	4.18	−8.49	—	—
			liq	2900	83.68	—	—	—	52.3
$Ca_3Al_2O_6$	205.4 ± 1.3	3589.5 ± 4.6	cubic	298	260.58	19.16	−50.25	—	—
$CaAl_2O_4$	114.0 ± 0.8	2325.9 ± 2.1		298	150.62	24.94	−33.30	—	—
$CaAl_4O_7$	177.8 ± 1.3	3999.1 ± 3.3		298	276.52	22.93	−74.48	—	—
$Ca_3B_2O_6$	183.7 ± 1.3	3424.6 ± 3.8	cryst	298	236.14	43.60	−54.48	—	—
			liq	1763	393.30	—	—	—	148.5
$Ca_2B_2O_5$	145.2 ± 0.8	2722.9 ± 5.0	α	298	183.05	48.12	−44.73	—	—
			β	804	218.78	10.04	—	—	4.6
			liq	1583	285.35	—	—	—	100.8
CaB_2O_4	105.9 ± 0.8	2027.1 ± 3.8	cryst	298	129.79	40.84	−33.76	—	—
			liq	1433	258.15	—	—	—	74.1
CaB_4O_7	134.7 ± 1.3	3340.9 ± 6.3	cryst	298	214.81	80.17	−71.80	—	—
			liq	1263	444.76	—	—	—	113.4
$CaCO_3$	92.7 ± 0.8	1206.9 ± 2.9	calcit	298	104.52	21.92	−25.94	—	—
$CaOCl_2$	113.0 ± 9.2	746.4		298	83.68	55.65	—	—	—
$CaCr_2O_4$	125.2	1829.7	cryst	298	169.66	13.26	−23.97	—	—
$Ca_2Fe_2O_5$	188.7 ± 1.3	2133.8 ± 5.4	cryst	298	248.61	—	−48.87	—	—
			liq	1750	310.45	—	—	—	151.0
$CaFe_2O_4$	145.2 ± 0.8	1479.5	cryst	298	164.93	19.92	−15.31	—	—
			liq	1489	229.70	—	—	—	108.4
$CaGeO_3$	87.4	1285.3 ± 5.4	tricl	298	120.50	16.11	−24.69	—	—
$CaOH(g)$	239.8	179.9	gas	298	211.71	17.57	−19.04	—	—
$Ca(OH)_2$	83.4 ± 0.4	−19.04 ± 1.3	hex	298	104.01	15.19	−18.74	—	—
$CaHfO_3$	99.2 ± 8.4	1779.9 ± 10.0	monocl	298	121.71	13.56	−17.78	—	—
$CaO.MgO$	66.3 ± 1.7	1243.1 ± 1.7		298	97.82	7.66	−18.24	—	—
$CaMg(CO_3)_2$	155.2 ± 0.4	2315.0 ± 5.0	dolom	298	156.98	80.33	−21.34	—	—

TABLE I—*continued*

Substance	$S_{298} \pm \delta S$ (J/deg mol)	$-\Delta H_{298} \pm \delta H$ (kJ/mol)	Phase	T K	$C_p = A + BT + C/T^2 + DT^2$ A	$B \times 10^3$ (J/deg mol)	$C \times 10^{-5}$	$D \times 10^6$	H_t (kJ/mol)
$CaMoO_4$	122.6 ± 1.0	1546.0 ± 3.3	tetrag	298	133.47	29.20	−22.34	—	—
$Ca(NO_3)_2$	193.2	936.8	cryst	298	122.88	154.01	−17.28	—	—
$CaNb_2O_6$	178.4 ± 13.0	2675.2 ± 17.2		298	214.64	20.92	−38.70	—	—
Ca_3P_2	123.8 ± 18.8	506.3 ± 25.1	cryst	298	107.95	28.03	—	—	—
$Ca_3(PO_4)_2$	236.0 ± 1.3	4117.1 ± 25.1	rhomb	298	201.84	163.51	−20.92	—	—
			monocl	1373	330.54	—	—	—	15.5
$Ca_2P_2O_7$	189.2 ± 0.4	3336.7 ± 20.9	α	298	221.88	61.76	−46.69	—	—
			β	1413	318.61	—	—	—	6.7
			liq	1626	405.01	—	—	—	100.8
$CaHPO_4$	111.4 ± 0.8	1814.2 ± 20.9		298	138.41	55.10	−40.38	—	—
Ca_2Pb	126.4 ± 7.1	209.2 ± 12.6	orth	298	62.43	22.01	—	—	—
CaS	56.5 ± 1.3	473.2 ± 3.8	cubic	298	50.63	3.70	−3.89	—	—
$CaS(g)$	232.4	−116.7 ± 20.9	gas	298	37.20	0.11	−2.28	—	—
$CaSO_4$	106.7 ± 1.7	1434.1 ± 4.2	orth	298	70.21	98.74	—	—	—
$CaSO_4.1/2H_2O$	130.5 ± 2.5	1575.5 ± 14.6		298	69.33	163.18	—	—	—
$CaSO_4.2H_2O$	194.1 ± 1.3	2021.4 ± 14.6	gypsum	298	91.38	317.98	—	—	—
Ca_3Sb_2	157.3 ± 18.8	728.0 ± 37.7	hex	298	105.02	29.50	—	—	—
$CaSe$	69.0 ± 6.3	368.2 ± 33.5	cubic	298	45.61	8.37	—	—	—
$CaSi$	45.2	150.6 ± 9.2	orth	298	42.68	15.06	—	—	—
Ca_3SiO_5	168.6 ± 1.3	2928.8 ± 7.9		298	208.57	36.07	−42.47	—	—
Ca_2SiO_4	120.5 ± 0.8	2328.4 ± 5.4	olivin	298	133.30	51.80	−19.41	—	—
			larnit	970	134.56	46.11	—	—	10.9
				1710	205.02	—	—	—	14.2
$Ca_2SiO_4(b)$	127.6 ± 0.8	2319.6 ± 5.4	larnit	298	145.90	40.75	−26.19	—	—
$Ca_3Si_2O_7$	210.9 ± 1.3	3942.6 ± 12.6	rankin	298	257.36	55.81	−53.64	—	—
$CaSiO_3$	83.1 ± 0.8	1635.1 ± 2.1	wollas	298	111.46	15.06	−27.28	—	—
			ps.wol	1463	108.16	16.48	−23.64	—	5.7
			liq	1817	146.44	—	—	—	82.8
$CaSiO_3(p)$	87.4 ± 0.8	1628.0 ± 2.5	ps.wol	298	108.16	16.48	−23.64	—	—
$CaO.Al_2O_3.2SiO_2$	202.5	4223.7 ± 5.9	anorth	298	297.06	43.39	−135.35	—	—
$CaO.Al_2O_3.SiO_2$	144.8	3293.2 ± 7.5	pyrox	298	233.22	21.13	−73.72	—	—
$2CaO.Al_2O_3.SiO_2$	198.3	3989.4 ± 5.9	gehlen	298	275.47	27.91	−78.20	—	—
$3CaO.Al_2O_3.3SiO_2$	241.4	6646.3 ± 17.6	gross	298	456.31	49.20	−131.42	—	—
$CaO.MgO.SiO_2$	110.5 ± 4.2	2263.1 ± 8.8	monti	298	150.62	32.01	−32.84	—	—

$CaO.MgO.2SiO_2$	143.1 ± 0.8	3203.3 ± 8.4	diops	298	186.02	123.76	−55.90	−43.93	—
			liq	1665	355.64	—	—	—	128.4
$2CaO.MgO.2SiO_2$	209.2	3876.9	akerm	298	251.88	47.24	−48.03	—	—
$2CaO.5MgO.8SiO_2.H_2O$	548.9	12358.7 ± 24.3	tremol	298	764.00	271.12	−168.20	—	—
$3CaO.MgO.2SiO_2$	253.1	4566.8 ± 15.9	merwin	298	305.01	50.38	−60.25	—	—
$Ca_4Ti_3O_{10}$	328.4 ± 10.5	5671.8 ± 18.8	orth	298	424.05	21.59	−82.42	—	—
$CaTiO_3$	93.7 ± 0.8	1660.6 ± 5.0	cubic	298	127.49	5.69	−27.99	—	—
			orth	1530	134.01	—	—	—	2.3
$CaO.TiO_2.SiO_2$	129.3	2598.7 ± 6.3	sphene	298	177.36	23.18	−40.29	—	—
			liq	1673	279.49	—	—	—	123.8
$CaUO_4$	121.1 ± 0.4	1998.3 ± 3.8		298	120.42	45.52	−9.46	—	—
$Ca_3V_2O_8$	274.9 ± 2.1	3777.7 ± 8.4		298	226.81	101.34	—	—	—
$Ca_2V_2O_7$	220.5 ± 1.7	3082.8 ± 6.7		298	177.82	121.00	—	—	—
CaV_2O_6	179.1 ± 1.7	2328.8 ± 5.4		298	135.23	119.16	—	—	—
$CaWO_4$	126.4 ± 0.8	1624.2 ± 12.1	scheel	298	134.56	20.67	−24.43	—	—
Ca_3WO_6	195.0 ± 14.6	2933.0 ± 16.3		298	236.52	29.71	−38.33	—	—
$CaZrO_3$	93.7 ± 8.4	1766.5 ± 9.6	cubic	298	119.24	12.05	−21.00	—	—
Cd	51.8	0.0	hex	298	22.05	12.55	0.14	—	—
			liq	594	29.71	—	—	—	6.2
Cd(g)	167.7	111.8	gas	298	20.79	—	—	—	—
Cd_3As_2	207.1 ± 10.5	38.1 ± 8.4	tetrag	298	136.19	11.92	−12.84	—	—
$CdBr_2$	138.8	315.3 ± 1.7	hex	298	79.96	21.09	−8.49	—	—
			liq	568	101.67	—	—	—	33.5
$CdBr_2(g)$	309.6 ± 3.3	140.0 ± 5.0	gas	298	65.40	—	−4.60	—	—
$CdCl_2$	115.3 ± 0.4	390.8 ± 1.3	hex	298	47.28	91.63	—	—	—
			liq	842	110.04	—	—	—	30.5
$CdCl_2(g)$	285.8 ± 2.9	194.6 ± 6.3	gas	298	63.30	—	−4.60	—	—
CdF_2	83.7 ± 6.3	700.4 ± 2.1	cubic	298	60.04	23.01	—	—	—
$CdF_2(g)$	265.3 ± 2.5	395.0 ± 4.2	gas	298	61.30	—	−6.90	—	—
CdI_2	158.3 ± 0.8	204.2 ± 2.5	hex	298	67.57	37.66	—	—	—
			β	595	89.96	—	—	—	—
			liq	661	102.09	—	—	—	20.7
$CdI_2(g)$	325.5 ± 3.8	60.2 ± 4.6	gas	298	66.23	—	−5.02	—	—
CdO	54.8 ± 1.7	258.4 ± 0.8	cubic	298	48.24	6.36	−5.31	—	—
$CdAl_2O_4$	125.1 ± 12.1	1916.9 ± 2.5	cubic	298	160.04	23.85	−33.05	—	—
$CdCO_3$	92.5 ± 5.0	751.9 ± 8.4	hex	298	43.10	131.80	—	—	—
$CdGa_2O_4$	139.7 ± 12.6	1355.6 ± 2.1	cubic	298	161.13	21.84	−25.90	—	—
$Cd(NO_3)_2$	207.9 ± 10.9	456.5 ± 4.6	cubic	298	179.91	—	—	—	—
			tetrag	431	213.38	—	—	—	2.9
CdS	69.0 ± 2.1	149.4 ± 2.9	hex	298	44.56	13.81	—	—	—
$CdSO_4$	123.1 ± 0.4	934.3 ± 1.3	rhomb	298	76.73	77.40	—	—	—

TABLE I—*continued*

Substance	$S_{298} \pm \delta S$ (J/deg mol)	$-\Delta H_{298} \pm \delta H$ (kJ/mol)	Phase	T K	$C_p = A + BT + C/T^2 + DT^2$ A	$B \times 10^3$ (J/deg mol)	$C \times 10^{-5}$	$D \times 10^6$	H_t (kJ/mol)
CdSb	95.6 ± 0.8	13.3 ± 1.3	orth	298	44.52	19.41	—	—	—
			liq	729	143.13	−75.44	—	—	35.6
CdSe	83.3 ± 2.1	144.8 ± 2.1	hex	298	46.82	9.33	—	—	—
$CdSeO_3$	138.1	576.6	cryst	298	78.45	54.39	—	—	—
$CdSiO_3$	97.5 ± 1.3	1189.3 ± 3.3	monocl	298	87.86	42.68	−10.46	—	—
CdTe	95.2 ± 0.8	100.8 ± 0.8	cubic	298	52.51	19.00	−7.36	—	—
			liq	1370	64.85	—	—	—	43.9
$CdTiO_3$	104.6 ± 8.4	1186.6 ± 2.9	hex	298	116.11	9.62	−18.20	—	—
Ce	56.9	0.0	fcc	298	22.38	15.06	—	—	—
			bcc	998	37.66	—	—	—	3.0
			liq	1071	37.66	—	—	—	5.2
Ce(g)	191.6	417.1	gas	298	24.89	15.02	−6.69	−3.77	—
$CeBr_3$	182.0	887.0	hex	298	94.73	24.52	−0.29	—	—
			liq	1006	152.72	—	—	—	51.9
Ce_2C_3	173.6 ± 15.9	176.6 ± 6.7	cubic	298	122.38	11.92	−20.08	—	—
CeC_2	90.0 ± 7.9	97.1 ± 7.1	tetrag	298	68.62	9.20	−9.20	—	—
$CeCl_3$	149.8 ± 7.5	1058.1 ± 1.7	hex	298	97.49	13.60	−5.02	—	—
			liq	1095	145.18	—	—	—	54.0
CeF_3	115.2 ± 0.4	1732.6 ± 8.4	hex	298	74.94	38.79	15.44	—	—
			liq	1705	125.02	—	—	—	58.6
CeH_2	55.8 ± 0.4	193.3	cubic	298	35.15	19.25	—	—	—
CeI_3	214.6 ± 14.6	649.8 ± 9.2	orth	298	80.12	60.75	3.77	—	—
			liq	1034	152.72	—	—	—	51.0
CeN	60.7 ± 8.4	326.4 ± 25.1	cubic	298	48.53	6.07	−7.24	—	—
Ce_2O_3	148.1 ± 0.4	1799.7 ± 1.3	cubic	298	139.33	11.63	−24.23	—	—
				1050	126.78	32.30	−9.29	—	—
				1800	154.81	—	—	—	—
CeO_2	62.3 ± 0.3	1090.4 ± 0.8	cubic	298	64.81	17.70	−7.61	—	—
$CeAlO_3$	95.4 ± 8.4	1800.8 ± 2.9	ortho	298	107.19	29.58	−18.87	—	—
$CeCrO_3$	109.2 ± 11.7	1533.0 ± 11.7		298	113.60	25.31	−12.43	—	—
CeS	78.2 ± 1.3	456.5 ± 8.4	cubic	298	52.54	13.55	−5.89	—	—
Ce_3S_4	255.2 ± 12.6	1652.7 ± 20.9	cubic	298	167.82	39.66	—	—	—
Ce_2S_3	180.3 ± 1.7	1188.3 ± 12.6	cubic	298	124.93	12.72	—	—	—
Ce_2O_2S	130.5 ± 7.5	1696.6 ± 15.1	hexag	298	124.89	23.64	−20.92	—	—
$Ce_2(SO_4)_3$	287.4 ± 23.0	3955.1 ± 4.6	cryst	298	221.75	198.74	—	—	—

$Cl_2(g)$	223.0	0.0	gas	298	36.99	0.71	−2.93	—	—
$Cl(g)$	165.1	−121.3	gas	298	23.79	−1.28	−1.40	—	—
$ClF(g)$	217.8 ± 0.2	50.8 ± 1.3	gas	298	36.30	1.00	−4.18	—	—
$ClF_3(g)$	281.5 ± 0.4	158.9 ± 1.3	gas	298	80.79	1.21	−14.69	—	—
$Cl_2O(g)$	267.9 ± 0.4	−81.4 ± 1.7	gas	298	56.78	0.73	−8.39	—	—
$ClO_2(g)$	257.1 ± 0.4	−93.3 ± 2.1	gas	298	55.56	1.53	−12.55	—	—
Co	30.0	0.0	hcp	298	18.12	23.14	−0.42	−0.08	—
			fcc	700	18.12	23.14	−0.42	−0.08	0.5
			fcc	1390	18.12	23.14	−0.42	−0.08	5.1
			liq	1768	40.50	—	—	—	16.2
$Co(g)$	179.4	−424.7	gas	298	27.03	−0.21	−3.59	—	—
CoAl	54.4	110.5 ± 8.4	cubic	298	42.68	12.55	—	—	—
			liq	1901	71.13	—	—	—	62.8
$CoBr_2$	133.9 ± 8.4	215.7 ± 2.1	hex	298	73.39	20.92	—	—	—
			cubic	648	92.05	—	—	—	—
$CoCl_2$	109.3 ± 0.4	312.5 ± 1.7	hex	298	82.09	6.74	−4.98	—	—
			liq	1013	99.16	—	—	—	44.8
$CoCl_2(g)$	298.3 ± 8.4	93.7 ± 8.4	gas	298	58.41	7.49	−0.73	−2.13	—
CoF_2	82.0 ± 0.4	671.5 ± 9.2	tetrag	298	64.02	15.56	—	—	—
CoF_3	94.6 ± 12.6	789.9 ± 12.6	hex	298	97.82	7.03	−7.49	—	—
CoI_2	153.1 ± 10.5	85.8 ± 10.5	hex	298	72.38	25.10	—	—	—
CoO	53.0 ± 0.4	237.7 ± 0.6	cubic	298	55.40	−6.44	—	7.11	—
Co_3O_4	114.3 ± 8.4	910.0 ± 4.6	cubic	298	140.75	17.28	−24.35	53.97	—
$CoAl_2O_4$	99.6 ± 2.9	1946.8 ± 3.8	cubic	298	165.69	18.83	−34.73	—	—
$CoCO_3$	87.9 ± 0.4	702.5 ± 12.6	hex	298	88.28	38.91	−17.99	—	—
$CoCr_2O_4$	126.8 ± 0.8	1432.2 ± 6.7	spinel	298	167.65	17.74	−13.97	—	—
$CoFe_2O_4$	142.7 ± 8.4	1088.7 ± 4.6	cubic	298	173.22	54.39	−32.76	—	—
				740	215.06	571.12	—	—	—
				784	657.72	−541.83	—	—	—
$Co(OH)_2$	93.3 ± 1.7	541.4 ± 6.3	hex	298	82.84	47.70	—	—	—
$Co(NO_3)_2$	177.0	421.3 ± 3.8	cryst	298	131.80	83.68	—	—	—
Co_2P	77.4	187.4 ± 17.2	rhomb	298	57.95	23.01	—	—	—
CoP	50.2	125.5 ± 18.0	rhomb	298	40.92	14.64	—	—	—
CoP_3	98.3	204.6 ± 31.8	cryst	298	93.72	25.10	—	—	—
$CoS_{0.89}$	51.5 ± 3.3	94.6 ± 4.2	cubic	298	40.25	15.52	—	—	—
Co_3S_4	184.1 ± 20.9	359.0 ± 25.1	cubic	298	143.30	76.57	—	—	—
CoS_2	69.0 ± 6.3	153.1 ± 8.4	cubic	298	60.67	25.31	—	—	—
$CoSO_4$	117.4 ± 1.3	888.3 ± 1.7	rhomb	298	122.59	40.29	−27.95	—	—
				964	168.62	—	—	—	2.1
CoSb	70.7 ± 5.4	41.8 ± 7.9	hex	298	42.26	25.94	—	—	—
$CoSeO_3$	128.0	577.4 ± 10.5	cryst	298	79.91	59.83	—	—	—
			liq	932	144.35	—	—	—	16.3

TABLE I—*continued*

Substance	$S_{298} \pm \delta S$ (J/deg mol)	$-\Delta H_{298}$ $\pm \delta$H (kJ/mol)	Phase	T K	$C_p = A + BT + C/T^2 + DT^2$ A	$B \times 10^3$ (J/deg mol)	$C \times 10^{-5}$	$D \times 10^6$	H_t (kJ/mol)
CoSi	42.7 ± 2.9	95.1 ± 7.9	cubic	298	49.16	12.09	−7.53	—	—
			liq	1733	87.36	—	—	—	69.2
$CoSi_2$	64.0 ± 5.0	98.7 ± 10.0	cubic	298	70.86	18.66	−9.92	—	—
			liq	1600	116.11	—	—	—	100.0
Co_2SiO_4	158.6 ± 4.2	1398.7 ± 4.6	orth	298	157.40	22.05	−26.69	—	—
			liq	1690	242.67	—	—	—	100.4
$CoTiO_3$	96.9 ± 3.3	1207.5 ± 3.8	trigon	298	123.47	9.71	−14.64	—	—
$CoWO_4$	126.4 ± 9.2	1142.2 ± 3.8	monocl	298	115.48	48.53	—	—	—
			β	986	122.38	41.92	—	—	1.9
Cr	23.6	0.0	bcc	298	21.76	8.98	−0.96	2.26	—
			liq	2130	39.33	—	—	—	17.2
Cr(g)	174.2	−397.5	gas	298	20.79	—	—	—	—
			gas	900	9.58	6.95	41.34	—	—
Cr_5B_3	137.7	248.9	tetrag	298	153.13	79.08	—	—	—
CrB	29.0	78.9orth	298	42.34	16.02	−10.04	—	—	
Cr_3B_4	91.8	281.2	orth	298	134.31	57.74	−39.12	—	—
CrB_2	32.9	125.5	hex	298	60.04	21.88	−18.62	—	—
$CrBr_2$	134.7 ± 7.1	302.1 ± 18.8	monocl	298	65.90	22.18	—	—	—
$Cr_{23}C_6$	612.1 ± 4.2	328.4 ± 37.7	cubic	298	683.25	209.20	−104.6	—	—
Cr_7C_3	201.0 ± 1.3	160.7 ± 16.7	hex	298	233.89	62.34	−38.07	—	—
Cr_3C_2	85.4 ± 0.4	85.4 ± 12.6	orth	298	123.26	25.90	−28.24	—	—
$CrCl_2$	115.3 ± 0.4	395.4 ± 10.5	orth	298	63.81	24.94	—	—	—
$CrCl_3$	124.7 ± 0.4	570.3 ± 10.9	monocl	298	79.50	41.21	—	—	—
CrF_2	85.8 ± 6.3	775.3 ± 11.7	monocl	298	61.63	19.50	−8.03	—	—
CrF_3	94.1 ± 0.4	1159.0 ± 8.4	hex	298	68.62	34.31	—	—	—
CrI_2	169.0	158.2 ± 5.9	monocl	298	66.94	22.59	—	—	—
CrI_3	199.6 ± 10.9	205.0 ± 6.7	hex	298	105.44	20.92	—	—	—
Cr_2N	65.3 ± 8.4	125.5 ± 12.6	hex	298	65.14	26.23	−6.15	—	—
CrN	37.7 ± 2.1	117.2 ± 8.4	cubic	298	41.17	16.32	—	—	—
Cr_2O_3	81.2 ± 1.3	1134.7 ± 7.5	rhomb	298	119.37	9.20	−15.65	—	—
CrO_3	73.2	587.0 ± 10.9	orth	298	71.76	87.86	−16.74	—	—
CrO_3(g)	270.7 ± 12.6	292.0	gas	298	82.55	—	−21.97	—	—
$Cr(CO)_6$	314.2 ± 6.3	1077.4 ± 4.2	orth	298	203.76	75.31	−19.87	—	—
CrO_2Cl_2(g)	329.7 ± 2.9	528.9 ± 11.7	gas	298	106.61	—	—	—	—

CrS	64.0 ± 6.3	155.6 ± 9.6	monocl	298	32.84	46.74	—	—	—
				450	51.64	4.90	—	—	0.3
$Cr_2(SO_4)_3$	258.8 ± 2.1	2931.3 ± 15.9		298	358.07	79.50	−89.75	—	—
Cr_3Si	85.8 ± 6.3	92.0 ± 16.7	cubic	298	82.22	42.38	−4.31	—	—
Cr_5Si_3	169.0 ± 12.6	211.3 ± 31.4	tetrag	298	198.57	49.29	25.61	—	—
				1300	586.85	−543.08	—	225.31	—
CrSi	43.7 ± 3.3	54.8 ± 8.4	cubic	298	52.01	8.74	−8.41	—	—
$CrSi_2$	58.6 ± 4.6	79.9 ± 12.6	hex	298	65.61	22.51	−7.74	—	—
			liq	1730	89.96	—	—	—	127.8
Cs	85.2	0.0	bcc	298	32.34	—	—	—	—
			liq	302	29.89	0.90	2.03	—	2.1
			liq	700	30.94	—	—	—	—
Cs(g)	175.5	−76.5	gas	298	20.79	—	—	—	—
Cs_2(g)	284.5	−113.8	gas	298	37.18	2.82	—	—	—
CsBr	113.0 ± 0.4	405.4 ± 0.8	cubic	298	50.38	8.54	—	—	—
			liq	911	77.40	—	—	—	23.6
CsCl	101.2 ± 0.2	442.7 ± 0.4	cubic	298	53.35	5.15	−2.09	—	—
				743	3.35	73.64	−3.77	—	3.8
			liq	918	57.99	17.91	—	—	20.3
CsCl(g)	256.0 ± 0.2	242.3 ± 4.2	gas	298	36.82	1.05	—	—	—
CsF	93.0	553.5 ± 1.3	cubic	298	45.81	18.83	—	—	—
			liq	976	74.06	—	—	—	21.8
CsF(g)	243.1 ± 0.2	361.1 ± 3.3	gas	298	37.34	0.61	−1.51	—	—
CsH	66.9 ± 8.4	54.0 ± 0.4	cubic	298	31.17	35.56	—	—	—
CsH(g)	215.1 ± 0.2	−116.9	gas	298	37.78	—	−5.61	—	—
CsI	122.2 ± 0.8	346.4 ± 8.4	cubic	298	29.54	43.35	8.08	—	—
			liq	905	72.38	—	—	—	25.5
CsI(g)	275.2 ± 0.2	153.1 ± 2.1	gas	298	38.91	—	−1.30	—	—
Cs_2O	146.9 ± 0.8	346.0 ± 2.9	hex	298	66.36	32.01	—	—	—
CsO_2	142.3 ± 12.6	286.2 ± 2.1	tetrag	298	72.38	30.96	—	—	—
$CsBO_2$	105.4 ± 0.6	976.8 ± 20.9	cubic	298	52.34	66.32	−2.33	—	—
Cs_2CO_3	204.0 ± 0.8	1136.4 ± 2.1	cryst	298	115.44	69.33	−10.96	—	—
$CsClO_4$	175.3 ± 0.8	437.2 ± 1.3	rhomb	298	30.54	259.41	—	—	—
			cubic	501	176.15	—	—	—	7.5
Cs_2CrO_4	228.6 ± 0.4	1429.3 ± 3.3	cryst	298	146.06	—	—	—	36.0
CsOH	98.7 ± 5.0	416.7 ± 1.3	α	298	52.97	64.02	—	—	—
			β	410	75.31	—	—	—	1.3
			gamma	493	82.42	—	—	—	6.1
			liq	616	81.59	—	—	—	4.6
CsOH(g)	254.7 ± 0.8	259.4 ± 12.6	gas	298	51.04	4.06	−2.22	—	—

TABLE I—*continued*

Substance	$S_{298} \pm \delta S$ (J/deg mol)	$-\Delta H_{298} \pm \delta H$ (kJ/mol)	Phase	T K	$C_p = A + BT + C/T^2 + DT^2$ A	$B \times 10^3$	$C \times 10^{-5}$ (J/deg mol)	$D \times 10^6$	H_t (kJ/mol)
Cs_2MoO_4	248.4	1514.6	rhomb	298	116.40	108.24	—	—	—
			hex	841	122.17	97.07	—	—	4.6
			liq	1230	210.04	—	—	—	31.8
$CsNO_2$	181.2 ± 4.2	369.4	cubic	298	91.76	—	—	—	10.9
$CsNO_3$	153.7 ± 0.4	505.8 ± 2.1	hex	298	62.76	110.46	—	—	—
			cubic	426	126.36	—	—	—	3.8
			liq	680	135.98	—	—	—	14.0
Cs_2SO_4	211.9 ± 0.4	1444.3 ± 1.3	orth	298	85.48	151.04	—	—	—
			hex	940	36.40	158.16	—	—	2.5
			liq	1286	207.11	—	—	—	36.4
Cs_2SiO_3	175.7 ± 10.5	1558.1 ± 15.1	cryst	298	113.72	74.64	−12.22	—	—
			liq	1100	194.56	—	—	—	39.7
Cs_2UO_4	219.7 ± 0.8	1926.3 ± 4.6		298	164.85	17.03	15.27	—	—
Cu	33.1	0.0	fcc	298	30.29	−10.71	−3.22	9.47	—
			liq	1358	32.84	—	—	—	13.3
Cu(g)	166.3	−337.4	gas	298	20.79	—	—	—	—
			gas	1300	23.97	−4.60	—	1.67	—
CuBr	96.1 ± 0.8	104.6 ± 1.3	cubic	298	56.74	9.37	−4.18	—	—
			hex	665	58.58	—	—	—	7.5
			cubic	743	58.99	—	—	—	1.5
CuCl	87.0 ± 2.9	137.2 ± 2.1	cubic	298	38.28	34.98	—	—	—
			liq	700	66.94	—	—	—	10.3
$CuCl_2$	108.1 ± 0.4	215.5 ± 4.2	monocl	298	67.03	17.57	—	—	—
CuF_2	77.4 ± 0.2	538.9 ± 8.4	monocl	298	72.01	19.96	−11.38	—	—
			liq	1100	100.42	—	—	—	55.2
CuF_2(g)	267.0	266.9	gas	298	55.15	2.28	−6.74	—	—
CuI	96.7 ± 1.3	68.0 ± 2.1	cubic	298	31.80	74.06	—	—	—
			hex	650	66.11	—	—	—	8.2
			cubic	690	68.62	—	—	—	1.9
			liq	873	90.79	—	—	—	10.9
CuI(g)	255.6	−153.1	gas	298	37.45	—	−0.92	—	—
CuCN	90.0 ± 0.4	−95.0 ± 2.1	cryst	298	59.66	24.48	−4.35	—	—
Cu_2O	92.9 ± 0.3	173.2 ± 2.1	cubic	298	58.20	23.97	−1.59	—	—
			liq	1515	100.42	—	—	—	64.0

CuO	42.6 ± 0.4	161.9 ± 1.7	monocl	298	46.44	11.55	−7.11	2.59	—
$CuAl_2O_4$	99.6 ± 2.9	1946.8 ± 4.6	cubic	298	155.64	34.10	−33.89	—	—
$CuCO_3$	87.9	596.2 ± 9.2		298	92.05	38.91	−17.99	—	—
$CuCr_2O_4$	130.5 ± 9.2	1293.3 ± 7.5	tetrag	298	166.31	20.92	−21.76	—	—
$CuFeO_2$	88.9 ± 1.0	513.0 ± 4.2	hex	298	97.99	7.53	−17.99	—	—
			β	1090	91.50	15.06	—	—	0.4
			liq	1470	126.69	—	—	—	64.4
$CuFe_2O_4$	146.7	969.0	tetrag	298	139.62	117.78	−23.43	—	—
			cubic	675	227.19	—	—	—	0.8
			cubic	795	166.02	41.00	—	—	—
$Cu(OH)_2$	87.0 ± 8.4	443.5 ± 5.4	orth	298	86.99	23.26	−5.40	—	—
Cu_3P	119.2	151.0 ± 12.6	hex	298	77.82	33.47	—	—	—
Cu_2S	120.9 ± 1.7	79.5 ± 1.7	orth	298	52.84	78.74	—	—	—
			hex	376	112.05	−30.75	—	—	3.8
			cubic	717	84.64	—	—	—	1.2
			liq	1402	89.12	—	—	—	9.6
CuS	66.5 ± 2.1	52.3 ± 4.2	hex	298	44.35	11.05	—	—	—
Cu_5FeS_4	362.3	380.3	tetrag	298	208.20	146.77	−5.65	—	—
			cubic	485	−143.55	1033.45	—	—	6.0
			cubic	540	189.03	117.49	−260.91	—	—
$CuFeS_2$	125.0	190.4	tetrag	298	86.99	53.56	−5.61	—	—
			cubic	830	−1441.97	1844.98	—	—	10.0
			cubic	930	172.46	—	—	—	—
$CuSO_4$	109.2 ± 0.4	771.4 ± 1.3	rhomb	298	73.43	152.84	−12.30	−71.59	—
$CuO.CuSO_4$	157.3	927.6	cryst	298	170.83	45.35	−39.25	—	—
$CuSO_4.H_2O$	145.1 ± 5.0	1082.8 ± 2.5	monocl	298	132.30	70.88	−18.62	—	—
$CuSO_4.3H_2O$	222.2 ± 5.0	1681.1 ± 3.3	monocl	298	204.30	71.80	−18.41	—	—
$CuSO_4.5H_2O$	301.2 ± 2.1	2276.5 ± 3.8	tricli	298	280.96	70.88	−18.58	—	—
Cu_2Sb	126.4	11.7 ± 1.7	cryst	298	68.53	27.61	—	—	—
Cu_2Se	129.7 ± 4.2	65.3 ± 6.3	tetrag	298	58.58	77.40	—	—	—
			cubic	395	84.10	—	—	—	6.8
$CuSe$	78.2 ± 6.3	41.8 ± 4.2	hex	298	54.81	—	—	—	—
			orth	326	62.76	—	—	—	1.4
$CuSe(g)$	264.6 ± 4.2	−309.6 ± 20.9	gas	298	37.40	—	−1.13	—	—
Cu_2Te	134.7 ± 6.3	41.8 ± 12.6	hex	298	59.83	53.56	—	—	—
			cubic	433	60.46	53.56	—	—	0.2
			delta	531	112.97	—	—	—	1.9
				590	133.89	—	—	—	1.0
				633	109.29	—	—	—	2.4
				841	87.86	—	—	—	2.0

TABLE I—*continued*

Substance	$S_{298} \pm \delta S$ (J/deg mol)	$-\Delta H_{298} \pm \delta H$ (kJ/mol)	Phase	T K	$C_p = A + BT + C/T^2 + DT^2$				H_t (kJ/mol)
					A	$B \times 10^3$ (J/deg mol)	$C \times 10^{-5}$ (J/deg mol)	$D \times 10^6$	
Dy	74.9	0.0	hcp	298	35.35	−21.88	−2.13	18.70	—
			bcc	1654	28.03	—	—	—	4.2
			liq	1685	49.92	—	—	—	11.0
Dy(g)	195.8	−290.4	gas	298	21.40	−2.12	−0.14	2.04	—
			gas	700	21.18	−1.99	—	2.29	—
			gas	1400	11.64	9.16	21.55	−1.37	—
$DyCl_3$	147.7 ± 11.3	995.8 ± 9.2	monocl	298	94.56	17.99	−1.42	—	—
			liq	924	144.77	—	—	—	25.5
$DyCl_3$(g)	383.3 ± 6.7	714.2 ± 11.3	gas	298	79.91	—	—	—	—
DyF_3	115.1 ± 10.5	1719.6 ± 7.5	orth	298	89.62	24.56	−0.38	—	—
			liq	1430	156.90	—	—	—	58.6
Dy_2O_3	149.8 ± 12.6	1862.7 ± 3.8	cubic	298	122.59	13.22	−8.49	—	—
			monocl	1590	148.11	—	—	—	0.9
			hex	2343	155.23	—	—	—	8.4
Er	73.1	0.0	hcp	298	28.37	−2.05	—	5.73	—
			bcc	1750	42.68	—	—	—	—
			liq	1795	38.70	—	—	—	19.9
Er(g)	193.9	−317.1	gas	298	20.79	—	—	—	—
			gas	500	22.12	−3.28	−0.64	2.27	—
			gas	1500	8.76	9.49	44.33	−1.20	—
$ErCl_3$	146.9 ± 4.2	992.4 ± 2.9	monocl	298	95.56	17.57	−1.05	—	—
			liq	1050	148.53	—	—	—	32.6
ErF_3	100.8 ± 1.3	1722.6 ± 6.3	orth	298	93.09	19.96	1.67	—	—
			hex	1390	135.02	—	—	—	29.5
			liq	1419	139.12	—	—	—	27.5
Er_2O_3	154.4 ± 0.6	1897.9 ± 4.2	cubic	298	115.27	29.29	−13.81	—	—
Eu	77.8	0.0	cubic	298	33.81	−19.33	−2.22	23.85	—
			liq	765	33.81	−19.33	−2.22	23.85	0.2
			liq	1096	38.07	—	—	—	9.2
Eu(g)	188.7	−175.3	gas	298	20.79	—	—	—	—
			gas	1200	22.42	−2.52	—	0.96	—
$EuCl_3$	125.5 ± 12.6	939.3 ± 4.6	hex	298	90.50	26.15	—	—	—
			liq	897	142.26	—	—	—	33.1
EuF_3	108.8	1619.2	orth	298	80.75	45.61	8.37	—	—
			hex	920	150.62	—	—	—	6.7

EuN	63.6 ± 0.8	217.6 ± 25.1	cubic	298	51.88	5.44	—	—	—
EuO	81.6 ± 4.2	589.9 ± 5.4	cubic	298	54.39	4.18	−6.15	—	—
Eu_2O_3	146.4	1662.7 ± 3.8	cubic	298	123.85	27.11	−8.70	—	—
			monocl	895	129.96	17.41	—	—	0.5
EuS	95.4 ± 0.8	451.9 ± 16.7	cubic	298	48.74	4.81	—	—	—
EuS(g)	272.0	−88.3 ± 16.7	gas	298	37.24	0.07	−1.91	—	—
F_2(g)	202.7	0.0	gas	298	35.90	1.34	−4.48	—	—
F(g)	158.6	−79.4	gas	298	21.69	−0.44	1.16	—	—
F_2O(g)	247.3 ± 0.4	18.4	gas	298	50.42	8.66	−8.59	−2.54	—
Fe	27.3	0.0	bcc	298	28.18	−7.32	−2.90	25.04	—
			bcc	800	−263.45	255.81	619.23	—	—
			bcc	1000	−641.91	696.34	—	—	—
			bcc	1042	1946.25	−1787.50	—	—	—
			bcc	1060	−561.95	334.13	2912.11	—	—
			fcc	1184	23.99	8.36	—	—	0.9
			bcc	1665	24.64	9.90	—	—	0.8
			liq	1809	46.02	—	—	—	13.8
Fe(g)	180.8	−415.9	gas	298	26.09	−1.40	—	—	—
			gas	400	30.93	−14.31	−0.99	5.95	—
			gas	1100	22.07	0.13	—	—	—
			gas	1600	24.83	−2.68	−26.61	1.11	—
			gas	3135	2.33	6.53	530.95	−0.14	—
Fe_2B	51.7	102.5	tetrag	298	78.87	14.14	−14.64	—	—
FeB	31.0	72.8	orth	298	49.96	10.00	−10.59	—	—
			liq	1863	89.96	—	—	—	62.6
$FeBr_2$	140.6 ± 2.5	248.9 ± 1.7	hex	298	73.60	22.26	—	—	—
			cubic	650	92.05	—	—	—	0.4
			liq	964	106.69	—	—	—	51.9
$FeBr_2$(g)	337.2	41.4	gas	298	59.87	3.14	—	—	—
Fe_3C	104.6 ± 4.2	−25.1 ± 4.2	cement	298	82.01	83.68	—	—	—
				480	107.32	12.55	—	—	0.8
$FeCl_2$	76.1 ± 2.5	341.8 ± 0.8	hex	298	79.24	8.70	−4.77	—	—
			liq	950	102.17	—	—	—	43.0
$FeCl_2$(g)	299.2 ± 4.2	141.0 ± 2.1	gas	298	59.50	3.31	−2.59	—	—
Fe_2Cl_4(g)	464.4 ± 12.6	431.4 ± 4.2	gas	298	130.21	3.26	−4.73	—	—
$FeCl_3$	142.3 ± 1.7	399.4 ± 1.0	hex	298	62.34	115.06	—	—	—
			liq	580	133.89	—	—	—	43.1
$FeCl_3$(g)	344.1	253.6 ± 4.2	gas	298	82.97	0.08	−4.73	—	—
Fe_2Cl_6(g)	537.0	654.8 ± 8.4	gas	298	182.67	0.13	−8.08	—	—
FeF_2	87.0 ± 0.4	707.1 ± 8.4	tetrag	298	74.64	8.03	−8.16	—	—
			liq	1373	98.32	—	—	—	51.9

TABLE I—*continued*

Substance	$S_{298} \pm \delta S$ (J/deg mol)	$-\Delta H_{298} \pm \delta H$ (kJ/mol)	Phase	T K	$C_p = A + BT + C/T^2 + DT^2$				H_t (kJ/mol)
					A	$B \times 10^3$ (J/deg mol)	$C \times 10^{-5}$	$D \times 10^6$	
Fe_4N	155.6 ± 7.5	11.1 ± 4.2	cubic	298	110.79	34.14	—	—	—
$Fe_{0.945}O$	60.1 ± 0.8	263.0 ± 2.5	cubic	298	48.79	8.37	−2.80	—	—
Fe_2O_3	87.4 ± 1.3	823.4 ± 3.3	hemat	298	98.28	77.82	−14.85	—	—
				950	150.60	—	—	—	—
				1050	132.67	7.36	—	—	—
Fe_3O_4	150.9 ± 1.7	1108.8 ± 4.2	magnet	298	91.55	202.00	—	—	—
				900	213.40	—	—	—	—
$FeAl_2O_4$	106.3 ± 0.8	1966.5 ± 2.5	cubic	298	155.31	26.15	−35.23	—	—
$FeCO_3$	92.9 ± 2.5	740.6 ± 3.3	sider	298	48.66	112.09	—	—	—
$Fe(CO)_5$(g)	445.2 ± 3.3	723.8 ± 8.4	gas	298	72.22	39.33	−37.24	—	—
FeOCl	76.6 ± 6.3	408.4 ± 8.4	rhomb	298	69.04	26.82	—	—	—
$FeCr_2O_4$	146.9 ± 1.7	1446.4 ± 5.0	spinel	298	163.01	22.34	−31.92	—	—
FeOOH	60.4 ± 0.8	558.1 ± 2.9	orth	298	49.37	83.68	—	—	—
$FeCl_2.2H_2O$	165.1	956.5 ± 6.3	monocl	298	217.57	—	—	—	—
$FeCl_2.4H_2O$	245.6	1552.3 ± 8.4	monocl	298	301.25	—	—	—	—
Fe_3Mo_2	146.4 ± 8.4	4.2 ± 8.4	homb	298	101.67	49.41	6.07	—	—
$FeMoO_4$	129.3 ± 1.7	1060.2 ± 12.1	monocl	298	125.52	33.47	−15.90	—	—
Fe_3P	101.6	164.0	tetrag	298	117.15	12.97	−17.78	—	—
Fe_2P	72.4	160.2	hex	298	76.78	17.03	−6.07	—	—
$FePO_4.2H_2O$	171.3 ± 1.3	1887.8 ± 9.2	streng	298	180.54	—	—	—	—
FeS	60.3 ± 0.4	100.4 ± 1.7	hex	298	−0.50	170.71	—	—	—
				411	72.80	—	—	—	2.4
				598	51.04	9.96	—	—	0.5
			liq	1461	71.13	—	—	—	32.3
FeS_2	52.9 ± 0.4	171.5 ± 2.1	pyrite	298	68.95	14.10	−9.87	—	—
$FeSO_4$	107.5 ± 1.3	927.6 ± 10.5	orth	298	122.01	37.82	−29.29	—	—
$Fe_2(SO_4)_3$	282.8 ± 2.1	2580.3 ± 5.0	hex	298	309.62	111.59	−64.22	—	—
$FeSe_{0.96}$	69.2 ± 4.2	66.9 ± 4.2	tetrag	298	54.31	21.13	−5.19	—	—
			hex	731	67.86	−12.07	—	—	9.6
$Fe_2(SeO_3)_3$	272.4 ± 18.8	1706.2 ± 14.6	cryst	298	226.98	123.43	−31.17	—	—
FeSi	37.7 ± 3.3	79.5 ± 3.3	cubic	298	44.56	15.15	−1.13	—	—
Fe_2SiO_4	145.2 ± 1.7	1471.1 ± 7.9	fayal	298	152.76	39.16	−28.03	—	—
			liq	1490	240.58	—	—	—	92.0

FeTi	52.7 ± 8.8	40.6 ± 1.7	cubic	298	53.01	9.62	−8.12	—	—
Fe_2Ti	74.5	87.4	hex	298	80.96	13.81	−13.18	—	—
Fe_2TiO_4	173.2	1501.2 ± 11.7	spinel	298	139.49	63.09	−14.23	—	—
$FeTiO_3$	105.9 ± 1.3	1237.6 ± 5.9	trigon	298	116.61	18.24	−20.04	—	—
			liq	1740	199.16	—	—	—	90.8
$FeWO_4$	131.6 ± 1.3	1182.8 ± 5.4	monocl	298	132.42	29.75	−23.93	—	—
Ga	40.8	0.0	orth	298	26.78	—	—	—	—
			liq	303	26.78	—	—	—	5.6
Ga(g)	168.9	−272.0	gas	298	31.76	−9.62	−3.05	2.18	—
GaAs	64.2 ± 0.4	74.1 ± 6.3	cubic	298	45.19	6.07	—	—	—
			liq	1511	58.99	—	—	—	87.9
$GaBr_3$	179.9 ± 11.7	386.6 ± 2.5	cryst	298	78.58	77.40	—	—	—
			liq	396	125.52	—	—	—	11.7
GaCl(g)	240.0 ± 0.2	81.8 ± 15.1	gas	298	37.99	—	−2.01	—	—
$GaCl_3$	135.1 ± 13.8	524.7 ± 4.6	cryst	298	118.41	—	—	—	—
			liq	351	128.03	—	—	—	11.5
GaI_3	203.8 ± 16.7	239.3 ± 12.1	orth	298	117.15	—	—	—	—
			liq	486	128.45	—	—	—	22.2
GaN	29.7 ± 7.5	109.6 ± 9.2	hex	298	38.07	9.00	—	—	—
Ga_2O_3	84.9 ± 0.4	1089.1 ± 4.6	monocl	298	112.55	15.48	−21.67	—	—
GaP	52.3 ± 0.8	102.5 ± 8.4	cubic	298	41.84	6.82	—	—	—
$Ga_2S(g)$	290.0 ± 8.4	−20.9 ± 33.5	gas	298	56.00	1.15	−9.25	—	—
GaS	57.7 ± 6.3	209.2 ± 18.8	hex	298	41.34	15.69	—	—	—
Ga_2S_3	142.3 ± 16.7	516.3 ± 12.6	cubic	298	90.50	47.28	—	—	—
GaSb	76.1 ± 0.8	43.9 ± 1.7	cubic	298	44.35	14.23	—	—	—
GaSe	70.3 ± 2.1	159.0 ± 12.6	hex	298	44.64	12.97	—	—	—
Ga_2Se_3	179.9 ± 16.7	408.8 ± 12.6	cubic	298	105.73	35.31	—	—	—
GaTe	85.4	125.5 ± 12.6	monocl	298	45.27	13.97	—	—	—
Ga_2Te_3	213.4 ± 16.7	274.9 ± 16.7	cubic	298	95.06	18.62	—	—	—
Gd	67.9	0.0	hcp	298	6.69	32.64	18.41	−8.37	—
			bcc	1533	28.45	—	—	—	3.9
			liq	1586	37.15	—	—	—	10.0
Gd(g)	194.2	−397.5	gas	298	30.91	−8.35	−0.95	2.15	—
			gas	1000	23.73	−4.00	23.20	2.55	—
			gas	2000	7.26	10.66	67.98	−0.94	—
$GdBr_3$	190.0	828.9	hex	298	101.96	5.77	−5.98	—	—
			liq	1058	139.33	—	—	—	38.2
$GdCl_3$	151.5 ± 2.5	1008.3 ± 3.3	orth	298	93.09	25.10	−2.47	—	—
			liq	875	139.33	—	—	—	40.7
GdF_3	114.8 ± 1.3	1699.1 ± 4.6	orth	298	102.01	6.49	−13.81	—	—
			hex	1348	130.83	—	—	—	6.0
			liq	1505	127.82	—	—	—	52.4

TABLE I—*continued*

Substance	$S_{298} \pm \delta S$ (J/deg mol)	$-\Delta H_{298} \pm \delta H$ (kJ/mol)	Phase	T K	$C_p = A + BT + C/T^2 + DT^2$ A	$B \times 10^3$ (J/deg mol)	$C \times 10^{-5}$	$D \times 10^6$	H_t (kJ/mol)
GdI_3	226.4	594.1	hex	298	106.61	0.88	−7.32	—	—
			β	1013	127.82	—	—	—	0.9
			liq	1204	155.85	—	—	—	54.0
Gd_2O_3	150.6 ± 0.4	1826.7 ± 3.3	cubic	298	119.20	12.95	−15.82	—	—
			monocl	1550	114.50	14.47	−10.84	—	1.7
GdOCl	95.4 ± 9.2	983.7 ± 3.8	tetrag	298	66.94	16.40	—	—	—
Ge	31.1	0.0	cubic	298	21.55	5.86	—	—	—
			liq	1210	27.61	—	—	—	36.9
Ge(g)	167.8	−372.0	gas	298	39.90	−15.95	−3.93	—	—
$GeBr_4$(g)	396.2 ± 1.7	298.7 ± 9.2	gas	298	107.36	0.67	−5.10	—	—
$GeCl_4$(g)	347.7 ± 1.3	494.8 ± 10.5	gas	298	106.86	0.92	−10.00	—	—
GeF_4(g)	301.8 ± 1.0	1190.2 ± 0.8	gas	298	97.61	8.37	−16.36	—	—
GeH_4(g)	217.1 ± 0.8	−90.8 ± 2.1	gas	298	62.55	22.09	−21.63	—	—
GeI_4(g)	428.9 ± 2.1	37.7 ± 10.5	gas	298	109.20	1.00	−4.60	—	—
GeO(g)	223.8 ± 0.4	30.7	gas	298	37.03	—	−5.65	—	—
GeO_2	39.7 ± 1.3	580.0 ± 1.5	tetrag	298	66.61	11.59	−17.74	—	—
			hex	1322	68.91	9.83	−17.70	—	21.1
GeP	61.1 ± 8.4	27.2 ± 10.5	cryst	298	45.40	11.30	−5.23	—	—
GeS	66.0 ± 0.8	76.1 ± 4.6	orth	298	41.80	20.13	—	—	—
			liq	938	60.67	—	—	—	20.9
GeS(g)	235.6 ± 0.4	−99.4 ± 5.9	gas	298	36.74	0.42	−2.85	—	—
GeS_2	87.4 ± 2.1	156.9 ± 12.6	orth	298	56.44	31.05	—	—	—
GeSe	78.2 ± 0.8	69.0 ± 10.5	orth	298	45.10	16.40	—	—	—
GeSe(g)	247.7 ± 2.1	−105.4 ± 18.8	gas	298	36.99	0.17	−1.76	—	—
$GeSe_2$	112.5 ± 3.3	113.0 ± 20.9	hex	298	62.89	27.74	—	—	—
GeTe	88.9 ± 1.0	48.5 ± 10.5	trig	298	48.12	12.55	—	—	—
			liq	997	60.67	—	—	—	46.9
GeTe(g)	255.6 ± 1.7	−145.6 ± 16.7	gas	298	37.36	—	−1.26	—	—
H(g)	114.6 ± 0.0	−218.0 ± 0.0	gas	298	20.79	—	—	—	—
H_2(g)	130.6 ± 0.0	0.0	gas	298	27.37	3.33	—	—	—
HBr(g)	198.6	36.3	gas	298	25.10	8.45	1.42	−1.46	—
HCl(g)	186.8	92.3	gas	298	24.80	7.99	1.88	−1.30	—
HF(g)	173.7 ± 0.0	273.3 ± 0.6	gas	298	26.90	3.43	1.09	—	—
HI(g)	206.5 ± 0.2	−26.4 ± 0.8	gas	298	25.90	8.49	0.66	−1.54	—
HCN(g)	201.7 ± 0.2	−135.1 ± 8.4	gas	298	46.28	5.02	−10.67	—	—

$H_2O(l)$	69.9	285.8	liq	298	75.44	—	—	—	—
$H_2O(g)$	188.7	241.8	gas	298	30.00	10.71	0.33	—	—
$H_2O_2(l)$	109.5 ± 0.5	187.8 ± 0.4	liq	298	89.33	—	—	—	—
$H_2O_2(g)$	232.9 ± 0.6	136.4 ± 0.8	gas	298	52.30	11.88	−11.88	—	—
HBO_2	49.0 ± 6.3	803.3 ± 1.3	cubic	298	52.09	34.48	6.95	—	—
$HBO_2(g)$	240.2 ± 1.7	564.0 ± 7.5	gas	298	51.76	12.01	−11.72	—	—
H_3BO_3	90.0 ± 0.8	1094.8 ± 1.3	tricli	298	81.38	—	—	—	—
$H_3BO_3(g)$	303.3 ± 2.5	1012.5 ± 20.9	gas	298	77.40	—	—	—	—
$HOF(g)$	226.6 ± 0.2	98.3 ± 4.2	gas	298	41.59	6.55	−6.80	—	—
$HNO_3(g)$	266.5	134.7	gas	298	66.11	32.05	−20.50	−6.86	—
$HCNO(g)$	238.1 ± 0.4	101.7 ± 8.4	gas	298	57.28	10.63	−13.81	—	—
H_3PO_4	110.5 ± 0.4	1266.9 ± 2.5	monocl	298	106.61	—	—	—	—
			liq	315	55.23	301.25	—	—	13.4
$H_2S(g)$	205.6 ± 0.4	20.5 ± 0.8	gas	298	28.54	20.61	−0.37	−3.70	—
$H_2S_2(l)$	200.8 ± 12.6	18.0 ± 0.8	liq	298	92.26	—	—	—	—
$H_2S_2(g)$	266.4	−15.7 ± 0.8	gas	298	51.38	16.19	−4.18	—	—
$H_2SO_4(l)$	156.9 ± 0.2	814.0 ± 0.8	liq	298	58.16	193.72	—	—	—
$H_2SO_4(g)$	298.7 ± 2.1	735.1 ± 8.4	gas	298	126.23	10.43	−43.46	—	—
$H_2Se(g)$	218.8 ± 0.4	−29.3 ± 1.3	gas	298	31.76	14.64	−1.30	—	—
$H_2Te(g)$	228.9 ± 2.1	−99.6 ± 1.3	gas	298	35.48	12.05	−3.10	—	—
Hf	43.5	0.0	hex	298	23.44	7.63	—	—	—
			bcc	2016	10.29	10.77	—	—	5.9
			liq	2504	33.47	—	—	—	27.2
Hf(g)	186.8	−620.1	gas	298	13.24	15.78	2.85	−3.47	—
			gas	2500	16.23	3.90	315.93	—	—
HfB_2	42.9 ± 0.4	328.9 ± 8.8	hex	298	73.35	7.82	−23.01	—	—
$HfBr_4$	240.6 ± 12.6	766.1 ± 2.1	cubic	298	108.70	63.43	—	—	—
$HfBr_4(g)$	427.6 ± 12.6	656.9 ± 16.7	gas	298	108.99	0.13	−4.85	—	—
HfC	39.5 ± 0.4	225.9 ± 6.3	cubic	298	42.34	12.13	−7.36	−2.43	—
$HfCl_4$	190.8 ± 2.5	990.4 ± 2.1	cubic	298	131.67	—	−9.96	—	—
$HfCl_4(g)$	377.2 ± 6.3	882.8 ± 6.3	gas	298	108.44	0.21	−8.54	—	—
HfF_4	113.0 ± 10.5	1930.5 ± 3.8	tetrag	298	133.47	3.14	−37.66	—	—
$HfF_4(g)$	327.6 ± 12.6	1694.5 ± 20.9	gas	298	120.25	—	−27.61	—	—
HfI_4	269.9 ± 12.6	493.7 ± 20.9	cubic	298	135.02	31.21	—	—	—
HfN	45.2 ± 1.3	373.6 ± 2.5	cubic	298	45.77	9.32	−6.69	—	—
HfO_2	59.4 ± 0.4	1117.5 ± 1.7	monocl	298	72.11	9.05	−12.94	—	—
			tetrag	1973	108.78	—	—	—	9.3
Hg	75.9	0.0	liq	298	30.38	−11.46	—	10.15	—
Hg(g)	174.8	−61.4	gas	298	27.66	—	—	—	—
HgBr	109.4	102.1	tetr	298	49.75	11.76	—	—	—
$HgBr_2$	170.3	169.5 ± 1.3	orth	298	66.59	29.29	—	—	—
			liq	428	102.09	—	—	—	17.9

TABLE I—*continued*

Substance	$S_{298} \pm \delta S$ (J/deg mol)	$-\Delta H_{298}$ $\pm \delta H$ (kJ/mol)	Phase	T K	$C_p = A + BT + C/T^2 + DT^2$				H_t (kJ/mol)
					A	$B \times 10^3$ (J/deg mol)	$C \times 10^{-5}$	$D \times 10^6$	
HgCl	95.8 ± 0.8	132.6 ± 0.8	tetrag	298	49.37	11.51	−1.80	—	—
$HgCl_2$	144.5 ± 0.4	230.1 ± 2.3	orth	298	70.00	20.29	−1.88	—	—
			liq	550	102.09	—	—	—	19.2
$HgCl_2$(g)	288.7 ± 1.7	149.0 ± 2.7	gas	298	62.13	0.13	−3.64	—	—
HgF	80.3 ± 8.4	242.7 ± 10.5	tetrag	298	50.00	10.92	−2.76	—	—
HgF_2	116.3	422.6	cubic	298	68.62	20.92	—	—	—
			liq	918	102.09	—	—	—	23.0
HgI	120.6	59.5 ± 0.8	tetrag	298	53.26	9.25	−2.76	—	—
			liq	563	68.20	—	—	—	13.6
HgI_2	170.7	105.4 ± 1.7	tetrag	298	77.40	—	—	—	—
			orth	403	84.52	—	—	—	2.7
			liq	525	104.60	—	—	—	18.8
HgO	70.2 ± 0.4	90.8 ± 0.8	orth	298	35.56	30.12	—	—	—
HgS	82.4 ± 2.1	53.3 ± 4.2	hex	298	43.76	15.56	—	—	—
			cubic	618	44.02	15.19	—	—	4.0
Hg_2SO_4	200.7 ± 0.4	743.1 ± 10.5	monocl	298	133.05	—	—	—	—
$HgSO_4$	140.2	704.2 ± 10.5	orth	298	58.58	146.44	—	—	—
HgSe	100.8 ± 3.3	43.5	cubic	298	48.95	15.48	—	—	—
$HgSeO_3$	162.3	365.3	cryst	298	83.68	61.92	—	—	—
HgTe	111.5 ± 1.7	32.2 ± 5.0	cubic	298	52.09	9.08	—	—	—
Ho	75.0	0.0	hcp	298	35.46	−22.07	−2.96	18.83	—
			bcc	1713	28.03	—	—	—	4.7
			liq	1745	43.93	—	—	—	12.1
Ho(g)	195.5	−300.8	gas	298	20.79	—	—	—	—
			gas	500	22.05	−2.95	−0.67	1.95	—
			gas	1500	7.88	9.72	54.75	−1.30	—
$HoCl_3$	147.7 ± 14.6	1006.3 ± 2.9	monocl	298	95.56	12.97	−0.96	—	—
			liq	993	148.66	—	—	—	30.5
HoF_3	110.9 ± 12.6	1714.2 ± 6.3	orth	298	106.86	0.50	−21.09	—	—
			orth	1343	163.34	−16.44	−239.07	—	—
			liq	1416	96.02	—	—	—	56.3
Ho_2O_3	158.2 ± 0.4	1881.1 ± 5.4	cubic	298	125.85	6.95	−11.51	—	—
I_2	116.1	0.0	orth	298	−50.65	246.91	27.97	—	—
			liq	387	80.67	—	—	—	15.8

$I_2(g)$	260.6	−62.4	gas	298	37.40	0.57	−0.62	—	—
I(g)	180.7	−106.8	gas	298	20.39	0.40	0.28	—	—
In	57.8	0.0		298	24.31	10.46	—	—	—
			liq	430	30.29	−1.38	—	—	3.3
In(g)	173.7	−240.4	gas	298	13.10	17.24	2.55	−2.97	—
				800	36.40	−5.23	−44.39	—	—
				1500	33.01	−5.15	—	0.59	—
				1900	23.81	—	85.23	−0.21	—
InAs	75.7 ± 0.8	57.7 ± 3.3	cubic	298	47.07	7.53	—	—	—
			liq	1215	59.83	—	—	—	77.0
InBr	112.1 ± 4.2	20.5 ± 8.4	orth	298	43.51	25.10	—	—	—
			liq	558	60.67	—	—	—	24.3
InBr(g)	259.5 ± 0.2	37.2	gas	298	37.57	0.42	−0.73	—	—
$InBr_3$	175.7 ± 12.6	410.9 ± 8.4	cryst	298	82.01	54.39	—	—	—
InCl	95.0 ± 6.3	186.2 ± 8.4	cubic	298	35.15	41.84	—	—	—
			β	393	58.58	—	—	—	6.9
			liq	498	62.76	—	—	—	9.2
$InCl_2$	122.2 ± 6.3	362.8 ± 16.7	orth	298	58.58	50.21	—	—	—
$InCl_3$	141.0 ± 8.4	537.2 ± 8.4	monocl	298	78.66	55.65	—	—	—
InI	123.8 ± 6.3	115.9 ± 8.4	orth	298	48.12	12.55	—	—	—
			liq	638	60.67	—	—	—	22.4
InI_3	203.3 ± 10.5	234.7 ± 8.4	monocl	298	164.01	—	—	—	—
			liq	480	135.98	—	—	—	20.1
InN	43.5 ± 5.9	138.1 ± 10.0	hex	298	38.07	12.13	—	—	—
In_2O_3	107.9 ± 3.3	925.9 ± 1.7	cubic	298	121.34	13.39	−30.12	—	—
InP	59.7 ± 0.4	75.3 ± 8.4	cubic	298	41.00	14.64	—	—	—
			β	910	55.23	—	—	—	0.4
			liq	1328	58.58	—	—	—	62.8
InS	71.0 ± 0.8	133.9 ± 10.5	orth	298	42.51	18.83	—	—	—
			liq	965	60.67	—	—	—	36.0
In_2S_3	163.6 ± 2.5	355.6 ± 20.9	cubic	298	128.95	3.26	−10.63	—	—
			tetrag	660	97.78	55.40	—	—	1.1
			gamma	1100	159.41	—	—	—	4.0
$In_2(SO_4)_3$	302.1 ± 2.1	2725.5 ± 3.3	hex	298	200.20	251.04	—	—	—
InSb	87.1 ± 0.8	30.5 ± 0.8	cubic	298	44.77	15.06	—	—	—
			liq	798	61.92	—	—	—	47.7
InSe	81.6 ± 4.2	118.0 ± 12.6	rhomb	298	45.44	16.32	—	—	—
In_2Se_3	201.3 ± 16.7	326.4 ± 16.7	hex	298	59.91	270.50	—	—	—
			hex	470	165.27	—	—	—	1.4
In_2Te	156.9 ± 10.5	79.9 ± 2.5	orth	298	56.48	36.82	—	—	—
InTe	105.7 ± 1.7	72.0 ± 1.7	tetrag	298	41.97	19.37	—	—	—
			liq	965	60.67	—	—	—	3.7

TABLE I—*continued*

Substance	$S_{298} \pm \delta S$ (J/deg mol)	$-\Delta H_{298}$ $\pm \delta H$ (kJ/mol)	Phase	T K	$C_p = A + BT + C/T^2 + DT^2$				H_t (kJ/mol)
					A	$B \times 10^3$ (J/deg mol)	$C \times 10^{-5}$	$D \times 10^6$	
In_2Te_3	234.3 ± 16.7	191.6 ± 4.2	cubic	298	110.88	41.84	—	—	—
			cubic	890	148.53	—	—	—	2.0
			liq	944	154.81	—	—	—	81.6
Ir	35.5	0.0	fcc	298	22.89	7.03	—	—	—
			liq	2720	41.84	—	—	—	26.1
Ir(g)	193.5	−669.4	gas	298	15.42	9.56	2.48	−1.54	—
$IrCl_3$	114.6	245.2	cryst	298	84.94	18.83	−4.18	—	—
IrF_6(g)	357.7 ± 1.7	543.9 ± 83.7	gas	298	120.92	—	—	—	—
IrO_2	51.0	249.4	tetrag	298	61.88	20.42	−10.96	—	—
Ir_2S_3	51.0	249.4	orth	298	110.29	32.97	−9.62	—	—
IrS_2	69.0	133.1	orth	298	68.58	15.77	−6.57	—	—
K	64.7	0.0	bcc	298	25.27	13.05	—	—	—
			liq	337	37.18	−19.12	—	12.30	2.4
K(g)	160.2	−89.0	gas	298	20.79	—	—	—	—
K_2(g)	249.6	−127.1	gas	298	37.41	2.06	−0.12	—	—
KBr	95.9 ± 0.4	393.6 ± 0.4	cubic	298	69.16	−45.56	−6.49	45.02	—
			liq	1007	69.87	—	—	—	25.5
KBr(g)	250.4 ± 0.2	179.9 ± 2.1	gas	298	37.40	0.86	−0.61	—	—
KCl	82.6 ± 0.2	436.6 ± 0.4	cubic	298	40.02	25.47	3.64	—	—
			liq	1045	73.60	—	—	—	26.3
KCl(g)	239.0 ± 0.2	214.6	gas	298	37.15	0.96	−0.84	—	—
$KCaCl_3$	196.6	1244.3	orth	298	137.28	4.03	−5.15	—	—
				600	123.60	−80.75	—	80.75	—
			liq	1014	178.66	—	—	—	41.8
KF	66.5 ± 0.4	566.1 ± 1.7	cubic	298	47.36	13.26	−1.97	—	—
			liq	1131	66.94	—	—	—	29.3
KF(g)	226.5 ± 0.2	324.3 ± 5.0	gas	298	37.24	0.75	−2.05	—	—
K_3AlF_6	284.5 ± 12.6	3326.3 ± 13.8	tetrag	298	238.61	41.00	−25.94	—	—
KBF_4	133.9	1884.1 ± 2.1	orth	298	65.35	162.59	0.73	—	—
			cubic	556	146.23	—	—	—	13.8
			liq	843	167.15	—	—	—	18.0
KH	51.5 ± 6.3	57.8 ± 0.2	cubic	298	37.80	26.78	−6.99	—	—
KH(g)	197.9 ± 0.2	−123.0 ± 3.3	gas	298	35.33	1.99	−4.60	—	—

KHF_2	104.3 ± 0.4	930.9 ± 1.3	tetrag	298	50.21	89.54	—	—	—
			cubic	470	100.25	—	—	—	12.0
			liq	512	104.60	—	—	—	6.6
KI	106.3 ± 0.4	327.8	cubic	298	38.83	28.91	4.94	—	—
			liq	954	72.38	—	—	—	24.0
KI(g)	258.2 ± 0.2	125.5 ± 2.1	gas	298	37.40	0.88	−0.42	—	—
KCN	127.8 ± 1.7	113.4	cubic	298	66.27	0.42	—	—	—
			liq	895	75.31	—	—	—	14.6
K_2O	94.1	363.2 ± 2.9	cubic	298	95.65	−4.94	−11.05	23.68	—
K_2O_2	113.0	495.0	rhomb	298	79.75	69.08	—	—	—
KO_2	122.6 ± 4.2	284.5 ± 2.5	tetrag	298	87.65	10.67	−11.97	—	—
KBO_2	80.0 ± 0.4	995.0 ± 8.4	hex	298	80.96	23.43	−18.66	—	—
			liq	1220	146.44	—	—	—	31.4
$K_2B_4O_7$	208.4 ± 6.3	3326.3 ± 12.6	cryst	298	138.57	279.57	−35.48	−131.75	—
			liq	1089	380.53	77.19	—	—	104.2
K_2CO_3	155.5 ± 0.4	1153.1 ± 4.2	monocl	298	97.95	92.09	−9.87	—	—
			liq	1173	209.20	—	—	—	27.9
$KClO_3$	143.1 ± 1.3	389.1 ± 2.9	monocl	298	97.49	60.67	−13.39	—	—
$KClO_4$	151.0 ± 1.3	427.2 ± 5.0	rhomb	298	138.49	62.76	−39.75	—	—
			cubic	574	138.49	62.76	−39.75	—	13.8
K_2CrO_4	200.2 ± 2.9	1390.3 ± 4.2	orth	298	123.72	74.89	—	—	—
			hex	939	148.53	50.21	—	—	10.0
			liq	1250	209.20	—	—	—	31.8
$KFeO_2$	87.9 ± 9.6	1238.5 ± 12.6	cryst	298	100.00	23.85	−11.72	—	—
KOH	78.9 ± 0.8	424.7 ± 1.3	monocl	298	43.30	72.59	—	—	—
			cubic	516	78.66	—	—	—	6.4
			liq	679	83.09	—	—	—	8.6
KOH(g)	236.3 ± 1.3	232.6 ± 10.5	gas	298	50.84	4.29	−3.18	—	—
KNO_3	132.8 ± 0.8	494.5 ± 1.7	rhomb	298	60.46	118.83	—	—	—
			hex	402	120.50	—	—	—	5.1
			liq	607	123.43	—	—	—	9.7
K_3PO_4	211.7 ± 6.3	1988.2 ± 7.5	cubic	298	164.85	—	—	—	—
K_2HPO_4	179.1 ± 1.3	1775.7 ± 2.9	cryst	298	164.56	41.46	−29.71	—	—
KH_2PO_4	134.9 ± 0.8	1568.0 ± 2.5	tetrag	298	144.52	38.95	−35.15	—	—
K_2S	115.1 ± 14.6	376.6 ± 12.6	cubic	298	66.94	25.94	—	—	—
			cubic	1050	142.34	—	—	—	10.5
			liq	1221	100.96	—	—	—	16.2
K_2SO_4	175.6 ± 1.0	1438.5 ± 1.3	orth	298	142.38	10.75	−19.08	80.75	—
			hex	857	114.06	81.59	—	—	8.6
			liq	1342	200.00	—	—	—	34.7

TABLE I—*continued*

Substance	$S_{298} \pm \delta S$ (J/deg mol)	$-\Delta H_{298} \pm \delta H$ (kJ/mol)	Phase	T K	$C_p = A + BT + C/T^2 + DT^2$ A	$B \times 10^3$ (J/deg mol)	$C \times 10^{-5}$	$D \times 10^6$	H_t (kJ/mol)
$K_2Si_2O_5$	190.6 ± 0.4	2508.7 ± 8.4	monocl	298	191.84	36.57	−37.15	—	—
			β	510	157.99	90.83	−9.96	—	1.3
			gamma	867	224.22	4.44	—	—	1.6
			liq	1318	275.31	—	—	—	35.2
K_2SiO_3	146.1 ± 0.8	1589.9 ± 6.3	orth	298	118.91	48.79	−14.14	—	—
			liq	1249	167.36	—	—	—	50.2
$K_2Si_4O_9$	265.7 ± 12.6	4315.8 ± 15.1	tricli	298	253.22	159.37	—	—	—
			β	865	391.37	16.19	—	—	3.3
			liq	1043	410.03	—	—	—	87.9
$KAl(SO_4)_2$	204.6 ± 1.3	2470.2 ± 3.3	hex	298	234.14	82.34	−58.41	—	—
$K_2O.Al_2O_3.2SiO_2$	266.1	4217.1	kaliop	298	239.58	—	—	—	—
$K_2O.Al_2O_3.4SiO_2$	368.2	6068.9	leucit	298	244.60	—	—	—	—
$K_2O.Al_2O_3.6SiO_2$	439.3	7914.0	microc	298	572.12	77.32	−190.46	—	—
$K_2O.Al_2O_3.6SiO_2$(a)	468.6	7908.2	andula	298	572.12	77.32	−190.46	—	—
$K_2O.Al_2O_3.6SiO_2$(s)	476.1	7903.2	sanidi	298	572.12	77.32	−190.46	—	—
K_2WO_4	175.7 ± 20.9	1581.6	monocl	298	113.39	125.52	—	—	—
			hex	650	194.56	—	—	—	10.5
			liq	1196	213.38	—	—	—	31.0
La	56.9	0.0	hex	298	26.44	2.33	—	—	—
			fcc	583	17.66	15.02	3.89	—	0.4
			bcc	1138	39.54	—	—	—	3.1
			liq	1191	34.31	—	—	—	6.2
La(g)	182.3	−430.0	gas	298	18.91	16.48	—	−5.15	—
				1400	32.51	—	−12.05	—	—
				1800	28.28	—	71.42	0.50	—
$LaAl_2$	98.7 ± 0.4	151.0 ± 33.5	cubic	298	69.45	14.23	—	—	—
$LaBr_3$	182.0	844.7 ± 1.7	hex	298	94.73	25.10	—	—	—
			liq	1062	144.35	—	—	—	54.0
$LaCl_3$	137.7 ± 0.8	1070.7 ± 1.3	hex	298	86.19	39.75	—	—	—
			liq	1131	125.52	—	—	—	54.4
$LaCl_3$(g)	364.4 ± 6.7	740.6 ± 6.7	gas	298	80.75	0.25	−2.93	—	—
LaF_3	113.4	1731.8 ± 5.0	hex	298	48.95	59.58	29.12	—	—
			liq	1766	152.72	—	—	—	50.2
LaF_3(g)	327.6 ± 2.9	1287.8	gas	298	82.68	—	−8.87	—	—
LaH_2	51.7 ± 0.4	201.3 ± 5.4	cubic	298	39.25	15.15	—	—	—

LaI_3	214.6	656.9 ± 10.5	cryst	298	97.15	19.81	−2.73	—	—
			liq	1062	151.75	—	—	—	56.1
LaN	60.5 ± 0.8	299.2 ± 16.7	cubic	298	45.52	7.28	—	—	—
La_2O_3	127.3 ± 0.4	1794.9 ± 2.9	cubic	298	120.71	12.89	−13.72	—	—
$LaAlO_3$	85.4 ± 10.5	1793.9 ± 3.3	orth	298	111.80	15.48	−21.13	—	—
				623	111.80	15.48	−21.13	—	0.1
				800	111.80	15.48	−21.13	—	0.3
LaOCl	82.8 ± 10.5	1018.8 ± 9.2	tetrag	298	70.50	12.26	−4.60	—	—
$La_3(MoO_4)_3$	389.9 ± 5.4	4323.7 ± 11.7	monocl	298	329.36	—	—	—	—
$LaPO_4$	121.3 ± 12.6	1912.5 ± 10.5	monocl	298	125.52	24.89	−27.82	—	—
La_2S_3	165.0 ± 0.8	1184.1 ± 12.6	rhomb	298	116.52	14.64	—	—	—
LaS	73.2	472.4 ± 5.0	cubic	298	46.48	5.44	—	—	—
LaS(g)	252.5 ± 0.4	−129.7	gas	298	37.07	0.17	−2.38	—	—
La_2Se_3	202.2 ± 0.8	933.0 ± 33.5	cubic	298	120.71	16.32	—	—	—
LaSe	81.2 ± 5.0	405.0	cubic	298	47.45	5.86	—	—	—
La_2Te_3	231.7 ± 0.8	784.5 ± 25.5	cubic	298	128.16	13.39	—	—	—
Li	29.1	0.0	bcc	298	1.30	56.27	6.02	—	—
			liq	454	31.21	−5.27	2.05	2.64	3.0
Li(g)	138.7	−159.3	gas	298	20.79	—	—	—	—
			gas	1700	22.64	−2.09	—	0.59	—
$Li_2(g)$	196.9	−211.3	gas	298	37.03	1.72	−1.34	0.59	—
			gas	1200	51.13	−4.94	−77.61	—	—
			gas	2000	61.09	−18.16	—	3.64	—
LiBr	74.1 ± 0.4	351.0 ± 1.0	cubic	298	30.25	41.38	7.20	—	—
			liq	823	65.27	—	—	—	17.7
LiBr(g)	224.3 ± 0.2	156.9 ± 2.5	gas	298	37.03	0.84	−3.01	—	—
Li_2C_2	58.6 ± 4.2	59.4 ± 8.4	cryst	298	101.84	10.21	−29.62	—	—
LiCl	59.3 ± 0.4	408.4 ± 0.4	cubic	298	41.42	23.39	—	—	—
			liq	883	73.39	−9.46	—	—	19.7
LiCl(g)	212.8 ± 0.4	195.4 ± 1.7	gas	298	36.80	0.89	−3.49	—	—
LiF	35.6 ± 0.2	615.0 ± 1.3	cubic	298	42.68	17.41	−5.31	—	—
			liq	1122	64.18	—	—	—	27.1
LiF(g)	200.2 ± 0.2	338.9 ± 1.3	gas	298	35.65	1.31	−4.27	—	—
$Li_2F_2(g)$	261.5	936.4	gas	298	81.59	2.26	−16.23	−0.54	—
$Li_3F_3(g)$	316.7	1404.2	gas	298	126.19	6.69	−22.93	−1.67	—
Li_3AlF_6	187.9 ± 0.4	3376.5 ± 5.9	rhomb	298	205.94	109.83	−32.30	—	—
			tetrag	748	284.51	—	—	—	2.1
			cubic	848	294.97	—	—	—	1.3
			eps	978	305.43	—	—	—	0.4
			liq	1057	359.82	—	—	—	86.2
Li_2BeF_4	130.5 ± 2.5	2272.7 ± 4.6	hex	298	92.05	147.95	−1.00	—	—
			liq	732	232.21	—	—	—	43.9

TABLE I—*continued*

Substance	$S_{298} \pm \delta S$ (J/deg mol)	$-\Delta H_{298} \pm \delta H$ (kJ/mol)	Phase	T K	$C_p = A + BT + C/T^2 + DT^2$				H_t (kJ/mol)
					A	$B \times 10^3$	$C \times 10^{-5}$ (J/deg mol)	$D \times 10^6$	
$LiBeF_3$	89.1 ± 4.2	1650.6 ± 5.4	cryst	298	54.39	125.52	—	—	—
			liq	650	158.99	—	—	—	27.2
LiH	20.6 ± 0.4	90.7 ± 0.4	cubic	298	16.40	52.72	−2.93	—	—
			liq	964	58.58	—	—	—	22.1
LiH(g)	170.8 ± 0.4	−139.3 ± 1.7	gas	298	28.29	10.30	−1.37	−2.63	—
LiI	86.7 ± 0.4	270.3 ± 5.4	cubic	298	42.26	28.07	—	—	—
			liq	742	63.18	—	—	—	14.6
LiI(g)	232.1 ± 0.2	88.3	gas	298	37.15	0.84	−2.55	—	—
Li_3N	62.6 ± 0.4	164.8 ± 1.3	hex	298	56.53	85.77	−6.07	—	—
Li_2O	37.7 ± 0.2	597.9 ± 1.0	cubic	298	69.79	17.66	−18.49	—	—
			liq	1726	97.07	—	—	—	43.1
Li_2O(g)	232.6 ± 4.6	165.7 ± 5.4	gas	298	60.46	0.84	−9.33	—	—
Li_2O_2	58.2 ± 6.3	633.9 ± 8.4	hex	298	52.22	76.15	—	—	—
$LiAlO_2$	53.3 ± 1.3	1188.7 ± 4.2	hex	298	92.34	12.18	−25.02	—	—
			liq	1883	133.89	—	—	—	87.9
Li_3AsO_4	173.2 ± 8.4	1623.4 ± 10.5	orth	298	161.92	—	—	—	—
$LiBO_2$	51.7 ± 0.4	1019.2 ± 1.7	monocl	298	59.41	50.00	−12.34	—	—
			liq	1117	144.35	—	—	—	33.8
$Li_2B_4O_7$	156.9 ± 2.9	3374.4 ± 8.8	tetrag	298	83.68	302.08	13.81	−84.31	—
			liq	1193	453.55	10.46	−53.97	3.77	120.5
LiB_3O_5	96.2 ± 3.8	2335.9 ± 5.0	orth	298	155.98	44.81	−23.01	—	—
$Li_2B_8O_{13}$	263.6	5947.6 ± 20.9	orth	298	377.82	219.66	−97.49	−78.66	—
Li_2CO_3	90.2 ± 0.4	1215.5 ± 4.2	monocl	298	56.82	138.07	—	—	—
			β	623	132.38	—	—	—	—
			gamma	683	14.35	180.75	—	—	2.5
			liq	1000	185.43	—	—	—	44.8
$LiClO_4$	119.5 ± 0.4	375.3 ± 1.3	cryst	298	137.24	44.64	−40.46	—	—
			liq	520	161.08	—	—	—	17.0
$LiFeO_2$	75.3 ± 0.8	769.0 ± 5.4	cubic	298	81.17	51.67	−12.55	—	—
LiOH	42.8 ± 0.2	484.9 ± 0.4	tetrag	298	50.17	34.48	−9.50	—	—
			liq	744	87.11	—	—	—	20.9
LiOH(g)	210.5 ± 2.1	234.3 ± 6.3	gas	298	50.54	4.23	−5.02	—	—
Li_2HfO_3	96.2 ± 8.4	1774.0 ± 5.4	monocl	298	134.72	34.39	−29.41	—	—
$LiNO_3$	88.7 ± 7.5	483.2 ± 1.3	hex	298	62.68	88.70	—	—	—
			liq	526	111.29	—	—	—	25.9

$LiNbO_3$	85.4± 4.2	1365.2	hex	298	114.93	24.35	−20.33	—	—
$LiPO_3$	72.4± 2.9	1254.8± 3.8	monocl	298	58.37	94.14	—	—	—
Li_2S	60.7± 8.4	446.9± 2.1	cubic	298	66.32	20.17	—	—	—
Li_2SO_4	114.0± 0.4	1437.2± 0.8	monocl	298	65.27	174.05	—	—	—
			cubic	848	607.52	−383.25	—	—	25.5
			liq	1130	207.94	—	—	—	9.3
Li_2Se	66.5± 8.4	420.1± 8.4	cryst	298	74.81	10.29	—	—	—
Li_2SiO_3	79.9± 0.8	1648.5±10.5	orth	298	126.90	28.24	−30.25	—	—
			liq	1474	167.36	—	—	—	28.0
$Li_2O.Al_2O_3.2SiO_2$	207.5	4230.0	eucryp	298	308.57	56.90	−87.86	—	—
				1300	259.41	100.42	—	—	2.5
$Li_2O.Al_2O_3.4SiO_2$	258.6	6092.7	α−spod	298	370.95	137.57	−83.68	—	—
$Li_2O.Al_2O_3.4SiO_2(\beta)$	308.8	6036.7	β−spod	298	414.38	91.21	−103.09	—	—
$LiTaO_3$	90.0± 8.4	1419.2	hexag	298	117.78	19.54	−20.92	—	—
Li_2Te	77.4	355.6	cryst	298	77.40	16.74	—	—	—
Li_2TiO_3	91.8± 0.8	1669.8±10.5	cubic	298	143.39	13.22	−33.47	—	—
			monocl	1485	125.52	33.47	—	—	11.5
			liq	1820	200.83	—	—	—	110.0
Li_2WO_4	113.0±12.6	1603.7	hex	298	101.67	106.27	—	—	—
			cubic	948	199.16	—	—	—	2.7
			liq	1013	205.02	—	—	—	28.5
Li_2ZrO_3	91.6± 8.4	1762.3± 7.1	cryst	298	133.26	32.97	−28.20	—	—
Lu	51.0	0.0	hex	298	27.41	−5.40	0.25	8.28	—
			liq	1936	47.91	—	—	—	19.2
Lu(g)	184.7	−427.6	gas	298	12.47	20.93	2.48	−7.15	—
			gas	800	30.63	−1.59	−26.41	−0.32	—
			gas	1700	28.24	−1.26	—	—	—
			gas	1936	38.64	−8.18	−72.20	1.32	—
LuF_3	94.8	1681.1	orth	298	89.12	19.25	−6.95	—	—
			hex	1230	106.44	—	—	—	25.1
			liq	1457	214.85	−0.54	−2347.20	—	30.3
$LuF_3(g)$	315.3	1246.4	gas	298	81.17	1.17	−9.71	—	—
Lu_2O_3	110.0± 0.4	1878.2± 7.5	cubic	298	112.63	23.01	−15.69	—	—
Mg	32.7	0.0	hcp	298	21.13	11.92	0.15	—	—
			liq	923	34.31	—	—	—	8.5
Mg(g)	148.5	−147.1	gas	298	20.79	—	—	—	—
MgB_2	36.0	92.0		298	49.79	22.72	−7.61	—	—
MgB_4	51.9	105.0		298	66.94	49.66	−10.33	—	—
MgB_{12}	89.5	221.8		298	151.46	—	—	—	—
$MgBr_2$	117.2± 5.0	524.3± 2.5	hex	298	75.65	12.55	−5.40	—	—
			liq	984	104.60	—	—	—	39.3
MgCe	105.6± 3.8	16.1± 3.3	cubic	298	44.56	25.86	—	—	—

TABLE I—*continued*

Substance	$S_{298} \pm \delta S$ (J/deg mol)	$-\Delta H_{298}$ $\pm \delta H$ (kJ/mol)	Phase	T K	$C_p = A + BT + C/T^2 + DT^2$ A	$B \times 10^3$ (J/deg mol)	$C \times 10^{-5}$	$D \times 10^6$	H_t (kJ/mol)
$MgCl_2$	89.6 ± 0.8	641.4 ± 0.8	rhomb	298	79.08	5.94	−8.62	—	—
			liq	987	92.47	—	—	—	43.1
Mg_2Cu	92.3 ± 1.3	29.5 ± 1.3	orth	298	61.09	29.71	—	—	—
			liq	839	96.65	—	—	—	36.0
$MgCu_2$	97.9 ± 1.7	35.1 ± 1.7	cubic	298	66.61	24.02	—	—	—
			liq	1092	95.40	—	—	—	41.8
MgF_2	57.2 ± 0.4	1124.2 ± 1.3	rutile	298	77.11	3.89	−14.94	—	—
			liq	1536	94.56	—	—	—	58.6
Mg_2Ge	72.9 ± 1.9	115.2 ± 1.7	cubic	298	73.43	17.99	−3.18	—	—
MgH_2	31.0 ± 4.2	76.1 ± 2.5		298	27.20	49.37	−5.86	—	—
MgI_2	129.7 ± 8.4	366.9 ± 6.7	hex	298	75.86	13.72	−4.48	—	—
			liq	907	104.60	—	—	—	29.3
Mg_3N_2	93.7 ± 7.5	461.5 ± 7.1	cubic	298	86.90	46.86	—	—	—
			β	823	83.97	44.60	—	—	0.5
			gamma	1061	119.24	—	—	—	0.9
MgO	26.9 ± 0.4	601.6 ± 0.4	cubic	298	48.99	3.43	−11.34	—	—
			liq	3105	60.67	—	—	—	77.8
$MgAl_2O_4$	88.7 ± 4.2	2312.9 ± 2.5	cubic	298	153.97	26.78	−40.92	—	—
			liq	2408	230.12	—	—	—	196.6
$MgCO_3$	65.9 ± 0.8	1095.8 ± 12.6	magnes	298	77.91	57.74	−17.41	—	—
$MgCr_2O_4$	105.9 ± 0.8	1777.8 ± 3.3	spinel	298	167.44	14.90	−40.08	—	—
$MgGeO_3$	70.5	1208.3 ± 5.0	cryst	298	119.79	14.73	−29.46	—	—
$Mg(OH)_2$	63.2 ± 0.4	925.1 ± 2.1	hex	298	46.82	102.93	—	—	—
$MgMoO_4$	118.8 ± 0.8	1400.4 ± 5.0	monocl	298	128.91	34.89	−24.85	−2.93	—
$Mg(NO_3)_2$	164.0	790.8		298	44.69	297.90	7.49	—	—
$Mg_3(PO_4)_2$	189.1 ± 0.8	3780.7 ± 16.7	cryst	298	121.46	335.77	1.09	−108.78	—
			liq	1621	462.12	7.74	—	—	121.3
Mg_2Pb	119.2 ± 12.6	48.1 ± 2.9	cubic	298	65.90	34.52	—	—	—
			orth	710	92.88	—	—	—	2.2
			liq	822	94.98	—	—	—	40.2
MgS	50.3 ± 0.4	345.6 ± 4.2	cubic	298	48.74	3.64	−3.79	—	—
$MgSO_4$	91.6 ± 0.8	1284.9 ± 12.6	cryst	298	106.44	46.28	−21.88	—	—
			liq	1400	158.99	—	—	—	14.6
Mg_3Sb_2	136.6 ± 14.6	300.0 ± 14.6	hex	298	112.97	40.17	—	—	—
			cubic	1198	160.67	—	—	—	73.3

Mg_2Si	63.8 ± 2.1	79.1 ± 7.9	cubic	298	68.45	23.01	−5.02	—	—
			liq	1350	93.30	—	—	—	97.9
Mg_2SiO_4	95.2 ± 0.8	2176.9 ± 2.5	forst	298	144.31	38.74	−32.84	−5.48	—
			liq	2163	205.02	—	—	—	71.1
$MgSiO_3$	67.9 ± 0.8	1548.5 ± 3.8	clino	298	92.05	33.05	−17.78	—	—
				903	120.33	—	—	—	0.7
				1258	122.42	—	—	—	1.6
			liq	1850	146.44	—	—	—	75.3
$2MgO.2Al_2O_3.5SiO_2$	407.1	9113.2	cord	298	626.34	91.21	−200.83	—	—
$3MgO.2SiO_2.2H_2O$	222.2	4364.7	serp	298	317.23	132.21	−73.55	—	—
$7MgO.8SiO_2.H_2O$	559.0	12088.0	anthop	298	832.62	142.67	−218.82	—	—
$3MgO.4SiO_2.H_2O$	260.7	5916.2	talc	298	353.00	175.27	−74.27	—	—
MgTe	74.5 ± 10.5	209.2 ± 25.1	hex	298	37.66	10.46	—	—	—
Mg_2TiO_4	115.1 ± 6.3	2164.4 ± 2.1	tet/cu	298	152.38	34.06	−30.54	—	—
			liq	2013	228.45	—	—	—	129.7
$MgTiO_3$	74.6 ± 1.3	1572.6 ± 2.1	trig	298	118.53	13.60	−27.91	—	—
			liq	1953	163.18	—	—	—	90.4
$MgTi_2O_5$	135.6 ± 6.3	2342.0 ± 2.9	orth	298	170.41	38.37	−31.30	—	—
			liq	1963	261.50	—	—	—	146.4
$MgUO_4$	131.9 ± 0.8	1856.9 ± 2.9		298	110.25	66.78	−23.43	—	—
MgV_2O_6	160.7 ± 2.9	2200.8 ± 5.4		298	231.29	−6.09	−64.77	−2.93	—
$Mg_2V_2O_7$	200.0 ± 3.8	2834.7 ± 8.4		298	284.60	4.06	−74.22	−5.82	—
$MgWO_4$	101.0 ± 0.8	1517.1 ± 2.9		298	115.02	42.30	−15.77	—	—
Mn	32.0	0.0	alpha	298	23.85	14.14	−1.57	—	—
			beta	990	34.85	2.76	—	—	2.2
			gamma	1360	25.23	14.90	−1.85	—	2.2
			delta	1410	46.44	—	—	—	1.8
			liq	1517	46.02	—	—	—	14.6
Mn(g)	173.6	−280.7	gas	298	20.79	—	—	—	—
			gas	1600	18.70	0.79	21.21	—	—
MnAs	77.1 ± 0.4	56.9 ± 4.2	hex	298	70.71	—	—	—	—
Mn_2B	66.4	91.6	tetrag	298	69.04	22.80	−16.57	—	—
MnB	36.0	70.7	orth	298	42.47	15.90	−10.13	—	—
Mn_3B_4	118.0	236.4	orth	298	122.59	44.77	−49.92	—	—
MnB_2	44.4	94.1	hex	298	64.43	16.99	−25.02	—	—
$MnBr_2$	138.1 ± 10.5	384.9 ± 8.4	hex	298	67.91	24.81	—	—	—
			liq	971	100.42	—	—	—	33.5
$MnCl_2$	118.2 ± 0.4	481.2 ± 0.8	hex	298	75.48	13.22	−5.73	—	—
			liq	923	94.56	—	—	—	37.7
$MnCl_2(g)$	295.4 ± 6.3	263.6 ± 2.1	gas	298	64.98	—	−6.15	—	—

TABLE I—*continued*

Substance	$S_{298} \pm \delta S$ (J/deg mol)	$-\Delta H_{298}$ $\pm \delta H$ (kJ/mol)	Phase	T K	$C_p = A + BT + C/T^2 + DT^2$ A	$B \times 10^3$ (J/deg mol)	$C \times 10^{-5}$	$D \times 10^6$	H_t (kJ/mol)
MnF_2	93.3 ± 0.4	846.8	tetrag	298	61.92	23.85	−1.97	—	—
			liq	1203	92.05	—	—	—	23.0
MnO	59.8 ± 0.8	384.9 ± 1.3	cubic	298	46.48	8.12	−3.68	—	—
			liq	2058	60.67	—	—	—	54.4
Mn_3O_4	154.8 ± 6.3	1387.4 ± 1.7	tetrag	298	144.77	45.44	−12.97	—	—
			cubic	1440	210.04	—	—	—	20.9
Mn_2O_3	110.5 ± 2.1	958.1 ± 2.1	orth	298	112.26	35.06	−13.71	—	—
MnO_2	53.1 ± 0.4	520.9 ± 1.3	tetrag	298	69.45	10.21	−16.23	—	—
$MnAl_2O_4$	103.8 ± 2.1	2100.4 ± 6.3	cubic	298	153.13	25.94	−32.22	—	—
$MnCO_3$	109.6 ± 1.3	881.6 ± 4.6	hex	298	79.83	50.21	—	—	—
$MnFe_2O_4$	154.0	1228.8 ± 5.0	cubic	298	145.60	45.27	−8.79	—	—
$MnMoO_4$	136.0	1191.6 ± 9.6	monocl	298	108.78	51.30	—	—	—
MnP	52.3 ± 9.2	96.2 ± 16.7	rhomb	298	44.94	10.46	—	—	—
MnS	80.3 ± 0.8	214.2 ± 2.1	cubic	298	47.70	7.53	—	—	—
			liq	1803	66.94	—	—	—	26.4
$MnS(g)$	238.9	−265.7 ± 12.6	gas	298	36.36	—	−3.01	—	—
MnS_2	99.9 ± 0.4	207.1	cubic	298	69.71	17.66	−4.35	—	—
$MnSO_4$	112.3 ± 1.3	1066.1 ± 4.6	orth	298	122.38	37.32	−29.50	—	—
Mn_2Sb	136.8 ± 6.3	32.6 ± 3.3	tetrag	298	66.94	29.29	—	—	—
$MnSb$	92.5 ± 4.2	27.2 ± 2.5	hex	298	46.02	20.29	—	—	—
$MnSe$	90.8 ± 1.7	155.6 ± 10.5	cubic	298	48.79	7.70	—	—	—
Mn_3Si	103.8 ± 1.3	110.5 ± 10.5	cubic	298	100.88	52.09	−14.73	—	—
Mn_5Si_3	238.5 ± 3.3	244.3 ± 20.9	hex	298	201.38	54.14	−19.58	—	—
			liq	1573	324.68	—	—	—	173.2
$MnSi$	46.2 ± 1.3	65.3 ± 5.4	cubic	298	49.33	12.76	−6.40	—	—
			liq	1548	78.66	—	—	—	59.4
$MnSi_{1.7}$	56.1 ± 1.0	75.7 ± 8.4	tetrag	298	71.92	4.60	−13.05	—	—
Mn_2SiO_4	163.2 ± 4.2	1730.9 ± 4.2	orth	298	159.08	19.50	−31.13	—	—
			liq	1620	243.09	—	—	—	88.7
$MnSiO_3$	102.5 ± 2.1	1320.9 ± 2.5	rhodon	298	110.54	16.23	−25.77	—	—
			liq	1564	151.67	—	—	—	66.9
$MnSn_2$	130.9 ± 3.8	27.4 ± 2.9	tetrag	298	92.38	1.26	−11.38	—	—
$MnTe$	93.7 ± 1.7	109.2 ± 3.8	hex	298	74.35	—	—	—	—
				307	56.69	2.76	—	—	—

$MnTe_2$	145.0 ± 0.4	123.4 ± 3.8	cubic	298	76.65	4.18	—	—	—
Mn_2TiO_4	170.3	1750.0	tetrag	298	168.20	17.41	−25.56	—	—
$MnTiO_3$	105.0 ± 0.8	1359.0	hex	298	121.67	9.29	−21.88	—	—
$MnWO_4$	140.6 ± 8.4	1305.8 ± 9.2	monocl	298	108.78	51.30	−34.73	8.87	—
Mo	28.6	0.0	bcc	298	28.52	−4.42	−3.62	4.28	—
			liq	2896	40.33	—	—	—	39.1
Mo(g)	181.8	−658.1	gas	298	20.79	—	—	—	—
			gas	1200	35.40	−13.68	−49.54	3.68	—
Mo_2B	52.6 ± 8.4	132.2 ± 25.1	tetrag	298	78.45	4.56	−19.37	2.80	—
			liq	2648	111.71	—	—	—	138.1
MoB	25.3 ± 8.4	123.8 ± 16.7	tetrag	298	45.40	5.82	−13.05	—	—
			liq	2873	80.75	—	—	—	55.2
Mo_2C	65.8 ± 0.4	52.3 ± 2.5	orth	298	64.27	23.60	−9.50	4.48	—
			hex	1500	79.50	8.37	—	—	21.8
			liq	2795	100.42	—	—	—	70.3
$MoCl_4$	159.0 ± 2.5	504.2 ± 20.9	monocl	298	120.50	—	—	—	—
			β	350	107.95	58.58	—	—	4.6
			liq	600	149.79	—	—	—	21.6
$MoCl_4$(g)	371.7	383.7 ± 16.7	gas	298	107.86	−0.42	−9.04	2.76	—
$MoCl_5$	238.5 ± 12.6	526.8 ± 9.2	monocl	298	165.69	—	—	—	—
			liq	470	88.28	—	—	—	18.8
$MoCl_5$(g)	398.1 ± 12.6	446.9 ± 9.2	gas	298	136.11	−0.75	−13.26	0.05	—
$MoCl_6$(g)	419.7 ± 16.7	460.2 ± 41.8	gas	298	157.53	0.01	−12.13	—	—
MoF_5	178.7 ± 12.6	1387.4 ± 4.2	monocl	298	154.81	—	—	—	—
			liq	319	155.64	—	—	—	12.6
MoF_5(g)	347.3 ± 8.4	1240.6 ± 5.4	gas	298	133.60	0.17	−27.66	0.08	—
Mo_2F_{10}(g)	530.9 ± 16.7	2694.5 ± 10.5	gas	298	264.01	13.97	−50.38	0.17	—
MoF_6(g)	350.6 ± 0.4	1556.9 ± 1.7	gas	298	153.76	1.59	−30.75	—	—
MoO_2	46.4 ± 0.4	587.9 ± 1.7	monocl	298	65.61	11.55	−12.13	—	—
MoO_2(g)	277.0 ± 8.4	8.4 ± 12.6	gas	298	55.65	0.44	−10.46	0.30	—
MoO_3	77.8 ± 0.8	745.2 ± 0.8	orth	298	75.19	32.64	−8.79	—	—
			liq	1074	126.90	—	—	—	48.5
MoO_3(g)	283.8 ± 4.2	345.2 ± 20.9	gas	298	74.89	6.95	−15.44	−1.46	—
$Mo(CO)_6$	327.2 ± 1.3	1059.8 ± 10.5	cubic	298	205.23	154.81	—	—	—
Mo_2S_3	115.0 ± 8.4	406.7 ± 8.4	monocl	298	110.29	32.97	−9.62	—	—
MoS_2	62.6 ± 0.4	275.7 ± 2.5	hex	298	71.65	7.45	−9.20	—	—
Mo_3Si	106.3 ± 1.3	116.3 ± 9.2	cubic	298	85.86	22.68	0.32	—	—
Mo_5Si_3	207.9 ± 9.2	310.0 ± 22.2	tetrag	298	183.34	35.02	−12.01	—	—
$MoSi_2$	65.1 ± 3.8	131.8 ± 8.4	tetrag	298	67.86	11.97	−6.57	—	—
N_2(g)	191.5	0.0	gas	298	30.42	2.54	−2.37	—	—
N(g)	153.2	−472.7	gas	298	20.79	—	—	—	—
			gas	1800	20.68	0.05	—	—	—

TABLE I—*continued*

Substance	$S_{298} \pm \delta S$ (J/deg mol)	$-\Delta H_{298} \pm \delta H$ (kJ/mol)	Phase	T K	$C_p = A + BT + C/T^2 + DT^2$ A	$B \times 10^3$ (J/deg mol)	$C \times 10^{-5}$	$D \times 10^6$	H_t (kJ/mol)
$NH_3(g)$	192.7	45.9	gas	298	37.32	18.66	−6.49	—	—
NH_4Br	111.3 ± 0.6	270.7 ± 1.0	cubic	298	124.68	−35.15	−23.85	—	—
				413	98.32	—	−33.89	—	3.6
NH_4Cl	95.0 ± 2.1	314.6 ± 0.4	cubic	298	38.87	160.25	—	—	—
				458	34.64	111.71	—	—	3.9
NH_4F	72.0 ± 0.2	463.6 ± 2.5	cubic	298	65.27	—	—	—	—
NH_4I	113.0	201.7	cubic	298	60.29	71.76	—	—	—
$N_2O(g)$	219.9 ± 0.2	−82.0 ± 0.6	gas	298	46.11	11.38	−10.04	−2.13	—
$NO(g)$	210.7 ± 0.2	−90.3 ± 0.4	gas	298	29.41	3.85	−0.59	—	—
$N_2O_3(g)$	309.2 ± 0.8	−82.8 ± 1.0	gas	298	80.10	17.82	−18.24	−3.51	—
$NO_2(g)$	239.9 ± 0.4	−33.1 ± 0.8	gas	298	35.69	22.91	−4.70	−6.33	—
				1500	53.76	1.28	—	—	—
$N_2O_4(g)$	304.0 ± 0.8	− 9.4 ± 1.7	gas	298	101.11	24.31	−28.58	−4.85	—
$N_2O_5(g)$	355.6 ± 10.5	−11.7 ± 1.7	gas	298	118.66	35.48	−29.20	−10.50	—
$NO_3(g)$	252.7	−71.1	gas	298	67.78	12.55	−22.51	−2.64	—
$NOBr(g)$	273.4	−82.1	gas	298	51.00	3.39	−6.28	—	—
$NOCl(g)$	261.6 ± 0.4	−52.4 ± 0.8	gas	298	42.80	8.91	−0.71	—	—
NH_4ClO_4	184.1	295.8 ± 5.0	orth	298	67.78	202.38	—	—	—
$NOF(g)$	248.0 ± 0.4	66.1 ± 3.3	gas	298	48.07	6.07	−7.61	—	—
NH_4NO_3	151.0 ± 0.4	365.4 ± 0.8	orth	298	140.16	—	—	—	—
			orth	305	59.41	196.65	—	—	1.7
			tetrag	357	150.62	—	—	—	1.3
			cubic	399	158.99	—	—	—	4.4
$(NH_4)_2SO_4$	220.1 ± 1.3	1180.3 ± 0.8	orth	298	103.55	280.75	—	—	—
Na	51.3	0.0	bcc	298	82.47	−369.32	—	627.60	—
			liq	371	37.51	−19.22	—	10.64	2.6
Na(g)	153.6	−107.5	gas	298	20.79	—	—	—	—
$Na_2(g)$	230.1	−142.3	gas	298	32.55	14.35	1.30	−7.74	—
			gas	1200	51.76	−9.58	−22.09	—	—
			gas	1900	69.83	−30.88	—	6.07	—
NaBr	86.9 ± 0.4	361.3 ± 0.4	cubic	298	47.91	13.31	—	—	—
			liq	1020	62.34	—	—	—	26.1
NaBr(g)	241.1 ± 0.4	144.3 ± 2.1	gas	298	36.61	1.19	—	—	—

NaCl	72.1 ± 0.4	411.3 ± 0.4	cubic	298	45.94	16.32	—	—	—
			liq	1074	77.78	−7.53	—	—	28.2
			liq	1500	66.94	—	—	—	—
NaCl(g)	229.7 ± 0.4	182.0 ± 0.8	gas	298	37.33	0.74	−1.59	—	—
$NaAlCl_4$	184.1 ± 8.4	1138.5 ± 3.3	rhomb	298	63.60	262.34	—	—	—
NaF	51.2 ± 0.2	572.8 ± 1.7	cubic	298	46.61	9.92	−2.13	—	—
			liq	1269	70.54	—	—	—	33.5
NaF(g)	217.6 ± 0.4	292.9	gas	298	37.15	0.77	−2.89	—	—
Na_3AlF_6	238.5 ± 1.7	3312.1 ± 5.0	monocl	298	172.26	158.45	—	—	—
			cubic	838	282.00	—	—	—	8.4
			cubic	1154	355.64	—	—	—	0.4
			liq	1284	396.22	—	—	—	107.5
$NaBF_4$	145.3 ± 0.8	1844.7 ± 3.3	orth	298	50.96	217.57	3.68	—	—
			monocl	516	152.63	—	—	—	6.7
			liq	679	165.35	—	—	—	13.6
NaH	40.0 ± 0.4	56.5 ± 0.4	cubic	298	26.15	42.80	−2.26	—	—
NaI	98.5 ± 0.4	287.9 ± 0.8	cubic	298	48.87	12.05	—	—	—
			liq	934	64.85	—	—	—	23.6
NaCN	115.7 ± 0.8	90.8	cubic	298	67.36	5.44	—	—	—
			liq	837	79.50	—	—	—	18.4
Na_2O	75.1 ± 0.4	415.1 ± 0.8	cubic	298	55.48	70.21	−4.14	−30.54	—
			β	1023	82.30	12.76	—	—	1.8
			gamma	1243	84.85	10.71	—	—	11.9
			liq	1405	104.60	—	—	—	47.7
Na_2O_2	94.8 ± 1.3	513.0 ± 6.3	hex	298	85.48	42.68	−8.03	—	—
			β	785	113.60	—	—	—	5.6
NaO_2	115.9 ± 1.3	260.7 ± 3.3	cubic	298	59.96	40.84	—	—	—
NaO(g)	228.4 ± 1.7	−102.5 ± 20.9	gas	298	37.24	0.92	−2.13	—	—
$NaAlO_2$	184.1 ± 0.8	1138.5 ± 5.9	orth	298	89.16	15.27	−17.95	—	—
			tetrag	740	89.16	15.27	−17.95	—	1.3
Na_3AsO_4	186.2 ± 5.4	1540.1 ± 6.7	rhomb	298	170.29	—	—	—	1.8
$NaBO_2$	73.5 ± 0.4	975.7 ± 2.5	hex	298	79.54	23.56	−18.41	—	—
			liq	1240	146.44	—	—	—	33.5
$Na_2B_4O_7$	189.5 ± 0.8	3284.9 ± 5.9	tricli	298	206.10	77.11	−37.49	—	—
			liq	1016	444.88	—	—	—	81.2
NaB_3O_5	116.1 ± 4.6	2299.1 ± 7.1	monocl	298	40.92	282.29	4.69	−95.81	—
$Na_2B_8O_{13}$	276.1 ± 16.7	5902.8 ± 10.0	monocl	298	345.18	226.35	−95.81	—	—
Na_2CO_3	135.0 ± 0.8	1129.7 ± 0.8	monocl	298	11.00	244.05	24.48	—	—
			hex	723	50.08	129.08	—	—	0.7
			liq	1130	189.54	—	—	—	29.7
$NaClO_4$	143.9 ± 0.4	377.8 ± 1.7	orth	298	138.49	54.10	−39.33	—	—

TABLE I—*continued*

Substance	$S_{298} \pm \delta S$ (J/deg mol)	$-\Delta H_{298} \pm \delta H$ (kJ/mol)	Phase	T K	$C_p = A + BT + C/T^2 + DT^2$ (J/deg mol) A	$B \times 10^3$	$C \times 10^{-5}$	$D \times 10^6$	H_t (kJ/mol)
$NaClO_3$	126.4 ± 4.2	357.7 ± 1.3	cubic	298	54.68	154.81	—	—	—
			liq	528	133.05	—	—	—	22.6
$NaCrO_2$	81.2 ± 5.0	876.5 ± 4.6	hex	298	94.56	15.06	−8.58	—	—
Na_2CrO_4	176.6 ± 0.4	1334.3 ± 9.2	orth	298	101.04	140.00	—	—	—
			hex	694	149.95	51.59	—	—	9.6
			liq	1070	204.60	—	—	—	24.3
$NaFeO_2$	88.3 ± 0.8	698.3 ± 5.9	orth	298	98.83	14.23	−16.74	—	—
				1270	98.83	14.23	−16.74	—	2.2
			liq	1620	125.52	—	—	—	49.4
$NaOH$	64.4 ± 0.4	425.9 ± 0.8	orth	298	71.76	−110.88	—	235.77	—
			cubic	572	85.98	—	—	—	7.2
			liq	596	89.45	−5.86	—	—	6.6
$NaHCO_3$	101.3 ± 0.8	949.1 ± 0.8	monocl	298	45.31	143.09	—	—	—
Na_2MoO_4	159.4 ± 1.3	1534.3 ± 11.3	cubic	298	142.80	58.32	−16.65	—	—
			rhomb	718	−215.48	506.26	—	—	21.8
			rhomb	866	−585.76	891.19	—	—	2.1
			delta	915	1105.41	−1204.99	—	—	8.3
			liq	962	212.97	—	—	—	21.4
$Na_2Mo_2O_7$	250.6 ± 2.1	2361.0 ± 12.6	orth	298	173.64	144.35	—	—	—
$NaNO_3$	116.3 ± 0.4	468.2 ± 1.3	hex	298	25.69	225.94	—	—	—
			liq	580	155.64	—	—	—	15.1
$NaPO_3$	95.5 ± 0.8	1220.1 ± 2.1	rhomb	298	52.93	115.06	—	—	—
Na_3PO_4	173.8 ± 0.8	1916.9 ± 2.1	tetrag	298	153.47	—	—	—	—
$Na_4P_2O_7$	270.3 ± 1.7	3166.5	rhomb	298	241.12	—	—	—	—
Na_2S	96.2 ± 14.6	366.1 ± 12.6	cubic	298	83.16	13.81	−0.25	—	—
Na_2SO_4	149.6 ± 0.4	1389.5 ± 1.3	orth	298	82.34	154.35	—	—	—
				522	145.06	54.60	—	—	10.8
			hex	980	142.67	59.33	—	—	0.3
			liq	1157	197.40	—	—	—	23.0
$Na_2SO_4.7H_2O$	411.7 ± 2.1	3456.8 ± 3.8	cryst	298	348.11	157.32	—	—	—
$Na_2SO_4.10H_2O$	585.8 ± 0.8	4328.8 ± 4.6	cryst	298	585.76	—	—	—	—
Na_2SO_3	146.0 ± 1.7	1095.0	hex	298	107.11	43.51	—	—	—
			liq	1184	182.00	—	—	—	25.9
Na_2SiF_6	207.1 ± 0.8	2912.9 ± 3.8	hex	298	182.00	72.17	−15.27	—	—
			liq	1120	276.14	—	—	—	99.6

Na_4SiO_4	195.8 ± 2.5	2101.2 ± 25.1	monocl	298	162.59	74.22	—	—	—
			liq	1393	259.41	—	—	—	57.7
Na_2SiO_3	113.8 ± 0.8	1563.1	orth	298	130.29	40.17	−27.07	—	—
			liq	1361	179.08	—	—	—	52.3
$Na_2Si_2O_5$	164.4 ± 2.9	2473.6 ± 5.0	monocl	298	185.69	70.54	−44.64	—	—
			orth	983	292.88	—	—	—	2.5
			liq	1147	261.08	—	—	—	35.6
$Na_2O.Al_2O_3.4SiO_2$	266.9	6039.6 ± 18.8	jadeit	298	403.00	95.56	−99.37	—	—
$Na_2O.Al_2O_3.6SiO_2$	420.1	7841.2 ± 10.0	l.alb	298	516.31	116.32	−125.60	—	—
			h.alb	973	565.59	81.67	−171.80	—	28.5
$Na_2O.Al_2O_3.2SiO_2$	248.5	4163.5	nephel	298	55.48	590.78	—	—	—
				467	224.18	134.22	—	—	—
				980	224.18	134.22	—	—	17.1
				1180	344.01	11.05	—	—	—
Na_2Te	96.2	334.7 ± 25.1	cubic	298	73.22	13.81	—	—	—
Na_2TiO_3	121.6 ± 0.4	1552.7 ± 2.1	cubic	298	135.19	46.78	−21.25	—	—
			β	560	108.57	71.13	—	—	1.7
			liq	1303	196.23	—	—	—	70.3
$Na_2Ti_2O_5$	173.8 ± 0.8	2513.7 ± 62.8	orth	298	230.71	9.62	−52.63	—	—
			liq	1258	286.60	—	—	—	112.3
$Na_2Ti_3O_7$	233.9 ± 0.8	3481.9 ± 4.2	monocl	298	294.76	20.63	−63.68	—	—
			liq	1401	393.92	—	—	—	155.2
Na_3UO_4	198.2 ± 0.8	2023.8 ± 2.9	cubic	298	188.91	25.19	−20.92	—	—
Na_2UO_4	166.1 ± 0.8	1889.5 ± 2.5	rhomb	298	162.55	25.90	−20.92	—	—
			β	1193	224.68	—	—	—	20.9
Na_3VO_4	189.5 ± 1.7	1841.4 ± 10.5	cubic	298	188.28	25.73	−27.66	—	—
$Na_4V_2O_7$	318.4 ± 1.7	3037.2 ± 10.5	cryst	298	323.42	28.87	−55.31	—	—
$NaVO_3$	113.8 ± 0.8	1191.6 ± 10.5	monocl	298	127.70	3.14	−27.61	—	—
Na_2WO_4	160.7 ± 1.3	1541.8	cubic	298	107.19	115.98	—	—	—
			β	864	209.20	—	—	—	34.4
			liq	969	209.20	—	—	—	23.8
Nb	36.6	0.0	bcc	298	27.78	−3.84	−2.55	3.60	—
			liq	2745	41.78	—	—	—	30.5
Nb(g)	186.1	−722.6	gas	298	32.13	−8.79	0.50	2.59	—
NbB_2	37.4 ± 0.8	175.3 ± 16.7	hex	298	46.99	38.53	−9.41	—	—
$NbBr_5$	305.4	556.1 ± 1.7	orth	298	116.78	130.54	—	—	—
			liq	540	184.10	—	—	—	35.6
$NbBr_5$(g)	449.2	443.5	gas	298	132.84	0.10	−6.28	—	—
Nb_2C	64.0 ± 0.4	195.0 ± 5.0		298	63.39	13.56	−3.47	—	—
$NbC_{0.7}$	31.9	116.7		298	39.92	6.49	−6.19	—	—
NbC	35.4 ± 0.4	141.4 ± 2.5	cubic	298	45.52	6.32	−9.04	—	—

TABLE I—*continued*

Substance	$S_{298} \pm \delta S$ (J/deg mol)	$-\Delta H_{298} \pm \delta H$ (kJ/mol)	Phase	T K	$C_p = A + BT + C/T^2 + DT^2$ A	$B \times 10^3$ (J/deg mol)	$C \times 10^{-5}$	$D \times 10^6$	H_t (kJ/mol)
$NbCl_4$	179.9 ± 10.5	694.5 ± 6.3	orth	298	133.47	—	−12.13	—	—
$NbCl_5$	214.2 ± 4.2	797.5 ± 4.2	monocl	298	148.53	—	—	—	—
			liq	479	184.10	—	—	—	33.9
$NbCl_5$(g)	404.0 ± 3.3	703.3 ± 8.4	gas	298	133.13	—	−12.55	—	—
$NbCr_2$	83.7 ± 0.4	20.9 ± 2.5	cubic	298	74.27	23.77	−7.36	—	—
NbF_5	157.3 ± 2.1	1813.8 ± 4.2	monocl	298	139.33	—	—	—	—
			liq	352	177.82	—	—	—	12.2
NbF_5(g)	382.8	1719.2 ± 8.4	gas	298	130.67	1.09	−24.69	—	—
Nb_2N	79.5	253.1 ± 7.1	hex	298	62.38	17.11	—	—	—
			hex	1000	70.75	8.74	—	—	—
NbN	43.9	236.4 ± 6.3	cubic	298	36.36	22.59	—	—	—
			hex	600	43.81	8.31	—	—	—
			hex	1643	62.76	—	—	—	4.2
			liq	2320	62.76	—	—	—	46.0
NbO	46.0 ± 8.4	419.7 ± 12.6	hex	298	42.97	8.87	−4.02	—	—
			liq	2210	62.76	—	—	—	85.4
NbO(g)	238.9 ± 3.8	−198.7 ± 20.9	gas	298	35.52	1.09	−4.56	—	—
NbO_2	54.5 ± 0.4	795.0 ± 8.4	tetrag	298	61.30	25.77	−10.21	—	—
			tetrag	1090	89.04	—	—	—	3.4
			tetrag	1200	83.05	—	—	—	—
			liq	2175	94.14	—	—	—	42.3
NbO_2(g)	272.0 ± 8.4	200.0 ± 20.9	gas	298	54.77	1.59	−10.17	—	—
Nb_2O_5	137.3 ± 1.3	1899.5 ± 4.2	orth	298	162.17	14.81	−30.63	—	—
			liq	1785	242.25	—	—	—	104.2
$NbOCl_2$	121.3	774.0 ± 8.4	cryst	298	96.23	16.74	−7.11	—	—
$NbOCl_3$	159.0 ± 10.5	880.7 ± 10.5	cryst	298	133.47	—	−12.13	—	—
$NbOCl_3$(g)	343.1	770.3	gas	298	107.95	—	−8.37	—	—
Nb_5Si_3	251.0	451.9 ± 108.8	tetrag	298	189.12	30.79	−15.06	—	—
$NbSi_2$	69.9 ± 10.5	138.1 ± 41.8	hex	298	63.18	15.36	−2.80	—	—
Nd	71.1	0.0	hcp	298	14.66	26.92	4.48	—	—
			bcc	1128	44.56	—	—	—	3.0
			liq	1289	48.79	—	—	—	7.1
Nd(g)	189.3	−327.6	gas	298	17.95	19.29	−0.79	−8.45	—
			gas	1000	28.79	1.42	−14.81	—	—

$NdCl_3$	153.0 ± 0.8	1040.6 ± 2.1	hex	298	82.84	55.06	—	—	—
			liq	1032	146.44	—	—	—	48.5
NdF_3	116.3 ± 8.4	1712.9 ± 7.5	hex	298	78.66	34.64	5.36	—	—
			liq	1650	133.89	—	—	—	54.8
NdH_2	58.9 ± 0.4	202.1	cubic	298	38.24	16.11	—	—	—
NdI_3	215.1	628.4	rhomb	298	90.86	35.82	−0.63	—	—
			β	847	117.40	—	—	—	13.8
			liq	1060	155.73	—	—	—	41.4
Nd_2O_3	158.6 ± 0.4	1808.3 ± 3.3	cubic	298	115.77	29.79	−11.88	—	—
			hex	1395	155.64	—	—	—	0.6
$NdOCl$	79.9 ± 7.9	1005.8 ± 6.7	tetrag	298	68.62	19.12	−3.97	—	—
NdS	74.1 ± 4.2	464.8 ± 5.0	tetrag	298	46.19	8.37	—	—	—
Nd_2S_3	185.4 ± 0.8	1125.5 ± 14.6	orth	298	118.53	13.35	—	—	—
Ni	29.9	0.0	fcc	298	11.17	37.78	3.18	—	—
			fcc	631	20.54	10.08	15.40	—	—
			liq	1728	38.91	—	—	—	17.2
Ni(g)	182.1	−430.1	gas	298	26.11	−1.30	−2.18	—	—
Ni_3Al	113.8	153.1 ± 4.6	cubic	298	88.49	32.22	—	—	—
NiAl	54.1	118.4 ± 4.2	cubic	298	41.84	13.81	—	—	—
			liq	1912	71.13	—	—	—	62.8
Ni_2Al_3	136.4	282.4 ± 14.6	hexag	298	106.06	34.31	—	—	—
$NiAl_3$	110.7	150.6 ± 6.3	ortho	298	84.10	35.15	—	—	—
NiAs	51.9	72.0	hex	298	44.10	12.97	—	—	—
Ni_3B	87.9	88.9	orth	298	95.40	26.36	−15.56	—	—
Ni_2B	66.3	63.8	tetrag	298	66.94	22.18	−12.05	—	—
Ni_4B_3	110.1	179.1	orth	298	156.06	49.16	−37.78	—	—
NiB	28.5	46.4	orth	298	42.97	14.64	−11.25	—	—
NiBi	88.3	7.7	hex	298	46.02	19.25	—	—	—
$NiBr_2$	120.9	213.8 ± 2.5	hex	298	69.04	19.66	—	—	—
$NiBr_2$(g)	320.9	−11.7	gas	298	68.70	−1.00	−4.94	—	—
			gas	1100	67.20	—	—	—	—
NiBr(g)	262.4	−184.1	gas	298	39.87	0.79	−3.51	—	—
$NiCl_2$	98.1 ± 0.4	304.6 ± 2.1	hex	298	73.22	13.22	−4.98	—	—
NiCl(g)	251.3	−179.9	gas	298	39.66	0.84	−4.10	—	—
$NiAl_2Cl_8$(g)	610.9	1543.9	gas	298	251.37	8.58	−25.19	—	—
NiF_2	73.6 ± 0.4	657.7 ± 2.1	tetrag	298	66.48	13.64	−5.73	—	—
NiF_2(g)	272.8 ± 2.5	330.0	gas	298	63.18	2.34	−10.13	—	—
NiF(g)	114.0	−104.6	gas	298	39.16	1.13	−6.15	—	—
NiH(g)	73.6	−393.3	gas	298	31.30	0.31	−2.72	—	—
NiI_2	154.0	78.2	hex	298	65.90	24.27	—	—	—
NiI(g)	270.1	−246.9	gas	298	40.21	0.46	−3.35	—	—

TABLE I—*continued*

Substance	$S_{298} \pm \delta S$ (J/deg mol)	$-\Delta H_{298} \pm \delta H$ (kJ/mol)	Phase	T K	$C_p = A + BT + C/T^2 + DT^2$ A	$B \times 10^3$ (J/deg mol)	$C \times 10^{-5}$ (J/deg mol)	$D \times 10^6$	H_t (kJ/mol)
NiO	38.0 ± 0.4	239.7 ± 1.3	cubic	298	−20.88	157.23	16.28	—	—
			β	525	58.07	—	—	—	—
			gamma	565	46.78	8.45	—	—	—
NiO(g)	241.2	−309.6	gas	298	42.33	−0.51	−0.78	—	—
$NiAl_2O_4$	98.3 ± 3.8	1920.5 ± 4.6	cubic	298	159.20	23.35	−30.75	—	—
$NiCO_3$	86.2 ± 0.4	696.2 ± 12.6	hex	298	88.70	38.91	−12.34	—	—
$Ni(CO)_4(g)$	415.5 ± 0.8	600.4 ± 4.2	gas	298	209.62	—	−53.56	—	—
$NiCr_2O_4$	129.7 ± 0.8	1375.7 ± 4.2	cubic	298	167.15	17.87	−21.05	—	—
$Ni(OH)_2(g)$	291.2	255.2	gas	298	87.07	5.56	−26.48	—	—
Ni_3P	106.3	220.1 ± 10.5	tetrag	298	77.82	33.47	—	—	—
Ni_5P_2	184.9	436.0 ± 16.7	cryst	298	135.14	56.48	—	—	—
Ni_2P	77.4	184.9 ± 16.7	hex	298	57.95	23.01	—	—	—
Ni_6P_5	276.1	627.6 ± 62.8	cryst	298	241.00	71.13	—	—	—
Ni_3S_2	133.9 ± 0.8	216.3 ± 5.0	hex	298	110.79	51.67	−7.53	—	—
			β	829	188.61	—	—	—	56.2
			liq	1062	191.79	—	—	—	19.7
NiS	53.0 ± 0.4	87.9 ± 6.3	rhomb	298	44.69	19.04	−2.89	—	—
			hex	652	34.39	28.66	—	—	6.4
			liq	1249	76.78	—	—	—	30.1
Ni_3S_4	186.6	301.2 ± 25.1	cubic	298	121.96	143.68	—	—	—
$NiSO_4$	101.3 ± 1.7	873.2 ± 1.3	orth	298	125.94	27.82	−32.64	—	—
NiSb	78.2	83.7	hex	298	46.23	11.63	—	—	—
$NiSe_2$	103.5 ± 0.4	64.6 ± 2.9	cubic	298	76.65	13.14	−4.60	—	—
$NiSeO_3$	103.3	567.4	cryst	298	79.50	59.83	—	—	—
NiSi	44.4 ± 3.3	89.5 ± 7.5	orth	298	48.74	6.15	−6.53	—	—
			liq	1265	79.50	—	—	—	43.0
$Ni_{.35}Si_{.65}$	22.0 ± 1.7	31.4 ± 4.2	cubic	298	25.02	3.68	−3.60	—	—
Ni_2SiO_4	110.0 ± 8.4	1400.8 ± 9.2	orth	298	185.10	19.87	−56.90	—	—
NiTe	80.0 ± 0.8	35.6	hex	298	54.48	6.78	−3.83	—	—
Ni_3Ti	138.1 ± 14.6	139.3 ± 8.4	hex	298	108.95	16.86	—	—	—
$NiTi_2$	83.7	80.8 ± 7.9	cubic	298	67.99	23.43	—	—	—
$NiTiO_3$	82.8 ± 4.2	1202.1 ± 6.3	trigon	298	115.10	15.98	−18.20	—	—
$NiWO_4$	118.0 ± 6.3	1128.8 ± 2.1	monocl	298	110.62	53.39	−4.39	—	—
$O_2(g)$	205.1	0.0	gas	298	29.96	4.18	−1.67	—	—

$O_3(g)$	238.8 ± 0.4	−142.3 ± 2.1	gas	298	44.35	15.61	−8.62	−4.35	—
$O(g)$	161.0	−249.2	gas	298	20.88	−0.05	0.97	—	—
Os	32.6	0.0	hcp	298	23.56	3.85	—	—	—
Os(g)	192.5	−788.0	gas	298	13.22	11.13	5.48	−1.59	—
			gas	3200	33.18	—	−229.83	0.21	—
OsO_2	51.9 ± 10.9	294.6 ± 9.6	tetrag	298	69.96	10.38	−14.18	—	—
OsO_4	136.8 ± 8.4	393.7 ± 8.4	monocl	298	151.46	—	—	—	—
			liq	314	157.74	—	—	—	14.3
$OsO_4(g)$	293.6 ± 0.4	336.2 ± 8.8	gas	298	85.98	20.42	−15.98	—	—
OsS_2	54.8 ± 8.4	146.9 ± 16.7	cubic	298	68.53	11.84	−8.79	—	—
$OsSe_2$	81.6 ± 12.6	120.1 ± 20.9	cubic	298	73.64	11.09	−4.18	—	—
P	41.1	0.0	white	298	19.12	15.82	—	—	—
			liq	317	26.32	—	—	—	0.7
P(r)	22.8 ± 0.2	17.4 ± 0.2	red	298	16.74	14.90	—	—	—
P(g)	163.1 ± 0.0	−316.5 ± 1.0	gas	298	20.79	—	—	—	—
$P_2(g)$	218.0 ± 0.2	−144.0 ± 0.4	gas	298	36.32	0.79	−4.14	—	—
$P_4(g)$	279.9 ± 0.3	−58.9 ± 0.3	gas	298	82.11	0.68	−13.44	—	—
$PBr_3(g)$	348.1 ± 0.4	145.9 ± 9.2	gas	298	74.43	0.18	−61.40	—	—
PCl_3	218.5 ± 1.0	320.9 ± 9.2	liq	298	131.38	—	—	—	—
$PCl_3(g)$	311.7 ± 0.5	288.7 ± 4.2	gas	298	82.37	0.41	−9.41	—	—
$PCl_5(g)$	364.0 ± 1.7	360.2 ± 9.6	gas	298	131.46	0.84	−17.78	—	—
$PF_3(g)$	272.8 ± 0.8	958.1	gas	298	67.43	14.99	−11.77	—	—
			gas	800	82.48	0.26	−32.67	—	—
$PF_5(g)$	300.7	1576.9 ± 1.3	gas	298	101.59	29.80	−22.84	—	—
			gas	800	131.08	0.69	−62.63	—	—
$PH_3(g)$	210.2 ± 0.4	− 5.4 ± 2.1	gas	298	26.30	40.48	−1.14	—	—
			gas	800	68.28	5.44	−90.29	—	—
PN(g)	211.0 ± 0.3	−104.6	gas	298	28.87	9.12	−1.61	−2.51	—
$P_4O_6(g)$	345.6 ± 1.3	2214.2	gas	298	182.42	48.12	−47.03	—	—
P_2O_5	114.4 ± 0.8	1505.0 ± 7.5	hex	298	74.89	162.34	−15.61	—	—
$P_4O_{10}(g)$	403.8	2834.2	gas	298	278.78	33.51	−9.12	—	—
$POBr_3(g)$	359.7 ± 0.4	406.6	gas	298	96.50	10.52	−8.70	—	—
$POCl_3$	222.5 ± 0.8	597.5 ± 2.1	liq	298	138.78	—	—	—	—
$POCl_3(g)$	325.4 ± 0.8	542.2 ± 2.1	gas	298	102.90	2.61	−17.26	—	—
P_4S_3	203.3 ± 0.4	154.4	orth	298	165.27	—	—	—	—
			hex	314	162.88	56.90	—	—	10.3
			liq	446	230.12	—	—	—	20.2
PS(g)	235.1 ± 0.8	−238.5 ± 9.6	gas	298	37.03	0.50	−1.80	—	—
P_2S_3	140.8 ± 8.4	121.3 ± 10.5	cryst	298	79.41	108.37	—	—	—
Pb	64.8	0.0	fcc	298	24.23	8.70	—	—	—
			liq	601	32.49	−3.10	—	—	4.8
			liq	1300	28.62	—	—	—	—

TABLE I—*continued*

Substance	$S_{298} \pm \delta S$ (J/deg mol)	$-\Delta H_{298} \pm \delta H$ (kJ/mol)	Phase	T K	$C_p = A + BT + C/T^2 + DT^2$ A	$B \times 10^3$ (J/deg mol)	$C \times 10^{-5}$	$D \times 10^6$	H_t (kJ/mol)
Pb(g)	175.3	−195.2	gas	298	20.79	—	—	—	—
			gas	900	17.70	−1.13	19.25	2.22	—
$PbBr_2$	161.1 ± 2.1	277.4 ± 2.5	orth	298	55.86	50.75	7.66	—	—
			liq	644	112.13	—	—	—	16.4
$PbCl_2$	136.0 ± 2.1	359.4 ± 0.8	orth	298	68.49	29.04	—	—	—
			liq	774	111.50	—	—	—	21.9
$PbCl_2$(g)	317.1 ± 2.9	174.1 ± 1.3	gas	298	58.03	0.33	−2.64	—	—
PbF_2	113.0 ± 8.4	677.0 ± 4.2	orth	298	74.29	18.83	−5.44	—	—
				716	74.29	18.83	−5.44	—	8.4
			liq	1103	109.20	—	—	—	14.7
PbF_2(g)	292.5 ± 3.3	435.1 ± 8.4	gas	298	57.40	0.54	−5.90	—	—
PbI_2	174.9 ± 0.4	175.3 ± 0.8	hex	298	72.17	18.07	—	—	—
			liq	683	108.57	—	—	—	23.4
PbO	66.3 ± 0.8	219.4 ± 0.8	r.tet	298	52.38	8.66	−8.20	—	—
			y.orth	762	45.27	12.80	−2.99	—	0.2
PbO(g)	239.9 ± 0.2	−70.3 ± 7.1	gas	298	36.53	0.77	−3.91	—	—
Pb_3O_4	212.0 ± 6.7	718.8 ± 6.3	tetrag	298	187.19	14.48	−33.39	9.37	—
PbO_2	71.8 ± 0.4	274.5 ± 2.9	tetrag	298	63.22	31.00	−8.95	−13.97	—
$PbCO_3$	131.0 ± 3.3	699.6 ± 5.4	orth	298	51.84	119.66	—	—	—
$PbMoO_4$	166.1 ± 2.1	1053.1 ± 5.0	tetrag	298	124.06	46.65	−16.53	—	—
$Pb(NO_3)_2$	218.0 ± 6.7	451.9 ± 3.3	cryst	298	125.94	149.37	−16.74	—	—
PbS	91.3 ± 1.7	98.3 ± 2.1	cubic	298	46.74	9.41	—	—	—
			liq	1387	66.94	—	—	—	18.8
PbS(g)	251.3 ± 0.2	−131.8 ± 6.3	gas	298	37.32	0.38	−2.05	—	—
$PbSO_4$	148.5 ± 0.8	920.0 ± 1.3	orth	298	74.18	102.51	−1.55	—	—
			monocl	1139	184.10	—	—	—	17.2
			liq	1443	179.91	—	—	—	40.2
PbSe	102.5 ± 2.1	100.0 ± 2.1	cubic	298	47.24	10.00	—	—	—
			liq	1350	62.76	—	—	—	49.4
PbSe(g)	263.6 ± 4.2	−126.4 ± 8.4	gas	298	37.40	—	−1.05	—	—
$PbSeO_3$	148.5	537.6	monocl	298	85.77	46.02	—	—	—
Pb_2SiO_4	186.6 ± 2.1	1369.0 ± 2.5	α	298	127.70	82.55	−13.68	—	—
			β	600	199.58	−1.55	−91.09	—	—
			liq	1018	189.12	—	—	—	53.6

$PbSiO_3$	109.6 ± 1.3	1147.7 ± 3.3	cryst	298	74.64	110.75	−11.46	−54.14	—
			liq	1037	130.12	—	—	—	34.5
PbTe	110.0 ± 2.1	68.6 ± 1.3	cubic	298	47.20	11.25	—	—	—
			liq	1197	62.76	—	—	—	57.3
$PbTiO_3$	111.9 ± 13.4	1194.7 ± 13.8	orth	298	120.29	17.91	−18.20	—	—
			β	763	109.08	22.80	−13.35	—	4.8
$PbWO_4$	167.8 ± 1.7	1121.3 ± 7.9	tetrag	298	120.00	41.25	—	—	—
Pd	37.9	0.0	fcc	298	23.78	7.39	—	—	—
			fcc	400	24.61	5.30	—	—	—
			fcc	1400	20.72	8.08	—	—	—
			liq	1825	34.73	—	—	—	17.6
Pd(g)	166.9	−377.0	gas	298	20.79	—	—	—	—
			gas	800	−6.74	20.17	74.81	—	—
			gas	2100	44.56	16.19	−949.01	−4.31	—
$PdCl_2$	103.8 ± 12.6	173.2 ± 12.6	hex	298	69.04	20.92	—	—	—
			liq	952	94.14	—	—	—	18.4
PdI_2	150.6 ± 10.5	63.6 ± 10.5	monocl	298	68.20	23.01	—	—	—
PdO	38.9 ± 4.2	115.5 ± 3.3	tetrag	298	45.31	7.03	−1.26	0.38	—
Pd_4S	180.7 ± 0.4	69.0 ± 2.1	tetrag	298	100.42	48.53	—	—	—
PdS	56.5 ± 6.3	70.7 ± 6.3	tetrag	298	41.71	17.20	−3.05	—	—
PdS_2	87.9 ± 12.6	78.2 ± 12.6	orth	298	68.58	15.77	−6.57	—	—
PdTe	89.6 ± 0.4	37.7 ± 16.7	hex	298	47.45	129.29	—	—	—
$PdTe_2$	126.6 ± 0.4	54.4 ± 20.9	hex	298	70.63	20.08	—	—	—
Pr	73.6	0.0	hcp	298	18.83	17.07	3.05	3.97	—
			bcc	1065	38.49	—	—	—	3.2
			liq	1191	42.97	—	—	—	6.9
Pr(g)	189.7	−355.6	gas	298	13.89	24.35	1.00	−10.42	—
			gas	1100	31.42	—	−27.15	−0.88	—
			gas	1600	30.46	−1.84	7.61	0.17	—
$PrCl_3$	153.3 ± 0.8	1057.7 ± 2.1	hex	298	88.37	45.61	−2.76	—	—
			liq	1059	133.89	—	—	—	50.6
$PrCl_3$(g)	379.5 ± 6.7	730.1 ± 5.0	gas	298	86.19	—	−3.35	—	—
PrF_3	117.2 ± 8.4	1712.1 ± 5.0	trigon	298	92.05	19.96	4.75	—	—
			liq	1672	130.75	—	—	—	57.3
PrF_3(g)	344.3 ± 4.2	1279.9 ± 11.7	gas	298	83.55	—	−6.95	—	—
PrH_2	56.8 ± 0.3	200.0 ± 3.3	cubic	298	35.48	18.83	—	—	—
PrI_3	226.4 ± 16.7	654.4 ± 16.7	orth	298	89.12	40.58	—	—	—
			liq	1011	143.09	—	—	—	52.3
Pr_2O_3	155.6 ± 2.1	1809.6	hex	298	119.66	17.78	−7.41	—	—
			cubic	2150	154.81	—	—	—	—
			liq	2570	152.30	—	—	—	92.0

TABLE I—*continued*

Substance	$S_{298} \pm \delta S$ (J/deg mol)	$-\Delta H_{298} \pm \delta H$ (kJ/mol)	Phase	T K	$C_p = A + BT + C/T^2 + DT^2$ A	$B \times 10^3$ (J/deg mol)	$C \times 10^{-5}$	$D \times 10^6$	H_t (kJ/mol)
PrS	77.8 ± 4.2	451.9 ± 25.1	cubic	298	51.97	4.39	—	—	—
Pt	41.5	0.0	fcc	298	24.69	5.02	−0.50	—	—
			fcc	1000	24.43	5.23	−0.75	—	—
			liq	2042	34.69	—	—	—	19.7
Pt(g)	192.3	−565.0	gas	298	36.32	−14.02	−5.86	—	—
			gas	500	31.21	−11.34	1.34	3.56	—
			gas	1700	18.66	1.38	33.64	—	—
$PtBr_4$	251.0	140.6 ± 16.7	rhomb	298	150.62	—	—	—	—
$PtCl_4$	205.0 ± 20.9	236.8 ± 16.7	cubic	298	146.44	—	—	—	—
PtO_2(g)	255.9 ± 10.5	−168.6 ± 10.5	gas	298	55.44	2.09	−11.51	—	—
PtS	55.1 ± 0.4	83.1 ± 2.5	tetrag	298	41.71	17.20	−3.05	—	—
PtS_2	74.7 ± 0.4	110.9 ± 2.5	hex	298	68.58	15.77	−6.57	—	—
Pt_5Se_4	336.8 ± 31.4	242.7 ± 37.7	monocl	298	188.28	78.45	—	—	—
PuN	64.9	299.2	cubic	298	44.89	15.48	—	—	—
Rb	76.8	0.0	bcc	298	31.71	—	—	—	—
			liq	313	40.88	−26.21	0.33	14.14	2.2
Rb(g)	170.0	−80.9	gas	298	20.79	—	—	—	—
Rb_2(g)	270.7	−118.4	gas	298	37.28	2.00	—	—	—
RbBr	110.0 ± 0.8	394.6 ± 0.8	cubic	298	49.37	10.67	—	—	—
			liq	965	66.94	—	—	—	23.3
RbCl	95.2 ± 0.4	435.1 ± 0.8	cubic	298	48.24	10.46	—	—	—
			liq	996	64.02	—	—	—	23.7
RbF	77.8 ± 2.1	555.6 ± 1.7	cubic	298	33.33	38.53	5.02	—	—
			liq	1050	59.41	—	—	—	23.0
RbH	58.6 ± 6.3	52.3 ± 0.4	cubic	298	27.70	40.67	—	—	—
RbI	118.8 ± 0.8	331.8 ± 5.0	cubic	298	49.08	11.30	—	—	—
			liq	929	66.53	—	—	—	22.0
Rb_2O	125.5 ± 4.2	338.9 ± 8.4	cubic	298	60.25	46.02	—	—	—
			hex	613	89.96	—	—	—	4.6
			liq	778	95.81	—	—	—	20.9
$RbBO_2$	94.6 ± 0.4	974.9 ± 20.9	cryst	298	58.37	50.63	—	—	—
Rb_2CO_3	181.3 ± 0.4	1133.0 ± 13.8	monocl	298	105.86	80.75	−10.88	—	—
RbOH	92.0 ± 8.4	418.8 ± 3.3	cryst	298	64.85	71.13	—	—	—
			β	508	79.50	—	—	—	5.4
			liq	658	83.68	—	—	—	8.9

Rb_2S	133.1 ± 12.6	361.1 ± 20.9	cubic	298	77.40	20.92	—	—	—
Rb_2SO_4	197.5 ± 0.4	1437.2 ± 1.3	orth	298	84.94	164.43	—	—	—
			hex	930	81.84	84.94	—	—	4.2
			liq	1341	207.11	—	—	—	38.4
Re	36.5	0.0	hcp	298	23.68	5.44	—	—	—
Re(g)	188.8	−775.0	gas	298	20.79	—	—	—	—
			gas	1400	29.75	−10.84	—	3.26	—
$ReBr_3$	200.8 ± 16.7	175.7 ± 5.9	monocl	298	80.33	68.20	—	—	—
$ReCl_3$	123.8 ± 0.4	263.6 ± 10.5	hex	298	105.48	27.61	−19.08	—	—
ReF_6(g)	363.6 ± 2.1	1353.5 ± 12.6	gas	298	155.64	1.30	−31.59	—	—
ReO_2	62.8 ± 4.6	432.6 ± 5.0	monocl	298	67.36	12.68	−12.93	—	—
ReO_3	80.8 ± 8.4	610.9 ± 12.6	cubic	298	108.78	—	—	—	—
Re_2O_7	207.1 ± 0.8	1248.5 ± 10.5	orth	298	121.96	184.10	−9.41	—	—
			liq	570	297.48	—	—	—	62.8
ReS_2	60.7 ± 8.4	178.7 ± 12.6	hex	298	68.58	15.77	−6.57	—	—
Re_2S_7	167.4 ± 20.9	451.5 ± 16.7	tetrag	298	184.10	50.21	—	—	—
Re_5Si_3	255.9 ± 16.7	157.3 ± 62.8	tetrag	298	190.79	45.19	−14.06	—	—
ReSi	55.4 ± 4.2	52.7 ± 20.9	cubic	298	52.59	9.62	−3.77	—	—
$ReSi_2$	74.1 ± 6.7	90.4 ± 31.4	tetrag	298	67.78	11.05	−6.11	—	—
Rh	31.5	0.0	fcc	298	20.79	13.43	0.33	−2.26	—
			liq	2236	41.84	—	—	—	21.5
Rh(g)	185.7	−553.1	gas	298	21.17	6.40	−2.13	−1.46	—
$RhCl_3$	126.8 ± 14.6	280.3 ± 12.6	monocl	298	105.44	27.61	−19.08	—	—
Rh_2O_3	106.3 ± 8.4	355.6	rhomb	298	86.78	57.74	—	—	—
RhO(g)	229.7 ± 10.5	−410.0 ± 50.2	gas	298	37.87	—	−5.44	—	—
Rh_3S_4	182.0 ± 25.1	357.7	cryst	298	145.60	60.25	−10.67	—	—
Rh_2S_3	125.5 ± 20.9	262.8	cryst	298	110.25	32.97	−9.62	—	—
Ru	28.5	0.0	hcp	298	18.58	9.29	2.80	—	—
			liq	2523	41.84	—	—	—	24.3
Ru(g)	186.4	−651.4	gas	298	23.93	2.01	−2.72	—	—
$RuCl_3$	127.6 ± 10.5	253.1 ± 14.6	hex	298	115.06	—	—	—	—
$RuCl_3$(g)	397.5 ± 20.9	−56.1 ± 18.8	gas	298	56.90	7.66	−3.01	—	—
$RuCl_4$(g)	374.5 ± 20.9	93.3 ± 16.7	gas	298	95.81	—	−10.46	—	—
RuF_5	161.1 ± 14.6	892.9 ± 1.7	cryst	298	163.18	—	—	—	—
			liq	358	182.00	—	—	—	18.8
RuO_2	58.6 ± 4.6	305.0 ± 6.3	tetrag	298	69.87	10.46	−14.85	—	—
RuO_4(g)	290.7 ± 0.4	184.1 ± 5.0	gas	298	101.80	3.05	−24.02	—	—
RuS_2	54.4 ± 12.6	205.9 ± 20.9	cubic	298	68.53	11.84	−8.79	—	—
S	32.0 ± 0.1	0.0	orth	298	23.51	—	—	—	—
			monocl	368	24.73	—	—	—	0.4
			liq	388	35.19	—	—	—	1.7

TABLE I—*continued*

Substance	$S_{298} \pm \delta S$ (J/deg mol)	$-\Delta H_{298} \pm \delta H$ (kJ/mol)	Phase	T K	$C_p = A + BT + C/T^2 + DT^2$ (J/deg mol)				H_t (kJ/mol)
					A	$B \times 10^3$	$C \times 10^{-5}$	$D \times 10^6$	
S(g)	167.7	−277.2 ± 0.4	gas	298	22.01	−0.42	1.51	—	—
			gas	2400	20.75	0.42	2.51	—	—
S_2(g)	228.1	−128.6	gas	298	33.53	6.02	−2.37	−2.09	—
			gas	1000	34.62	3.67	−6.07	−0.43	—
S_3(g)	276.0	−146.4	gas	298	57.60	0.32	−6.33	—	—
S_4(g)	325.9	−188.3	gas	298	82.55	3.14	−7.11	—	—
			gas	718	82.42	3.14	−7.11	—	—
S_5(g)	320.9	−111.0	gas	298	106.48	0.85	−16.45	—	—
S_6(g)	353.9	−98.7	gas	298	131.75	0.67	−17.20	—	—
S_7(g)	394.1	−102.9	gas	298	155.91	1.09	−23.34	—	—
			gas	718	155.91	1.09	−23.34	—	—
S_8(g)	423.1	−96.8	gas	298	181.18	0.92	−22.79	—	—
S_2Cl_2	223.8 ± 4.2	58.2 ± 2.1	liq	298	124.26	—	—	—	—
S_2Cl_2(g)	327.2 ± 0.4	16.7 ± 4.2	gas	298	80.79	3.04	−7.76	—	—
SCl_2(g)	281.5 ± 0.4	17.6 ± 3.3	gas	298	57.53	0.36	−6.00	—	—
SF_5(g)	301.2 ± 8.4	908.3 ± 15.9	gas	298	126.19	3.18	−33.47	—	—
SF_6(g)	291.6 ± 0.4	1220.5 ± 1.7	gas	298	146.69	5.90	−46.02	—	—
SO(g)	221.8 ± 0.2	− 5.0 ± 2.1	gas	298	33.35	2.85	−3.60	—	—
SO_2(g)	248.1	296.8	gas	298	49.96	4.77	−10.46	—	—
SO_3	122.6 ± 8.4	438.5 ± 0.8	liq	298	179.91	—	—	—	—
SO_3(g)	256.7 ± 0.2	395.8 ± 0.8	gas	298	70.00	6.61	−19.35	—	—
$SOCl_2$(g)	307.9 ± 0.4	212.3 ± 3.8	gas	298	74.27	7.20	−8.70	—	—
SO_2Cl_2(g)	311.0 ± 0.4	354.8 ± 2.1	gas	298	97.49	4.81	−19.46	—	—
SO_2F_2(g)	283.5 ± 0.4	758.6 ± 8.4	gas	298	96.02	5.69	−28.49	—	—
Sb	45.5	0.0	rhomb	298	23.05	7.28	—	—	—
			liq	904	31.38	—	—	—	19.9
Sb(g)	180.2	−264.6	gas	298	20.79	—	—	—	—
			gas	800	17.83	1.91	9.68	—	—
Sb_2(g)	254.8	−231.2	gas	298	37.36	0.04	−0.91	—	—
			gas	1300	37.28	0.06	—	—	—
			gas	1860	37.39	—	—	—	—
Sb_4(g)	350.0	−206.5	gas	298	82.91	0.17	−1.80	—	—
			gas	1300	82.81	0.14	—	—	—
			gas	1860	62.14	—	—	—	—

$SbBr_3$	210.0 ± 4.2	259.4 ± 14.6	orth	298	112.97	—	—	—	—
			liq	370	125.52	—	—	—	14.6
$SbBr_3$(g)	372.4 ± 0.8	180.7	gas	298	83.26	—	−2.68	—	—
$SbCl_3$	183.3 ± 3.8	381.2 ± 1.7	orth	298	43.10	225.94	—	—	—
			liq	346	123.85	—	—	—	13.0
$SbCl_3$(g)	338.5 ± 1.3	312.1 ± 3.3	gas	298	83.01	—	−4.98	—	—
$SbCl_5$(g)	401.7 ± 1.0	389.1 ± 5.4	gas	298	132.21	−9.83	—	—	—
SbF_3	127.2	915.5 ± 16.7	orth	298	107.11	—	—	—	—
			liq	564	127.61	—	—	—	21.3
SbH_3(g)	233.0 ± 0.4	145.1 ± 0.8	gas	298	50.50	18.70	−13.18	—	—
SbI_3	215.5 ± 1.7	100.4 ± 27.6	hex	298	71.13	88.70	—	—	—
			liq	443	143.51	—	—	—	17.6
Sb_2O_3	132.7 ± 4.2	716.1 ± 3.3	cubic	298	75.31	97.49	—	—	—
			orth	845	92.05	66.11	—	—	8.1
			liq	929	156.90	—	—	—	54.8
SbO_2	63.6 ± 4.2	453.5 ± 2.5	cubic	298	47.28	33.89	—	—	—
Sb_2O_5	125.1 ± 8.4	1007.5 ± 4.6	cubic	298	69.04	230.12	—	—	—
SbOCl	107.5 ± 9.2	380.7 ± 16.7	monocl	298	67.99	21.97	—	—	—
Sb_2S_3	107.53 ± 3.3	205.0	orth	298	101.84	60.54	—	—	—
			liq	823	167.36	—	—	—	47.9
Sb_2Se_3	212.1 ± 3.3	127.6 ± 1.3	orth	298	118.74	20.92	—	—	—
			liq	888	171.54	—	—	—	53.8
Sb_2Te_3	246.4 ± 3.3	56.5 ± 1.3	rhomb	298	112.88	53.14	—	—	—
			liq	892	196.65	—	—	—	99.0
Sc	34.8	0.0	hex	298	24.74	1.33	0.35	5.10	—
			cubic	1610	44.22	—	—	—	4.0
			liq	1814	44.35	—	—	—	14.1
Sc(g)	300.2	−379.1	gas	298	20.92	—	—	—	—
			gas	1600	28.12	−8.74	—	2.68	—
$ScCl_3$	121.3	918.8 ± 3.3	rhomb	298	95.65	15.40	−7.28	—	—
			liq	1240	143.43	—	—	—	67.4
ScF_3	89.1	1648.9 ± 8.4	hex	298	98.58	3.22	−13.16	—	—
			liq	1825	88.87	—	—	—	62.8
ScN	29.7 ± 4.2	313.8 ± 16.7	cubic	298	45.81	5.44	−9.20	—	—
Sc_2O_3	77.0 ± 0.4	1908.3 ± 3.3	cubic	298	99.79	22.22	−11.09	—	—
Se	42.3	0.0	hex	298	17.89	25.10	—	—	—
			liq	493	35.15	—	—	—	5.9
Se(g)	176.6	−235.4	gas	298	21.46	1.51	−0.92	—	—
Se_2(g)	246.9	−138.2	gas	298	44.60	−2.66	−2.50	—	—
Se_3(g)	314.9	−176.1	gas	298	58.15	3.04	−2.21	—	—
Se_4(g)	379.1	−183.3	gas	298	83.16	0.03	−2.51	—	—

TABLE I—*continued*

Substance	$S_{298} \pm \delta S$ (J/deg mol)	$-\Delta H_{298} \pm \delta H$ (kJ/mol)	Phase	T K	$C_p = A + BT + C/T^2 + DT^2$				H_t (kJ/mol)
					A	$B \times 10^3$ (J/deg mol)	$C \times 10^{-5}$ (J/deg mol)	$D \times 10^6$	
$Se_5(g)$	385.3	−138.1	gas	298	107.93	0.09	−5.91	—	—
$Se_6(g)$	433.5	−135.1	gas	298	132.90	0.07	−5.92	—	—
$Se_2Cl_2(g)$	353.8	21.8 ± 8.4	gas	298	82.42	1.55	−4.52	—	—
$SeCl_2(g)$	295.6	33.5 ± 12.6	gas	298	57.95	0.13	−3.95	—	—
$SeCl_4$	194.6 ± 16.7	188.7 ± 6.3	monocl	298	133.89	—	—	—	—
$SeF_4(g)$	296.4	811.7 ± 41.8	gas	298	101.04	3.68	−27.53	—	—
$SeF_5(g)$	326.4	940.6 ± 14.6	gas	298	127.61	2.80	−29.29	—	—
$SeF_6(g)$	313.5 ± 0.8	1117.1 ± 1.3	gas	298	151.13	3.60	−37.40	—	—
$SeO(g)$	233.9	−62.3 ± 20.9	gas	298	34.94	1.51	−3.68	—	—
SeO_2	66.7 ± 1.7	225.1 ± 2.1	tetrag	298	69.58	3.89	−11.05	—	—
$SeO_2(g)$	264.8	107.9 ± 8.4	gas	298	52.84	3.10	−9.90	—	—
Si	18.8	0.0	diam	298	23.93	2.22	−4.14	—	—
			liq	1685	27.20	—	—	—	50.2
$Si(g)$	167.9	−450.0	gas	298	19.82	1.00	2.01	—	—
$Si_2(g)$	229.7	−585.8	gas	298	37.25	0.46	−0.28	—	—
$Si_3(g)$	267.8	−636.0	gas	298	61.99	0.13	−0.73	—	—
$SiBr_2(g)$	303.3 ± 2.1	42.7 ± 0.8	gas	298	57.78	—	−4.64	—	—
$SiBr_4$	278.2 ± 1.3	457.3 ± 8.4	liq	298	146.44	—	—	—	—
$SiBr_4(g)$	379.2 ± 0.8	415.5 ± 16.7	gas	298	107.40	0.31	−9.33	—	—
SiC	16.5 ± 0.2	66.9 ± 6.3	cubic	298	42.59	8.37	−16.61	−1.26	—
$SiCl_2(g)$	281.8 ± 4.2	167.4 ± 4.2	gas	298	57.57	0.38	−5.65	—	—
$SiCl_4(g)$	330.8 ± 0.4	662.7 ± 1.3	gas	298	106.52	0.75	−14.73	—	—
$SiF_2(g)$	256.5 ± 0.8	619.2	gas	298	54.76	1.91	−9.99	—	—
$SiF_4(g)$	282.6 ± 0.4	1615.0 ± 0.8	gas	298	91.46	13.26	−19.66	—	—
$SiH(g)$	197.9 ± 0.4	−368.2 ± 20.9	gas	298	28.33	4.84	—	—	—
			gas	700	30.29	3.65	—	—	—
$SiH_4(g)$	204.6 ± 0.2	−34.3 ± 2.1	gas	298	13.87	111.03	−0.49	−40.31	—
$Si_2H_6(g)$	274.3 ± 2.1	−80.3 ± 1.3	gas	298	53.93	136.40	−10.33	−44.77	—
$SiH_2Cl_2(g)$	286.6 ± 0.4	320.5 ± 12.6	gas	298	85.40	10.04	−23.64	—	—
$SiI_2(g)$	317.1 ± 2.5	−92.5 ± 8.4	gas	298	58.07	0.06	−3.08	—	—
SiI_4	258.2 ± 3.3	189.5 ± 16.7	cubic	298	81.96	87.45	—	—	—
			liq	394	147.49	41.30	—	—	19.7
$SiI_4(g)$	413.4 ± 1.3	125.1 ± 7.5	gas	298	106.98	1.00	−6.02	—	—

Si_3N_4	113.0 ± 8.4	744.8 ± 12.6	hex	298	76.36	109.04	−6.53	−27.07	—
SiO(g)	211.5 ± 0.4	98.3 ± 8.4	gas	298	29.83	8.24	−2.05	−2.28	—
SiO_2	41.5 ± 0.4	910.9 ± 1.7	α−qtz	298	43.93	38.83	−9.69	—	—
			β−qtz	847	58.91	10.04	—	—	0.7
SiO_2(cr)	43.4 ± 0.8	908.3 ± 1.7	α−cris	298	46.90	31.51	−10.08	—	—
			β−cris	540	71.63	1.88	−39.06	—	1.3
			liq	2000	86.19	—	—	—	9.6
$SiOF_2$(g)	271.2	966.5	gas	298	64.27	20.23	−14.64	−5.77	—
SiP	32.6 ± 9.6	61.9 ± 14.6	cryst	298	42.89	10.88	−5.65	—	—
SiS(g)	223.7 ± 2.1	−105.9 ± 12.6	gas	298	36.42	0.74	−4.04	—	—
SiS_2	80.3 ± 4.2	213.4 ± 20.9	orth	298	74.14	12.38	—	—	—
			liq	1363	91.00	—	—	—	20.9
SiSe(g)	235.1 ± 1.3	−202.9 ± 20.9	gas	298	36.74	0.42	−2.89	—	—
Sm	69.5	0.0	hcp	298	27.57	29.58	−5.61	−10.46	—
			bcc	1191	46.94	—	—	—	3.1
			liq	1346	50.21	—	—	—	8.6
Sm(g)	182.9	−206.7	gas	298	30.33	—	—	—	—
SmC_2	77.8 ± 8.4	97.9 ± 8.4	tetrag	298	68.62	11.30	−7.95	—	—
$SmCl_2$	127.6 ± 6.3	816.3 ± 8.4	cryst	298	77.40	16.74	—	—	—
$SmCl_2$(g)	315.5	500.4 ± 17.6	gas	298	65.19	−8.58	—	—	—
$SmCl_3$	113.0 ± 12.6	1028.4 ± 3.3	hex	298	82.26	47.70	0.75	—	—
Sm_2O_3	151.0 ± 0.4	1822.6 ± 3.3	monocl	298	128.66	19.41	−17.99	—	—
			monocl	1195	154.39	—	—	—	1.0
SmOCl	100.4 ± 12.6	1000.4 ± 10.0	tetrag	298	70.71	22.38	−5.73	—	—
SmS	81.2 ± 4.2	431.0 ± 41.8	cubic	298	59.33	3.18	−1.92	—	—
SmSe	94.1 ± 6.3	443.5 ± 6.3	cubic	298	59.83	3.22	−1.84	—	—
Sn	51.2	0.0	tetrag	298	21.59	18.16	—	—	—
			liq	505	32.84	−6.28	—	—	7.2
			liq	700	28.45	—	—	—	—
Sn(g)	168.4	−301.2	gas	298	8.31	31.51	3.16	—	—
			gas	500	−1.49	66.33	4.21	−32.18	—
			gas	1000	50.14	−10.73	−63.12	—	—
			gas	1400	29.20	−1.36	91.76	—	—
$SnBr_2$	149.8 ± 11.7	260.2 ± 6.7	cryst	298	114.64	33.05	—	—	—
			liq	505	99.58	—	—	—	18.0
$SnBr_4$(g)	405.0 ± 1.7	348.1	gas	298	107.91	—	−4.06	—	—
$SnCl_2$	131.8	331.0	rhomb	298	50.63	83.68	—	—	—
			liq	520	96.23	—	—	—	12.6
$SnCl_4$	259.0 ± 3.3	528.9 ± 14.6	liq	298	164.85	—	—	—	—
$SnCl_4$(g)	364.8 ± 1.3	489.1	gas	298	106.98	0.84	−7.82	—	—
SnH_4(g)	228.7 ± 0.4	−162.8 ± 2.1	gas	298	51.80	37.66	−11.30	—	—

TABLE I—*continued*

Substance	$S_{298} \pm \delta S$ (J/deg mol)	$-\Delta H_{298}$ $\pm \delta H$ (kJ/mol)	Phase	T K	$C_p = A + BT + C/T^2 + DT^2$ A	$B \times 10^3$ (J/deg mol)	$C \times 10^{-5}$	$D \times 10^6$	H_t (kJ/mol)
SnI_2	164.8	152.1 ± 3.3	cryst	298	70.29	29.29	—	—	—
			liq	593	94.56	—	—	—	18.8
$SnI_2(g)$	344.8 ± 1.7	− 0.8	gas	298	61.92	—	−4.60	—	—
SnI_4	270.3	206.3 ± 5.0	cubic	298	147.28	—	—	—	—
			liq	418	162.34	—	—	—	19.2
$SnI_4(g)$	446.4 ± 1.7	112.2	gas	298	108.37	—	−2.64	—	—
SnO	57.2 ± 0.4	280.7 ± 0.8	tetrag	298	39.96	14.64	—	—	—
SnO(g)	232.0 ± 0.2	−20.9 ± 9.6	gas	298	35.23	1.34	−3.51	—	—
SnO_2	52.3 ± 1.3	577.6 ± 0.4	tetrag	298	73.89	10.04	−21.59	—	—
$Sn(NO_3)_2$	208.8 ± 9.2	456.1 ± 18.8	cryst	298	209.20	—	—	—	—
SnS	77.0 ± 0.8	107.9 ± 2.1	orth	298	35.69	31.30	−3.77	—	—
			orth	875	32.55	15.65	—	—	0.7
			liq	1153	74.89	—	—	—	31.6
SnS(g)	242.3 ± 0.8	−112.5 ± 4.2	gas	298	36.94	0.33	−2.30	—	—
Sn_3S_4	243.5 ± 16.7	370.3 ± 20.9	tetrag	298	150.96	62.34	—	—	—
Sn_2S_3	164.4	263.6 ± 20.9	orth	298	107.03	43.93	—	—	—
SnS_2	87.4 ± 0.8	153.6	hex	298	64.89	17.57	—	—	—
$SnSO_4$	138.6 ± 0.8	1014.6 ± 31.4	rhomb	298	118.20	108.78	—	—	—
SnSe	98.1	94.1 ± 3.8	orth	298	46.65	19.96	—	—	—
SnSe(g)	255.2 ± 1.3	−120.9 ± 5.0	gas	298	37.36	—	−1.30	—	—
$SnSe_2$	118.0	124.7 ± 8.4	hex	298	62.05	31.63	—	—	—
SnTe	98.7	60.7 ± 0.8	cubic	298	48.95	10.13	—	—	—
			liq	1079	63.60	—	—	—	45.2
Sr	52.3	0.0	fcc	298	27.70	−3.68	−1.38	19.20	—
			bcc	820	37.66	—	—	—	0.8
			liq	1042	35.15	—	—	—	8.2
Sr(g)	164.5	−164.0	gas	298	20.79	—	—	—	—
$SrBr_2$	143.5 ± 4.2	718.0 ± 1.7	α	298	75.10	13.47	—	—	—
			β	918	115.06	—	—	—	12.2
			liq	930	116.40	—	—	—	10.1
$SrBr_2(g)$	323.4 ± 8.4	407.1 ± 12.6	gas	298	62.34	8.74	−1.30	−0.08	—
SrC_2	71.1 ± 8.4	84.5 ± 16.7	tetrag	298	68.62	11.30	−7.95	—	—
$SrCl_2$	114.8 ± 0.2	828.9 ± 2.5	cubic	298	113.72	−99.70	−14.90	92.88	—
			β	1000	123.01	—	—	—	6.0
			liq	1147	104.60	—	—	—	16.2

$SrCl_2(g)$	316.2 ± 5.0	473.2 ± 6.3	gas	298	58.16	20.04	−2.13	−0.21	—
SrF_2	82.1	1217.1 ± 2.9	cubic	298	69.37	16.07	−3.47	−0.59	—
			cubic	1475	69.37	16.07	−3.47	−0.59	23.7
			liq	1750	99.04	—	—	—	29.7
$SrF_2(g)$	291.6 ± 2.1	766.1 ± 4.2	gas	298	57.99	0.08	−4.52	—	—
SrH_2	49.8 ± 6.7	179.9 ± 5.9	cryst	298	33.47	22.59	—	—	—
SrI_2	159.1 ± 0.8	561.5 ± 2.1	cryst	298	69.83	27.41	−0.05	—	—
			liq	811	110.04	—	—	—	19.7
$SrI_2(g)$	339.4 ± 8.4	274.9 ± 6.3	gas	298	62.34	0.02	−0.84	—	—
SrO	55.5 ± 0.4	592.0 ± 3.8	cubic	298	50.75	6.07	−6.28	—	—
SrO(g)	230.0 ± 0.4	13.4 ± 16.7	gas	298	36.40	0.67	−3.10	—	—
SrO_2	59.0 ± 10.0	633.5 ± 15.1	tetrag	298	73.97	18.41	—	—	—
$SrAl_2O_4$	108.8 ± 20.9	2338.9 ± 16.7	α	298	177.19	4.94	−53.01	—	—
			β	932	146.11	29.29	—	—	1.9
SrB_4O_7	144.3 ± 6.3	3332.6 ± 26.8	ortho	298	165.69	142.26	−35.69	—	139.7
$SrCO_3$	97.1 ± 1.7	1220.1 ± 8.4	orth	298	88.78	35.90	−15.48	—	—
			hex	1197	142.26	—	—	—	19.7
$Sr(OH)_2$	97.1 ± 8.4	968.9 ± 9.2	cryst	298	35.23	133.05	—	—	—
$SrHfO_3$	113.0 ± 12.1	1783.6 ± 12.1	monocl	298	122.17	13.81	−19.87	—	—
$SrMoO_4$	128.9 ± 5.0	1549.3 ± 10.0	tetrag	298	134.14	29.37	−23.01	—	—
SrS	68.2 ± 2.9	452.7 ± 16.3	cubic	298	54.31	5.27	−6.49	—	—
$SrSO_4$	117.2 ± 8.4	1453.1 ± 17.6	celest	298	91.21	55.65	—	—	—
Sr_2SiO_4	149.8 ± 8.4	2302.9 ± 3.3	cryst	298	154.39	31.38	−31.38	—	—
$SrSiO_3$	96.2 ± 8.4	1633.4 ± 3.8	cryst	298	116.73	11.09	−29.29	—	—
Sr_2TiO_4	159.0	2287.8 ± 9.6	tetrag	298	160.87	16.07	−19.54	—	—
$SrTiO_3$	108.4 ± 1.3	1670.7 ± 7.9	cubic	298	118.11	8.54	−19.16	—	—
$SrWO_4$	133.9 ± 5.0	1621.3 ± 20.5	cryst	298	120.67	36.11	—	—	—
$SrZrO_3$	108.8 ± 8.4	1767.3 ± 14.6	cryst	298	121.25	12.22	−21.42	—	—
Ta	41.5	0.0	bcc	298	23.77	7.41	−0.42	−2.72	—
			bcc	1400	47.82	−22.34	—	6.57	—
			liq	3293	41.84	—	—	—	33.9
Ta(g)	185.1	−781.6	gas	298	17.95	11.21	−0.56	−1.72	—
TaB_2	46.0 ± 4.2	209.2	hex	298	59.45	18.83	−15.06	—	—
			liq	3250	125.52	—	—	—	83.7
$TaBr_5$	303.3 ± 18.8	598.3 ± 6.3	orth	298	119.16	122.59	—	—	—
			liq	543	184.10	—	—	—	47.3
Ta_2C	81.6 ± 4.2	208.4 ± 9.2	hex	298	66.44	13.93	−8.58	—	—
TaC	42.3 ± 0.4	142.7 ± 3.8	cubic	298	43.30	8.16	−7.95	—	—
			liq	4100	66.94	—	—	—	104.6
$TaCl_3$	154.8 ± 8.4	552.3 ± 3.8	cubic	298	96.23	16.32	−7.11	—	—

TABLE I—*continued*

Substance	$S_{298} \pm \delta S$ (J/deg mol)	$-\Delta H_{298}$ $\pm\delta H$ (kJ/mol)	Phase	T K	$C_p = A + BT + C/T^2 + DT^2$ A	$B \times 10^3$	$C \times 10^{-5}$ (J/deg mol)	$D \times 10^6$	H_t (kJ/mol)
$TaCl_4$	192.5 ± 8.4	705.0 ± 4.6	monocl	298	133.47	—	−12.13	—	—
$TaCl_4(g)$	382.8 ± 10.5	566.9 ± 8.8	gas	298	107.95	—	−8.37	—	—
$TaCl_5$	221.8 ± 6.3	859.0 ± 4.2	monocl	298	147.90	—	—	—	—
			liq	490	180.75	—	—	—	35.1
$TaCl_5(g)$	412.9 ± 3.3	764.8 ± 8.4	gas	298	132.51	0.27	−11.09	—	—
$TaCr_2$	88.1 ± 0.6	27.0 ± 2.5	cubic	298	73.85	22.80	−7.20	—	—
TaF_5	169.9 ± 15.9	1903.7 ± 2.1	monocl	298	133.89	—	—	—	—
			liq	369	177.82	—	—	—	12.6
$TaF_5(g)$	352.9 ± 16.7	1820.0 ± 16.7	gas	298	130.67	1.09	−24.69	—	—
$TaFe_2$	106.7 ± 5.4	57.7 ± 7.5	hex	298	66.94	26.36	—	—	—
Ta_2N	74.5 ± 9.2	272.4 ± 5.0	hex	298	70.50	17.66	−7.11	—	—
TaN	41.8 ± 1.3	252.3 ± 2.5	cubic	298	55.27	2.72	−12.64	—	—
			liq	3360	62.76	—	—	—	66.9
$TaO_2(g)$	279.5 ± 1.7	184.1 ± 20.9	gas	298	54.81	0.17	−10.13	—	—
Ta_2O_5	143.1 ± 1.3	2046.4 ± 4.2	orth	298	161.29	21.42	−29.20	—	—
			liq	2058	242.67	—	—	—	120.1
TaS_2	75.3	354.0 ± 16.7	hex	298	69.87	—	—	—	—
Ta_5Si_3	280.7 ± 26.8	334.7 ± 33.5	tetrag	298	179.70	39.12	−8.91	—	—
$TaSi_2$	75.3 ± 6.3	119.2 ± 12.6	hex	298	73.26	7.70	−9.08	—	—
Tb	73.3	0.0	hcp	298	16.78	18.70	6.02	—	—
			bcc	1591	27.74	—	—	—	5.0
			liq	1633	46.44	—	—	—	10.8
Tb(g)	203.1	−388.7	gas	298	25.46	−2.85	—	—	—
			gas	500	19.50	5.87	4.15	—	—
			gas	1000	21.23	6.50	−9.08	−1.07	—
TbF_3	90.4 ± 1.3	1707.9 ± 5.0	orth	298	97.82	19.66	−12.07	—	—
$TbCl_3$	147.7 ± 7.5	998.7 ± 3.3	orth	298	94.10	25.75	−3.10	—	—
			orth	783	123.93	—	—	—	14.2
			liq	855	144.47	—	—	—	19.5
Tb_2O_3	156.9 ± 8.4	1865.2 ± 7.5	cubic	298	108.78	41.42	−8.37	—	—
TbO_2	82.8	971.5 ± 3.8	cubic	298	64.81	17.70	−7.59	—	—
Te	49.5	0.0	hex	298	19.16	21.97	—	—	—
			liq	723	37.66	—	—	—	17.5
Te(g)	182.6	−211.7	gas	298	19.41	1.84	0.75	—	—

$Te_2(g)$	260.6	−159.1	gas	298	34.64	6.61	−0.26	—	—
$TeCl_4$	200.8 ± 20.9	323.8	cryst	298	138.49	—	—	—	—
			liq	497	230.12	—	—	—	18.9
$TeCl_4(g)$	376.6 ± 20.9	205.9 ± 20.9	gas	298	96.71	0.15	−5.49	—	—
$TeF_4(g)$	324.1 ± 12.6	948.1	gas	298	104.14	2.08	−20.13	—	—
$TeF_6(g)$	335.9 ± 1.7	1369.0 ± 0.8	gas	298	152.08	3.10	−31.71	—	—
$TeO(g)$	240.6 ± 0.8	−74.5	gas	298	35.31	1.34	−34.73	—	—
TeO_2	70.1 ± 0.8	316.3 ± 2.1	tetrag	298	65.19	15.06	−7.95	—	—
			liq	1004	112.63	2.18	—	—	29.1
$TeO_2(g)$	274.9 ± 4.2	52.3 ± 8.4	gas	298	54.77	2.43	−11.84	—	—
Th	53.4	0.0	fcc	298	25.10	8.37	−0.23	−0.22	—
			bcc	1633	15.69	11.97	—	—	3.6
			liq	2028	46.02	—	—	—	13.8
$Th(g)$	190.0	−597.1	gas	298	8.62	21.21	5.86	−3.20	—
$ThBr_4$	228.0 ± 6.3	965.7 ± 6.3	orth	298	127.61	15.06	−6.15	—	—
			tetrag	693	127.61	15.06	−6.15	—	4.2
			liq	952	171.54	—	—	—	54.4
$ThBr_4(g)$	429.7 ± 12.6	766.5 ± 10.5	gas	298	110.16	—	−5.02	—	—
ThC	58.0 ± 0.8	126.4 ± 10.5	cubic	298	42.89	7.36	—	—	—
ThC_2	68.7 ± 0.4	122.2 ± 6.3	monocl	298	63.89	12.09	−9.25	—	—
			monocl	1700	83.68	—	—	—	6.3
			cubic	1763	83.68	—	—	—	6.3
$ThCl_4$	190.4 ± 5.0	1186.6 ± 2.9	orth	298	120.29	23.26	−6.15	—	—
			tetrag	679	120.29	23.26	−6.15	—	5.0
			liq	1043	163.18	—	—	—	61.5
$ThCl_4(g)$	390.8 ± 8.4	974.9 ± 9.2	gas	298	108.07	—	−6.15	—	—
ThF_4	142.0 ± 0.4	2111.2 ± 8.4	monocl	298	111.92	24.48	−7.55	—	—
			liq	1383	152.72	—	—	—	43.9
$ThF_4(g)$	341.8 ± 10.5	1775.3 ± 11.7	gas	298	108.37	—	−13.60	—	—
ThH_2	59.4 ± 4.2	143.5 ± 4.2	bctet	298	52.76	11.09	—	—	—
ThI_4	265.7 ± 10.5	664.4 ± 5.9	monocl	298	140.16	12.97	−6.15	—	—
			liq	839	177.40	—	—	—	48.1
$ThI_4(g)$	472.0 ± 10.5	469.4	gas	298	108.11	—	−1.30	—	—
ThN	57.3 ± 1.3	378.7 ± 10.5	cubic	298	48.12	9.41	−5.86	—	—
Th_3N_4	182.8 ± 2.1	1305.4 ± 20.9	hex	298	164.56	26.11	−22.30	—	—
$ThO(g)$	240.0 ± 0.4	28.5 ± 2.1	gas	298	37.91	—	−5.98	—	—
ThO_2	65.2 ± 0.4	1226.7 ± 1.3	cubic	298	69.29	9.33	−9.18	—	—
$ThOBr_2$	131.0 ± 10.5	1129.7 ± 12.6	cryst	298	98.45	12.22	−7.66	—	—
$ThOCl_2$	113.6 ± 7.9	1236.0 ± 5.0	orth	298	98.32	11.21	−9.25	—	—
$ThOI_2$	158.6 ± 6.3	992.4 ± 5.0	cryst	298	99.50	11.13	−7.70	—	—
Th_2N_2O	124.3 ± 6.3	1288.7 ± 25.1	hex	298	116.90	18.74	−15.73	—	—

TABLE I—*continued*

Substance	$S_{298} \pm \delta S$ (J/deg mol)	$-\Delta H_{298}$ $\pm \delta H$ (kJ/mol)	Phase	T K	$C_p = A + BT + C/T^2 + DT^2$ (J/deg mol) A	$B \times 10^3$	$C \times 10^{-5}$	$D \times 10^6$	H_t (kJ/mol)
$ThRe_2$	123.6 ± 5.0	174.1 ± 5.0	hex	298	70.96	23.56	—	—	—
Th_7Rh_3	389.1	757.3	hex	298	247.27	81.59	−5.19	—	—
ThRh	64.9	232.6 ± 6.3	orth	298	40.00	16.82	−1.05	—	—
$ThRh_3$	120.9	321.7 ± 10.5	cubic	298	77.49	34.81	−2.13	—	—
$ThRh_5$	199.8	316.7 ± 12.6	cryst	298	114.93	52.80	−3.22	—	—
ThRu	61.3	121.8 ± 7.5	orth	298	43.05	10.88	—	—	—
Th_7Ru_3	378.7	505.4	hex	298	198.74	77.40	—	—	—
ThS	69.8 ± 0.4	399.6 ± 6.3	cubic	298	50.12	5.46	−3.59	—	—
Th_2S_3	171.5 ± 11.7	1083.2 ± 12.6	orth	298	121.96	15.06	−3.77	—	—
ThS_2	96.2 ± 0.8	627.6 ± 33.5	orth	298	67.03	11.00	—	—	—
$Th(SO_4)_2$	165.3 ± 20.9	2541.4 ± 20.9	cryst	298	104.60	230.96	—	—	—
Th_3Si_2	163.2 ± 12.6	284.9 ± 20.9	tetrag	298	124.26	28.45	−9.79	—	—
ThSi	58.2 ± 12.6	128.0 ± 12.6	orth	298	49.41	10.29	−4.64	—	—
Th_3Si_5	213.4 ± 20.9	486.2 ± 37.7	hex	298	196.10	35.82	−22.26	—	—
$ThSi_2$	82.0 ± 8.4	174.1 ± 16.7	hex	298	73.35	12.76	−8.79	—	—
Ti	30.7	0.0	hex	298	24.94	6.57	−1.63	1.34	—
			bcc	1155	30.84	−8.87	—	6.44	4.2
			liq	1943	41.84	—	—	—	16.7
Ti(g)	180.2	−473.6	gas	298	22.26	−3.26	2.76	2.13	—
TiB	34.7	160.2 ± 16.7	orth	298	54.06	−0.04	−21.63	—	—
TiB_2	28.5 ± 0.4	315.9 ± 7.5	hex	298	56.38	25.86	−17.47	−3.35	—
			liq	3498	108.78	—	—	—	108.8
$TiBr_2$	108.4	405.4	trigon	298	76.09	10.75	−0.51	—	—
$TiBr_2(g)$	308.8	179.1	gas	298	60.28	2.13	−0.62	—	—
$TiBr_3$	176.6	551.9 ± 6.7	trigon	298	−10.79	284.34	34.18	−119.2	—
$TiBr_3(g)$	359.0	376.6	gas	298	88.20	−1.20	−7.49	—	—
$TiBr_4$	243.5	619.7 ± 2.1	cubic	298	80.92	169.62	—	—	—
			liq	311	151.88	—	—	—	12.9
$TiBr_4(g)$	398.5	551.5	gas	298	107.76	0.17	−6.36	—	—
TiC	24.7 ± 0.4	184.5 ± 4.6	cubic	298	48.43	3.16	−1.36	1.23	—
$TiCl_2(g)$	87.4 ± 12.6	515.5 ± 16.7	gas	298	60.12	2.22	−2.76	—	—
$TiCl_3$	139.7 ± 1.3	721.7 ± 5.0	trigon	298	95.81	11.05	−1.80	—	—
$TiCl_3(g)$	316.7 ± 4.2	539.3 ± 6.3	gas	298	87.24	−0.71	−12.93	—	—

$Ti_2Cl_6(g)$	482.0 ± 16.7	1248.1 ± 16.7	gas	298	182.59	0.21	−11.55	—	—
$TiCl_4$	252.4 ± 0.8	804.2 ± 1.3	liq	298	142.80	8.70	−0.17	—	—
$TiCl_4(g)$	353.1 ± 4.2	763.2 ± 3.8	gas	298	107.19	0.47	−10.54	—	—
$TiCr_2$	86.5	− 0.3	cubic	298	72.43	24.77	−6.90	—	—
TiF_4	134.0	1649.3 ± 3.8	cryst	298	123.30	36.23	−17.64	—	—
$TiF_4(g)$	314.8 ± 2.1	1551.4 ± 4.2	gas	298	104.27	1.98	−18.04	—	—
TiI_4	246.0 ± 6.7	377.4 ± 3.3	α	298	78.24	158.99	—	—	—
			cubic	379	148.11	—	—	—	9.9
			liq	428	156.48	—	—	—	19.8
TiN	30.3	338.1	cubic	298	49.83	3.93	−12.38	—	—
TiO	34.7 ± 2.1	542.7 ± 12.6	monocl	298	44.22	15.06	−7.78	—	—
			cubic	1265	49.58	12.55	—	—	3.4
Ti_2O_3	77.2 ± 0.4	1520.9 ± 8.4	trigon	298	31.80	213.38	—	—	—
			β	470	147.70	3.43	−47.51	—	1.1
Ti_3O_5	129.4 ± 1.7	2459.1 ± 4.2	monocl	298	231.04	−24.77	−61.25	—	—
			orth	450	158.99	50.21	—	—	13.1
TiO_2	50.6 ± 0.4	944.0 ± 0.8	rutile	298	73.35	3.05	−17.03	—	—
$TiO_2(a)$	49.9 ± 0.4	941.4 ± 2.9	anatas	298	76.36	0.84	−20.08	—	—
$TiOCl(g)$	263.6	244.3	gas	298	60.50	0.96	−8.37	—	—
$TiOCl_2(g)$	320.9	545.6	gas	298	81.46	0.88	−8.93	—	—
$TiOF(g)$	250.6	433.0	gas	298	59.73	1.35	−10.74	—	—
$TiOF_2(g)$	284.5	924.7	gas	298	79.98	1.63	−16.18	—	—
TiS	56.5 ± 8.4	272.0 ± 29.3	hex	298	45.90	7.36	—	—	—
$TiS(g)$	246.4 ± 2.1	−330.5	gas	298	36.99	0.22	−2.95	—	—
TiS_2	78.2 ± 6.3	407.1 ± 33.5	hex	298	33.81	114.39	—	—	—
			hex	420	62.72	21.51	—	—	—
Ti_5Si_3	218.0 ± 14.2	579.5 ± 58.6	hex	298	196.44	44.77	−20.08	—	—
TiSi	49.0 ± 7.5	129.7 ± 14.6	orth	298	48.12	11.42	−5.44	—	—
$TiSi_2$	61.1 ± 9.2	133.9 ± 20.9	orth	298	70.42	17.57	−9.04	—	—
Tl	64.2	0.0	hcp	298	15.65	25.27	2.80	—	—
			β	507	20.92	20.92	—	—	0.4
			liq	577	30.12	—	—	—	4.3
$Tl(g)$	180.8	−181.0	gas	298	20.79	—	—	—	—
			gas	700	23.60	−4.56	−3.17	2.08	—
TlBr	122.6 ± 0.4	172.7 ± 0.8	cubic	298	46.32	20.71	—	—	—
			liq	733	105.65	−37.82	—	—	16.4
TlCl	111.5 ± 0.4	204.2 ± 1.7	cubic	298	46.02	16.74	—	—	—
			liq	702	59.41	—	—	—	15.9
TlF	95.7 ± 0.4	325.5 ± 4.6	rhomb	298	54.81	—	—	—	—
			tetrag	356	55.44	—	—	—	0.4
			liq	595	67.28	—	—	—	13.9

TABLE I—*continued*

Substance	$S_{298} \pm \delta S$ (J/deg mol)	$-\Delta H_{298} \pm \delta H$ (kJ/mol)	Phase	T K	$C_p = A + BT + C/T^2 + DT^2$ A	$B \times 10^3$ (J/deg mol)	$C \times 10^{-5}$	$D \times 10^6$	H_t (kJ/mol)
TlI	127.7 ± 0.4	123.7 ± 1.3	rhomb	298	49.25	13.47	—	—	—
			cubic	451	32.30	47.11	—	—	0.8
			liq	715	71.96	—	—	—	14.7
TlI(g)	276.6 ± 1.3	−18.6	gas	298	38.62	—	−2.09	—	—
Tl_2O	134.3 ± 11.3	167.4 ± 7.5	hex	298	56.07	41.84	—	—	—
			liq	852	94.98	—	—	—	30.3
Tl_2O_3	137.2 ± 13.8	390.4 ± 5.9	cubic	298	131.88	3.56	−22.26	—	—
Tl_2S	159.0 ± 12.6	95.0 ± 4.6	hex	298	71.55	29.29	—	—	—
			liq	730	99.58	—	—	—	23.0
Tl_2SO_4	200.8 ± 13.8	933.7 ± 0.8	orth	298	100.42	125.52	—	—	—
			liq	905	205.02	—	—	—	23.8
Tl_2Se	173.6 ± 7.5	93.3 ± 2.1	tetrag	298	69.75	32.64	—	—	—
TlSe	102.9 ± 1.7	61.1 ± 1.0	tetrag	298	39.33	35.15	—	—	—
Tl_2Te	174.1 ± 12.6	80.3 ± 5.4	cryst	298	67.99	27.20	—	—	—
Tm	74.0	0.0	hcp	298	19.83	13.01	3.22	−1.88	—
			bcc	1780	37.24	—	—	—	—
			liq	1818	41.42	—	—	—	16.9
Tm(g)	190.0	−232.2	gas	298	20.79	—	—	—	—
			gas	900	22.52	−2.24	−3.18	0.83	—
$TmCl_3$	147.3 ± 11.7	988.3 ± 2.5	monocl	298	95.60	11.72	−1.26	—	—
			liq	1101	148.53	—	—	—	34.9
TmF_3	95.0 ± 1.3	1694.5	orth	298	100.12	10.50	−16.40	—	—
			hex	1325	97.86	—	—	—	30.3
			liq	1431	140.33	—	—	—	28.9
Tm_2O_3	139.7 ± 0.4	1888.7 ± 5.9	cubic	298	129.70	3.26	−14.31	—	—
			cubic	1680	133.89	—	—	—	1.3
U	50.2	0.0	orth	298	25.10	2.38	—	23.68	—
			tetrag	941	42.93	—	—	—	2.8
			cubic	1049	38.28	—	—	—	4.8
			liq	1408	48.66	—	—	—	9.2
U(g)	199.7	−531.4	gas	298	16.74	6.32	4.98	1.55	—
UAl_2	97.5 ± 10.5	98.7 ± 5.4	cubic	298	75.31	10.46	—	—	—
UAl_3	136.0	114.2 ± 16.7	cubic	298	100.42	13.39	—	—	—
UAl_4	145.6 ± 12.6	130.5 ± 16.7	tetrag	298	119.24	33.47	—	—	—

UAs_2	206.7 ± 0.4	251.0 ± 12.6	tetrag	298	79.96	—	—	—	—
UB_2	55.1 ± 0.4	147.7 ± 12.6	hex	298	107.03	−40.25	−35.48	24.52	—
UB_4	71.1 ± 8.4	245.6 ± 20.9	tetrag	298	112.13	29.08	−37.24	—	—
UB_{12}	139.7	433.0	tetrag	298	308.57	15.31	−1.28	—	—
UBr_4	236.4 ± 10.5	802.1 ± 10.5	monocl	298	136.82	20.50	−11.30	—	—
			liq	792	163.18	—	—	—	55.2
UC	59.0 ± 0.6	97.5 ± 3.8	cubic	298	143.59	−1.26	−8.70	4.39	—
U_2C_3	138.4 ± 0.4	181.6 ± 7.5	cubic	298	125.10	12.80	−15.52	—	—
UC_2	68.8 ± 1.0	88.3 ± 5.0	tetrag	298	69.04	8.54	−9.41	—	—
			cubic	2038	123.01	—	—	—	10.9
UCl_3	159.0 ± 0.8	861.9 ± 3.3	hex	298	87.03	32.43	4.39	—	—
			liq	1114	129.70	—	—	—	46.4
UCl_3(g)	356.9	569.0 ± 16.7	gas	298	91.00	—	−9.62	—	—
UCl_4	197.3 ± 0.8	1018.8 ± 2.1	tetrag	298	113.80	35.86	−3.31	—	—
			liq	863	162.34	—	—	—	50.0
UCl_4(g)	389.9 ± 6.3	828.4 ± 7.1	gas	298	112.97	—	−9.04	—	—
U_2Cl_8	623.8 ± 25.1	1778.6 ± 25.1	gas	298	248.53	—	−19.87	—	—
UCl_5	246.9 ± 10.5	1041.4 ± 2.1	monocl	298	140.04	35.44	−5.36	—	—
			liq	600	186.69	—	—	—	35.6
U_2Cl_{10}(g)	633.0 ± 10.5	1960.2	gas	298	287.44	—	−28.87	—	—
UCl_6	285.8 ± 1.7	1068.2 ± 2.1	hex	298	173.43	35.06	−7.41	—	—
			liq	451	213.97	—	—	—	20.9
UCl_6(g)	432.6	987.8 ± 8.4	gas	298	158.03	—	−12.34	—	—
UF_3	117.2 ± 8.4	1507.1 ± 3.8	hex	298	85.77	31.38	1.05	—	—
UF_3(g)	331.8 ± 4.2	1036.8 ± 25.1	gas	298	83.68	—	−8.66	—	—
UF_4	151.7 ± 0.4	1919.6 ± 4.6	monocl	298	107.53	29.29	−0.25	—	—
			β	1118	150.21	—	—	—	16.3
			liq	1309	165.27	—	—	—	30.7
UF_4(g)	349.4 ± 8.4	1600.4	gas	298	110.88	—	−18.12	—	—
UF_5	188.3 ± 12.6	2072.3 ± 5.0	tetrag	298	125.52	30.21	−1.97	—	—
			liq	621	166.61	—	—	—	46.9
UF_5(g)	377.0 ± 8.4	1937.2 ± 20.9	gas	298	126.78	—	−15.48	—	—
UF_6	227.8 ± 0.6	2197.9 ± 3.3	orth	298	52.72	384.93	—	—	—
			liq	337	198.32	—	—	—	19.2
UF_6(g)	376.6 ± 4.2	2148.1 ± 2.5	gas	298	151.04	5.44	−20.38	—	—
UFe_2	104.6	32.2 ± 1.3	cubic	298	69.87	29.29	—	—	—
			liq	1502	138.07	—	—	—	67.8
UH_3	63.7 ± 0.4	127.0 ± 0.4	cubic	298	30.38	42.34	5.65	—	—
UI_4	265.7 ± 16.7	510.4 ± 8.4	cryst	298	149.37	9.96	−15.90	—	—
			liq	793	165.69	—	—	—	23.6
UN	62.5 ± 0.4	294.6 ± 5.0	cubic	298	55.73	4.98	−8.79	—	—

TABLE I—*continued*

Substance	$S_{298} \pm \delta S$ (J/deg mol)	$-\Delta H_{298} \pm \delta H$ (kJ/mol)	Phase	T K	$C_p = A + BT + C/T^2 + DT^2$ A	$B \times 10^3$ (J/deg mol)	$C \times 10^{-5}$	$D \times 10^6$	H_t (kJ/mol)
UO_2	77.0	1085.0	cubic	298	80.33	6.78	−16.57	—	—
U_4O_9	334.1 ± 0.6	4510.4 ± 7.1	cubic	298	356.27	35.44	−66.40	—	—
U_3O_8	282.5 ± 0.5	3574.8 ± 2.5	orth	298	282.42	36.94	−49.96	—	—
UO_3	96.2 ± 0.4	1223.8 ± 1.3	monocl	298	93.30	10.88	−10.88	—	—
$UOBr_2$	157.6 ± 0.4	977.8 ± 8.4	cryst	298	110.58	13.68	−14.85	—	—
$UOBr_3$	205.0 ± 16.7	947.7 ± 16.7	cryst	298	130.54	20.50	−13.81	—	—
UO_2Br_2	169.5 ± 12.6	1132.6 ± 12.6	cryst	298	117.95	17.53	−10.71	—	—
$UOCl$	102.5 ± 8.4	833.9 ± 4.2	cryst	298	75.81	14.35	−8.28	—	—
$UOCl_2$	138.3 ± 0.4	1069.4 ± 2.7	orth	298	98.95	14.64	−7.41	—	—
$U_2O_5Cl_5$	326.4 ± 8.4	2197.4 ± 4.2	cryst	298	234.30	35.56	−22.68	—	—
$UOCl_3$	169.9 ± 8.4	1140.1 ± 9.2	cryst	298	122.59	20.92	−11.92	—	—
UO_2Cl	112.5 ± 8.4	1169.4 ± 8.4	cryst	298	90.12	22.26	−7.74	—	—
$(UO_2)_2Cl_3$	276.1 ± 8.4	2404.5 ± 2.1	cryst	298	225.94	35.56	−29.29	—	—
UO_2Cl_2	150.6 ± 0.4	1242.6 ± 1.3	orth	298	115.23	18.20	−11.42	—	—
			liq	851	159.83	—	—	—	44.1
$UO_2Cl_2(g)$	372.4	972.8	gas	298	102.42	3.59	−13.22	—	—
UO_2F_2	135.6 ± 0.4	1651.4 ± 2.1	hex	298	122.93	8.62	−19.87	—	—
UP	78.2 ± 0.6	262.3 ± 12.6	cubic	298	57.36	−5.77	−5.44	8.79	—
US	78.0 ± 0.4	318.0 ± 6.7	cubic	298	52.84	6.53	−3.79	—	—
US_2	110.4 ± 0.4	526.3 ± 6.3	hex	298	71.80	9.62	—	—	—
US_3	138.5 ± 0.4	548.1 ± 33.5	monocl	298	95.60	—	—	—	—
UO_2SO_4	154.8 ± 8.4	1845.1 ± 2.1	orth	298	112.47	108.78	—	—	—
USe	96.5 ± 0.4	275.7 ± 14.6	cubic	298	52.89	6.40	—	—	—
U_3Si	173.8 ± 0.8	134.7 ± 16.7	tetrag	298	142.97	—	−31.38	—	—
U_3Si_2	197.5 ± 8.4	170.7 ± 6.3	tetrag	298	169.37	2.43	−35.19	—	—
USi	66.5 ± 4.2	83.7 ± 4.2	cubic	298	64.56	1.63	−13.01	—	—
USi_2	82.0 ± 5.0	129.7 ± 2.5	tetrag	298	89.62	4.06	−16.82	—	—
USi_3	106.3 ± 6.3	130.5 ± 2.9	cubic	298	113.18	6.44	−20.67	—	—
V	30.9	0.0	bcc	298	24.14	6.19	−1.38	−0.71	—
			bcc	900	25.90	−0.13	—	4.08	—
			liq	2183	47.28	—	—	—	23.0
$V(g)$	182.2	−514.2	gas	298	25.10	—	—	—	—
VBr_2	125.5 ± 12.6	364.4 ± 25.1	hex	298	73.64	12.55	—	—	—
VBr_3	142.3 ± 12.6	446.0 ± 25.1	hex	298	92.05	32.22	—	—	—

$VBr_4(g)$	334.7 ± 16.7	351.5 ± 33.5	gas	298	107.74	0.84	−7.32	—	—
$VC_{0.88}$	25.1 ± 0.8	101.9 ± 2.1	cubic	298	36.36	13.31	−7.11	—	—
VCl_2	97.1 ± 1.3	461.5 ± 8.4	hex	298	72.17	11.38	−2.97	—	—
VCl_3	131.0 ± 1.7	581.2 ± 2.1	hex	298	96.23	16.40	−7.03	—	—
VCl_4	257.4 ± 2.1	569.9 ± 2.5	liq	298	161.71	—	—	—	—
$VCl_4(g)$	366.1	525.9	gas	298	99.16	8.37	−5.44	—	—
VF_4	121.3 ± 11.3	1403.3 ± 16.7	hex	298	95.19	39.75	—	—	—
$VF_5(g)$	320.9 ± 0.8	1433.9 ± 12.6	gas	298	130.46	0.63	−28.74	—	—
V_3Ge	114.4	141.2	cubic	298	111.50	4.56	−16.78	—	—
V_5Ge_3	109.6	331.8	tetrag	298	226.35	7.28	−33.56	—	—
VN	37.3 ± 0.4	218.0 ± 5.0	cubic	298	45.77	8.79	−9.25	—	—
VO	33.1 ± 0.4	431.8 ± 2.1	cubic	298	50.21	11.84	−13.51	—	—
V_2O_3	92.9 ± 2.9	1218.8 ± 1.7	hex	298	112.97	19.29	−14.98	—	—
VO_2	47.1 ± 1.0	713.8 ± 1.0	monocl	298	73.01	2.43	−14.98	—	—
			tetrag	341	74.68	7.11	−16.53	—	4.3
			liq	1818	106.69	—	—	—	56.9
V_2O_5	130.5 ± 0.4	1550.2 ± 1.7	orth	298	141.00	42.68	−23.43	—	—
			liq	952	190.37	—	—	—	66.9
$VOCl_3$	241.8 ± 3.3	735.5 ± 5.9	liq	298	150.62	—	—	—	—
$VOCl_3(g)$	342.7 ± 0.8	696.2 ± 5.9	gas	298	108.99	—	−17.15	—	—
V_3Si	97.9 ± 1.7	172.4 ± 5.0	cubic	298	90.37	16.74	−7.15	—	—
V_5Si_3	178.2 ± 4.2	431.0 ± 16.7	tetrag	298	210.66	15.36	−33.56	—	—
VSi_2	59.0 ± 2.1	122.2 ± 33.5	hex	298	67.78	14.98	−6.61	—	—
			liq	1953	119.24	—	—	—	152.7
V_3Sn	124.7	66.9	cubic	298	104.60	9.08	−8.83	—	—
W	32.6	0.0	bcc	298	23.68	4.06	−0.47	−0.33	—
			bcc	1000	29.41	−2.95	−12.40	2.13	—
			bcc	2000	126.40	−58.68	−879.21	11.17	—
			liq	3693	54.02	—	—	—	50.0
$W(g)$	173.8	−851.0	gas	298	10.26	17.54	3.28	24.02	—
			gas	600	−46.25	157.86	29.82	−73.35	—
			gas	1000	116.14	−65.43	−227.29	13.25	—
			gas	2000	25.46	−3.52	370.70	1.24	—
W_2B	66.9	66.9 ± 9.2	tetrag	298	77.11	6.15	−13.05	—	—
WB	33.1	66.1	tetrag	298	50.42	3.10	−15.69	—	—
W_2C	81.6 ± 4.2	26.4 ± 2.5	hex	298	89.75	10.88	−14.56	—	—
WC	34.7 ± 2.1	40.6 ± 1.7	hex	298	43.39	8.62	−9.33	−1.03	—
WCl_4	198.3	443.5	rhomb	298	106.48	77.86	1.38	−18.62	—
$WCl_4(g)$	379.1	336.0	gas	298	107.40	0.46	−7.78	—	—
WCl_5	217.6	514.2 ± 37.7	monocl	298	124.43	109.91	−1.38	—	—

TABLE I—*continued*

Substance	$S_{298} \pm \delta S$ (J/deg mol)	$-\Delta H_{298}$ $\pm \delta H$ (kJ/mol)	Phase	T K	$C_p = A + BT + C/T^2 + DT^2$ A	$B \times 10^3$ (J/deg mol)	$C \times 10^{-5}$	$D \times 10^6$	H_t (kJ/mol)
$WCl_5(g)$	405.4	412.5	gas	298	131.38	1.42	−10.29	—	—
$W_2Cl_{10}(g)$	711.3	868.6	gas	298	281.71	0.52	−16.53	—	—
WCl_6	238.5	594.1	rhomb	298	125.52	167.36	—	—	—
			β	450	209.20	—	—	—	4.2
			gamma	503	188.28	—	—	—	15.5
			liq	555	200.83	—	—	—	6.7
$WCl_6(g)$	419.2	493.7	gas	298	157.53	0.21	−12.26	—	—
$WF_6(g)$	341.0 ± 0.4	1721.7 ± 2.5	gas	298	154.39	2.76	−31.38	—	—
WO_3	75.9 ± 1.3	842.7 ± 2.9	monocl	298	91.29	10.46	−19.00	—	—
			orth	603	77.40	25.10	—	—	1.4
			tetrag	1013	99.79	—	—	—	1.9
			tet2	1173	105.86	—	—	—	1.2
			tet3	1490	112.97	—	—	—	0.5
			liq	1745	131.80	—	—	—	73.4
$WO_3(g)$	286.2 ± 12.6	292.9 ± 25.1	gas	298	70.92	13.39	−12.30	−3.89	—
WO_2	50.5 ± 1.7	589.5 ± 6.3	monocl	298	77.40	−7.03	−18.49	8.87	—
$WO_2(g)$	285.3	−76.6	gas	298	49.08	10.00	−7.32	−2.89	—
$W(CO)_6$	331.8	946.8 ± 8.4	rhomb	298	164.43	261.92	—	—	—
$WOCl_4$	172.8	671.1	tetrag	298	115.06	104.60	—	—	—
			liq	483	182.00	—	—	—	33.5
$WOCl_4(g)$	377.0	573.2	gas	298	127.90	2.62	−20.71	—	—
WOF_4	175.7 ± 4.2	1394.5 ± 62.8	monocl	298	125.52	167.36	—	—	—
			liq	379	182.00	—	—	—	5.0
$WOF_4(g)$	334.7 ± 4.2	1336.8 ± 62.8	gas	298	125.69	4.18	−27.70	—	—
$WO2I_2(g)$	377.0 ± 8.4	430.1 ± 10.5	gas	298	102.55	4.60	−10.59	—	—
WS_2	64.9 ± 6.3	259.4 ± 16.7	hex	298	68.62	15.61	−8.66	—	—
W_5Si_3	229.7 ± 13.8	135.1 ± 20.9	tetrag	298	179.66	39.16	−8.91	—	—
WSi_2	64.0 ± 5.0	92.9 ± 9.2	tetrag	298	67.82	11.05	−6.11	—	—
Y	44.4	0.0	hcp	298	23.39	7.95	1.21	—	—
			bcc	1751	35.02	—	—	—	5.0
			liq	1795	39.79	—	—	—	11.4
Y(g)	179.4	−423.0	gas	298	26.15	−6.74	2.34	2.30	—
YCl_3	140.2	1000.0 ± 3.8	monocl	298	104.73	3.22	−12.13	—	—
			liq	994	177.57	—	—	—	31.5

$YCl_3(g)$	351.5 ± 6.7	702.9 ± 8.8	gas	298	83.68	—	−5.23	—	—
YF_3	109.6 ± 6.3	1718.4 ± 3.3	orth	298	99.41	7.45	−5.69	—	—
			hex	1350	122.38	—	—	—	32.4
			liq	1428	133.68	—	—	—	27.9
$YF_3(g)$	316.7 ± 2.9	1263.1 ± 13.0	gas	298	85.52	—	−13.39	—	—
YH_2	38.4 ± 0.4	221.8 ± 4.2	cubic	298	32.80	21.25	−4.18	—	—
YH_3	41.9 ± 0.4	265.7	hex	298	38.74	15.48	—	—	—
YI_3	207.1	616.7 ± 31.4	hex	298	100.92	11.51	−7.41	—	—
YN	37.7 ± 5.0	299.2 ± 20.9	cubic	298	45.61	6.49	−7.32	—	—
Y_2O_3	99.2 ± 0.4	1905.0 ± 5.0	cubic	298	109.62	20.08	−11.30	—	—
Yb	59.8	0.0	β	298	22.84	14.52	—	−5.44	—
			gamma	1068	36.11	—	—	—	1.8
			liq	1092	36.82	—	—	—	7.7
Yb(g)	172.8	−152.0	gas	298	20.79	—	—	—	—
$YbCl_2$	122.6 ± 2.1	799.1 ± 8.4	ortho	298	67.99	20.92	—	—	—
$YbCl_3$	135.1 ± 12.6	961.1 ± 3.3	monocl	298	94.68	9.33	−1.88	—	—
			liq	1148	121.34	—	—	—	—
YbN	62.8 ± 0.8	359.8 ± 20.9	cubic	298	46.44	8.37	—	—	—
Yb_2O_3	133.1 ± 0.4	1814.6 ± 3.3	cubic	298	128.66	19.46	−17.15	—	—
Zn	41.6	0.0	hcp	298	22.38	10.04	—	—	—
			liq	693	31.38	—	—	—	7.3
Zn(g)	160.9	−130.4	gas	298	20.79	—	—	—	—
Zn_3As_2	165.3 ± 11.7	134.7 ± 10.5	tetrag	298	112.76	41.84	—	—	—
$ZnBr_2$	136.0 ± 4.2	329.7 ± 2.5	tetrag	298	52.72	43.51	—	—	—
			liq	675	113.80	—	—	—	15.6
$ZnCl_2$	111.5 ± 0.4	415.1 ± 0.8	tetrag	298	59.83	37.66	—	—	—
			liq	591	100.83	—	—	—	16.1
$ZnCl_2(g)$	276.6 ± 2.1	265.7 ± 2.5	gas	298	61.71	—	−4.31	—	—
ZnF_2	73.7 ± 0.4	764.4 ± 1.7	tetrag	298	62.30	11.36	—	—	—
			liq	1148	94.14	—	—	—	41.8
ZnI_2	161.5 ± 4.2	208.2 ± 0.8	tetrag	298	85.14	11.46	−12.55	—	—
Zn_3N_2	140.2 ± 11.7	22.6 ± 8.4	cubic	298	79.50	94.14	—	—	—
ZnO	43.6 ± 0.4	350.5 ± 0.6	hex	298	48.99	5.10	−9.12	—	—
$ZnAl_2O_4$	87.0 ± 2.1	2071.1 ± 2.5	cubic	298	166.52	15.48	−46.02	—	—
$ZnCO_3$	82.4 ± 1.3	818.0 ± 1.3	hex	298	38.91	138.07	—	—	—
$ZnCr_2O_4$	116.3 ± 1.3	1548.1 ± 4.6		298	167.36	14.23	−25.10	—	—
$ZnFe_2O_4$	153.3 ± 0.8	1179.1 ± 5.9	cubic	298	22.38	7.32	−48.53	—	—
$Zn(OH)_2$	77.0 ± 0.4	645.4	rhomb	298	74.27	—	—	—	—
Zn_3P_2	150.6 ± 10.5	159.0 ± 14.6	tetrag	298	126.23	26.07	−15.23	—	—
			liq	1513	155.64	—	—	—	58.6

TABLE I—*continued*

Substance	$S_{298} \pm \delta S$ (J/deg mol)	$-\Delta H_{298} \pm \delta H$ (kJ/mol)	Phase	T K	$C_p = A + BT + C/T^2 + DT^2$				H_t (kJ/mol)
					A	$B \times 10^3$ (J/deg mol)	$C \times 10^{-5}$ (J/deg mol)	$D \times 10^6$	
ZnP_2	60.2 ± 8.4	101.7 ± 12.6	tetrag	298	71.25	16.74	−2.72	—	—
			liq	1255	91.21	—	—	—	37.2
ZnS	57.7 ± 0.4	205.2 ± 2.5	cubic	298	49.25	5.27	−4.85	—	—
			hex	1293	49.45	4.44	−4.35	—	13.4
ZnS(g)	238.1 ± 2.1	−202.1 ± 16.7	gas	298	37.45	0.08	−1.67	—	—
$ZnSO_4$	110.5 ± 0.4	981.4 ± 1.7	rhomb	298	76.36	76.15	—	—	—
$ZnSO_4.H_2O$	145.5	1301.5 ± 2.5	monocl	298	127.61	87.03	—	—	—
$ZnSO_4.2H_2O$	192.5 ± 6.3	1596.0	cryst	298	168.11	102.72	—	—	—
$ZnSO_4.6H_2O$	363.6 ± 2.5	2779.0 ± 2.5	monocl	298	316.10	140.42	—	—	—
$ZnSO_4.7H_2O$	388.7 ± 2.1	3078.6 ± 2.5	rhomb	298	334.09	151.13	—	—	—
$ZnO.2ZnSO_4$	264.4	2320.4 ± 5.0	cryst	298	201.71	157.40	−9.12	—	—
ZnSb	82.6 ± 1.7	19.0 ± 1.3	rhomb	298	44.69	17.32	—	—	—
			liq	820	62.76	—	—	—	30.8
ZnSe	71.1 ± 6.3	163.2 ± 10.5	cubic	298	50.17	5.77	—	—	—
$ZnSeO_3$	98.3 ± 8.4	652.3 ± 3.8	cryst	298	77.19	55.23	—	—	—
			liq	894	140.16	—	—	—	46.4
Zn_2SiO_4	131.4 ± 1.3	1643.1 ± 3.3	hex	298	144.89	36.94	−30.29	—	—
ZnTe	77.8 ± 3.3	119.2 ± 1.7	cubic	298	44.10	18.74	—	—	—
Zn_2TiO_4	144.8 ± 2.9	1649.8 ± 2.9	tetrag	298	166.61	23.18	−32.17	—	—
$ZnWO_4$	129.7 ± 8.4	1230.9 ± 7.9	monocl	298	121.71	33.51	−9.12	—	—
Zr	39.0	0.0	hcp	298	22.84	8.95	−0.67	—	—
			bcc	1136	21.51	6.57	36.69	—	3.9
			liq	2128	33.47	—	—	—	18.8
Zr(g)	181.3	−601.2	gas	298	23.00	3.05	3.64	—	—
ZrB_2	35.9 ± 0.4	323.8 ± 6.3	hex	298	64.22	9.41	−16.57	—	—
$ZrBr_4$	224.7	759.8 ± 2.5	cubic	298	133.01	4.74	−8.58	—	—
$ZrBr_4$(g)	414.6	643.5 ± 7.1	gas	298	107.91	0.08	−4.81	—	—
ZrC	33.3 ± 1.7	207.1 ± 3.8	cubic	298	51.13	3.39	−12.97	—	—
$ZrCl_4$	181.4 ± 2.5	980.3 ± 1.7	cubic	298	124.98	14.14	−8.37	—	—
$ZrCl_4$(g)	367.7 ± 0.4	868.6 ± 2.1	gas	298	107.45	0.29	−8.26	—	—
ZrF_4	104.6 ± 0.4	1911.3 ± 3.3	monocl	298	117.49	16.74	−17.24	—	—
ZrF_4(g)	318.4	1674.0	gas	298	105.60	1.19	−16.15	—	—
ZrI_4	260.2 ± 4.2	488.7 ± 6.3	cubic	298	130.79	9.12	−5.10	—	—
ZrN	38.9 ± 1.3	368.2 ± 2.5	cubic	298	46.44	7.03	−7.20	—	—

ZrO(g)	227.3	−58.6	gas	298	26.28	14.90	—	—	—
ZrO_2	50.4 ± 0.4	1100.8 ± 2.1	monocl	298	69.62	7.53	−14.06	—	—
			tetrag	1450	74.48	—	—	—	5.9
			liq	2950	87.86	—	—	—	87.0
ZrS_2	78.2 ± 12.6	577.4 ± 20.9	hex	298	64.27	15.06	—	—	—
$ZrSiO_4$	84.5 ± 1.7	2034.7 ± 1.7	tetrag	298	131.71	16.40	−33.81	—	—

Table II. *Thermochemical data of binary metallic systems*
(N_2 is the atomic fraction of the second element named in the first column)
Ag. Silver alloys: partial and integral excess Gibbs energies of solution in J/g–atom

System	Components	Thermo-Chemical Function	Temp. (K)	$N_2=$ 0.0	0.1	0.3	0.5	0.7	0.9
⟨Ag–Al⟩	⟨Ag⟩ ⟨Al⟩	$\Delta \bar{G}^{E Al}$	722	−23,430.4	−23,514.1	−13,095.9	—	—	—
{Ag–Al}	{Ag} {Al}	ΔH	1273	0	−3138.0	−6401.5	−3941.3	−1129.7	280.3
	{Al}	ΔH_{Al}	1273	−35,564.0	−27,417.8	−8978.9	5522.9	2259.4	405.8
⟨Ag–Au⟩	⟨Ag⟩ ⟨Au⟩	ΔH	800	0	−1799.1	−4045.9	−4648.4	−3765.6	−1548.1
	⟨Au⟩	$\Delta \bar{H}_{Au}$	800	−20,292.4	−15,882.5	−8941.2	−4221.7	−1397.5	−142.3
{Ag–Au}	{Ag} {Au}	ΔH	1350	0	−1853.5	−4322.1	−5146.3	−4322.1	−1853.5
	{Au}	$\Delta \bar{H}_{Au}$	1350	−20,585.3	−16,673.2	−10,083.4	−5146.3	−1853.5	−205.0
{Ag–Bi}	{Ag} {Bi}	$\Delta \bar{H}$	1000	0	—	698.7	1221.7	1711.3	1046.0
	{Bi}	$\Delta \bar{H}_{Bi}$	1000	—	—	2334.7	2820.0	1832.6	272.0
⟨Ag–Cd⟩	⟨Ag⟩ ⟨Cd⟩	ΔH	673	0	−2523.0	−6100.3	—	−6443.4	—
	⟨Cd⟩	$\Delta \bar{H}_{Cd}$	673	−26,777.6	−23,417.8	−15,413.9	—	—	—
{Ag–Cd}	{Ag} {Cd}	ΔG^E	1223	0	−2179.9	−4568.9	−4744.7	−3389.0	−1196.6
	{Cd}	ΔG^E_{Cd}	1223	—	−18,978.6	−8660.9	−2970.6	−552.3	0
{Ag–Cu}	{Ag} {Cu}	ΔH	1423	0	1903.7	3853.5	4242.6	3468.5	1472.8
	{Cu}	$\Delta \bar{H}_{Cu}$	1423	23,012.0	15,690.0	7481.0	3765.6	1435.1	159.0
{Ag–Ga}	{Ag} {Ga}	ΔH	1000	0	—	−3631.7	−2669.4	−853.5	230.1
	{Ga}	$\Delta \bar{H}_{Ga}$	1000	—	—	−3610.8	1464.4	1719.6	322.2
{Ag–Ge}	{Ag} {Ge}	ΔH	1250	0	−682.0	393.3	2234.3	2686.1	1376.5
	{Ge}	$\Delta \bar{H}_{Ge}$	1250	−12,552.0	−2133.8	7322.0	5481.0	2058.5	255.2

⟨Ag–Hg⟩	⟨Hg⟩	ΔG^E_{Hg}	500	−12,259.1	−9706.9	−1230.1	—	—	—
{Ag–In}	{Ag} {In}	ΔH	1100	0	—	−5648.4	−4133.8	−1974.8	−510.4
	{In}	ΔH_{In}	—	−18,828.0	—	−8255.0	1782.4	656.9	92.05
⟨Ag–Mg⟩	⟨Ag⟩ ⟨Mg⟩	ΔH	773	0	−4644.2	—	−18,409.6	—	—
	⟨Mg⟩	$\Delta \bar{H}_{Mg}$	773	−46,860.8	−38,430.0	—	—	—	—
⟨Ag–Mn⟩	⟨Mn⟩β	$\Delta \bar{G}^E_{Mn}$	1150	836.8	3422.5	7154.6	—	—	—
{Ag–Pb}	{Ag} {Pb}	ΔH	1273	0	1054.4	2912.1	3702.8	2974.8	1146.4
	{Pb}	$\Delta \bar{H}_{Pb}$	1273	10,460.0	10,460.0	8004.0	3681.9	920.5	46.02
⟨Ag–Pd⟩	⟨Ag⟩ ⟨Pd⟩	ΔH	1200	0	−2778.2	−5510.3	−5020.8	−3263.5	−1046.0
	⟨Pd⟩	$\Delta \bar{G}^E_{Pd}$	1200	−24,518.2	−17,070.7	−4769.8	962.3	878.6	87.86
{Ag–S}	$\frac{1}{2}(S_2)$	ΔH_S	1400	−72,676.1	−85,897.5	−123,218.8	—	—	—
⟨Ag–Sb⟩	⟨Sb⟩	$\Delta \bar{G}^E_{Sb}$	600	−3405.8	−5589.8	—	—	—	—
{Ag–Sb}	{Ag} {Sb}	ΔH	1250	0	−2108.7	−2635.9	−62.76	1004.2	656.9
	{Sb}	$\Delta \bar{H}_{Sb}$	1250	−25,104.0	−16,828.0	6405.7	4435.0	1472.8	167.4
{Ag–Si}	{Ag} {Si}	ΔH	1473	0	−259.4	782.4	2627.6	3581.5	2129.7
	{Si}	$\Delta \bar{H}_{Si}$	1473	−4991.5	100.42	6995.6	6548.0	3577.3	698.7
{Ag–Sn}	{Ag} {Sn}	ΔH	1250	0	−2242.6	−2677.8	−949.8	246.9	347.3
	{Sn}	$\Delta \bar{H}_{Sn}$	900	—	—	—	2510.4	795.0	175.7
	{Sn}	$\Delta \bar{G}^E_{Sn}$	1250	4142.2	−1397.5	−5158.9	−4288.6	−2309.6	−322.2
{Ag–Tl}	{Ag} {Tl}	ΔH	—	—	723.8	1882.8	2485.3	2288.6	1171.5
	{Tl}	$\Delta \bar{H}_{Tl}$	975	—	—	5104.5	3075.2	1372.4	267.8
⟨Ag–Zn⟩	⟨Ag⟩ ⟨Zn⟩	ΔH	873	0	−1673.6	−8472.6	−3983.2	−3891.1ε	—
	⟨Zn⟩	$\Delta \bar{H}_{Zn}$	873	−11,824.0	−14,213.0	−9405.6	−7886.8	−209.2ε	—
{Ag–Zn}	{Ag} {Zn}	ΔG^E	1023	—	—	−4267.7	−4619.1	−3598.2	−1338.9
	{Zn}	$\Delta \bar{G}^E_{Zn}$	1023	—	—	−10,878.4	−4204.9	−949.8	−33.47

Ag. Silver alloys: partial and integral excess entropies of solution in J/deg g–atom

System	Components	Thermo-Chemical Function	Temp. (K)	$N_2=$ 0.0	0.1	0.3	0.5	0.7	0.9
{Ag–Al}	{Ag} {Al}	ΔS^E	1273	0	−0.05	0.08	1.56	2.00	1.15
	{Al}	$\Delta \bar{S}^E_{Al}$	1273	−1.31	0	1.31	5.43	1.74	0.29
⟨Ag–Au⟩	⟨Ag⟩ ⟨Au⟩	ΔS^E	800	0	−0.51	−1.20	−1.44	−1.21	−0.52
	⟨Au⟩	$\Delta \bar{S}^E_{Au}$	800	−5.67	−4.61	−2.81	−1.44	−0.52	−0.06
{Ag–Au}	{Ag} {Au}	ΔS^E	1350	0	−0.62	−1.54	−1.96	−1.75	−0.79
	{Au}	$\Delta \bar{S}^E_{Au}$	1350	−6.63	−5.76	−3.96	−2.26	−0.90	−0.11
{Ag–Bi}	{Ag} {Bi}	ΔS^E	1000	0	—	0.72	0.41	0.55	0.30
	{Bi}	$\Delta \bar{S}^E_{Bi}$	1000	—	—	−0.45	0.28	0.70	0.00
⟨Ag–Cd⟩	⟨Ag⟩ ⟨Cd⟩	ΔS^E	673	0	−0.36	−0.84	—	−1.46ε	—
	⟨Cd⟩	$\Delta \bar{S}^E_{Cd}$	673	−6.22	−2.15	−4.67	—	—	—
{Ag–Cu}	{Au} {Cu}	ΔS^E	1423	0	0.44	0.64	0.51	0.34	0.13
	{Cu}	$\Delta \bar{S}^E_{Cu}$	1423	5.98	3.08	0.46	0.10	0.08	0.00
{Ag–Ga}	{Ag} {Ga}	ΔS^E	1000	0	—	1.51	2.01	2.60	1.69
	{Ga}	$\Delta \bar{S}^E_{Ga}$	1000	—	—	1.97	3.85	2.87	0.50
{Ag–Ge}	{Ag} {Ge}	ΔS^E	1250	0	−0.03	0.95	1.95	1.95	0.87
	{Ge}	$\Delta \bar{S}^E_{Ge}$	1250	−2.68	1.77	5.29	3.36	1.15	0.07
{Ag–In}	{Ag} {In}	ΔS^E	1100	0	—	−3.64	−2.54	−1.26	−0.44
	{In}	$\Delta \bar{S}^E_{In}$	1100	—	—	−5.63	1.41	0.09	−0.00
⟨Ag–Mg⟩	⟨Ag⟩ ⟨Mg⟩	ΔS^E	773	0	0.56	—	−0.38	—	—
	⟨Mg⟩	$\Delta \bar{S}^E_{Mg}$	773	7.04	13.33	—	—	—	—
{Ag–Pb}	{Ag} {Pb}	ΔS^E	1273	0	0.67	1.57	1.83	1.28	0.39
	{Pb}	$\Delta \bar{S}^E_{Pb}$	1273	8.87	5.58	3.74	1.44	0.03	−0.04
⟨Ag–Pd⟩	⟨Ag⟩ ⟨Pd⟩	ΔS^E	1200	0	−0.85	−1.59	−1.83	−1.69	−0.70
{Ag–S}	$\frac{1}{2}(S_2)$	$\Delta \bar{S}^E_S$	1400	−46.86	−51.59	−72.30	—	—	—
{Ag–Sb}	{Ag} {Sb}	ΔS^E	1250	0	0.28	1.30	2.54	2.28	0.96
	{Sb}	$\Delta \bar{S}^E_{Sb}$	1250	4.23	2.24	8.38	3.43	0.97	0.11
{Ag–Si}	Si	$\Delta \bar{S}^E_{Si}$	1250	−5.40	−2.99	−0.10	1.00	0.90	0.29
{Ag–Sn}	{Ag} {Sn}	ΔS^E	1250	0	−0.67	−0.18	0.79	0.86	0.19
	{Sn}	$\Delta \bar{S}^E_{Sn}$	900	—	—	—	1.40	0.14	0.09
{Ag–Tl}	{Ag} {Tl}	ΔS^E	975	0	—	0.59	0.56	0.37	0.22
	{Tl}	$\Delta \bar{S}^E_{Tl}$	975	—	—	0.88	0.25	0.12	0.10
⟨Ag–Zn⟩	⟨Ag⟩ ⟨Zn⟩	ΔS^E	873	0	−0.40	0.46	1.57	1.07ε	—
	⟨Zn⟩	$\Delta \bar{S}^{E}_{Zn}$	873	−0.13	0.03	1.75	−1.55	2.98ε	—

AI. Aluminium alloys: partial and integral heats or excess Gibbs energies of solution in J/g–atom

System	Components	Thermo-Chemical Function	Temp. (K)	$N_2 =$ 0.0	0.1	0.3	0.5	0.7	0.9
⟨Al–Ag⟩	⟨Ag⟩	$\Delta \bar{G}^E_{Ag}$	722	2953.9	—	—	—	−2205.0	−58.58
{Al–Ag}	{Al} {Ag}	ΔH	1273	0	280.3	−1129.7	−3941.3	−6405.7	--3133.8
	{Ag}	$\Delta \bar{H}_{Ag}$	1273	7322.0	−849.4	−9029.1	−13,409.7	−5301.1	−435.1
{Al–Au}	{Al} {Au}	ΔH	1338	0	−7719.5	−24,246.3	−33,764.9	−29,903.0	−12,200.5
	{Au}	$\Delta \bar{H}_{Au}$	1338	−69,726.4	−82,106.8	−74,994.0	−42,224.9	−13,363.7	−857.7
{Al–Be}	{Be}	$\Delta \bar{H}$	1600	—	—	—	21,388.6	10,250.8	1589.9
{Al–Bi}	{Al} {Bi}	ΔH	—	0	2225.9	6326.2	7146.3	4882.7	1841.0
	{Bi}	$\delta \bar{H}_{Bi}$	1173	22,217.0	—	—	—	644.3	129.7
{Al–Cu}	{Al} {Cu}	ΔH	1373	0	−1920.5	−6045.9	−9050.0	−8284.3	−3347.2
	{Cu}	$\Delta \bar{H}_{Cu}$	1373	−17,677.4	−20,292.4	−19,560.2	−13,681.7	−2845.1	−272.0
{Al–Fe}	{Fe}	$\Delta \bar{H}_{Fe}$	971	−128,448.8	—	—	—	—	—
{Al–Ga}	{Al} {Ga}	ΔH	1023	0	297.1	619.2	656.9	502.1	205.0
	{Ga}	$\Delta \bar{H}_{Ga}$	1023	3481.1	2510.4	1213.4	472.8	167.4	12.55
{Al–Ge}	{Al} {Ge}	ΔH	1200	0	−1138.0	−3050.1	−3882.8	−3234.2	−1255.2
	{Ge}	$\Delta \bar{H}_{Ge}$	1200	−11,547.8	−11,121.1	−8200.6	−4171.4	−1146.4	−37.66
{Al–In}	{Al} {In}	ΔH	1173	0	2443.5	5192.3	5711.2	4548.0	1903.7
	{In}	$\Delta \bar{H}_{In}$	1173	28,451.2	20,945.1	10,485.1	4757.2	1677.8	200.8
{Al–Mg}	{Al} {Mg}	ΔH	1073	0	−1305.4	−3008.3	−3372.3	−2569.0	−962.3
	{Mg}	$\Delta \bar{H}_{Mg}$	1073	−14,560.3	−11,853.3	−6610.7	−2681.9	−656.9	−33.47
{Al–Pb}	{Al} {Pb}	ΔH	>1650	0	3866.0	8911.9	9355.4	6443.4	2443.5
	{Pb}	$\Delta \bar{H}_{Pb}$	1200	41,714.5	—	—	—	—	—
{Al–Si}	{Al} {Si}	ΔH	—	0	−1020.9	−2447.6	−3104.5	−2769.8	−1255.2
	{Si}	$\Delta \bar{H}_{Si}$	—	−10,460.0	−9100.2	−6276.0	−3598.2	−1405.8	−175.7
{Al–Sn}	{Al} {Sn}	ΔH	973	0	1941.4	3765.6	4041.7	3188.2	1322.1
	{Sn}	$\Delta \bar{H}_{Sn}$	973	24,455.5	15,397.1	7071.0	3204.9	1142.2	138.1
⟨Al–Zn⟩	⟨Al⟩ ⟨Zn⟩	ΔH	653	0	1364.0	2991.6	3451.8	—	—
	⟨Zn⟩	$\Delta \bar{H}_{Zn}$	653	15,656.5	11,924.4	6292.7	3786.5	—	—
{Al–Zn}	{Al} {Zn}	ΔH	1000	0	941.4	2158.9	2569.0	2200.8	979.1
	{Zn}	$\Delta \bar{H}_{Zn}$	1000	10,619.0	8347.1	4928.8	2640.1	606.7	150.6

Al. Aluminium alloys: partial and integral excess entropies of solution in J/deg g–atom

System	Components	Thermo-Chemical Function	Temp. (K)	$N_2=$ 0.0	0.1	0.3	0.5	0.7	0.9
{Al–Ag}	{Al} {Ag}	ΔS^E	1273	0	1.15	2.01	1.56	0.08	−0.05
	{Ag}	$\Delta \bar{S}^E_{Ag}$	1273	14.71	8.88	2.62	−2.31	−0.46	−0.06
{Al–Au}	{Al} {Au}	ΔS^E	1338	0	1.36	0.52	−0.12	0.18	0.53
	{Au}	$\Delta \bar{S}^E_{Au}$	1338	23.58	5.97	−4.15	−0.24	0.96	0.23
{Al–Be}	{Be}	$\Delta \bar{S}^E_{Be}$	1600	—	—	—	10.47	4.82	0.74
{Al–Bi}	{Bi}	$\Delta \bar{S}^E_{Bi}$	1173	−6.38	—	—	—	−0.31	0.08
{Al–Cu}	{Al} {Cu}	ΔS^E		0	1.15	2.57	3.40	3.84	2.05
	{Cu}	$\Delta \bar{S}^E_{Cu}$	1373	13.55	9.81	6.30	4.77	3.64	0.38
{Al–Ga}	{Al} {Ga}	ΔS^E	1023	0	0.18	0.35	0.33	0.23	0.09
	{Ga}	$\Delta \bar{S}^E_{Ga}$	1023	2.18	1.46	0.59	0.15	0.05	0.00
{Al–Ge}	{Al} {Ge}	ΔS^E	1200	0	0.36	0.55	0.44	0.29	0.11
	{Ge}	$\Delta \bar{S}^E_{Ge}$	1200	4.92	2.60	0.41	0.10	0.05	0.00
{Al–In}	{Al} {In}	ΔS^E	1173	0	0.40	0.69	0.59	0.37	0.13
	{In}	$\Delta \bar{S}^E_{In}$	1173	5.02	3.10	0.83	0.13	0	0.03
{Al–Mg}	{Al} {Mg}	ΔS^E	1073	0	0.00	−0.44	−0.83	−0.82	−0.36
	{Mg}	$\Delta \bar{S}^E_{Mg}$	1073	1.26	−1.07	−2.23	−1.35	−0.46	−0.03
{Al–Si}	{Al} {Si}	ΔS^E	—	0	0.23	0.47	0.25	−0.05	−0.22
	{Si}	$\Delta \bar{S}^E_{Si}$	—	2.55	1.90	0.42	−0.54	−0.59	−0.13
{Al–Sn}	{Al} {Sn}	ΔS^E	973	0	0.67	1.13	1.21	1.05	0.50
	{Sn}	$\Delta \bar{S}^E_{Sn}$	973	9.39	4.63	1.96	1.11	0.54	0.13
⟨Al–Zn⟩	⟨Al⟩ ⟨Zn⟩	ΔS^E	653	0	0.48	1.06	1.31	—	—
	⟨Zn⟩	$\Delta \bar{S}^E_{Sn}$	653	5.97	4.22	2.13	2.19	—	—
{Al–Zn}	{Al} {Zn}	ΔS^E	1000	0	0.34	0.72	0.95	0.89	0.33
	{Zn}	$\Delta \bar{S}^E_{Zn}$	1000	4.21	2.72	1.69	1.30	0.44	−0.02

Au. Gold alloys: partial and integral heats or excess Gibbs energies of solution in J/g–atom

System	Components	Thermo-Chemical Function	Temp. (K)	$N_2=$ 0.0	0.1	0.3	0.5	0.7	0.9
⟨Au–Ag⟩	⟨Ag⟩⟨Au⟩	ΔH	800	0	−1552.3	−3761.4	−4648.4	−4045.9	−1794.9
	⟨Ag⟩	$\Delta \bar{H}_{Ag}$	800	−16,903.4	−14,242.3	−9288.5	−5079.4	−1949.7	−230.1
{Au–Ag}	{Au}{Ag}	ΔH	1350	0	−1853.5	−4322.1	−5146.3	−4322.1	−1853.5
	{Ag}	$\Delta \bar{H}_{Ag}$	1350	−20,585.3	−16,673.2	−10,087.6	−5146.3	−1853.5	−205.0
{Au–Al}	{Au}{Al}	ΔH	1338	0	−12,200.5	−29,903.0	−33,764.9	−24,246.3	−7719.5
	{Al}	$\Delta \bar{H}_{Al}$	1338	−128,549.2	−114,239.9	−68,492.1	−25,300.6	−2497.8	543.9
{Au–Bi}	{Au}{Bi}	ΔH	973	0	—	418.4	627.6	502.1	188.3
	{Bi}	$\Delta \bar{H}_{Bi}$	973	—	—	1769.8	698.7	121.34	8.37
⟨Au–Cd⟩	⟨Au⟩⟨Cd⟩	ΔH	700	0	−5987.3	−14,196.3α_2	−18,577.0	−15,313.4ε	—
	⟨Cd⟩	$\Delta \bar{H}_{Cd}$	700	−65,270.4	−55,241.4	−39,287.8α_2	−20,313.3	−17,572.8ε	—
{Au–Cd}	{Au}{Cd}	ΔH	1000	0	—	—	−13,192.2	−10,769.6	−4334.6
	{Cd}	$\Delta \bar{H}_{Cd}$	1000	—	—	—	−13,275.8	−3916.2	−322.2
⟨Au–Co⟩	⟨Co⟩β	$\Delta \bar{H}_{Co}$	1150	40,124.6	37,141.4	—	—	—	—
⟨Au–Cu⟩	⟨Au⟩⟨Cu⟩	ΔH	800	0	−1213.4	−3573.1	−5108.7	−4974.8	−1841.0
	⟨Cu⟩	$\Delta \bar{H}_{Cu}$	800	−11,631.5	−12,367.9	−10,920.2	−7384.8	−2644.3	121.34
{Au–Cu}	{Au}{Cu}	ΔH	1550	0	−1435.1	−3510.4	−4368.1	−3828.4	−1707.1
	{Cu}	$\Delta \bar{H}_{Cu}$	1550	−15,585.4	−13,234.0	−8744.6	−4836.7	−1878.6	−221.8
⟨Au–Fe⟩	⟨Au⟩⟨Fe⟩α	ΔH	1123	0	2209.2	4882.7	6552.1	—	—
	⟨Fe⟩α	$\Delta \bar{H}_{Fe}$	1123	25,480.6	18,472.4	12,008.1	9665.0	—	—

System	Components	Thermo-Chemical Function	Temp. (K)	$N_2=$ 0.0	0.1	0.3	0.5	0.7	0.9
{Au–Fe}	{Au} {Fe}	ΔH		0	5857.6	14,058.2	17,154.4	—	—
	{Fe}	$\Delta \bar{H}_{Fe}$	1473	64,852.0	53,534.3	34,112.2	18,325.9	—	—
{Au–Ga}	{Au} {Ga}	ΔH	1030	0	−7886.8* ÷	−15,690.0	−15,982.9	−11,380.5	−4050.1
	{Ga}	$\Delta \bar{H}_{Ga}$	1030	−93,721.6*	−65,270.4*	−28,158.3	−9581.4	−2071.1	−121.34
{Au–Ge}	{Au} {Ge}	ΔH	1423	0	−2133.8	−4518.7	−4686.1	−3472.7	−1255.2
⟨Au–Hg⟩	⟨Hg⟩	$\Delta \bar{H}_{Hg}$	500	−5313.7	−2217.5	—	—	—	—
{Au–Mn}	{Au} {Mn}	ΔG	1535	0	−3502.0	−11,321.9	−17,593.7	−14,254.9	−4686.1
	{Mn}	ΔG^E_{Mn}	1535	−35,564.0	−36,262.7	−40,283.6	−23,723.3	−1828.4	217.6
⟨Au–Ni⟩	⟨Au⟩ ⟨Ni⟩	ΔH	1150	0	2092.0	5635.8	7560.5	6727.9	2690.3
	⟨Ni⟩	$\Delta \bar{H}_{Ni}$	1150	21,505.8	20,267.3	15,765.3	9351.2	2991.6	117.15
{Au–Pb}	{Au} {Pb}	ΔH	1200	0	−435.1	−790.8	−698.7	−305.4	259.4
	{Pb}	$\Delta \bar{H}_{Pb}$	1200	—	−3046.0	−1138.0	−92.05	548.1	343.1
⟨Au–Pd⟩	⟨Au⟩ ⟨Pd⟩	ΔH	—	0	−4079.4	−7656.7	−7447.5	−4736.3	−1439.3
	⟨Pd⟩	$\Delta \bar{H}_{Pd}$	—	−46,609.8	−32,258.6	−12,907.6	−3200.8	58.58	142.3
⟨Au–Pt⟩	⟨Au⟩ ⟨Pt⟩	ΔH	—	0	0	1108.8	2656.8	3326.3	1903.7
	⟨Pt⟩	$\Delta \bar{H}_{Pt}$	—	−2656.8	2133.8	6485.2	6276.0	3096.2	447.7
{Au–Sn}	{Au} {Sn}	ΔH	823	0	—	−10,321.9	−11,572.9	−8744.6	−3179.8
	{Sn}	$\Delta \bar{H}_{Sn}$	823	—	—	−22,685.6	−9301.0	−1857.7	−92.05
{Au–Te}	⟨Au⟩ {Te}	ΔH	737	0	—	—	4225.8	2757.3	1058.6
{Au–Tl}	{Tl}	$\Delta \bar{H}_{Tl}$	973	—	—	—	205.0	−12.55	−8.37
{Au–Zn}	{Au} {Zn}	ΔH	1080	—	0	−17,698.3	−22,744.2	−18,773.6	−7309.4
	{Zn}	$\Delta \bar{H}_{Zn}$	1080	—	—	−50,208.0	−25,187.7	−6025.0	−393.3

Au. Gold alloys: partial and integral excess entropies of solution in J/degg–atom

System	Components	Thermo-Chemical Function	Temp. (K)	$N_2=$ 0.0	0.1	0.3	0.5	0.7	0.9
⟨Au–Ag⟩	⟨Au⟩⟨Ag⟩	ΔS^E	800	0	−0.51	−1.20	−1.44	−1.21	−0.52
	⟨Ag⟩	$\Delta \bar{S}^E_{Ag}$	800	−5.81	−4.69	−2.81	−1.43	−0.51	−0.05
{Au–Ag}	{Au}{Ag}	ΔS^E	1350	0	−0.79	−1.75	−1.96	−1.54	−0.62
	{Ag}	$\Delta \bar{S}^E_{Ag}$	1350	−9.05	−6.94	−3.72	−1.66	−0.51	−0.05
{Au–Al}	{Au}{Al}	ΔS^E	1338	0	0.53	0.18	−0.12	0.52	1.36
	{Al}	$\Delta \bar{S}^E_{Al}$	1338	8.37	3.22	−1.64	0.01	2.52	0.85
{Au–Bi}	{Au}{Bi}	ΔS^E	973	0	—	1.03	1.25	1.02	0.44
	{Bi}	$\Delta \bar{S}^E_{Bi}$	973	—	—	2.55	1.23	0.38	0.06
⟨Au–Cd⟩	⟨Au⟩⟨Cd⟩	ΔS^E	700	0	−2.87α	−6.64α_2	−6.59β	−6.32ε	—
		$\Delta \bar{S}^E_{Cd}$	700	−34.97	−23.92	−15.94α_2	−7.87	−15.44ε	—
⟨Au–Co⟩	⟨Co⟩β	Δ	1150	8.37	14.74	—	—	—	—
⟨Au–Cu⟩	⟨Au⟩⟨Cu⟩	ΔS^E	800	0	0.20	0.34	0.29	0.13	0.25
	⟨Cu⟩	$\Delta \bar{S}^E_{Cu}$	800	2.56	1.51	0.39	−0.14	0.10	0.22
{Au–Cu}	{Au}{Cu}	ΔS^E	1550	0	0.47	1.00	1.06	0.79	0.30
	{Cu}	$\Delta \bar{S}_{ECu}$	1550	5.46	4.03	1.96	0.76	0.18	0.01
⟨Au–Fe⟩	⟨Au⟩⟨Fe⟩α	ΔS^E	1123	0	1.72	2.97	3.18	—	—
	⟨Fe⟩α	21.76	12.68	4.63	3.23	—	—		
{Au–Fe}	{Au}{Fe}	ΔS^E	1473	0	4.56	9.64	11.06	—	—
	{Fe}	$\Delta \bar{S}^E_{Fe}$	1473	53.40	39.12	19.05	7.36	—	—
{Au–Ga}	{Au}{Ga}	ΔS^E	1400	0	0	0.25	1.13	1.26	0.25
	{Ga}	$\Delta \bar{S}_{EGa}$	1400	−4.44	0	3.26	2.62	1.03	0.13
{Au–Ge}	{Au}{Ge}	ΔS^E	1423	0	−0.38	−0.54	0.13	0.54	0.67
⟨Au–Hg⟩	⟨Hg⟩	$\Delta \bar{S}^E_{Hg}$	500	−8.37	−4.77	—	—	—	—
⟨Au–Ni⟩	⟨Au⟩⟨Ni⟩	ΔS^E	1150	0	0.58	1.92	2.79	2.35	0.67
	⟨Ni⟩	$\Delta \bar{S}^E_{Ni}$	1150	5.08	6.34	6.33	3.57	0.53	−0.16
{Au–Pb}	{Au}{Pb}	ΔS^E	1200	0	0.64	1.49	1.78	1.66	1.09
	{Pb}	$\Delta \bar{S}^E_{Pb}$	1200	—	5.76	3.49	1.78	1.36	0.39
⟨Au–Pd⟩	⟨Au⟩⟨Pd⟩	ΔS^E	—	0	−1.13	−2.64	−3.14	−2.64	−1.13
	⟨Pd⟩	$\Delta \bar{S}^E_{Pd}$	—	−12.55	−10.17	−6.15	−3.14	−1.13	−0.13
⟨Au–Pt⟩	⟨Au⟩⟨Pt⟩	ΔS^E	—	0	−1.21	−2.43	−2.30	−1.46	−0.44
	⟨Pt⟩	$\Delta \bar{S}^E_{Pt}$	—	−14.64	−10.08	−3.97	−0.94	—	—
{Au–Sn}	{Au}{Sn}	ΔS^E	823	0	—	2.08	1.31	0.77	0.39
	{Sn}	$\Delta \bar{S}^E_{Sn}$	823	—	—	−0.11	−0.64	0.38	0.05
{Au–Tl}	{Au}{Tl}	ΔS^E_{Tl}	973	—	—	—	2.67	0.87	0.08
{Au–Zn}	{Au}{Zn}	ΔS^E	1080	0	—	−0.60	−2.06	−2.12	−0.74
	{Zn}	$\Delta \bar{S}^E_{Zn}$	1080	—	—	−6.57	−4.73	−0.71	0.04

Co. Cobalt alloys: partial and integral heats or excess Gibbs energies of solution in J/g–atom

System	Components	Thermo-Chemical Function	Temp. (K)	$N_2=$ 0.0	0.1	0.3	0.5	0.7	0.9
{Co–Al}	{Co} {Al}	ΔG^E_{Al}	1873	−82,650.7	−68,596.7	−42,706.1	—	—	—
⟨Co–Cr⟩	⟨Co⟩fcc ⟨Cr⟩bcc	ΔH	1473	0	313.8	2133.8	—	7112.8	2991.6
	⟨Cr⟩	$\Delta \bar{H}_{Cr}$	1473	2803.3	4037.6	13,137.8	tab¢—	—	502.1
⟨Co–Cu⟩	⟨Cu⟩	ΔG^E_{Cu}	1300	0	—	—	—	—	—
⟨Co–Fe⟩	⟨Co⟩γ⟨Fe⟩γ	ΔH	1473	0	−439.3	−1129.7	−1313.8	−974.9	−355.6
	⟨Fe⟩γ	$\Delta \bar{H}_{Fe}$	1473	−4539.6	−4167.3	−2832.6	−1087.8	−221.8	−16.74
{Co–Fe}	{Co} {Fe}	ΔH	1873	0	−916.3	−2125.5	−2426.7	−1907.9	−761.5
	{Fe}	$\Delta \bar{H}_{Fe}$	1873	—	−8313.6	−4769.8	−2092.0	−610.9	−58.58
{Co–Ge}	⟨Co⟩ {Ge}	ΔH	1287	0	—	—	−2493.7	−4435.0	−1652.7
⟨Co–Mn⟩	⟨Co⟩ ⟨Mn⟩β	ΔG^E	1023	0	−2138.0	−4715.4	−5606.6	−10,397.2	−4979.0
	⟨Mn⟩β	$\Delta \bar{G}^E_{Mn}$	1023	−24,581.0	−18,639.7	−10,384.7	−6234.2	−2899.5	−150.6
⟨Co–Mo⟩	⟨Mo⟩	$\Delta \bar{G}^E_{Mo}$	1350	16,736.0	14,309.3	—	—	—	—
Co–Ni							this system is nearly ideal in the solid and liquid states		
⟨Co–Pt⟩	⟨Co⟩ ⟨Pt⟩	ΔG^E	1273	0	−4464.3	−10,422.3	−12,405.6	−10,422.3	−4464.3
{Co–Si}	{Co} {Si}	ΔH	1873	0	−14,518.5	−38,953.0	−46,651.6	−34,183.3	−12,091.8
	{Si}	$\Delta \bar{H}_{Si}$	1873	−146,440.0	−142,674.4	−102,089.6	−36,819.2	−5857.6	−125.5
{Co–Sn}	{Co} {Sn}	ΔH	1773	0	−1589.9	−2301.2	−1046.0	543.9	836.8

Co. Cobalt alloys: partial and integral excess entropies of solution in J/deg g–atom

System	Components	Thermo-Chemical Function	Temp. (K)	$N_2=$ 0.0	0.1	0.3	0.5	0.7	0.9
⟨Co–Cr⟩	⟨Co⟩ ⟨Cr⟩	ΔS^E	1473	0	0.27	1.78	—	6.19	3.79
	⟨Cr⟩	$\Delta \bar{S}_{Cr}$	1473	1.76	2.95	11.02	—	—	2.28
⟨Co–Fe⟩	—	—	—	—	—	—	—	—	—
{Co–Fe}	—	—	—	—	—	—	—	—	—
{Co–Si}	{Co} {Si}	ΔS^E	1873	0	−2.55	−7.24	−9.20	−7.20	−2.68
	{Si}	$\Delta \bar{S}_{Si}$	1873	−25.36	−26.32	−20.00	−9.00	−1.84	−0.08

Cr. Chromium alloys: partial and integral heats or excess Gibbs energies of solution in J/g–atom

System	Components	Thermo-Chemical Function	Temp. (K)	$N_2=$ 0.0	0.1	0.3	0.5	0.7	0.9
⟨Cr–Co⟩	⟨Cr⟩bcc ⟨Co⟩fcc	ΔS^E	1473	0	2991.6	7112.8	—	2133.8	313.8
	⟨Co⟩	$\Delta \bar{S}^E_{Co}$	1473	0	25,438.7	—	—	−2585.7	−100.42
⟨Cr–Fe⟩	⟨Cr⟩⟨Fe⟩	ΔS^E	1550	0	2259.4	5271.8	6276.0	5271.8	2259.4
	⟨Fe⟩	$\Delta \bar{S}^E_{Fe}$	1550	25,104.0	20,334.2	12,301.0	6276.0	2259.4	251.0
{Cr–Fe}	{Cr}{Fe}	ΔS^E	1750–2150	0	1903.7	4393.2	5209.1	4372.3	1924.6
	{Fe}	$\Delta \bar{S}^E_{Fe}$	1750–2150	20,995.3	17,154.4	10,167.1	5188.2	1924.6	251.0
⟨Cr–Mo⟩	⟨Cr⟩⟨Mo⟩	ΔH	1400	0	3117.1	6506.1	7217.4	5627.5	2071.1
⟨Cr–Ni⟩	⟨Cr⟩⟨Ni⟩	ΔH	—	0	4866.0	8987.2	6359.7	1405.8	−878.6
	⟨Ni⟩	$\Delta \bar{H}_{Ni}$	—	58,040.4	40,225.0	10,296.8	−8070.9	−4698.6	−878.6
{Cr–Ni}	{Cr}{Ni}	ΔG^E	—	0	−1410.0	−1715.4	−2732.2	−3104.5	−1401.6
	{Ni}	$\Delta \bar{G}^E_{Ni}$	—	—	−2615.0	−3569.0	−5125.4	−2104.6	−146.4
⟨Cr–V⟩	⟨Cr⟩⟨V⟩	ΔH	1550	0	−1828.4	−3259.3	−1903.7	−983.2	−849.4
	⟨V⟩	$\Delta \bar{H}_V$	1550	−21,798.6	−15,037.3	−3004.1	3882.8	−3305.4	−112.97

Cr. Chromium alloys: partial and integral excess entropies of solution in J/deg g–atom

System	Components	Thermo-Chemical Function	Temp. (K)	$N_2=$ 0.0	0.1	0.3	0.5	0.7	0.9
⟨Cr–Co⟩	⟨Cr⟩bcc	ΔS^E	1473	0	3.79	6.19	—	1.78	0.27
	⟨Co⟩fcc	—		(38.07)	17.32	—	—	2.19	0.27
⟨Cr–Fe⟩	⟨Co⟩	$\Delta \bar{S}^E_{Fe}$	1473	0	1.11	2.49	2.88	2.32	0.93
	⟨Cr⟩⟨Fe⟩	ΔS^E	1550	—	8.85	5.75	2.54	0.87	0.10
	⟨Fe⟩	$\Delta \bar{S}^E_{Fe}$	1550	—					
{Cr–Fe}	{Cr}{Fe}	ΔS^E	1750–2150–	0	1.11	2.49	2.88	2.32	0.93
	{Fe}	$\Delta S^{E\bar{D}}_{Fe}$	1750–2150	—	8.85	5.75	2.54	0.87	0.10
⟨Cr–Mo⟩	⟨Cr⟩⟨Mo⟩	ΔS^E	1200	0	0.86	1.76	1.92	1.38	0.36
⟨Cr–Ni⟩	⟨Cr⟩⟨Ni⟩	ΔS^E	—	0	2.77	4.87	4.05	1.79	0.36
	⟨Ni⟩	ΔS^E_{Ni}	—	—	21.80	6.40	−2.11	−1.22	0.06
⟨Cr–V⟩	⟨Cr⟩⟨V⟩	ΔS^E	1550	0	−0.11	0.13	1.01	0.89	0.02
	⟨V⟩	$\Delta \bar{S}^E_V$	1550	−1.79	−0.51	2.20	3.61	−1.92	−0.04

Cu. Copper alloys: partial and integral heats or excess Gibbs energies of solution in J/g–atom

System	Components	Thermo-Chemical Function	Temp. (K)	$N_2=$ 0.0	0.1	0.3	0.5	0.7	0.9
{Cu–Ag}	{Cu} {Ag}	ΔH	1423	0	1472.8	3468.5	4242.6	3853.5	1903.7
	{Ag}	$\Delta \bar{H}_{Ag}$	1423	16,317.6	11,556.2	11,556.2	8221.6	2297.0	368.2
{Cu–Al}	{Cu} {Al}	ΔH	1373	0	−3347.2	−8284.3	−9050.0	−6045.9	−1920.5
	{Al}	$\Delta \bar{H}_{Al}$	1373	−36,087.0	−31,024.4	−20,970.2	−4414.1	−251.0	121.34
⟨Cu–Au⟩	⟨Cu⟩ ⟨Au⟩	ΔH	800	0	−1841.0	−4974.8	−5108.7	−3573.1	−1213.4
	⟨Au⟩	$\Delta \bar{H}_{Au}$	800	−16,401.3	−19,497.4	−10,405.6	−2832.6	−422.6	29.29
{Cu–Au}	{Cu} {Au}	ΔH	1550	0	−1707.1	−3828.4	−4368.1	−3510.4	−1435.1
	{Au}	$\Delta \bar{H}_{Au}$	1550	−19,351.0	−15,062.4	−8376.4	−3895.3	−1267.8	−125.5
{Cu–Bi}	{Cu} {Bi}	ΔH	1200	0	—	5121.2	5355.5	4401.6	1958.1
	{Bi}	$\Delta \bar{H}_{Bi}$	1200	—	—	8903.6	4414.1	1887.0	322.2
⟨Cu–Fe⟩	⟨Cu⟩ ⟨Fe⟩	ΔH	1323	0	3907.9	9623.2	12,049.9	10,614.8	4761.4
{Cu–Fe}	{Cu} {Fe}	ΔH	1823	0	3363.9	7489.4	8920.3	8020.7	3974.8
	{Fe}	$\Delta \bar{H}_{Fe}$	1823	38,911.2	29,196.0	16,853.2	9547.9	4820.0	744.8
{Cu–Ge}	{Cu} {Ge}	ΔH	1423	0	−2133.8	−4476.9	−4686.1	−3472.7	−1255.2
{Cu–In}	{Cu} {In}	ΔH	1073	0	—	−3133.8	−1004.2	96.23	276.1
	{In}	$\Delta \bar{H}_{In}$	1073	—	($N_2=0.2$) −5970.6	1945.6	3468.5	974.9	167.4
{Cu–Mg}	{Cu} {Mg}	ΔH	1100	0	—	−9560.4	−10,313.6	−8087.7	−3799.1
	{Mg}	$\Delta \bar{H}_{Mg}$	1100	—	($N_2=0.2$) −26,857.1	−18,911.7	−7836.6	−3230.0	−870.3

⟨Cu–Mn⟩	⟨Cu⟩⟨Mn⟩γ	ΔH	1100	0	1004.2	2543.9	3196.6	4221.7	3012.5
	⟨Mn⟩γ	$\Delta \bar{G}^E_{Mn}$	1100	−3681.9	−338.9	3054.3	3648.4	2598.3	—
{Cu–Mn}	{Cu}{Mn}	ΔG^E	1500	0	−472.8	405.8	1991.6	3020.8	2146.4
	{Mn}	$\Delta \bar{G}^E_{Mn}$	1500	−8368.0	−1489.5	5924.5	5585.6	3648.4	857.7
⟨Cu–Ni⟩	⟨Cu⟩⟨Ni⟩	ΔH	773	0	543.9	1422.6	1861.9	1945.6	1192.4
	⟨Ni⟩	$\Delta \bar{H}_{Ni}$	773	6192.3	5376.4	3681.9	2510.4	1694.5	376.6
{Cu–Ni}	{Cu}{Ni}	ΔG^E	—	0	1129.7	2845.1	3493.6	3075.2	1318.0
	{Ni}	$\Delta \bar{G}^E_{Ni}$	—	12,426.5	10,250.8	6568.9	3493.6	1318.0	146.4
{Cu–Pb}	{Cu}{Pb}	ΔH	1473	0	2958.1	5949.6	6723.7	5598.2	2443.5
	{Pb}	$\Delta \bar{H}_{Pb}$	1473	36,066.1	24,225.4	11,786.3	6464.3	2397.4	305.4
⟨Cu–Pd⟩	⟨Cu⟩⟨Pd⟩	ΔH	1350	0	−4204.9	−10,209.0	−10,698.5	−8819.9	−3782.3
	⟨Pd⟩	$\Delta \bar{H}_{Pd}$	1350	−45,605.6	−38,802.4	−22,530.8	−8606.5	−3727.9	−418.4
⟨Cu–Pt⟩	⟨Cu⟩⟨Pt⟩	ΔH	1350	0	−5138.0	−10,384.7	−11,087.6	−8280.1	−2882.8
	⟨Pt⟩	$\Delta \bar{H}_{Pt}$	1350	−61,755.8	−42,584.8	−19,589.5	−9075.1	−991.6	−20.92
{Cu–Sb}	{Cu}{Sb}	ΔH	1190	—	—	−5623.3	−2916.2	−631.8	343.1
	{Sb}	$\Delta \bar{H}_{Sb}$	1190	—	(−29,288.0)	−276.1	4573.1	1945.6	338.9
{Cu–Si}	{Cu}{Si}	ΔG^E	1760	0	−5773.9	−11,798.9	−11,757.0	−7865.9	−2677.8
	{Si}	$\Delta \bar{G}^E_{Si}$	1760	−60,542.5	−51,295.8	−20,417.9	−5941.3	−502.1	−41.84
{Cu–Sn}	{Cu}{Sn}	ΔH	1400	0	−2786.5	−3907.9	−1987.4	−405.8	217.6
	{Sn}	$\Delta \bar{H}_{Sn}$	1400	−33,472.0	−21,894.9	1054.4	2849.3	1301.2	205.0
{Cu–Tl}	{Cu}{Tl}	ΔH	1573	0	2707.0	6949.6	8577.2	6619.1	2271.9
	{Tl}	$\Delta \bar{H}_{Tl}$	1573	28,158.3	25,790.2	17,936.8	8497.7	1083.7	−46.02
⟨Cu–Zn⟩	⟨Cu⟩⟨Zn⟩	ΔH	773	0	−2623.4	−7338.7	—	—	—
	⟨Zn⟩	$\Delta \bar{H}_{Zn}$	773	−23,012.0	−28,673.0	−17,363.6	—	—	—
{Cu–Zn}	{Zn}	$\Delta \bar{G}^E_{Zn}$	—	—	—	—	−3631.7	−631.8	4.18
{Cu–O}	$\frac{1}{2}${O_2}	$\Delta \bar{H}_O$	1673	−86,608.8	−95,395.2	−146,858.4	—	—	—

Cu. Copper alloys: partial and integral excess entropies of solution in J/deg g–atom

System	Components	Thermo-Chemical Function	Temp. (K)	$N_2=$ 0.0	0.1	0.3	0.5	0.7	0.9
{Cu–Ag}	{Cu} {Ag}	ΔS^E	1423	0	0.13	0.34	0.51	0.64	0.44
	{Ag}	$\Delta \bar{S}^E_{Ag}$	1423	1.35	1.23	0.94	0.92	0.73	0.15
{Cu–Al}	{Cu} {Al}	ΔS^E	1373	0	2.05	3.84	3.40	2.57	1.15
	{Al}	$\Delta \bar{S}^E_{Al}$	1373	24.43	17.10	4.30	2.03	0.98	0.18
⟨Cu–Au⟩	⟨Cu⟩ ⟨Au⟩	ΔS^E	800	0	0.25	0.13	0.29	0.34	0.20
	⟨Au⟩	$\Delta \bar{S}^E_{Au}$	800	5.31	0.48	0.21	0.73	0.32	0.05
{Cu–Au}	{Cu} {Au}	ΔS^E	1550	0	0.30	0.79	1.06	1.00	0.47
	{Au}	$\Delta \bar{S}^E_{Au}$	1550	3.04	2.85	2.20	1.37	0.58	0.08
{Cu–Bi}	{Cu} {Bi}	ΔS^E	1200	0	—	1.87	1.82	1.55	0.75
	{Bi}	$\Delta \bar{S}^E_{Bi}$	1200	—	—	2.61	1.40	0.79	0.18
⟨Cu–Fe⟩	⟨Fe⟩ ⟨Cu⟩	ΔS^E	1323	0	1.13	2.64	3.14	2.64	1.13
{Cu–Fe}	{Cu} {Fe}	ΔS^E		0	0.16	0.46	0.68	0.77	0.55
	{Fe}	$\Delta \bar{S}^E_{Fe}$	1823	1.72	1.50	1.49	1.01	0.87	0.18
{Cu–Ge}	{Cu} {Ge}	ΔS^E	14523	0	−0.50	0.50	0.08	0.54	0.54
{Cu–In}	{Cu} {In}	ΔS^E	1073	0	—	−0.24	0.67	0.55	0.36
	{In}	$\Delta \bar{S}^E_{In}$	1073	0	($N_2=0.2$) −0.40	2.30	1.68	−0.03	0.01
{Cu–Mg}	{Cu} {Mg}	ΔS^E	1100	0	—	−2.87	−3.23	−2.92	−1.88
	{Mg}	$\Delta \bar{S}^E_{Mg}$	1100	0	($N_2=0.2$) −7.50	−5.74	−3.03	−2.18	0.77
⟨Cu–Mn⟩	⟨Cu⟩ ⟨Mn⟩γ	ΔS^E	1100	0	1.08	2.06	1.81	2.03	—
⟨Cu–Ni⟩	⟨Cu⟩ ⟨Ni⟩	ΔS^E		0	−0.36	−0.79	−0.97	−0.64	−0.04
	⟨Ni⟩	$\Delta \bar{S}^E_{Ni}$	—	−3.22	−2.93	−1.97	−0.77	0.17	0.18
{Cu–Pb}	{Cu} {Pb}	ΔS^E	1473	0	0.72	0.98	1.05	0.94	0.46
	{Pb}	$\Delta \bar{S}^E_{Pb}$	1473	10.67	4.59	1.16	1.18	0.52	0.09
⟨Cu–Pd⟩	⟨Cu⟩ ⟨Pd⟩	ΔS^E	1350	0	−0.63	−2.67	−3.54	−4.13	−2.23
	⟨Pd⟩	$\Delta \bar{S}^E_{Pd}$	1350	−4.28	−7.90	−8.96	−5.94	−3.47	−0.46
⟨Cu–Pt⟩	⟨Cu⟩ ⟨Pt⟩	ΔS^E	1350	0	−0.40	−0.75	−0.87	0.52	0.16
	⟨Pt⟩	$\Delta \bar{S}^E_{Pt}$	1350	−5.25	−3.00	−1.46	−1.37	1.11	0.20
{Cu–Sb}	{Cu} {Sb}	ΔS^E	1190	—	—	1.16	2.42	2.38	1.18
	{Sb}	$\Delta \bar{S}^E_{Sb}$	1190	—	1.46	6.64	4.08	1456.0	0.22
{Cu–Sn}	{Cu} {Sn}	ΔS^E	1400	0	1.05	1.67	2.28	2.05	1.03
	{Sn}	$\Delta \bar{S}^E_{Sn}$	1400	16.76	6.18	4.26	2.60	1.16	0.22
{Cu–Tl}	{Cu} {Tl}	ΔS^E	1573	0	0.26	1.20	1.94	1.56	0.46
	{Tl}	$\Delta \bar{S}^E_{Tl}$	1573	1.28	3.56	4.69	2.75	0.04	−0.07
⟨Cu–Zn⟩	⟨Cu⟩ ⟨Zn⟩	ΔS^E	773	0	−0.08	−1.48	—	—	—
	⟨Zn⟩	$\Delta \bar{S}^E_{Zn}$	773	5.44	−5.92	−3.95	—	—	—

Fe. Iron alloys: partial and integral heats or excess Gibbs energies of solution in J/g–atom

System	Components	Thermo-Chemical Function	Temp. (K)	$N_2=$ 0.0	0.1	0.3	0.5	0.7	0.9
{Fe–Al}	{Fe} {Al}	ΔH	1873	0	−6108.6	−18,949.3	—	—	—
	{Al}	$\Delta\bar{H}_{Al}$	1873	−64,224.4	−58,032.1	−44,120.3	—	—	—
⟨Fe–Au⟩	⟨Fe⟩α⟨Au⟩	ΔH	1123	0	—	—	6066.8	4811.6	2217.5
	⟨Au⟩	$\Delta\bar{H}_{Au}$	1123	—	—	—	3439.2	1828.4	401.7
{Fe–Au}	{Fe} {Au}	ΔH	1383	0	—	—	4309.5	3598.2	1631.8
⟨Fe–Co⟩	⟨Fe⟩ ⟨Co⟩γ	ΔH	1473	0	−355.6	−974.9	−1313.8	−1129.7	−439.3
	⟨Co⟩γ	$\Delta\bar{H}_{Co}$	1473	−3861.8	−3535.5	−2790.7	−1548.1	−376.6	−25.10
{Fe–Co}	{Fe} {Co}	ΔH	1873	0	−761.5	−1907.9	−2426.7	−2125.5	−916.3
	{Co}	$\Delta\bar{H}_{Co}$	1873	—	−7104.4	−4937.1	−2765.6	−991.6	−96.23
⟨Fe–Cr⟩	⟨Fe⟩ ⟨Cr⟩	ΔH	1550	0	2259.4	5271.8	6276.0	5271.8	2259.4
	⟨Cr⟩	$\Delta\bar{H}_{Cr}$	1550	25,104.0	20,334.2	12,301.0	6276.0	2259.4	251.0
{Fe–Cr}	{Fe} {Cr}	ΔH	—	0	1924.6	4372.3	5209.1	4393.2	1903.7
	{Cr}	$\Delta\bar{H}_{Cr}$	—	21,610.4	16,987.0	10,083.4	5230.0	1924.6	209.2
⟨Fe–Cu⟩	⟨Fe⟩ ⟨Cu⟩	ΔH	1323	0	4761.4	10,614.8	12,049.9	9623.2	3907.9
{Fe–Cu}	{Fe} {Cu}	ΔH	1823	0	3974.8	8020.7	8920.3	7485.2	3363.9
	{Cu}	$\Delta\bar{H}_{Cu}$	1823	47,572.1	33,041.0	15,489.2	8292.7	3472.7	493.7
{Fe–Ge}	⟨Fe⟩ {Ge}	ΔH	1287	0	—	—	8953.8	4853.4	1841.0
⟨Fe–Ir⟩	⟨Fe⟩γ⟨Ir⟩	ΔG^E	1473	0	−5296.9	−12,685.9	−14,003.8	−10,732.0	−4225.8
	⟨Ir⟩	$\Delta\bar{G}^E_{Ir}$	1473	−56,316.6	−49,203.8	−28,129.0	−10,782.2	−3163.1	−297.1
⟨Fe–Mn⟩	⟨Fe⟩γ⟨Mn⟩γ	ΔH	1450	0	−1313.8	−6200.7	−6916.2	−6091.9	−3330.5
	⟨Mn⟩γ	$\Delta\bar{H}_{Mn}$	1450	−13,999.7	−12,351.2	−9585.5	−6602.4	−3217.5	−468.6

System	Components	Thermo-Chemical Function	Temp. (K)	$N_2 =$ 0.0	0.1	0.3	0.5	0.7	0.9
{Fe–Mn}	{Fe} {Mn}	ΔG^E	1863	0	397.5	924.7	1100.4	924.7	397.5
	{Mn}	$\Delta \bar{G}^E_{Mn}$	1863	4393.2	3560.6	2154.8	1100.4	397.5	46.02
⟨Fe–Ni⟩	⟨Fe⟩γ⟨Ni⟩	ΔH	1200	0	−962.3	−1945.6	−3702.8	−4184.0	−2075.3
	⟨Ni⟩	$\Delta \bar{H}_{Ni}$	1200	—	−4309.5	−8828.2	−5983.1	−3933.0	−334.7
{Fe–Ni}	{Fe} {Ni}	ΔH	1873	0	−1025.1	−2928.8	−4602.4	−4853.4	−2426.7
	{Ni}	$\Delta \bar{H}_{Ni}$	1873	−10,041.6	−9761.3	−9723.6	−7213.2	−3832.5	−205.0
⟨Fe–Pd⟩	⟨Fe⟩γ⟨Pd⟩	ΔH	1273	0	1702.9	−2836.8	−9443.3	−12,020.6	−6317.8
	⟨Pd⟩	$\Delta \bar{H}_{Pd}$	1273	36,107.9	907.9	−26,078.9	−23,597.8	−10,116.9	−1188.3
{Fe–Pd}	{Fe} {Pd}	ΔH	1873	0	1029.3	1941.4	2217.5	2000.0	916.3
	{Pd}	$\Delta \bar{H}_{Pd}$	1873	21,966.0	6652.6	3899.5	2301.2	1146.4	117.15
⟨Fe–Pt⟩	⟨Fe⟩γ⟨Pt⟩	ΔG^E	1123	0	−5723.7	−13,982.9	$-18,175.3_2$	$-13,706.8_3$	−6656.7
	⟨Pt⟩	$\Delta \bar{G}^E_{Pt}$	1123	−75,814.1	−49,136.9	−37,530.5	$-18,911.7_2$	-3577.3_3	0
{Fe–Pt}	{Fe} {Pt}	ΔG^E	1880	0	−7447.5	−21,505.8	−29,170.8	—	—
	{Pt}	$\Delta \bar{G}^E_{Pt}$	1880	−66,944.0	−78,002.3	−62,521.5	−34,656.1	—	—
{Fe–S}	(S_2)	$\Delta \bar{H}_{S_2}$	—	−253,132.0	−251,458.4	−247,692.8	−240,161.6	—	—
{Fe–Si}	{Fe} {Si}	ΔH	1873	0	−12,761.2	−33,388.3	−37,865.2	−28,032.8	−10,250.8
	{Si}	$\Delta \bar{H}_{Si}$	1873	−131,377.6	−125,101.6	−81,169.6	−27,823.6	−5606.6	−376.6
{Fe–Sn}	{Fe} {Sn}	ΔG^E	1820	0	1535.5	4238.4	—	—	—
	{Sn}	$\Delta \bar{G}^E_{Sn}$	1820	15,581.2	15,087.5	12,405.6	—	—	—
⟨Fe–V⟩	⟨Fe⟩α⟨V⟩	ΔH	1600	0	−1301.2	−3589.9	−5242.6	−5535.4	−2213.3
	⟨V⟩	$\Delta \bar{H}_V$	1600	−12,928.6	−12,510.2	−10,593.9	−8535.4	−2200.8	−276.1
{Fe–V}	{Fe} {V}	ΔF^E	2193	0	−941.4	−2761.4	−4481.1	−5188.2	−2736.3
	{V}	$\Delta \bar{G}^E_V$	2193	−10,669.2	−10,020.7	−8660.9	−7740.4	−4518.7	−502.1

Fe. Iron alloys: partial and integral excess entropies of solution in J/deg g–atom

System	Components	Thermo-Chemical Function	Temp. (K)	$N_2=$ 0.0	0.1	0.3	0.5	0.7	0.9
{Fe–Al}	{Fe} {Al}	ΔS^E	1873	0	−1.13	−3.51	—	—	—
	{Al}	$\Delta \bar{S}^E_{Al}$	1873	−10.56	−11.76	−11.32	—	—	—
{Fe–Au}	{Fe} {Au}	ΔS^E	1383	—	—	—	—	2.51	1.59
⟨Fe–Co⟩	—	—	—	—	—	—	—	—	—
{Fe–Co}	—	—	—	—	—	—	—	—	—
⟨Fe–Cr⟩	⟨Fe⟩ ⟨Cr⟩	ΔS^E	1550	0	0.93	2.32	2.88	2.49	1.11
	⟨Cr⟩	$\Delta \bar{S}^E_{Cr}$	1550	10.25	8.41	5.69	3.22	1.09	0.21
{Fe–Cr}	{Fe} {Cr}	ΔS^E	1750–2150	0	0.93	2.32	2.88	2.49	1.11
	{Cr}	$\Delta \bar{S}^E_{Cr}$	1750–2150	10.25	8.41	5.69	3.22	1.09	0.21
⟨Fe–Cu⟩	⟨Fe⟩γ⟨Cu⟩	ΔS^E	1323	0	1.13	2.64	3.14	2.64	1.13
	⟨Cu⟩	$\Delta \bar{S}^E_{Cu}$	1323	12.55	10.17	6.15	3.14	1.13	0.13
{Fe–Cu}	{Fe} {Cu}	ΔS^E	1823	0	0.16	0.46	0.68	0.77	0.55
	{Cu}	$\Delta \bar{S}^E_{Cu}$	1823	7.36	3.84	0.56	0.35	0.02	0.01
⟨Fe–Mn⟩	⟨Fe⟩γ⟨Mn⟩γ	ΔS^E	1450	0	−1.24	−3.17	−4.26	−4.11	−2.05
	⟨Mn⟩γ	$\Delta \bar{S}^E_{Mn}$	1450	−13.26	−11.55	−8.59	−5.65	−2.68	−0.39
⟨Fe–Ni⟩	⟨Fe⟩γ⟨Ni⟩	ΔS^E	1200	0	−0.25	−0.29	−0.81	−1.04	−0.56
	⟨Ni⟩	$\Delta \bar{S}^E_{Ni}$	1200	—	1.22	−1.79	−1.13	−1.31	−0.26
{Fe–Ni}	{Fe} {Ni}	ΔS^E	1873	0	−0.21	−0.64	−1.10	−1.24	−0.74
	{Ni}	$\Delta \bar{S}^E_{Ni}$	1873	−2.13	−2.13	−2.35	−1.76	−1.16	−0.18
⟨Fe–Pd⟩	⟨Fe⟩γ⟨Pd⟩	ΔS^E	1273	0	2.56	3.42	2.88	1.93	0.69
	⟨Pd⟩	$\Delta \bar{S}^E_{Pd}$	1273	35.79	16.92	2.79	0.88	0.29	0.00
{Fe–Pd}	{Fe} {Pd}	ΔS^E	1873	0	−0.12	−0.47	−0.56	−0.41	−0.14
	{Pd}	$\Delta \bar{S}^E_{Pd}$	1873	3.15	−2.15	−1.27	−0.46	−0.05	−0.00
{Fe–S}	(S_2)	$\Delta S^E_{S_2}$	—	−53.35	−48.95	−43.05	−70.21	—	—
{Fe–Si}	$Fe} {Si}	ΔS^E	1873	0	−2.09	−6.90	−8.62	−7.15	−2.79
	{Si}	$\Delta \bar{S}^E_{Si}$	1873	−17.51	−23.72	−19.71	−8.97	−2.55	−0.17
⟨Fe–V⟩	⟨Fe⟩α⟨V⟩	ΔS^E	1600	0	−0.53	−1.38	−1.62	−1.10	−0.13
	⟨V⟩	$\Delta \bar{S}^E_{V}$	1600	14.23	8.64	2.31	−1.43	−0.52	−0.18

Mg. Magnesium alloys: partial and integral heats or excess Gibbs energies of solution in J/g–atom

System	Components	Thermo-Chemical Function	Temp. (K)	$N_2 =$ 0.0	0.1	0.3	0.5	0.7	0.9
{Mg–Al}	{Ag} {Al}	ΔH	1073	0	−962.3	−2569.0	−3372.3	−3012.5	−1305.4
	{Al}	$\Delta \bar{H}_{Al}$	1073	−9865.9	−9317.8	−7033.3	−4066.8	−1464.4	−133.9
{Mg–Bi}	{Mg} {Bi}	ΔH	1100	0	−7740.4	−18,828.0	−18,953.5	−12,552.0	−3138.0
	{Bi}	$\Delta \bar{H}_{Bi}$	1100	—	−69,872.8	−35,020.1	−10,920.2	2887.0	418.4
{Mg–Ca}	{Mg} {Ca}	ΔH	1110	0	(−4937.1)	−7949.6	−7154.6	−4686.1	−1673.6
	{Ca}	$\Delta \bar{H}_{Ca}$	1110	—	—	−9874.2	−2552.2	−418.4	−62.76
⟨Mg–Cd⟩	⟨Mg⟩ ⟨Cd⟩	ΔH	543	0	−1661.0	−4460.1	−5531.2	−4217.5	−1422.6
	⟨Cd⟩	$\Delta \bar{H}_{Cd}$	543	−16,861.5	−16,192.1	−12,029.0	−5146.3	−761.5	71.13
{Mg–Cd}	{Mg} {Cd}	ΔH	923	0	−2025.1	−4610.8	−5610.7	−4698.6	−1924.6
	{Cd}	$\Delta \bar{H}_{Cd}$	923	−23,313.2	−17,685.8	−10,899.3	−5861.8	−1845.1	−150.6
{Mg–Cu}	{Mg} {Cu}	ΔH	1100	0	−3799.1	−8087.7	−13,100.1	−11,987.2	—
		$\Delta \bar{H}_{Cu}$	1100	−49,203.8	−30,149.9	−19,422.1	−12,786.3	−5552.2	—
{Mg–In}	{Mg} {In}	ΔH	923	0	−2807.5	−6778.1	−6870.1	−4736.3	−1807.5
	{In}	$\Delta \bar{H}_{In}$	923	−29,664.6	−26,384.3	−14,372.0	−3158.9	−920.5	−117.15
{Mg–Li}	{Mg} {Li}	ΔH	1000	0	−1518.8	−4610.8	−5368.1	−3585.7	−1092.0
	{Li}	$\Delta \bar{H}_{Li}$	1000	−12,635.7	−16,694.2	−12,560.4	−3288.6	−46.02	108.78
{Mg–Ni}	{Ni}	$\Delta \bar{G}^E_{Ni}$	1000	−41,936.2	−21,162.7	—	—	—	—
{Mg–Pb}	{Mg} {Pb}	ΔH	973	0	−3953.9	−9581.4	−9388.9	−6150.5	−2071.1
	{Pb}	$\Delta \bar{H}_{Pb}$	973	−39,748.0	−38,212.5	−19,058.1	−3895.3	−288.7	0
{Mg–Sb}	{Mg} {Sb}	ΔH	1100	0	−21,966.0	−52,718.4	−46024.0	−21,338.4	−6903.6
		$\Delta \bar{H}_{Sb}$	1100	—	−183,677.6	−107,528.8	51,044.8	2510.4	−83.68
{Mg–Sn}	{Mg} {Sn}	ΔG^E	1073	0	−6723.7	−13,765.4	−14,162.8	−10,083.4	−3564.8
	{Sn}	$\Delta \bar{G}^E_{Sn}$	1073	−79,307.7	−56,768.5	−25,501.5	−8660.9	−1644.3	−29.29
{Mg–Tl}	{Mg} {Tl}	ΔH	923	0	−2769.8	−11,129.4	−12,581.3	−10,091.8	−4518.7
	{Tl}	$\Delta \bar{H}_{Tl}$	923	−26,568.4	−26,568.4	−12,936.9	−2928.8	0	117.15
{Mg–Zn}	{Mg} {Zn}	ΔH	923	0	−2092.0	−4853.4	−6380.6	−5857.6	−2887.0
	{Zn}	$\Delta \bar{H}_{Zn}$	923	−21,756.8	−18,409.6	−11,798.9	−8033.3	−3389.0	−648.5

Mg. Magnesium alloys: partial and integral excess entropies of solution in J/deg g–atom

System	Components	Thermo-Chemical Function	Temp. (K)	$N_2=$ 0.0	0.1	0.3	0.5	0.7	0.9
{Mg–Al}	{Mg} {Al}	ΔS^E	1073	0	−0.36	−0.82	−0.83	−0.44	0.00
	{Al}	$\Delta \bar{S}^E_{Al}$	1073	−3.86	−3.29	−1.66	−0.31	0.32	0.12
{Mg–Bi}	{Mg} {Bi}	ΔS^E	1100	0	−4.20	−9.48	−8.47	−4.96	−1.90
	{Bi}	$\Delta \bar{S}_{Bi}$	1100	—	−39.85	−11.46	−8.68	0.96	−0.48
{Mg–Ca}	{Mg} {Ca}	ΔS^E	1110	0	(−1.30)	−2.68	−2.41	−1.57	−0.54
	{Ca}	$\Delta \bar{S}^E_{Ca}$	1110	—	—	−3.68	−0.75	−0.13	0
⟨Mg–Cd⟩	⟨Mg⟩ ⟨Cd⟩	ΔS^E	543	0	−0.33	−1.06	−1.30	−0.61	0.09
	⟨Cd⟩	$\Delta \bar{S}^E_{Cd}$	543	−3.17	−3.49	−3.33	−0.59	0.77	0.21
{Mg–Cd}	{Mg} {Cd}	ΔS^E	923	0	−0.57	−0.94	−1.05	−0.88	−0.38
	{Cd}	$\Delta \bar{S}^E_{Cd}$	923	−8.05	−4.00	−1.51	−1.05	−0.38	−0.04
{Mg–Cu}	{Mg} {Cu}	ΔS^E	1100	0	−1.88	−2.92	−3.23	−2.87	—
	{Cu}	$\Delta \bar{S}^E_{Cu}$	1100	−28.89	−11.84	−4.64	−3.43	−1.65	—
{Mg–In}	{Mg} {In}	ΔS^E	923	0	0.22	−0.71	−0.41	0.35	0.41
	{In}	$\Delta \bar{S}^E_{In}$	923	6.87	−1.32	−3.08	1.87	0.95	0.27
{Mg–Li}	{Mg} {Li}	ΔS^E	1000	0	−0.50	−1.41	−0.77	−0.83	−1.00
	{Li}	$\Delta \bar{S}^E_{Li}$	1000	−3.46	−5.81	−2.51	3.10	2.60	0.44
{Mg–Pb}	{Mg} {Pb}	ΔS^E	973	0	0.57	−0.62	−0.58	−0.23	−0.20
	{Pb}	$\Delta \bar{S}^E_{Pb}$	973	14.42	−0.63	−3.67	0.27	−0.06	−0.12
{Mg–Sb}	{Mg} {Sb}	ΔS^E	1100	0	−10.92	−22.43	−12.30	−1.67	0.21
	{Sb}	$\Delta S^E{}_{Sb}$	1100	—	−72.80	−2.93	25.90	4.06	4.23
{Mg–Tl}	{Mg} {Ti}	ΔS^E	923	0	−0.12	−0.22	0.73	1.18	0.77
	{Tl}	$\Delta \bar{S}^E_{Tl}$	923	1.81	−1.88	2.22	2.64	1.36	0.22
{Mg–Zn}	{Mg} {Zn}	ΔS^E	923	0	−1.05	−2.15	−2.78	—	−1.19
	{Zn}	$\Delta \bar{S}^E_{Zn}$	923	−8.54	−7.74	−5.40	−3.60	−1.34	−0.21

Na. Sodium alloys: partial and integral heats or excess Gibbs energies of solution in J/g–atom

System	Components	Thermo-Chemical Function	Temp. (K)	$N_2=$ 0.0	0.1	0.3	0.5	0.7	0.9
{Na–Bi}	{Na} {Bi}	ΔH	1173	0	−16,652.3	−41,337.9	−30,375.8	−18,242.2	−4497.8
	{Bi}	$\Delta \bar{H}_{Bi}$	1173	—	−182,422.4	−8493.5	−7991.4	3138.0	1046.0
{Na–Cd}	{Na} {Cd}	ΔH	673	0	−142.3	−1217.5	−3138.0	−4606.6	−3092.0
	{Cd}	$\Delta \bar{H}_{Cd}$	673	−230.1	−2694.5	−6681.8	−7907.8	−5422.5	−669.4
{Na–Cs}	{Na} {Cs}	ΔH	384	0	606.7	995.8	916.3	619.2	217.6
	{Cs}	$\Delta \bar{H}_{Cs}$	384	4895.3	3577.3	1581.6	368.2	66.94	12.55
{Na–Ga}	{Na} {Ga}	ΔG^E	823	0	1096.2	1255.2	301.2	−878.6	−870.3
	{Ga}	$\Delta \bar{G}^E_{Ga}$	823	—	4811.6	−949.8	−2920.4	−2008.3	−343.1
{Na–Hg}	{Na} {Hg}	ΔH	648	0	−5230.0	−16,192.1	−25,438.7	−25,396.9	−8409.8
	{Hg}	$\Delta \bar{H}_{Hg}$	648	−49,371.2	−53,889.9	−53,973.6	−42,383.9	−7949.6	0
{Na–In}	{Na} {In}	ΔH	713	0	−1874.4	−5744.6	−8472.6	−7388.9	−2811.6
	{In}	$\Delta \bar{H}_{In}$	713	−17,572.8	−19,518.4	−18,200.4	−12,552.0	−1928.8	−133.9
{Na–K}	{Na} {K}	ΔH	384	0	318.0	677.8	740.6	560.7	213.4
	{K}	$\Delta \bar{H}_{K}$	384	3640.1	2719.6	1380.7	564.8	150.6	8.37
{Na–Pb}	{Na} {Pb}	ΔH	700	0	−6753.0	−16,384.5	−16,246.5	−11,246.6	−3895.3
	{Pb}	$\Delta \bar{H}_{Pb}$	700	−64,852.0	−69,036.0	−27,765.0	−8640.0	−1359.8	0
{Na–Rb}	{Na} {Rb}	ΔH	384	0	598.3	1188.3	1230.1	907.9	334.7
	{Rb}	$\Delta \bar{H}_{Rb}$	384	6527.0	4665.2	2104.6	828.4	205.0	12.55
{Na–Sn}	{Na} {Sn}	ΔH	773	0	−6694.4	−17,489.1	−19,371.9	−12,091.8	−3535.5
{Na–Tl}	{Na} {Tl}	ΔH	673	0	−3276.1	−8811.5	−11,443.2	−9154.6	−3363.9
	{Tl}	$\Delta \bar{H}_{Tl}$	673	−33,472.0	−31,823.5	−24,376.0	−11,911.8	−2502.0	−33.47

Na. Sodium alloys: partial and integral excess entropies of solution in J/deg g–atom

System	Components	Thermo-Chemical Function	Temp. (K)	$N_2=$ 0.0	0.1	0.3	0.5	0.7	0.9
{Na–Bi}	{Na} {Bi}	ΔS^E	1173	0	−4.87	−13.37	−6.05	−1.32	1.82
	{Bi}	$\Delta \bar{S}^E_{Bi}$	1173	0	−65.69	19.96	3.35	6.90	1.55
{Na–Cd}	{Na} {Cd}	ΔS^E	673	0	−0.67	−2.09	−3.77	−4.64	−2.90
	{Cd}	ΔS^E_{Cd}	673	−7.24	−6.67	−7.56	−7.21	−4.56	−0.50
{Na–Cs}	{Na} {Cg}	ΔS^E ΔS^E	384	0	0	0	0	0	0
{Na–Hg}	{Na} {Hg}	ΔS^E	648	0	−2.76	−10.50	−18.66	−19.83	−4.23
	{Hg}	ΔS^E_{Hg}	648	−21.84	−31.97	−40.12	−37.20	−1.21	1.13
{Na–In}	{Na} {In}	ΔS^E	713	0	−2.09	−4.98	−6.53	−5.32	−1.82
	{In}	$\Delta \bar{S}^E_{In}$	713	−23.43	−18.79	−12.44	−9.22	−0.33	0
{Na–K}	{Na} {K}	ΔS^E	384	0	0	0	−0.21	−0.46	−0.46
	{K}	$\Delta \bar{S}^E_{K}$	384	0	0	−0.54	−0.79	−0.67	−0.40
{Na–Pb}	{Na} {Pb}	ΔS^E	700	0	−2.94	−7.42	−5.90	−3.15	−0.83
	{Pb}	$\Delta \bar{S}^E_{Pb}$	700	−24.33	−34.06	−6.97	0.42	0.97	0.11
{Na–Rb}	{Na} {Rb}	ΔS^E ΔS^E	384	0	0	0	0	0	0
{Na–Sn}	{Na} {Sn}	ΔS^E		0	−3.28	−8.60	−0.93	−5.13	−2.11
{Na–Tl}	{Na} {Tl}	ΔS^E		0	−1.62	−3.84	−4.45	−2.69	−0.50
	{Tl}	$\Delta \bar{S}^E_{Tl}$	673	−18.02	−14.61	−9.23	−2.39	0.86	0.29

Ni. Nickel alloys: partial and integral heats or excess Gibbs energies of solution in J/g–atom

System	Components	Thermo-Chemical Function	Temp. (K)	$N_2=$ 0.0	0.1	0.3	0.5	0.7	0.9
⟨Ni–Al⟩	⟨Ni⟩⟨Al⟩	ΔH	298	0	−15,271.6	—	−58,785.2	—	—
	⟨Ni⟩⟨Al⟩	ΔG^E	1273	0	−12,656.6	—	−46,881.7	—	—
	⟨Al⟩	$\Delta \bar{G}^E_{Al}$	1273	−144,933.8	−109,746.3	—	−41,066.0	—	—
⟨Ni–Au⟩	⟨Ni⟩⟨Au⟩	ΔH	1150	0	2690.3	6727.9	7560.5	5635.8	2092.0
	⟨Au⟩	$\Delta \bar{H}_{Au}$	1150	27,698.1	25,832.0	15,447.3	5765.6	1292.9	75.31
{Ni–Au}	{Ni}{Au}	ΔH	1369	0	—	—	2150.6	1807.5	774.0
Ni–Co					this system is nearly ideal in the solid and liquid states				
⟨Ni–Cr⟩	⟨Ni⟩⟨Cr⟩	ΔH	—	0	−878.6	1405.8	6359.7	8987.2	4866.0
	⟨Cr⟩	$\Delta \bar{H}_{Cr}$	—	−17,572.8	−878.6	15,648.2	20,794.5	8426.6	937.2
{Ni–Cr}	{Ni}{Cr}	ΔG^E	—	0	−1401.6	−3104.5	−2732.2	−1715.4	−1410.0
	{Cr}	$\Delta \bar{G}^E_{Cr}$	—	−15,376.2	−12,706.8	−5439.2	−338.9	−920.5	−1276.1
⟨Ni–Cu⟩	⟨Ni⟩⟨Cu⟩	ΔH	—	0	1192.4	1945.6	1861.9	1422.6	543.9
	⟨Cu⟩	$\Delta \bar{H}_{Cu}$	—	16,485.0	8660.9	2531.3	1213.4	460.2	12.55
{Ni–Cu}	{Ni}{Cu}	ΔG^E	—	0	1318.0	3075.2	3493.6	2845.1	1129.7
	{Cu}	$\Delta \bar{G}^E_{Cu}$	—	14,644.0	11,840.7	7175.6	3493.6	1255.2	125.5
⟨Ni–Fe⟩	⟨Ni⟩⟨Fe⟩γ	ΔH	1200	0	−2075.3	−4184.0	−3702.8	−1945.6	−962.3
	⟨Fe⟩γ	$\Delta \bar{H}_{Fe}$	1200	−24,267.2	−17,572.8	−4769.8	−1422.6	1004.2	−585.8
{Ni–Fe}	{Ni}{Fe}	ΔH	1873	0	−2426.7	−4853.4	−4602.4	−2928.8	−1025.1
	{Fe}	$\Delta \bar{H}_{Fe}$	—	−32,216.8	−22,648.0	−7246.7	−1987.4	−16.74	−54.39
{Ni–Ge}	{Ni}{Fe}	ΔH	1287	0	—	—	−54,768.6	−36,191.6	−12,342.8
⟨Ni–Mn⟩	⟨Ni⟩⟨Mn⟩β	ΔH	1050	0	−5418.3	−12,970.4	−14,213.0	−6828.3	—
	⟨Mn⟩β	$\Delta \bar{H}_{Mn}$	1050	−60,145.0	−48,890.0	−29,526.5	−7656.7	4686.1	—
⟨Ni–Pd⟩	⟨Ni⟩⟨Pd⟩	ΔH	1273	0	548.1	351.5	−535.6	−1179.9	−845.2
	⟨Pd⟩	$\Delta \bar{H}_{Pd}$	1273	8468.4	3054.3	−2200.8	−2778.2	−1606.7	−280.3
{Ni–Pd}	{Ni}{Pd}	ΔH	1873	0	682.0	1066.9	1196.6	987.4	422.6
	{Pd}	$\Delta \bar{H}_{Pd}$	1873	11,673.4	4158.9	1937.2	1121.3	410.0	50.21
⟨Ni–Pt⟩	⟨Ni⟩⟨Pt⟩	ΔH	298	0	−2723.8	−7811.5	−9263.4	−7238.3	−3188.2
	⟨Ni⟩⟨Pt⟩	$\Delta G^E 1625$	0	−3564.8	−7782.2	−8619.0	−6698.6	−2640.1	
	⟨Pt⟩	$\Delta \bar{G}^E_{Pt}$	1625	−40,919.5	−31,057.8	−16,263.2	−7008.2	−2058.5	−175.7
{Ni–Si}	{Ni}{Si}	ΔH	1873	0	−18,786.2	−52,885.8	−55,354.3	−36,735.5	12,761.2
	{Si}	$\Delta \bar{H}_{Si}$	1873	−188,280.0	−187,024.8	−140,164.0	−23,430.4	−3430.9	−167.4
{Ni–Sn}	{Ni}{Sn}	ΔH	1773	0	−10,460.0	−18,409.6	−17,154.4	−10,460.0	−4602.4
⟨Ni–Zn⟩	⟨Ni⟩⟨Zn⟩	ΔG^E	1100	0	−3652.6	−9020.7	−12,535.3	—	—
	⟨Zn⟩	$\Delta \bar{G}^E_{Zn}$	1100	−38,969.8	−34,170.7	−22,878.1	−14,581.2	—	—

Ni. Nickel alloys: partial and integral excess entropies of solution in J/deg g–atom

System	Components	Thermo-Chemical Function	Temp. (K)	$N_2=$ 0.0	0.1	0.3	0.5	0.7	0.9
⟨Ni–Au⟩	⟨Au⟩⟨Ni⟩	ΔS^E	1150	0	0.67	2.35	2.79	1.92	0.58
	⟨Au⟩	$\Delta \bar{S}^E_{Au}$	1150	4.77	8.19	6.59	2.02	0.03	−0.06
⟨Ni–Cr⟩	⟨Ni⟩⟨Cr⟩	ΔS^E	—	0	0.36	1.79	4.05	4.87	2.77
	⟨Cr⟩	$\Delta \bar{S}^E_{Cr}$	—	−2.82	3.10	8.82	10.20	4.21	1.36
⟨Ni–Cu⟩	⟨Ni⟩⟨Cu⟩	ΔS^E	—	0	−0.04	−0.64	−0.97	−0.79	−0.36
	⟨Cu⟩	$\Delta \bar{S}^E_{Cu}$	—	4.70	−2.03	−2.55	−0.98	−0.29	−0.07
⟨Ni–Fe⟩	⟨Ni⟩⟨Fe⟩γ	ΔS^E	1200	0	−0.56	−1.04	−0.81	0.29	−0.25
	⟨Fe⟩γ	$\Delta \bar{S}^E_{Fe}$	1200	−5.02	−3.30	−0.88	−0.50	0.36	−0.41
{Ni–Fe}	{Ni}{Fe}	ΔS^E	1873	0	−0.74	−1.24	−1.10	−0.64	−0.21
	{Fe}	$\Delta \bar{S}^E_{Fe}$	1873	−9.20	−5.86	−1.42	−0.42	0.08	0
⟨Ni–Mn⟩	⟨Ni⟩⟨Mn⟩β	ΔS^E	1050	0	−0.69	−2.30	−2.07	1.22	—
	⟨Mn⟩γ	$\Delta \bar{S}^E_{Mn}$	1050	−7.11	−6.91	−6.38	3.35	6.28	—
⟨Ni–Pd⟩	⟨Ni⟩⟨Pd⟩	ΔS^E	1273	0	0.69	1.57	1.79	1.38	0.49
	⟨Pd⟩	$\Delta \bar{S}^E_{Pd}$	1273	7.85	6.17	3.46	1.54	0.33	−0.01
{Ni–Pd}	{Ni}{Pd}	ΔS^E	1873	0	−0.01	−0.28	−0.36	−0.28	−0.11
	{Pd}	$\Delta \bar{S}^E_{Pd}$	1873	1.90	−1.08	−0.90	−0.33	−0.08	−0.00
⟨Ni–Pt⟩	⟨Ni⟩⟨Pt⟩	ΔS^E	1625	0	0.52	−0.02	−0.40	−0.33	−0.34
{Ni–Si}	{Ni}{Si}	ΔS^E	1873	0	−2.72	−10.25	−11.59	−7.70	−2.30
	{Si}	$\Delta \bar{S}^E_{Si}$	1873	−23.01	−31.30	−33.47	−6.44	−0.17	0.33

Pb. Lead alloys: partial and integral heats or excess Gibbs energies of solution in J/g–atom

System	Components	Thermo-Chemical Function	Temp. (K)	$N_2=$ 0.0	0.1	0.3	0.5	0.7	0.9
{Pb–Ag}	{Pb} {Ag}	ΔH	1273	0	1146.4	2974.8	3702.8	2912.1	1054.4
	{Ag}	$\bar{H}_{Ag}$	1273	11,715.2	11,045.8	7761.3	3727.9	728.0	4.18
{Pb–Al}	{Pb} {Al}	ΔH	—	0	2443.5	6665.1	9355.4	8907.7	3866.0
{Pb–Au}	{Pb} {Au}	ΔH	1200	0	259.4	−305.4	−698.7	−790.8	−435.1
	{Au}	$\Delta\bar{H}_{Au}$	1200	221.8	−477.0	−2301.2	−1301.2	−640.2	−146.4
{Pb–Bi}	{Pb} {Bi}	ΔH	700	0	−334.7	−878.6	−1108.8	−899.6	−334.7
	{Bi}	$\Delta\bar{H}_{Bi}$	700	−3514.6	−3263.5	−2301.2	−1108.8	−292.9	−20.92
{Pb–Cd}	{Pb} {Cd}	ΔH	773	0	857.7	2100.4	2656.8	2447.6	1242.6
	{Cd}	$\Delta\bar{H}_{Cd}$	773	9330.3	7891.0	5280.2	3142.2	1477.0	288.7
{Pb–Cu}	{Pb} {Cu}	ΔH	1473	0	2443.5	5598.2	6723.7	5949.6	2958.1
	{Cu}	ΔH_{Cu}	1473	27,614.4	21,689.9	13,058.3	−6987.3	3447.6	598.3
{Pb–Ga}	{Pb} {Ga}	ΔH	923	0	1635.9	3297.0	3832.5	3585.7	1882.8
		$\Delta\bar{H}_{Ga}$	923	20,208.7	13,305.1	6673.5	4259.3	2410.0	431.0
{Pb–Hg}	{Pb} {Hg}	ΔH	600	0	−142.3	−75.31	196.6	464.4	384.9
	{Hg}	$\Delta\bar{H}_{Hg}$	600	−2322.1	−723.8	652.7	962.3	728.0	154.8
⟨Pb–In⟩	⟨In⟩ ⟨Pb⟩	ΔH	315	0	418.4	1046.0	1297.0	1129.7	0
{Pb–In}	{Pb} {In}	ΔH	315	0	326.4	778.2	949.8	815.9	355.6
	{In}	$\Delta\bar{H}_{In}$	673	3999.9	2974.8	1878.6	1000.0	376.6	41.84
{Pb–K}	{Pb} {K}	ΔH	848	0	−5154.7	−14,359.5	−19,756.8	−17,694.1	−6221.6
	{K}	$\Delta\bar{H}_{K}$	848	−51,530.1	−50,910.9	−41,250.1	−29,288.0	−5238.4	259.4
{Pb–Mg}	{Pb} {Mg}	ΔH	973	0	−2071.1	−6150.5	−9388.9	−9577.2	−3953.9
	{Mg}	$\Delta\bar{H}_{Mg}$	973	−20,920.0	−20,710.8	−19,823.8	−14,878.3	−5514.5	−146.4
{Pb–Na}	{Pb} {Na}	ΔH	700	0	−3895.3	−11,246.6	−16,246.5	−16,384.5	−6753.0
	{Na}	$\Delta\bar{H}_{Na}$	700	−38,492.8	−39,036.7	−34,308.8	−23,848.8	−11,506.0	167.4
{Pb–Pt}	{Pb} {Pt}	ΔG^E	1273	0	−3033.4	−7301.1	−8957.9	—	—
	{Pt}	$\Delta\bar{G}^E_{Pt}$	1273	−33,296.3	−27,656.2	−17,957.7	−9225.7	—	—
{Pb–Sb}	{Pb} {Sb}	ΔH	905	0	0	−33.47	−66.94	−50.21	−8.37
	{Sb}	$\Delta\bar{H}_{Sb}$	905	133.9	−75.31	−188.3	−92.05	12.55	8.37
{Pb–Sn}	{Pb} {Sn}	ΔH	1050	0	543.9	1192.4	1368.2	1146.4	502.1
	{Sn}	$\Delta\bar{H}_{Sn}$	1050	6276.0	4677.7	2573.2	1305.4	510.4	62.76
⟨Pb–Tl⟩	⟨Pb⟩ ⟨Tl⟩β	ΔH	523	0	−468.6	−1309.6	−1937.2	−2066.9	—
	⟨Tl⟩β	$\Delta\bar{H}_{Tl}$	523	−4895.3	−4497.8	−3974.8	−2928.8	−1732.2	—
{Pb–Tl}	{Pb} {Tl}	ΔH	773	0	−326.4	−836.8	−1054.4	−882.8	−368.2
	{Tl}	$\Delta\bar{H}_{Tl}$	773	−3430.9	−3112.9	−2175.7	−1121.3	−359.8	−33.47
{Pb–Zn}	{Pb} {Zn}	ΔH	—	0	2238.4	5690.2	7531.2	7301.1	3891.1
	{Zn}	$\Delta\bar{H}_{Zn}$	—	—	20,961.8	15,083.3	9790.6	4769.8	753.1

Pb. Lead alloys: partial and integral excess entropies of solution in J/deg g–atom

System	Components	Thermo-Chemical Function	Temp. (K)	$N_2 =$ 0.0	0.1	0.3	0.5	0.7	0.9
{Pb–Ag}	{Pb} {Ag}	ΔS^E	1273	0	0.39	1.28	1.83	1.57	0.67
	{Ag}	$\Delta \bar{S}^{EAg}$	1273	3.31	4.29	4.20	2.22	0.64	0.13
{Pb–Au}	{Pb} {Au}	ΔS^E	1200	0	1.09	1.66	1.78	1.49	0.64
	{Au}	$\Delta \bar{S}^E_{Au}$	1200	9.81	7.40	2.36	1.78	0.64	0.07
{Pb–Bi}	{Pb} {Bi}	ΔS^E	700	0	0.08	0.21	0.21	0.17	0.13
	{Bi}	$\Delta \bar{S}^E_{Bi}$	700	1.09	0.85	0.42	0.17	0.13	0.04
{Pb–Cd}	{Pb} {Cd}	ΔS^E	773	0	0.19	0.52	0.71	0.73	0.46
	{Cd}	$\Delta \bar{S}^E_{Cd}$	773	1.95	1.88	1.44	0.98	0.60	0.18
{Pb–Cu}	{Pb} {Cu}	ΔS^E	1473	0	0.46	0.94	1.05	0.98	0.72
	{Cu}	$\Delta \bar{S}_{ECu}$	1473	5.58	3.84	1.93	0.93	0.90	0.30
{Pb–Hg}	{Pb} {Hg}	ΔS^E	600	0	−0.42	−0.96	−1.16	−0.98	−0.46
	{Hg}	$\Delta \bar{S}^E_{Hg}$	600	−4.94	−3.66	−2.27	−1.19	−0.41	−0.10
{Pb–In}	{Pb} {In}	ΔS^E	673	0	0.25	0.50	0.51	0.35	0.13
	{In}	$\Delta \bar{S}^E_{In}$	673	3.46	2.05	0.92	0.32	0.04	0
{Pb–K}	{Pb} {K}	ΔS^E	848	0	−1.87	−5.49	−8.88	−9.79	−3.41
	{K}	$\Delta \bar{S}^E_{K}$	848	−18.83	−18.47	−17.33	−20.92	−3.82	0.28
{Pb–Mg}	{Pb} {Mg}	ΔS^E	973	0	−0.20	−0.23	−0.58	−0.62	0.57
	{Mg}	$\Delta \bar{S}^E_{Mg}$	973	−3.87	−0.90	−0.62	−1.42	0.68	0.70
{Pb–Na}	{Pb} {Na}	ΔS^E	700	0	−0.83	−3.15	−5.90	−7.42	−2.94
	{Na}	$\Delta \bar{S}^E_{Na}$	700	−7.17	−9.33	−12.76	−12.20	−7.61	0.51
{Pb–Sb}	{Pb} {Sb}	ΔS^E	905	0	0.19	0.40	0.44	0.38	0.18
	{Sb}	$\Delta \bar{S}^E_{Sb}$	905	2.23	1.60	0.81	0.41	0.20	0.03
{Pb–Sn}	{Pb} {Sn}	ΔS^E	1050	0	−0.77	−1.27	−1.02	−0.56	−0.15
	{Sn}	$\Delta \bar{S}^E_{Sn}$	1050	−9.98	−5.86	−1.31	0.04	0.15	0.03
⟨Pb–Tl⟩	⟨Pb⟩ ⟨Tl⟩β	ΔS^E	523	0	−0.58	−1.25	−1.50	−1.38	—
	⟨Tl⟩β	$\Delta \bar{S}^E_{Tl}$	523	−6.96	−4.84	−2.74	−1.42	−0.95	—
{Pb–Tl}	{Pb} {Tl}	ΔS^E	773	0	−0.23	−0.50	−0.54	−0.42	−0.26
	{Tl}	$\Delta \bar{S}^E_{Tl}$	773	−2.54	−2.05	−0.99	−0.41	−0.16	−0.15
{Pb–Zn}	{Pb} {Zn}	ΔS^E	—	0	0.77	1.99	2.76	2.87	1.69
	{Zn}	$\Delta \bar{S}^E_{Zn}$	—	9.20	7.30	5.52	3.93	2.05	0.29

Pt. Platinum alloys: partial and integral heats or excess Gibbs energies of solution in J/g–atom

System	Components	Thermo-Chemical Function	Temp. (K)	N_2= 0.0	0.1	0.3	0.5	0.7	0.9
⟨Pt–Au⟩	⟨Pt⟩⟨Au⟩	ΔH	—	0	1861.9	3326.3	2644.3	1108.8	0
	⟨Au⟩	$\Delta \bar{H}_{Au}$	—	23,807.0	14,999.6	3891.1	−669.4	−1192.4	−251.0
⟨Pt–Co⟩	⟨Pt⟩⟨Co⟩	ΔG^E	1273	0	−4464.3	−10,418.2	−12,405.6	−10,418.2	−4464.3
	⟨Co⟩	$\Delta \bar{G}^E_{Co}$	1273	−49,622.2	−40,195.7	−24,313.2	−12,405.6	−4464.3	−497.9
⟨Pt–Cr⟩	⟨Pt⟩⟨Cr⟩	ΔG^E	1773	0	−13,137.8	−30,375.8	—	—	—
	⟨Cr⟩	$\Delta \bar{G}^E_{Cr}$	1773	−146,858.4	−117,570.4	−69,663.6	—	—	—
⟨Pt–Cu⟩	⟨Pt⟩⟨Cu⟩	ΔH	1350	0	−2887.0	−8284.3	−11,087.6	−10,384.7	−5138.0
	⟨Cu⟩	$\Delta \bar{H}_{Cu}$	1350	−28,744.1	−28,635.3	−25,288.1	−13,095.9	−6443.4	−979.1
⟨Pt–Fe⟩	⟨Pt⟩⟨Fe⟩	ΔG^E	1123	0	−6652.6	−13,706.8γ_3	−18,175.3γ_2	−13,982.9	−5723.7
	⟨Fe⟩γ	$\Delta \bar{G}^E_{Fe}$	1123	−66,584.2	−66,584.2	−37,338.0γ_3	−17,430.5γ_2	−3886.9	−1882.8
⟨Pt–Ni⟩	⟨Pt⟩⟨Ni⟩	ΔG^E	1625	0	−2640.1	−6694.4	−8619.0	−7782.2	−3564.8
	⟨Pt⟩⟨Ni⟩	ΔH	298	0	−3188.2	−7238.3	−9263.4	−7811.5	−2723.8
	⟨Ni⟩	$\Delta \bar{G}^E_{Ni}$	1625	−28,032.8	−24,794.4	−17,531.0	−10,229.9	−4142.2	−510.4
⟨Pt–Pd⟩	⟨Pt⟩⟨Pd⟩	ΔH	—	0	−1221.7	−3234.2	−4317.9	−4016.6	−1887.0
	⟨Pd⟩	$\Delta \bar{H}_{Pd}$	—	−12,635.7	−11,736.1	−8911.9	−5481.0	−2301.2	−292.9

Pt. Platinum alloys: partial and integral excess entropies of solution in J/deg g–atom

System	Components	Thermo-Chemical Function	Temp. (K)	N_2= 0.0	0.1	0.3	0.5	0.7	0.9
⟨Pt–Au⟩	⟨Pt⟩⟨Au⟩	ΔS^E	—	0	−0.44	−1.46	−2.30	−2.43	−1.21
	⟨Au⟩	$\Delta \bar{S}^E_{Au}$	—	−3.77	−4.81	−5.04	−3.66	−1.72	−0.24
⟨Pt–Cu⟩	⟨Pt⟩⟨Cu⟩	ΔS^E	1350	0	0.16	−0.52	−0.87	−0.75	−0.40
	⟨Cu⟩	$\Delta \bar{S}^E_{Cu}$	1350	3.87	−0.18	−4.33	−0.37	−0.44	−0.11
⟨Pt–Ni⟩	⟨Pt⟩⟨Ni⟩	ΔS^E	1625	0	−0.34	−0.33	−0.40	−0.02	0.52
⟨Pt–Pd⟩	⟨Pt⟩⟨Pd⟩	ΔS^E	—	0	−0.69	−1.81	−2.49	−2.23	−1.05
		$\Delta \bar{S}^E_{Pd}$	—	−7.11	−6.57	−4.98	−3.03	−1.28	—

REFERENCES

Chapters I, II, III and IV

[1884] E. Warburg, *Wied. Ann. Phys.* 1884 **21** 622.
[1898] M. Herschkowitsch, *Z. phys. Chem.* 1898 **27** 123.
[1899] A. H. Jouniaux, *Compt. rend.* 1899 **129** 283; 1901 **132** 1270; 1903 **136** 1006.
[1901] T. J. Baker, *Proc. Roy. Soc.* 1901 **68** 9.
[1902] (a) H. Pélabon, *Ann. Chim. Phys.* (7) 1902 **25** 365.
(b) T. W. Richards, *Z. physik. Chem.* 1902 **40** 597; 1904 **49** 15.
[1905] F. Haber and A. Moser, *Z. Elektrochem.* 1905 **11** 593.
[1907] A. Sieverts, *Z. phys. Chem.* 1907 **60** 129.
[1908] (a) M. Bodenstein, *Z. Elektrochem.* 1908 **14** 544.
(b) F. M. G. Johnson, *Z. phys. Chem.* 1908 **61** 457.
(c) J. J. van Laar, *Z. Phys. Chem.* 1908 **63** 216; 1908 **64** 257.
(d) J. Thomsen, *Thermochemistry*, London, 1908.
[1909] (a) M. Knudsen, *Ann. Physik.* 1909 **28** 75; 1909 **29** 179.
(b) G. Masing, *Z. anorg. Chem.* 1909 **62** 265.
[1913] I. Langmuir, *Phys. Rev.* (2) 1913 **2** 329.
(b) G. Tammann, *Z. phys. Chem.* 1913 **85** 273.
(c) H. von Wartenberg, *Z. Elektrochem.* 1913 **19** 482.
[1917] N. Parravano and P. De Cesaris, *Gazz. chim. Ital.* 1917 **47** 144.
[1919] O. Ruff and B. Bergdahl, *Z. anorg. allg. Chem.* 1919 **106** 76.
[1920] F. G. Keyes and W. A. Felsing, *J. Amer. Chem. Soc.* 1920 **42** 246.
[1921] R. Walter, *Z Metallkunde* 1921 **13** 225.
[1922] (a) W. Biltz and G. Hohorst, *Z. anorg. Chem.* 1922 **121** 1.
(b) W. Leitgebel, *Z. anorg. allg. Chemie* 1922 **202** 305.
(c) T. W. Richards and J. B. Conant, *J. Amer. Chem. Soc.* 1922 **44** 601.
[1923] W. Biltz and C. Haase, *Z. anorg. Chem.* 1923 **129** 141.
[1924] (a) W. Biltz, W. Wagner, H. Pieper and W. Holverscheidt, *Z. anorg. Chem.* 1924 **134** 25.
(b) E. D. Eastman and R. M. Evans, *J. Amer. Chem. Soc.* 1924 **46** 888.
(c) P. H. Emmett and J. F. Shultz, *J. Amer. Chem. Soc.* 1929 **51** 3249.
[1925] (a) F. Neumann, *Z. anorg. Chem.* 1925 **145** 193.
(b) W. H. Rodebush and A. O. Dixon, *Phys. Rev.* 1925 **26** 85.
[1926] (a) K. Jellinek and K. Uhloht, *Z. physik. Chem.* 1926 **119** 161.
(b) K. Jellinek and K. Uhloht, *Z. anorg. allg. Chem.* 1926 **151** 157.
(c) C. H. M. Jenkins, *Proc. Roy. Soc.* (A) 1926 **110** 227.
(d) R. N. Pease and R. S. Cook, *J. Amer. Chem. Soc.* 1926 **48** 1199.
(e) F. E. Poindexter, *Phys. Rev.* 1926 **28** 208.
[1927] (a) W. Fischer and W. Biltz, *Z. anorg. Chem.* 1927 **166** 290; *ibid.* 1928 **176** 81.
(b) R. W. Millar, *J. Amer. Chem. Soc.* 1927 **49** 3003.
(c) R. Schenck, *Z. anorg. Chem.* 1927 **164** 145.
(d) H. von Wartenberg and S. Aoyama, *Z. Elektrochem.* 1927 **33** 144.
[1928] (a) W. Biltz and F. Meyer, *Z. anorg. allg. Chem.* 1928 **176** 23.
(b) K. Jellinek and A. Rudat, *Z. anorg. allg. Chem.* 1928 **175** 281.
(c) W. A. Roth and P. Chall, *Z. Elektrochem.* 1928 **34** 185.
[1929] (a) P. H. Emmett and J. F. Shultz, *J. Amer. Chem. Soc.* 1929 **51** 3249.
(b) L. L. Hirst and A. R. Olson, *J. Amer. Chem. Soc.* 1929 **51** 2398.

(c) K. Jellinek and G. A. Rosner, *Z. physik. Chem.* (A) 1929 **143** 51; 1931 **152** 67.
(d) R. Lorenz and H. Velde, *Z. anorg. Chem.* 1929 **183** 81.
[1930] (a) W. Biltz and R. Juza, *Z. anorg. Chem.* 1930 **190** 161.
(b) P. H. Emmett, S. B. Hendricks and S. Brunauer, *J. Amer. Chem. Soc.* 1930 **52** 1456.
(c) E. Lange and J. Monheim, *Z. phys. Chem.* 1930 **149** 51.
(d) C. G. Maier, *J. Phys. Chem.* 1930 **43** 2866.
(e) F. Oppenheimer, *Z. anorg. Chem.* 1930 **189** 297.
(f) W. A. Roth, H. Umbach and P. Chall, *Arch. Eisenhüttenwes.* 1930 **4** 87.
[1931] (a) W. Leitgebel, *Z. anorg. allg. Chem.* 1931 **202** 305.
(b) R. Schenck, F. Kurzen and H. Wesselkock, *Z. anorg. Chem.* 1931 **203** 159.
(c) M. Volmer, *Z. physik. Chem., Bodenstein Festb.* 1931 863.
[1932] (a) R. Brunner, *Z. Elektrochem.* 1932 **38** 55.
(b) W. Fischer and R. Gewehr, *Z. anorg. allg. Chem.* 1932 **209** 17.
(c) W. Fischer and O. Rahlfs, *Z. anorg. allg. Chem.* 1932 **205** 1.
(d) H. Haraldsen, *Skr. norske Vidensk Akad. I. Math.-naturw. Kl.* 1932 No. **9**.
(e) B. Neumann, C. Kröger and H. Kunz, *Z. anorg. Chem.* 1932 **207** 133.
(f) W. A. Roth and H. Troitzsch, *Arch. Eisenhüttenwes.* 1932/33 **6** 79.
[1933] (a) F. Ishikawa and M. Watanabe, *Sci. Rep. Tôhoku Univ.* 1933 **22** 393.
(b) M. Watanabe, *Sci. Rep. Tôhoku Univ.* 1933 **22** 407, 902.
[1934] (a) E. Bauer and R. Brunner, *Helv. chim. acta* 1934 **17** 958.
(b) W. Biltz, G. Rohlffs and H. U. von Vogel, *Z. anorg. Chem.* 1934 **220** 113.
(c) J. Fischer, *Z. anorg. allg. Chem.* 1934 **219** 1.
(d) W. Franke, R. Juza, K. Meisel and W. Biltz, *Z. anorg. Chem.* 1934 **218** 346.
(e) G. Grube and E. A. Rau, *Z. Elektrochem.* 1934 **40** 352.
(f) G. Naeser, *Mitt. KWI Eisenforsch Düsseldorf* 1934 **16** 1.
[1936] (a) H. Moser, *Phys. Z.* 1936 **37** 737.
(b) J. H. McAteer and H. Seltz, *J. Amer. Chem. Soc.* 1936 **58** 2081.
(c) K. K. Kelley, *US Bur. Mines, Bull.* No. 393, 1936.
(d) O. Kubaschewski, *Z. Elektrochem.* 1936 **42** 5.
(e) C. Robert, *Helv. Phys. Acta* 1936 **9** 405.
(f) C. Sykes and F. W. Jones, *J. Inst. Met.* 1936 **59** 257.
[1937] (a) F. Adcock, *J. Iron Steel Inst.* 1937, *No.* 1, 281P.
(b) R. Fricke and W. Zerrweck, *Z. Elektrochem.* 1937 **43** 52.
(c) E. A. Guggenheim, *Trans. Faraday Soc.* 1937 **33** 151, *Proc. Roy. Soc.* (*A*) 1935 **148** 304.
(d) A. L. Marshall, R. W. Dornte and F. J. Norton, *J. Amer. Chem. Soc.* 1937 **59** 1161.
(e) A. Ölander, *Z. Metallk.* 1937 **29** 361.
(f) G. Scatchard, *Trans. Faraday Soc.* 1937 **33** 160.
[1938] (a) British Standards Institution, *Bomb Calorimeter Thermometers*, BSS No. 791, London, 1938.
(b) Z. Shibata and H. Kitagawa, *J. Fac. Sci. Hokk. Univ.* Ser. III 1938 **2** 223.
(c) N. Swindells and C. Sykes, *Proc. Roy. Soc.* (*A*) 1938 **168** 237.
[1939] (a) R. G. Bates, *J. Amer. Chem. Soc.* 1939 **61** 1040.
(b) W. Biltz and M. Heimbrecht, *Z. anorg. Chem.* 1939 **241** 349.
(c) W. Fischer, R. Gewehr and H. Wingchen, *Z. anorg. allg. Chem.* 1939 **242** 161.
(d) G. Grube and M. Flad, *Z. Elektrochem.* 1939 **45** 835.
(e) G. Grube and K. Ratsch, *Z. Elektrochem.* 1939 **45** 838.
(f) R. Hargreaves, *J. Inst. Metals* 1939 **64** 115.
(g) O. Kubaschewski and A. Walter, *Z. Elektrochem.* 1939 **45** 630, 732.
(h) F. Weibke and U. von Quadt, *Z. Elektrochem.* 1939 **45** 715.
(i) R. Schenck and P. von der Forst, *Z. anorg. allg. Chem.* 1939 **241** 145.
(j) R. Schenck, K. Meyer and K. Mayer, *Z. anorg. Chem.* 1939 **243** 17, 259.
[1940] (a) T. W. Dakin and D. T. Ewing, *J. Amer. Chem. Soc.* 1940 **62** 2280.
(b) R. Fricke and F. Blaschke, *Z. Elektrochem.* 1940 **46** 46.
(c) K. Hauffe, *Z. Elektrochem.* 1940 **46** 348.
(d) W. A. Roth, U. Wolff and O. Fritz, *Z. Elektrochem.* 1940 **46** 42.
(e) J. C. Southard, *Ind. Eng. Chem.* 1940 **32** 442.
(f) E. Zintl, *Z. anorg. Allg. Chem.* 1940 **245** 1.
[1941] (a) H. Hohmann and H. Bommer, *Z. anorg. Chem.* 1941 **248** 383.
(b) O. Kubaschewski, *Z. Elektrochem.* 1941 **47** 623.
(c) W. A. Roth, H. Berendt and G. Wirths, *Z. Elektrochem.* 1941 **47** 185.
(d) A. Schneider and E. K. Stoll, *Z. Elektrochem.* 1941 **47** 519.
(e) F. Weibke and H. Matthes, *Z. Elektrochem.* 1941 **47** 421.
[1942] (a) R. Fricke and E. Dönges, *Z. anorg. Chem.* 1942 **250** 202.

(b) G. Grube and M. Flad, *Z. Elektrochem.* 1942 **48** 377.
(c) E. A. Gulbransen and K. F. Andrew, *Trans. Electrochem. Soc.* 1942 **81** 327.
[1943] (a) A. Knappwost, *Z. Elektrochem.* 1943 **49** 1.
(b) E. Scheil, *Z. Elektrochem.* 1943 **49** 242.
(c) A. Schneider and U. Esch, *Z. Elektrochem.* 1943 **49** 55.
(d) C. H. Shomate, *J. Amer. Chem. Soc.* 1943 **65** 785.
(e) C. H. Shomate and E. H. Huffmann, *J. Amer. Chem. Soc.* 1943 **65** 1625.
(f) H. von Wartenberg, *Z. anorg. Chem.* 1943 **252** 141.
[1945] F. E. Young, *J. Amer. Chem. Soc.* 1945 **67** 257.
[1946] (a) L. S. Darken and R. W. Gurry, *J. Amer. Chem. Soc.* 1945 **67** 1398; 1946 **68** 798.
(b) K. Wohl, *Trans. AICE* 1946 **42** 215.
[1947] W. A. Roth, *Thermochemie*, Berlin: Göschen, 1947.
[1948] (a) Discussion Faraday Soc. No. 4 1948.
(b) M. A. Fineman and W. E. Wallace, *J. Amer. Chem. Soc.* 1948 **70** 4165.
(c) P. V. Gel'd and M. I. Kochnev, *Zh. Priklad. Khim.* 1948 **21** 1249.
(d) C. F. Goodeve, *Disc. Faraday Soc.* 1948 **4** 9.
(e) P. Gross, C. S. Campbell, P. J. C. Kent and D. Levi, *Disc. Faraday Soc.* 1948 No. 4 206.
(f) O. Kubaschewski and O. Huchler, *Z. Elektrochem.* 1948 **52** 170.
(g) B. A. Rose, G. J. Davis and H. J. T. Ellingham, *Disc. Faraday Soc.* 1948 No. **4** 154.
[1949] (a) M. N. Dastur and J. Chipman, *J. Metals* 1949 **1** 441.
(b) M. N. Dastur and N. A. Gokcen, *J. Metals* 1949 **1** 665.
(c) W. F. Giauque, *J. Amer. Chem. Soc.* 1949 **71** 3192.
(d) G. Grube and H. Speidel, *Z. Elektrochem.* 1949 **53** 339, 541.
(e) K. Hauffe and A. L. Vierk, *Z. Elektrochem.* 1949 **53** 151.
(f) K. K. Kelley, *US Bur. Mines, Bull.* 476, 1949.
(g) O. J. Kleppa, *J. Amer. Chem. Soc.* 1949 **71** 3275; 1950 **72** 3346; 1951 **73** 385; 1952 **74** 6037, 6052.
(h) O. Kubaschewski, *Trans. Faraday Soc.* 1949 **45** 931.
(i) O. Kubaschewski and O. von Goldbeck, *Trans. Faraday Soc.* 1949 **45** 948.
(j) G. W. Murphy, *Rev. Sci. Instr.* 1949 **20** 372.
(k) M. Rey, *Compt. rend.* 1949 **228** 473, 545.
(l) F. D. Richardson, *J. Iron Steel Inst.* 1949 **163** 382.
(m) T. Rosenqvist, *J. Metals* 1949 **1** 451.
(n) H. von Wartenberg, *Z. Elektrochem.* 1949 **53** 343.
[1950] (a) C. R. Barber, *J. Sci. Instr.* 1950 **27** 47.
(b) L. S. Darken, *J. Amer. Chem. Soc.* 1950 **72** 2909.
(c) E. A. Gulbransen and K. F. Andrew, *J. Electrochem. Soc.* 1950 **97** 383.
(d) A. W. Herbenar, C. A. Siebert and O. S. Duffendack, *J. Metals* 1950 **2** 323.
(e) H. H. Kellogg, *Trans. AIME* 1950 **188** 862.
(f) O. Kubaschewski and O. von Goldbeck, NPL Rep., June 1950.
(g) H. M. Schadel and C. E. Birchenall, *J. Metals* 1950 **2** 1134.
(h) H. M. Schadel, G. Derge and C. E. Birchenall, *J. Metals* 1950 **2** 1282.
(i) H. Schäfer and R. Hörnle, *Z. anorg. allg. Chem.* 1950 **263** 261.
(j) R. Schuhmann, *Acta met.* 1950 **3** 219.
(k) C. W. Sherman, H. I. Elvander and J. Chipman, *J. Metals* 1950 **2** 334; 1952 **4** 597.
(l) A. L. Vierk, *Z. Elektrochem.* 1950 **54** 436.
[1951) (a) J. F. Elliott and J. Chipman, *Trans Faraday Soc.* 1951 **47** 139.
(b) B. E. Hopkins, G. C. H. Jenkins and H. E. N. Stone, *J. Iron Steel Inst.* 1951 **168** 377.
(c) J. A. Kitchener and S. Ignatowicz, *Trans. Faraday Soc.* 1951 **47** 1278.
(d) O. Kubaschewski and R. Hörnle, *Z. Metalk.* 1951 **42** 129.
(e) W. M. Latimer, *J. Amer. Chem. Soc.* 1951 **73** 1480.
(f) E. Whalley, *J. Chem. Phys.* 1951 **19** 509.
[1952] (a) J. B. Bookey, *J. Iron Steel Inst.* 1952 **172** 61.
(b) L. S. Brooks, *J. Amer. Chem. Soc.* 1952 **74** 227.
(c) T. M. Buck, W. E. Wallace and R. M. Rulon, *J. Amer. Chem. Soc.* 1952 **74** 136.
(d) P. Chiche, *Ann. Chimie* 1952 **7** 361.
(e) H. Flood and K. Grjotheim, *J. Iron and Steel Inst.* 1952 **171** 64.
(f) N. A. Gokcen and J. Chipman, *J. Metals* 1952 **4** 171.
(g) E. J. Huber, C. E. Holley and E. H. Meierkord, *J. Amer. Chem. Soc.* 1952 **74** 3406.
(h) J. Lumsden, *Thermodynamics of Alloys*, Inst. Metals, 1952.
(i) N. C. Tombs and A. J. E. Welch, *J. Iron Steel Inst.* 1952 **172** 69.
(j) C. Wagner, *Thermodynamics of Alloys*, Addison-Wesley, Cambridge, Mass., 1952.
[1953] (a) H. Flood, T. Forland and K. Grjotheim, *Phys. Chem. of Melts*, I.M.M. London 1953 p. 46.

(b) C. B. GRIFFITH and M. W. MALLETT, *J. Amer. Chem. Soc.* 1953 **75** 1823.
(c) J. A. HALL, (a) *Fundamentals of Thermometry.*
(b) *Practical Thermometry*, Institute of Physics, London, 1953.
(d) J. MOREAU, *Compt. Rend.* 1953 **236** 85.
(e) F. D. RICHARDSON and W. E. DENNIS, *J. Iron Steel Inst.* 1953 **175** 257.
[1954] A. W. SEARCY and R. D. FREEMAN, *J. Chem. Phys.* 1954 **22** 762.
[1955] (a) P. GROSS, C. HAYMAN and D. L. LEVI, *Trans. Faraday Soc.* 1955 **51** 626; 1954 **50** 477.
(b) O. KUBASCHEWSKI and W. A. DENCH, *Acta met.* 1955 **3** 339.
(c) W. OELSEN, K. H. RIESKAMP and O. OELSEN, *Arch. Eisenhüttenw.* 1955 **26** 253.
(d) F. D. RICHARDSON and L. E. WEBB, *Trans. IMM* 1955 **64** 529.
(e) A. W. SEARCY and R. D. FREEMAN, *J. Amer. Chem. Soc.* 1955 **76** 5229.
[1956] (a) J. L. BARTON and H. BLOOM, *J. Phys. Chem.* 1956 **60** 1413; 1959 **63** 1785.
(b) P. HERASYMENKO, *Acta met.* 1956 **4** 1.
(c) MCCOSKEY and D. C. GINNINGS, *J. Res. Nat. Bur. Stand.* 1956 **57** 67.
(d) J. P. MORRIS and G. R. ZELLARS, *J. Metals* 1956 **8** 1086.
(e) E. K. RIDEAL and R. LITTLEWOOD, *Trans. Faraday Soc.* 1956 **52** 1598.
[1957] (a) G. R. FITTERER, *Proc. Open Hearth Conf.* Pittsburgh, 1957 **40** 281.
(b) J. HARPER and A. E. WILLIAMS, *Extraction and Refining of the Rarer Metals*, Inst. Min. Metall. 1957 p. 143.
(c) K. KIUKKOLA and C. WAGNER, *J. Electrochem. Soc.* 1957 **104** 379.
(d) J. L. MEIJERING, *Acta Met.* 1957 **5** 257.
(e) M. SHIMOJI and K. NIWA, *Acta Met.* 1957 **5** 496.
(f) R. W. URE, *J. Chem. Phys.* 1957 **26** 1363.
[1958] (a) C. B. ALCOCK and F. D. RICHARDSON, *Acta Met.* 1958 **6** 385; 1960 **8** 882.
(b) I. BACKHURST, *J. Iron Steel Inst.* 1958 **189** 124.
(c) N. CUSACK and P. KENDALL, *Proc. Phys. Soc. (London)* 1958 **72** 898.
(d) W. A. FISCHER and A. HOFFMANN, *Arch. Eisenhüttenw.* 1958 **29** 339.
(e) R. HULTGREN, P. NEWCOMBE, R. L. ORR and L. WARNER, *Met. Chem.* NPL: Sympos. No. 9 1958, HMSO, 1959, paper 1H.
(f) O. KUBASCHEWSKI, *Metal Chemistry*, NPL Symposium No. 9, 1958, HMSO, 1959, paper 3C.
(g) O. KUBASCHEWSKI, *Trans. Faraday Soc.* 1958 **54** 814.
(h) R. E. MACHOL and E. F. WESTRUM, *J. Amer. Chem. Soc.* 1958 **80** 2950.
(i) M. B. PANISH, *et al.*, *J. Phys. Chem.* 1958 **62** 980.
(j) A. SCHNEIDER and G. HEYMER, *Metal Chemistry*, NPL Symposium No. 9, 1958, HMSO, 1959, paper 4A.
(k) F. E. WITTIG, *Metal Chemistry*, NPL Sympos. No. 9, 1958, HMSO, 1959, paper 1A.
[1959] (a) E. W. DEWING and F. D. RICHARDSON, *Trans. Faraday Soc.* 1959 **55** 611.
(b) G. GATTOW and A. SCHNEIDER, *Angew. Chem.* 1959 **71** 189.
(c) U. MERTENS, *J. Phys. Chem.* 1959 **63** 443.
(d) A. S. PASHINKIN and A. V. NOVOSELOVA, *Russ. J. Inorg. Chem.* 1959 **4** 1229.
(e) J. N. PRATT and A. T. ALDRED, *J. Sci. Instrum.* 1959**36** 465.
(f) R. SPEISER, A. J. JACOBS and J. W. SPRETNAK, *Trans. AIME* 1959 **215** 185.
(g) G. R. ZELLARS, S. L. PAYNE, J. P. MORRIS and R. L. KIPP, *Trans. AIME* 1959 **215** 181.
[1960] (a) C. B. ALCOCK and G. W. HOOPER, *Proc. Roy. Soc. (A)* 1960 **254** 551.
(b) J. BERKOWITZ and W. A. CHUPKA, *Trans. N.Y. Acad. Sci.* 1960 **79** 1073.
(c) E. BONNIER and R. CABOZ, *Compt. Rend.* 1960 **250** 527.
(d) L. H. DREGER and J. L. MARGRAVE, *J. Phys. Chem.* 1960 **64** 1323.
(e) O. J. KLEPPA, *J. Phys. Chem.* 1960 **64** 1937.
(f) F. KOHLER, *Monatsh. Chemie* 1960 **91** 738.
(g) O. KUBASCHEWSKI and G. HEYMER, *Acta Met.* 1960 **8** 416.
(h) O. KUBASCHEWSKI and G. HEYMER, *Trans. Faraday Soc.* 1960 **56** 473.
(i) O. KUBASCHEWSKI, G. HEYMER and W. A. DENCH, *Z. Elektrochem.* 1960 **64** 801.
(j) D. T. PETERSON and R. KONTRIMAS, *J. Phys. Chem.* 1960 **64** 362.
(k) H. SCHÄFER and A. TEBBEN, *Z. anorg. allg. Chem.* 1960 **304** 317.
[1961] (a) R. BENZ and C. WAGNER, *J. Phys. Chem.* 1961 **65** 1308.
(b) Y. M. NESTEROVA, A. S. PASHINKIN and A. V. NOVOSELOVA, *Russ. J. Inorg. Chem.* 1961 **6** 1031.
(c) H. F. RAMSTAD and F. D. RICHARDSON, *Trans. Met. Soc. AIME* 1961 **221** 1021.
[1962] (a) C. B. ALCOCK and P. GRIEVESON, *J. Inst. Metals* 1962 **90** 304.
(b) A. FERRIER and M. OLETTE, *Compt. rend.* 1962 **254** 4293.
(c) D. L. HILDENBRAND and W. F. HALL, *J. Phys. Chem.* 1962 **66** 754.
(d) K. L. KOMAREK and M. SILVER, *Thermodynamics of Nuclear Materials*, Proc. IAEA Sympos., Vienna, 1962, p. 749.
(e) H. R. LARSON and J. F. ELLIOTT, *Trans. Met. Soc. AIME* 1962 **66** 497.
(f) O. NYQUIST, K. W. K. LANGE and J. CHIPMAN, *Trans. Met. Soc. AIME* 1962 **224** 714.

(g) A. RAHMEL, *Z. Elektrochem.* 1962 **66** 363.
(h) R. A. RAPP and F. MAAK, *Acta met.* 1962 **10** 63.
(i) T. N. REZUKHINA and Z. V. PROSHINA, *Russ. J. Phys. Chem.* 1962 **36** 333.
(j) H. SCHMALZRIED, *Z. Elektrochem.* 1962 **66** 572.
(k) H. A. SKINNER (ed.) *Experimental Thermochemistry*, Vol. 2, Interscience, New York–London, 1962.
(l) E. G. WOLFF and C. B. ALCOCK, *Trans. Brit. Ceram. Soc.* 1962 **61** 667.
[1963] (a) E. CALVET and H. PRATT, *Recent Progress in Microcalorimetry*, Pergamon, Oxford, 1963.
(b) W. A. DENCH, *Trans. Faraday Soc.* 1963 **59** 1279.
(c) W. A. DENCH and O. KUBASCHEWSKI, *J. Iron Steel Inst.* 1963 **201** 140.
(d) T. R. INGRAHAM and H. H. KELLOGG, *Trans. Met. Soc. AIME* 1963 **227** 1419.
(e) M. H. RAND and O. KUBASCHEWSKI, *The Thermochemical Properties of Uranium Compounds*, Oliver & Boyd, Edinburgh, 1963.
(f) R. A. RAPP, *Trans. AIME* 1963 **227** 371.
(g) D. M. SPEROS and R. L. WOODHOUSE, *J. Phys. Chem.* 1963 **67** 2164.
(h) M. WEINSTEIN and J. F. ELLIOTT, *Trans. Met. Soc. AIME* 1963 **227** 382.
[1964] (a) C. B. ALCOCK and T. N. BELFORD, *Trans. Faraday Soc.* 1964 **60** 822.
(b) S. ARONSON, *Compounds of Interest in Nuclear Reactor Technology*, 1964, AIME Special Report No. 13 247.
(c) R. F. BREBRICK and A. J. STRAUSS, *J. Chem. Phys.* 1964 **40** 3230; **41** 197.
(d) J. J. EGAN, *J. Phys. Chem.* 1964 **68** 978.
(e) P. FESCHOTTE and O. KUBASCHEWSKI, *Trans. Faraday Soc.* 1964 **60** 1941.
(f) H. H. KELLOGG, *Trans Met. Soc. AIME* 1964 **230** 1622.
(g) Y. MATSUSHITA and K. GOTO, *Tetsu-to-Hagane* (overseas) 1964 **4** 128.
[1965] (a) TERMICHESKIE KONSTANTY VESHCHESTV, V. P. GLUSHKO and V. A. MEDVEDEV (eds), Akademiya Nauk SSSR, Moscow, 1965 onward.
(b) O. KUBASCHEWSKI and T. G. CHART, *J. Inst. Metals* 1965 **93** 329.
(c) P. A. RICE and D. V. RAGONE, *J. Chem. Phys.* 1965 **42** 701; 1966 **45** 4141.
(d) A. W. SMITH, F. W. MESZAROS and C. D. AMATA, *J. Amer. Ceram. Soc.* 1965 **49** 240.
(e) B. C. H. STEELE and C. B. ALCOCK, *Trans. AIME* 1965 **233** 1359.
(f) G. W. TOOP, *Trans. Met. Soc. AIME* 1965 **233** 850.
[1966] (a) C. B. ALCOCK, *Nature* 1966 **209** 351.
(b) C. B. ALCOCK, J. B. CORNISH and P. GRIEVESON, *Thermodynamics*, Proc. IAEA Symposium, 1966, **I** 211.
(c) S. ARONSON and A. AUSKERN, *Thermodynamics*, Vol. **I**, 1966, IAEA, Vienna, p. 165.
(d) R. P. BURNS, *J. Chem. Phys.* 1966 **44** 3307.
(e) J. B. DARBY JR, R. KLEB and O. J. KLEPPA, *Rev. Sci. Instr.* 1966 **37** 164.
(f) A. KANT, *J. chem. Phys.* 1966 **44** 2450.
(g) M. PELEG and C. B. ALCOCK, *Sci. Instrum.* 1966 **43** 558.
[1967] (a) C. B. ALCOCK and J. H. E. JEFFES, *Trans. Inst. Min. Metall.* 1967 **76** C246.
(b) G. R. BELTON and R. F. FRUEHAN, *J. Phys. Chem.* 1967 **71** 1403.
(c) W. A. FISCHER, *Z. Naturforsch.* 1967 **22A** 1575.
(d) R. J. FRUEHAN, L. J. MARTONIK and E. T. TURKDOGAN, *Trans. AIME* 1967 **239** 1276
(e) G. K. JOHNSON, E. GREENBERG, J. L. MARGRAVE and W. N. HUBBARD, *J. Chem. Eng. Data* 1967 **12** 137.
(f) H. H. KELLOGG, *Applications of Fundamental Thermodynamics to Metallurgical Processes* (ed. G. R. FITTERER), Gordon and Breach, 1967 p. 357.
(g) O. KUBASCHEWSKI, *Proc. 1st Conf. on Thermodynamic Properties of Materials*, Pittsburgh, 1964, Gordon and Breach, New York, 1967 p. 11.
(h) O. KUBASCHEWSKI and L. E. H. STUART, *J. Chem. Eng. Data* 1967 **12** 418.
(i) A. KUBIK and C. B. ALCOCK, *Metal Sci. J.* 1967 **1** 19.
(k) J. L. MARGRAVE (ed.), *The Characterization of High Temperature Vapors*, Wiley: New York, 1967.
(l) O. REPETYLO, M. OLETTE and P. KOZAKEVITCH, *J. Metals* 1967 (May), p. 45.
(m) F. E. RIZZO, R. E. BIDWELL and D. F. FRANK, *Trans. AIME* 1967 **239** 593.
(n) P. J. SPENCER and J. N. PRATT, *Brit. J. Appl. Phys.* 1967 **18** 1473.
(o) J. W. WARD, R. N. R. MULFORD and R. L. BIVINS, *J. Chem. Phys.* 1967 **47** 1718.
(p) J. W. WARD, R. N. R. MULFORD and M. KAHN, *J. Chem. Phys.* 1967 **47** 1710.
(q) R. C. WILHOIT, *J. Chem. Educ.* 1967 **44** A571; A629; A685; A853.
[1968] (a) C. AFFORTIT and R. LALLEMENT, *Rev. Int. Hautes Temper. et Refract.* 1968.
(b) G. G. CHARETTE and S. N. FLENGAS, *J. Electrochem. Soc.* 1968 **115** 796.
(c) G. G. CHARETTE and S. N. FLENGAS, *Can. Met. Quart.* 1968 **7** 191.
(d) R. J. FRUEHAN, *Trans. Met. Soc. AIME* 1968 **242** 2007.
(e) C. GATELLIER and M. OLETTE, *C. r. Acad. Sci.* 1968 **266** 1133.
(f) J. H. E. JEFFES and C. B. ALCOCK, *J. Materials Science* 1968 **3** 635.

(g) P. A. G. O'Hare, J. L. Settle, H. M. Feder and W. N. Hubbard, *Thermodynamics of Nuclear Materials*, IAEA, Vienna, 1968, p. 265.
(h) R. J. Ruka, J. E. Bauerle and L. Dykstra, *J. Electrochem. Soc.* 1968 **115** 497.
[1969] (a) A. Cezairliyan, *High Temp.–High Press.* 1969 **1** 517.
(b) W. A. Dench and O. Kubaschewski, *High Temp.–High Press.* 1969 **1** 357.
(c) P. Fellner and C. Krohn, *Can. Met. Quart.* 1969 **8** 275.
(d) F. Foerster and H. Richter, *Radex Rundschau* 1969 (2), p. 518.
(e) S. W. Gilby and G. R. St Pierre, *Trans. Met. Soc. AIME* 1969 **245** 1749.
(f) K. Goto, T. Ito and M. Someno, *Trans. AIME* 1969 **245** 1662.
(g) M. Hoch, High Temps.–High Pressures 1969 **1** 531.
(h) H. Kleykamp, *Ber. Bunsenges. Phys. Chem.* 1969 **73** 354.
(i) F. Müller and O. Kubaschewski, *High Temp.–High Press.* 1969 **1** 543.
(j) A. Neckel and S. Wagner, *Ber. Bunsenges. Phys Chem.* 1969 **73** 210.
(k) T. Ostvold and O. J. Kleppa, *Inorg. Chem.* 1969 **8** 78.
(l) E. K. Storms, *High Temp. Sci.* 1969 **1** 456.
(m) Y. D. Tretyakov and R. A. Rapp. *Trans. AIME* 1969 **242** 1235.
[1970] (a) C. B. Alcock, R. Sridhar and R. C. Svedberg, *J. Chem. Thermodyn.* 1970 **2** 255.
(b) G. Boureau and P. Gerdanian, *High Temp.–High Press.* 1970 2 681.
(c) A. Cezairliyan, M. S. Morse, H. A. Berman and C. W. Beckett, *J. Res. Nat. Bureau Standards* 1970 **74A** 65.
(d) V. Ya. Chekhovskoi, A. E. Sheindlin and B. Ya. Berezin, *High Temp.–High Press.* 1970 **2** 301.
(e) G. R. Fitterer and L. A. Pugliese, *Met. Trans.* 1970 **1** 1997.
(f) G. M. Foley, *Rev. Sci. Instr.* 1970 **41** 827.
(g) M. Hoch, *Rev. Int. Hautes Temper. et Refract.* 1970 **7** 242.
(h) M. Hillert, *Phase Transformation*, ASM, Metals Park, Ohio, 1970, 181.
(i) M. Hillert and L. J. Staffanson, *Acta Chem. Scand.* 1970 **24** 3618.
(j) E. Huber and C. E. Holley, *Physicochemical Measurements in Metals Research* (ed. R. A. Rapp), Wiley, 1970, p. 243.
(k) L. Kaufman and H. Bernstein, *Computer Calculation of Phase Diagrams*, Academic Press, New York, 1970.
(l) Z. Kozuka and C. S. Samis, *Met. Trans.* 1970 **1** 871.
(m) O. Kubaschewski, A. Cibula and D. C. Moore, *Gases and Metals*, Metals and Metallurgy Trust, Iliffe, 1970, p. 17.
(n) J. F. Martin, F. Müller and O. Kubaschewski, *Trans. Faraday Soc.* 1970 **66** 1065.
(o) B. Predel and U. Mohs, *Arch. Eisenhüttenws.* 1970 **41** 61.
(p) R. A. Rapp and D. A. Shores, *Physicochemical Measurements in Metals Research* (R. A. Rapp, ed.), Interscience, 1970 2 123.
(q) F. R. Sale, *J. Sci. Instr.* 1970 **3** 653.
(r) K. Schwerdtfeger and R. T. Turkdogan, *Physicochemical Measurements in Metals Research* (A. Rapp, ed.), Interscience, 1970 **I** 321.
(s) G. Sodeck, P. Etner and A. Neckel, *High Temp. Sci.* 1970 **2** 311.
(t) S. Zador and C. B. Alcock, *J. Chem. Thermodyn.* 1970 **2** 9.
[1971] (a) I. Ansara, *Metallurgical Chemistry*, Proc. Symp. Brunel Univ. and NPL, 1971 (ed. O. Kubaschewski), HMSO, 1972, p. 403.
(b) B. Ya. Berezin, V. Ya. Chekhovskoi and A. E. Sheindlin, *High Temp.–High Press.* 1971 **3** 287.
(c) A. Cezairliyan, *J. Res. Nat. Bureau Stds* 1971 **75C** 7.
(d) T. G. Chart and O. Kubaschewski, *Metallurgical Chemistry*, Proc. Symp. Brunel Univ. and NPL Symposium, 1971, HMSO, 1972, p. 567.
(e) E. L. Flores-Magon and M. Vikram Rao, *Met. Trans.* 1971 **2** 2471.
(f) T. L. Francis, F. E. Phelps and G. Maczura, *Bull. Amer. Ceram. Soc.* 1971 **54** 414.
(g) P. Gerdanian, *Thermochimie*, Coll. Int. CNRS, No. 201, Marseille, 1971, CNRS, 1972, p. 259.
(h) M. L. Kapoor and M. G. Frohberg, *Arch. Eisenhüttenw.* 1971 **42** 5.
(i) O. J. Kleppa, *Thermochimie*, Coll. Int. CNRS, No. 201, Marseille, 1971, CNRS, 1972, p. 119
(j) O. Kubaschewski and J. F. Counsell, *Monatsh. Chemie* 1971 **102** 1924.
(k) M. Laffitte, *Metallurgical Chemistry*, Proc. Symp. Brunel Univ. and NPL, 1971, HMSO, 1972, p. 3.
(l) M. Laffitte, *Thermochimie*, Coll. Int. CNRS, No. 201, Marseille, 1971, CNRS, 1972, p. 135.
(m) E. Lugscheider and G. Jangg, *Z. Metallkunde* 1971 **62** 548.
(n) M. J. O'Neill and A. P. Gray, *Thermal Analysis* (ed. H. G. Wiedemann). ICTA 1971, Birkhäuser Verlag Basel and Stuttgart, 1972, p. 279.
(o) T. N. Rezukhina and B. S. Pokarev, *J. Chem. Thermodyn.* 1971 **3** 369.
(p) K. Schwitzgebel, P. S. Lowell, T. B. Parsons and K. J. Sladek, *J. Chem. Eng. Data* 1961 **16** 418.
(q) W. Slough, *Metallurgical Chemistry*, Proc. Symp. Brunel Univ. and NPL, 1971, p. 311, HMSO, 1972.

(r) F. E. STAFFORD, *High Temp.-High Press.* 1971 **3** 213.
(s) M. S. WHITTINGHAM and R. A. HUGGINS, *J. Chem. Phys.* 1971 **54** 414.
(t) I. WYNNE-JONES and L. J. MILES, *Proc. Brit. Ceram. Soc.* 1971 **19** 161.
[1972] (a) C. B. ALCOCK and J. C. CHAN, *Can. Met. Quart.* 1972 **11** 559.
(b) M. CARBONEL, C. BERGMAN and M. LAFFITTE, *Thermochimie Colloques Internationaux du CNRS* 1972, No. 201, 311.
(c) A. CEZAIRLIYAN, *High Temp. Sci.* 1972 **4** 248.
(d) J. V. DAVIES and H. O. PRITCHARD, *J. Chem. Thermodynamics* 1972 **4** 9.
(e) M. HOCH, *High Temps.-High Pressures* 1972 **4** 493, 659.
(f) K. T. JACOB and C. B. ALCOCK, *Acta Met.* 1972 **20** 221.
(g) K. C. MILLS, K. KINOSHITA and P. GRIEVESON, *J. Chem. Thermodynamics* 1972 **4** 581.
(h) W. SLOUGH, *Metallurgical Chemistry*, HMSO, 1972, p. 30; (ed. O. KUBASCHEWSKI), (i) P. J. SPENCER, F. H. HAYES and O. KUBASCHEWSKI, *Rev. chim. miner.* 1972 **9** 13.
(j) M. LE VAN, *Bull. Soc. Chim. France* 1972 **No. 2** 579.
(k) D. L. VIETH and M. J. POOL, Report No. 72-30, Dept. Mat. Sci. and Met. Eng., Univ. Cincinnati, Ohio, 1972.
(l) W. W. WENDLANDT, *J. Chem. Educ.* 1972 **49** A571; A623; A671.
[1973] (a) J. C. CHAN, C. B. ALCOCK and K. T. JACOB *Can. Met. Quart.* 1973 **12** 439.
(b) J. P. HAGER, S. M. HOWARD and J. H. JONES, *Met. Trans.* 1973 **4** 2383.
(c) R. HULTGREN, R. L. ORR, P. D. ANDERSON and K. K. KELLEY, *Selected Values of the Thermodynamic Properties of Metals and Alloys*, Wiley, New York, 1973.
(d) R. HULTGREN, *et al.*, *Selected Values of the Thermodynamic Properties of Binary Alloys*, ASM, Metals Park, Ohio, 1973.
(e) K. T. JACOB and C. B. ALCOCK, *Acta Met.* 1973 **21** 1011.
(f) B. V. JOGLEKAR, P. S. NICHOLSON and W. W. SMELTZER, *Can. Met. Quart.* 1973 **12** 155.
(g) K. L. KOMAREK, *Z. Metallkunde* 1973 **64** 406.
(h) W. A. KRIVSKY, *Met. Trans.* 1973 **4** 1439.
(i) F. N. MAZANDARANY and R. D. PEHLKE, *Met. Trans.* 1973 **4** 2067.
(j) A. R. MIEDEMA, *J. Less-Common Met.* 1973 **32** 117.
(k) K. C. MILLS and K. KINOSHITA, *J. Chem. Thermodynamics* 1973 **5** 129.
(l) W. SLOUGH, *National Physical Laboratory Report Chem.* No. 25, October 1973.
(m) P. J. SPENCER, F. H. HAYES and L. ELFORD, *Proc. Sympos. Chem. Metall. Iron and Steel*, Iron and Steel Inst. 1973, p. 322.
[1974] (a) C. W. BALE and A. D. PELTON, *Metall. Trans.* 1974 **5** 2323.
(b) G. BOUREAU and P. GERDANIAN, *Can. Met. Quart.* 1974 **13** 339.
(c) M. GAUNE-ESCARD and J.-P. BROS, *Can. Met. Quart.* 1974 **13** 335.
(d) M. HOCH, *Proc. IAEA Symp. on Thermodynamics of Nuclear Materials*, Vienna, Oct. 21–25, 1974
(e) M. KANNO, *J. Nucl. Mat.* 1974 **51** 24.
(f) O. KUBASCHEWSKI and T. G. CHART, *J. Chem. Thermodyn.* 1974 **6** 467.
(g) V. A. LEVITSKII and Y. Y. SCOLIS, *J. Chem. Thermodyn.* 1974 **6** 1181.
(h) A. MCLEAN, *High Temp.-High Press.* 1974 **6** 21.
(i) K. C. MILLS, *National Physical Laboratory*, DCS Note 20, March 1974.
(j) T. N. REZUKHINA, T. F. SISLOVA, L. I. HOLOKHONOVA and E. G. IPPOLITOV, *J. Chem. Thermodyn.* 1974 **6** 883.
(k) F. D. RICHARDSON, *Physical Chemistry of Melts in Metallurgy*, Academic Press, London, 1974.
(l) G. K. SIGWORTH and J. F. ELLIOTT, *Metal Sci.* 1974 **8** 298.
[1975] (a) D. J. FRAY and B. SAVORY, *J. Chem. Thermodyn.* 1975 **7** 485.
(b) K. T. JACOB and C. B. ALCOCK, *Met. Trans. B* 1975 **6** 215.
(c) A. R. MIEDEMA, R. BOOM and F. R. DE BOER, *J. Less-Common Met.* 1975 **41** 283.
(d) B. V. SLOBODIN, I. I. MILLER and A. A. FOTIEV, *Russ. J. Phys. Chem.* 1975 **49** 511.
[1976] (a) ALLEGRET, Diplome These, Grenoble (1976).
(b) R. BOOM, F. R. DE BOER and A. R. MIEDEMA, *J. Less-Common Met.* 1976 **45** 237.
(c) R. BOOM, F. R. DE BOER and A. R. MIEDEMA, *J. Less-Common Met.* 1976 **46** 271.
(d) E. FROMM and E. GEBHARDT (eds), *Gase und Kohlenstoff in Metallen*, Springer, Berlin, 1976
(e) K. HACK, Diplomarbeit, 1976. Lehrstuhl für theoretische Hüttenkunde, Tech. Hochscule, Aachen.
(f) M. HOCH and T. VENARDAKIS, *Rev. Int. Hautes Temper. et Refract.* 1976 **13** 75.
(g) M. HOCH and T. VENARDAKIS, *Ber. Bunsenges.* 1976 **80** 770.
(h) A. R. MIEDEMA, *J. Less-Common Met.* 1976 **46** 67.
[1977] (a) I. BARIN and O. KNACKE, *Thermochemical Properties of Inorganic Substances*, Springer, Berlin, 1973; supplement with O. KUBASCHEWSKI, 1977.
(b) O. KUBASCHEWSKI and J. GRUNDMANN, *Ber. Bunsenges.* 1977 **81** 1239.
(c) O. KUBASCHEWSKI and H. ÜNAL, *High Temps.-High Pressures* 1977 **9** 361.

(d) H-L. LUKAS, E. TH. HENIG and B. ZIMMERMANN, *CALPHAD* 1977 **1** 225.
(e) A. NAVROTSKY, *Phys. Chem. Minerals* 1977 **2** 89.
[1978] (a) I. ANSARA, C. BERNARD, L. KAUFMAN and P. SPENCER, *CALPHAD* 1978 **2** 1.
(b) P. J. SPENCER and O. KUBASCHEWSKI, *CALPHAD* 1978 **2** 147.
[1979] (a) I. ARPSHOFEN, B. PREDEL, E. SCHULTHEIß and M. HOCH, *Thermochim. Acta* 1979 **33** 197.
(b) O. J. KLEPPA, *Calculation of Phase Diagrams and Thermochemistry of Alloy Phases*, (ed. Y. A. CHANG and J. F. SMITH), AIME, Warrendale, 1979, p. 213.
(c) O. KUBASCHEWSKI and C. B. ALCOCK, *Metallurgical Thermochemistry*, 5th edition, Pergamon, Oxford, 1979.
(d) O. KUBASCHEWSKI and K. HACK, *Z. Metallkunde* 1979 **70** 789.
(e) H.-D. NÜSSLER, Doktorarbeit, Lehrstuhl für Theoretische Hüttenkunde, RWTH Aachen, 1979.
(f) M. J. POOL, B. PREDEL and E. SCHULTHEISS, *Thermochimica Acta* 1979 **28** 349.
(g) J RICHTER and W. VREULS, *Ber. Bunsenges. Phys. Chem.* 1979 **83** 1023.
(h) P. J. SPENCER and I. BARIN, *Materials in Eng. Applications*, 1979 **1** 167.
[1980] (a) I. BARIN, B. FRASSEK, R. GALLAGHER and P. J. SPENCER, *Erzmetall* 1980 **33** 226–231.
(b) K. J. DICKENS, O. J. KLEPPA and H. YOKOKAWA, *Rev. Sci. Instr.* 1980 **51** 675.
(c) T. HOSTER and O. KUBASCHEWSKI, *Thermochimica Acta* 1980 **40** 15.
(d) H. OPPERMANN, G. STÖVER, L. M. CHRIPLOWITSCH and I. E. PAUKOV, *Z. anorg. allg. Chem.* 1980 **461** 173.
[1981] (a) J. C. GACHON, M. NOTIN and J. HERTZ, *Thermochim. Acta* 1981 **48** 155.
(b) G. HATEM, P. GAUNE, J-P. BROS, F. GEHRINGER and E. HAYER. *Rev. Sci. Instr.* 1981 **52** 585.
(c) H. IWAHARA, T. ESAKA, H. UCHIDA and N. MAEDA, *Solid State Ionics* 1981 **3/4** 359.
(d) W. L. WORRELL, *Solid State Ionics* 1981 **3/4** 559.
[1982] (a) R. KRACHLER, H. IPSER and K. L. KOMAREK, *Z. Metallkunde* 1982, **73** 731.
(b) The NBS Tables of Chemical Thermodynamic Properties, (ed. D. D. WAGMAN, *et al.*), *J. Phys. Chem. Ref. Data* 1982 **11** Supplement No. 2.
[1984] G. ERIKSSON and K. HACK, *CALPHAD* 1984 **8** 15–24.
(b) W. HEMMINGER and C. HÖHNE, *Calorimetry: Fundamentals and Practice*, Part 1, (ed. L. M. NEIL), Verlag Chemie, Weinheim, 1984.
(c) B. JANSSON, Ph.D. Thesis, Div. of Phys. Met., Royal Inst. Technology, Stockholm, 1984.
(d) O. KUBASCHEWSKI, *High Temps.-High Pressures* 1984 **16** 197.
(e) S. C. MRAW and O. J. KLEPPA, *J. Chem. Thermodyn.* 1984 **16** 865.
[1985] (a) M. W. M. HISHAM and S. W. BENSON, *J. Phys. Chem.* 1985 **89** 1905.
(b) M. W. M. HISHAM and S. W. BENSON, *J. Phys. Chem.* 1985 **89** 3417.
(c) JANAF Thermochemical Tables, 3rd edition, *J. Phys. Chem. Ref. Data* 1985 **14** Supplement No. 1.
(d) B. SUNDMAN, B. JANSSON and J-O. ANDERSSON, *CALPHAD* 1985 **9** 153.
[1986] (a) N. A. GOKCEN, *Statistical Thermodynamics of Alloys*. Plenum Press, New York, 1986.
(b) M. W. M. HISHAM and S. W. BENSON, *J. Phys. Chem.* 1986 **90** 885.
(c) F. PELLICANI, F. VILLETTE and J. DUBOIS, *Scan Inject 4*, Lulea, Sweden, 1986.
(d) D. PETTIFOR, *New Scientist*, 29 May, 1986, p. 48.
(e) N. SAUNDERS and A. P. MIODOWNIK, *J. Mater. Res.* 1986 **1** 38.
(f) D. ZIEGLER and A. NAVROTSKY, *Geochim. Cosmochim. Acta* 1986 **50** 2461.
(g) E. ZIMMERMAN, Doktorarbeit, Lehrstuhl für Theoretische Hüttenkunde, RWTH Aachen, 1986.
[1987] (a) I. ANSARA and B. SUNDMAN, *Computer Handling and Dissemination of Data*, (ed. P. GLAESER), Elsevier, CODATA 1987.
(b) K. T. JACOB, M. IWASE and Y. WASEDA, *Solid State Ionics* 1987 **23** 245.
(c) M. W. M. HISHAM and S. W. BENSON, *J. Phys. Chem.* 1987 **91** 3631.
(d) M. W. M. HISHAM and S. W. BENSON, *J. Phys. Chem.* 1987 **91** 5998.
(e) M. W. M. HISHAM and S. W. BENSON, *J. Phys. Chem. Ref. Data* 1987 **16** 467.
(f) M. W. M. HISHAM and S. W. BENSON, *J. Chem. Eng. Data* 1987 **32** 243.
(g) J. KLEIN and F. MÜLLER, *High Temps.-High Pressures* 1987 **19** 201.
[1988] (a) B. CHEYNET, *Proc. Metall. Soc. of CIM*, Montreal '88, 1988 **11** 87.
(b) A. T. DINSDALE, S. M. HODSON, T. I. BARRY and J. R. TAYLOR, *Proc. Metall. Soc. of CIM*, Montreal '88, 1988 **11** 59.
(c) M. W. M. HISHAM and S. W. BENSON, *J. Phys. Chem.* 1988 **92** 6107.
(d) T. IIDA and R. L. GUTHRIE, *The Physical Properties of Liquid Metals*, Oxford Science Publ., 1988.
(e) H. OPPERMANN and R. KRAUSZE, *J. Less-Common Metals* 1988 **137** 217.
(f) B. SUNDMAN, *Proc. Metall. Soc. of CIM*, Montreal '88, 1988 **11** 75.
(g) W. T. THOMPSON, G. ERIKSSON, A. D. PELTON and C. W. BALE, *Proc. Metall. Soc. of CIM*, Montreal '88 1988 **11** 87.
[1989] (a) J.-P. BROS, *J. Less-Common Met.* 1989 **154** 9.
(b) O. J. KLEPPA and L. TOPOR, *Thermochim. Acta* 1989 **139** 291.
(c) W. WILSMANN and F. MÜLLER, *Thermochim. Acta* 1989 **151** 309.

[1990] (a) C. W. BALE and G. ERIKSSON, *Can. Met. Quart.* 1990 **29** 105.
(b) G. ERIKSSON and K. HACK, *Metall. Trans. B* 1990 **21B** 1013.
(c) M. HOCH, private communication, 1990.
(d) S. AN MEY, K. HACK, K. ITAGAKI and P. J. SPENCER, *CALPHAD* 1990 **14** 377.
(e) G. QI, K. ITAGAKI, S. AN MEY and P. J. SPENCER, *Z. Metallkunde* 1990 **14** 127.
(f) P. J. SPENCER and H. HOLLECK, *High Temperature Sci.* 1990 **27** 295.
(g) T. TANAKA, N. A. GOKCEN, Z. MORITA and P. J. SPENCER, *Anales de Fisica* 1990 **86B** 104.
(h) T. TANAKA, N. A. GOKCEN and Z. MORITA, *Z. Metallkunde* 1990 **81** 49.
(i) C. B. ALCOCK and B. LI, Solid State Ionics 1990 **39** 245.
[1991] (a) M. KOWALSKI, *Lehrstuhl für Theoretische Hüttenkunde*, RWTH Aachen, private communication, 1991.
(b) H. OPPERMANN and K. WITTE, *Z. anorg. allg. Chem.* 1991 **593** 200.
(c) A. D. PELTON, *Materials Science and Technology* 1991 **5** Chapter 5.
(d) H. STOLTEN, Doktorarbeit, *Lehrstuhl für Theoretische Hüttenkunde*, RWTH Aachen, 1991.
[1992] (a) M. FAN and F. MÜLLER, RWTH AACHEN, to be published.
(b) P. J. SPENCER and D. NEUSCHÜTZ, *Chem. Eng. Technol.* 1992 **15** 119.
(c) T. TANAKA, N. A. GOKCEN, D. NEUSCHÜTZ, P. J. SPENCER and Z. MORITA, *Steel Research.* 1991 **62** 385.

REFERENCES

References to Chapter V

1. L. BREWER, Review of the Fusion and Evaporation Data of Metal Halides, *Natl Nuclear Sci. Ser., Div. IV*, 1949 **19B** 193.
2. L. BREWER, L. A. BROMLEY, P. W. GILLES and N. L. LOFGREN, Review of Thermochemical Properties of Metal Carbides, Nitrides and Sulphides, *ibid.* p. 40; corresponding review of the metal halides, *ibid.* p. 76.
3. L. BREWER and R. H. LAMOREAUX, *Atomic Energy Review*, No. **7**, Thermochemical Properties of Molybdenum and its Compounds, Int. At. En. Agency, Vienna (1980).
4. T. G. CHART, Assessment of Thermochemical Data for Transition Metal–silicon Systems, High Temp.-High Press. 1973 **5** 241; *Met. Sci.* 1974 **8** 344; 1975 **9** 504.
5. T. G. CHART and K. HACK, Evaluation of the Thermochemical Properties of Transition Metal-boron Systems, Comm. des Communautes Europeennes, Recherche CECA No. 7210-CA/3/303, Rep. no. **5–9**, Nov. 1981.
6. CODATA, *J. Chem. Therm.* 1978 **10** 903.
7. E. H. P. CORDFUNKE and P. A. G. O'HARE, *The Chemical Thermodynamics of Actinide Elements and Compounds*, Part **3**, IAEA, Vienna (1978).
8. H. F. FISCHMEISTER, L_t and L_f of Sulphates reviewed, *Z. Phys. Chem.* 1956 **7** 91.
9. V. P. GLUSHKO and V. A. MEDVEDEV (Ed.), *Termicheskie Konstanty Veschestv*, Part I: O.H. Halogens, Inert Gases (1965); Part II: Chalcogenides (1966); Part III: N, P, As, Sb Bi (1968); Part IV: C, Si, Ge, Sn, Pb (1970); Part V: B, Al, Ga, Ir, Tl (1971); Part VI: Zn, Cd, Hg, Ag, Au, Fe, Co, Ni, Ru, Rh, Pd, Os, Ir, Pt (1972); Part VIII: Sc, Y, Lanthanides, Actinides (1978); Part X: Alkaline Metals (1981); Akademyia Nauk, Moscow.
10. V. P. GLUSHKO and L. V. GURVICH, *Termodinamicheskie Svoistva Individualnikh Veshestv.*, Vols **I–IV**, Nauka, Moscow (1979), and in English Vols **I** and **II**, Hemisphere, NY (1989).
11. R. HULTGREN, R. L. ORR, P. D. ANDERSON and K. K. KELLEY, *Selected Values of Thermodynamic Properties of Metals and Alloys*, J. Wiley, New York 1974.
12. JANAF, Thermochemical Tables (D. R. STULL, *et al.*), US Dept of Commerce Natl Bur. Stand., US Government Printing Office, 1971.
13. JANAF, Thermochemical Tables (M. W. CHASE, *et al.*), *J. Phys. Chem. Ref. Data* 1974 **3** No. 2, 311; 1975 **4** No. 1, 1; 1978 **7** 793; 1982 **11** 695.
14. J. H. E. JEFFES and H. MCKERELL, Thermodynamic Properties of the Metal Hydrides, *J. Iron Steel Inst.* 1964 **202** 666.
15. M. KH. KARAPYETYANTS and M. L. KARAPYETYANTS, *Osnovye Termodynamicheskie Konstantly, Neorganicheskikh I Organicheskikh Veshchestv*, Khimya, Moscow (1968).
16. K. K. KELLEY, Evaluation of Vapour Pressures and Heats of Evaporation of Inorganic Substances, *US Bur. Mines Bull.* 1935 **383**.
17. K. K. KELLEY, Evaluation of the Heats of Fusion of Metals and Metal Compounds from Data on Freezing Point Lowering, *US Bur. Mines Bull.* 1936 **393**.
18. K. K. KELLEY, Revision of the Entropies of Inorganic Substances, *US Bur. Mines Bull* 1950 **477**.
19. K. K. KELLEY, Evaluation of High-temperature Heat Capacities of Inorganic Compounds, *US Bur. Mines Bull.* 1949 **476**; 1960 **584**.
20. E. G. KING, A. D. MAH and L. B. PANKRATZ, *INCRA-Ser. Metallurgy of Cu, Copper Res. Ass. and US Bur. Mines* 1973 **2**.
21. O. KNACKE, O. KUBASCHEWSKI and K. HESSELMANN, *Thermochemical Properties of Inorganic Substances*, 2nd Edition, Springer-Verlag, 1991.
22. K. KOMAREK (Ed.), *Atomic Energy Reviews: Thermochemical Properties* **8** (P. J. Spencer): Hafnium 1981; **9** (O. Kubaschewski): Titanium 1983, Int. At. En. Agency, Vienna.
23. O. KUBASCHEWSKI and J. A. CATTERALL, *Thermodynamic Data of Alloys*, Pergamon Press, 1956.
24. O. KUBASCHEWSKI (Ed.), *Atomic Energy Reviews, Thermochemical Properties*, **1** (M. H. Rand): Plutonium

1966; **2** (V. I. Lavrentev and Ya. I. Gerassimov): Niobium 1968; **3** (Ya. I. Gerassimov and V. I. Lavrentev): Tantalum 1972; **4** (P. J. Spencer): Beryllium 1973; **5** (M. H. Rand): Thorium 1975; **6** (C. B. Alcock and K. T. Jacobs): Zirconium 1976, Int. At. En. Agency, Vienna.

25. O. KUBASCHEWSKI, The Thermodynamic Properties of Double Oxides, *High Temp.-High Press.* 1972 **4** 1.
26. A. D. MAH, Thermodynamic Properties of Vanadium Compounds, *US Bur. Mines Rep. Inv.* 1966, 6727.
27. A. D. MAH and L. B. PANKRATZ, Thermodynamic Properties of Nickel and its Inorganic Compounds, *US Dept Int. Bur. Mines Bull.* 1976, 688.
28. K. C. MILLS, *Thermodynamic Data for Inorganic Sulphides, Selenides and Tellurides*, Butterworths, London, 1974.
29. National Bureau of Standards (F. D. ROSSINI, *et al.*), Selected Values of Chemical Thermodynamic Properties, Circular 500, Washington, 1952.
30. National Bureau of Standards (D. D. WAGMAN, *et al.*), Selected Values of Chemical Thermodynamic Properties, Technical Notes, 270-3/7, Washington, 1968–73.
31. National Bureau of Standards, *High Temperature Properties and Decomposition of Inorganic Salts*, Part 1: Sulfates (NSRDS-NBS **7** 1966); Part 2: Carbonates (NSRDS-NBS **30** 1969).
32. National Bureau of Standards, *Technical News Bull.* (Jan. 1971) *CODATA Bull.* **10** (Dec. 1973); *CODATA Sp. Rep.* **3** (Sept. 1975).
33. F. L. OETTING, *Chem. Rev.* 1967 **67** 261.
34. F. L. OETTING, M. H. RAND and R. J. ACKERMANN, *The Chemical Thermodynamics of Actinides Elements and Compounds*, Part 1, Int. At. En. Agency, Vienna (1976).
35. L. B. PANKRATZ, Thermodynamic Properties of Elements and Oxides, *US Bur. Mines* 1982 **672**.
36. L. B. PANKRATZ, Thermodynamic Properties of Halides, *ibid.* 1984 **674**.
37. L. B. PANKRATZ, J. M. STUVE and N. A. GOKCEN, Thermodynamic Data for Mineral Technology, *ibid.* 1984 **677**.
38. L. B. PANKRATZ, A. D. MAH and S. W. WATSON, Thermodynamic Properties of Sulfides, *ibid.* 1987 **689**.
39. M. H. RAND and O. KUBASCHEWSKI, *The Thermochemical Properties of Uranium Compounds*, Oliver and Boyd, Edinburgh, 1963.
40. F. D. RICHARDSON, Review of ΔG (formation) of Metal Carbides, *J. Iron Steel Inst.* 1953 **175** 33.
41. H. L. SCHICK, *Thermodynamics of Certain Refractory Compounds*, Academic Press, New York, 1966.
42. W. SLOUGH and G. P. JONES, Thermodynamic Data for Borate Systems, NPL (Teddington), *DCS Rep. Chem.* 1974 **31**.
43. J. F. SMITH, Vanadium Assessment, *Bull Alloy Phase Diagrams* 1981 **2** 2.
44. P. J. SPENCER, The Thermodynamic Properties of (a) alkaline metal alloys and (b) silicates, NPL (Teddington), *DCS Rep.* (a) 1971 **10** and (b) 1973 **21**.
45. D. R. STULL, Vapour Pressures of Inorganic Substances, *Ind. Eng. Chem.* 1947 **39** 540.
46. F. WEIBKE and O. KUBASCHEWSKI, *Thermochemie der Legierungen*, Springer, Berlin, 1943.
47. R. W. WYCKOFF, *Crystal Structures*, Interscience, London, 1948.

Subject Index

Activity coefficient, definition 31
Activity definition 31
Adiabatic calorimeter, high temperature versions 76–8, 90–2
Adiabatic calorimeter, principle of 66, 74–5
Adsorption spectrum method for vapour pressure measurement 98
Alcock–Richardson model of ternary interaction 60
Alloy thermodynamics by EMF measurements 153–4
Amalgams, liquid—EMF studies of 143
Aqueous solution calorimetry 84–9
Aqueous solution electrolytes 143
Atomistics and solution thermodynamics 47–50

Beta-alumina solid electrolyte 160
Bismuth vapour pressure measurement 99
Boiling point method for vapour pressure 105–8
Boltzmann calculation of entropy change 48
Bonnier, ternary Gibbs–Duhem approximation 58
Bourdon manometer for vapour pressure measurement 97
Bunsen calorimeter, diphenyl ether 70–1

Calculation of activities using equilibrium diagrams 50–1
Calorimetric methods 65–94
Carbide equilibria by EMF measurements 159
Carbide stability by H_2–CH_4 equilibria 113
Chemical potential 29
Chloride, electrolyte for, preparation 143–4
Chlorination calorimetry 84
Clausing coefficient 120
Clausius–Clapeyron equation 7
Combustion calorimetry 83–4
Cooling curve, determination of specific heat 79
Coordination effects in stability of compounds 193
Crompton–Richards fusion entropy rule 170

Dalton's rule 27
Darken, ternary Gibbs–Duhem calculation 57
Debye equation 14, 15
Deoxidation constant 203–5
Determination of water equivalent 68
Dew-point method for vapour pressure 100
Differential scanning calorimetry 81–3
Differential thermal analysis 79–80
Differential thermal analysis, high temperature 81
Dispersed phase electrolytes 161
Drop calorimeter 70
Drop calorimeter, high temperature 71–2
Drop calorimeter, levitation 72, 74
Duhem–Margules equation 35
Dulong and Petit's rule 164

Electromotive forces 140–62
Elements, standard entropies 171–2
Enthalpies of double compounds 183, 185–7
Enthalpies of formation, estimation 179–95
Enthalpies of formation of compounds 182–3, 184–5
Enthalpies of formation of oxy-compounds 187–9
Enthalpies of sublimination and stabilities of compounds 193–4
Enthalpy of formation and volume contraction 189–91
Enthalpy of fusion, calculation from phase diagrams 45–6
Enthalpy of reaction 4
Enthalpy of transformation 11
Entropies of fusion data 171
Entropies of gases and molecular weight 176–7
Entropies of mixing of non-metallic solution phases 177–9
Entropy 13–17
Entropy of mixing and volume change 54–5
Entropy of mixing, covalent bonds effect 55–6
Equilibria with a gaseous phase 95–140
Equilibrium constant 3–4
Estimation of enthalpies and entropies of transformation 169–170
Estimation of enthalpies of fusion 169–170
Estimation of entropies and entropy changes 171–7
Estimation of heat capacities 163–9

Examples of applications 202–256
assessment of standard values 241–53
calculation of metallurgical equilibrium diagrams 253–6
ceramics 218–27
chemical and physical vapour phase deposition 227–35
corrosion 231–5
environmental and energy problems 235–41
iron and steelmaking 203–12
non-ferrous metallurgy 212–18
Excess integral entropy, relation to enthalpy of mixing 54

Fischer transportation method 110
Flood, exchange reaction equilibria 178
Flow rate effect on evaporation 109
Fluoride electrolytes 159

Gas electrodes in O_2 concentration cells 154–5
Gases, solubilities in metals measurement 105
Gibbs–Duhem application in mass spectrometry 132–5
Gibbs–Duhem equation 34
Gibbs energy 17–27
Gibbs energy change, approximate calculation in 24
Gibbs energy change, from EMF measurements 18
Gibbs energy change, from thermal data 19
Gibbs energy definition 17
Gibbs energy relation to equilibrium constant 3, 18
Gibbs–Helmholtz equation 3, 33
Glass electrolytes 148–8
Guggenheim series expansion for solution data 38

Heat capacity change, average values for 168–9
Heat capacity deviation 23–4
Heat capacity for double molecules 164
Heat capacity equation 12
Heat content 5
Heat content and enthalpy of formation 4–8
Heat content equation 12
Heat content temperature effect 10
Heat flow calorimeter, principle of 66
Henry's law 32, 45–6
Hess's principle 6
High speed pulse calorimeter 78–9
Hoch estimation of heat capacities 166–7
Hoch estimation of heat capacities for liquids 167–8
Hydroxide electrolytes, EMF studies with 146

Inductively heated Knudsen cell 130–1
Inorganic compounds, entropy estimation 172–4
Integral excess Gibbs functions 31
Integral Gibbs energies 29
Integral properties, relation to partial pressures 30
Interaction coefficient 61–3
Ion beam detection in mass spectrometry 127, 129
Isoperibol calorimeter, principle of 66
Isopiestic method 102–4
Isopiestic technique for O_2 in titanium 101
Isothermal calorimeter, principle of 66

Kellogg estimation of solid compounds' heat capacities 164–5
Kirchhoff's equation 9
Knudsen cell–mass spectrometer combination method 124–33
Knudsen cells, various designs 131–2
Knudsen effusion method 121–133
Kohler, ternary Gibbs–Duhem approximation 57
Krivsky method for decarburisation of alloys 206

Lambert cosine law for evaporation 122–3
Langmuir apparatus for free evaporation 122–3
Langmuir vaporisation method 133–6
Latimer estimation of entropies 172–4
Le Chatelier's principle 3
Legendre polynomials 39
Le Van's estimation of enthalpies of oxy-acids 187–8
Levitation calorimeter 72–4
Levitation melting and transpiration technique 139–40
Liquid electrodes in oxide EMF studies 156–8

Manometric methods for vapour pressure 97–8
Margules, ternary Gibbs–Duhem approximation 59
Mass action law 1–3
Measurement of temperature 66–8
Melting point and heats of formation 180
Membrane manometer for vapour pressures 98
Mercury-in-glass thermometers 67
Miedema method for intermetallic compounds 198–9
Mills, values for compounds entropies 172–5
Mole fraction definition 2
Molten chloride electrolytes, EMF studies with 144–6
Molten salt electrolytes 143–6
Molybdenum bell manometer for vapour pressure measurement 97

Neumann–Kopp rule 49, 167
Nitride stability by H_2–NH_3 equilibria 113
Non-regular solutions 53–6

"Ofchen" calorimeter for enthalpies of formation 89
Order–disorder transformation 53

Oxide stability by H_2–H_2O equilibria 113–8
Oxy-compounds, enthalpies of formation 187–9
Oxygen in titanium by isopiestic technique 101
Oxygen partial Gibbs energies in titanium 42–3
Oxygen solutions in iron 40, 42–44

Partial excess Gibbs functions 37
Partial molar Gibbs energy 27–30
Partial molar properties, relation to integral properties 30, 34–44
Partial thermodynamic properties of Cu–Zn alloys 41
Permeability of oxide electrolytes 156
Perovskite electrolytes for H_2 cells 161
Pettifor version of Periodic Table 180–2
Phases of variable composition 27–30
Phosphides dissociation measurement 103
Phosphorus value in vapour pressure measurements 104
Platinum resistance thermometers 67–8

Radiation methods for vapour pressure 98–100
Radioactive isotopes in vapour pressure measurements 121–2
Radiofrequency technique for gas equilibrium 116–17
Raoult's law 32, 44–5
Reaction entropy, from EMF measurements 16
Reaction entropy, from equilibrium constant 17
References 349–59
Regular solution definition 50–3
Regular solution order–disorder transformation 53
Regular solution spinodal decomposition 51–2
Richter's rules, standard entropy estimation 174–5
Ruff boiling point method 106–8

Second law of thermodynamics 13
Seebeck coefficient in non-isothermal cells 152–3
Sensing electrode in EMF measurements 161–2
Shimoji–Niwa free volume theory 199–200
Shutter, use in mass spectrometry 126
Sievert's law 46
Sievert's method for gas solubility 104–5
Silver, glass electrolyte in EMF studies 147–8
SiO formation in Fe–Si–C alloy oxidation 207–8
Solid electrolytes 146–62
Solid fluoride electrolytes 159–61
Solid oxide electrolytes 148–58
Solid oxide electrolytes, semiconduction in 149
Solutions 27–63
Spinels, entropy of formation 177, 179
Spinodal decomposition 51–2
Stability diagram for intermediate compounds 196
Statistical probability 49
Sulphate dissociation equilibria 119–20
Sulphide equilibria by EMF studies 156, 161–2
Sulphide equilibria by H_2–H_2S equilibria 118–19
Sulphur valve for vapour pressure measurement 102

Tanaka, free volume theory, data for alloys 200–1
Tandem calorimeter 80–1
Temkin's rule 178
Temperature dependence of enthalpy of reaction 8–12
Temperature gradient effects in vapour pressure measurement 123–4
Temperature measuring devices 67–8
Tensi-eudiometer method for vapour pressure 102–4
Ternary solutions 56–63
Thermal diffusion in gas mixtures 116–18
Thermistors 68
Thermochemical data, elements and compounds 258–323
Thermochemical data, metallic alloys 324–48
Thermocouples 68
Thermodynamic functions 3–27
Thermodynamic probability 48
Thermodynamics of alloys, estimation 197–201
Thermometers, range of application 67–8
Third law of thermodynamics 14
Tian–Calvet calorimeter 92–5
Time of flight mass spectrometer 126
Toop, ternary Gibbs–Duhem approximation 57
Torsion cell for vapour pressure measurement 137
Torsion effusion 136–7
Transport number in electrolytes 141–2
Transportation method for vapour pressure 108–12
Trouton–Pictet evaporation rule 169

Unal estimation of heat capacities 165–6

van't Hoff's isobar 6
Vaporisation rate effects of partial pressure 112
Vapour pressure and partial Gibbs energies 30–4
Vapour pressure curves of binary systems 37
Vapour transpiration method 138–40

Water equivalent 65, 68–78

Zero point entropy, of alloys 16